CARBOHYDRATES

Carbohydrates
Structures, Syntheses and Dynamics

Edited by

Paul Finch

Centre for Chemical Sciences,
Royal Holloway,
University of London, U.K.

SPRINGER-SCIENCE+BUSINESS MEDIA, B.V.

Library of Congress Catalog Card Number: 98-67965.

ISBN 978-90-481-4033-6 ISBN 978-94-015-9281-9 (eBook)
DOI 10.1007/978-94-015-9281-9

Printed on acid-free paper

Contents

Contributors ix

Preface xi

1 Monosaccharides: geometry and dynamics **1**

ALFRED D. FRENCH and PAUL FINCH

1.1 Introduction 1
1.1.1 How to describe preferred sugar structures 1
1.1.2 Flexibility 3
1.1.3 Prediction of structure 4
1.1.4 Conventional orientation of sugar ring drawings 8
1.2 Methods for determination of monosaccharide geometry 9
1.2.1 Diffraction 9
1.2.2 Vibrational spectroscopy 11
1.2.3 Nuclear magnetic resonance spectroscopy 11
1.2.4 Optical rotation 12
1.2.5 Boltzmann weighting 12
1.2.6 Computer modeling methods 13
1.3 Shapes of acyclic carbohydrates 17
1.4 Shapes of carbohydrate rings 18
1.4.1 Size and energy 18
1.4.2 Characteristic furanose shapes 18
1.4.3 Characteristic pyranose shapes 23
1.4.4 Pseudorotation in furanose rings 24
1.4.5 Puckering of furanose rings 24
1.4.6 Puckering of pyranose rings 26
1.4.7 Seven-membered rings 26
1.4.8 Energy surfaces 28
1.5 Detailed factors that determine the shapes of sugar rings 28
1.6 Likely shapes for monosaccharides 31
1.7 Effects of condensed phases (solvents and crystals) 33
1.8 Dynamic behavior of monosaccharides 36
1.8.1 Molecular dynamics studies 36
1.8.2 Mutarotation 37
References 41
Software 46

2 Chemical synthesis of monosaccharides **47**

KARL J. HALE and ANTHONY C. RICHARDSON

2.1 The asymmetric synthesis of monosaccharides from achiral starting materials 47
2.2 Asymmetric synthesis of monosaccharides via chemical homologation and degradation 64
2.3 Monosaccharide mimetics and pseudo sugars 75
2.3.1 Preparative methods for 5-thioaldoses 75

2.3.2 Preparative methods for 5-amino-5-deoxy-aldoses 79
2.3.3 Preparative methods for monosaccharides containing a phosphorus atom in the ring 80
2.3.4 Preparative methods for carba-monosaccharides 81
2.4 Recent methodological developments in the chemical modification and manipulation of monosaccharide derivatives 91
2.4.1 Protecting groups 91
2.4.2 New methodology for the synthetic manipulation of monosaccharides 96
References 104

3 Carbohydrate sulphates **107**

RUTH FALSHAW, RICHARD H. FURNEAUX and GEORGE C. SLIM

3.1 Introduction 107
3.2 Natural occurrence and biological role 107
3.2.1 Sulphated galactans 107
3.2.2 Sulphated fucans 110
3.2.3 Sulphated mannans 111
3.2.4 Sulphated heteropolysaccharides 111
3.2.5 Glycosaminoglycans 114
3.2.6 Glycoproteins 124
3.2.7 Miscellaneous sulphated carbohydrates 126
3.2.8 Differentiation of sulphate and phosphate groups in biological systems 128
3.3 Chemical synthesis and modification 128
3.3.1 Sulphation reagents 128
3.3.2 Sulphated free sugars and their glycosides 129
3.3.3 Sulphated polysaccharides 131
3.3.4 Cyclic sulphates 133
3.3.5 Sulphate diesters 133
3.4 Chemical stability 134
3.4.1 Alkali treatment and nucleophilic displacement reactions 134
3.4.2 Acid hydrolysis 135
3.4.3 Methanolysis 135
3.4.4 Solvolytic desulphation 136
3.4.5 Other methods of sulphate cleavage 136
3.4.6 Effect of sulphate esters on other substituents 136
3.5 Methods of analysis 137
3.5.1 Separation techniques 137
3.5.2 Spectroscopic methods 137
3.5.3 Chemical methods 139
3.6 Conformational effects of sulphation 140
References 141

4 Conjugation of monosaccharides – synthesis of glycosidic linkages in glycosides, oligosaccharides and polysaccharides **150**

STEFAN OSCARSON

4.1 Introduction 150
4.2 Chemical synthesis of *O*-glycosides 151
4.2.1 Glycosylation methods 151
4.2.2 Anomeric configuration 157
4.2.3 Strategies 162
4.2.4 Protecting groups 165
4.2.5 Additional remarks 171
4.3 Chemical synthesis of *S*-, *N*- and *C*-glycosides 172
4.3.1 *S*-glycosides 172
4.3.2 *N*-glycosides 173

4.3.3 *C*-glycosidic compounds 175
4.4 Enzymatic synthesis of *O*-glycosides 178
4.4.1 Glycosyl transferases 178
4.4.2 Glycosyl hydrolases 181
4.4.3 Other enzymes 182
References 182

5 **Chemistry of glycopeptides** **187**
HORST KUNZ, BIRGIT LÖHR and JÖRG HABERMANN
5.1 Introduction 187
5.2 Saccharide-peptide linkages of natural glycoproteins 188
5.2.1 The *N*-glycosyl asparagine linkage 188
5.2.2 *O*-glycosidic linkages of glycoproteins 188
5.3 Problems of isolation and structure elucidation of glycoproteins 190
5.4 Synthesis of glycosyl amino acids 192
5.4.1 The glucosamine asparagine linkage 192
5.4.2 *O*-glycosyl serine and threonine linkages 198
5.5 Synthesis of oligosaccharides and glycopeptides 207
5.5.1 The Thomsen–Friedenreich antigen disaccharide 207
5.5.2 Sialyl T_N disaccharide 208
5.5.3 β-Mannosyl chitobiose asparagine 210
5.5.4 Lewis^X antigen glycopeptides 211
5.5.5 Sialyl Lewis^X glycopeptides: chemical and enzymatic synthesis 213
5.6 Solid-phase synthesis of glycopeptides 216
5.7 Conclusions 221
References 221

6 **Oligosaccharide geometry and dynamics** **228**
JOHN BRADY
6.1 Introduction 228
6.2 Experimental characterization of oligosaccharide conformations 232
6.2.1 X-ray and electron diffraction 232
6.2.2 Nuclear magnetic resonance of oligosaccharides 235
6.2.3 Other experimental methods 236
6.3 Disaccharide energetics 237
6.3.1 Quantum mechanical studies 237
6.3.2 Empirical energy calculations 239
6.4.4 Conformational fluctuations and dynamics 246
6.5 Hydration of oligosaccharides 247
6.6 Conformational free energies 251
6.7 Conclusions 254
References 254

7 **Shapes and interactions of polysaccharide chains** **258**
SERGE PÉREZ and MILOU KOUWIJZER
7.1 Introduction 258
7.2 Ordered structures of polysaccharides 259
7.2.1 Structural features 259
7.2.2 Diffraction methods 266
7.2.3 Structural features of some crystalline polysaccharides 272
7.3 Secondary and tertiary structures of polysaccharides in solutions and gels 277
7.3.1 Chirooptical methods 278
7.3.2 Nuclear magnetic resonance spectroscopy 278
7.3.3 Light, X-ray and neutron scattering 279

7.3.4 Rheology 280
7.3.5 Marine polysaccharides 282
7.3.6 Plant polysaccharides 286
7.3.7 Microbial polysaccharides 288
References 290

8 Chemistry of polysaccharide modification and degradation 294

MANSSUR YALPANI

8.1 Introduction 294
8.2 Non-selective chemical modifications 295
8.2.1 Acylations 296
8.2.2 Alkylations 299
8.2.3 Oxidations 300
8.3 Selective chemical modifications 301
8.3.1 Acylations 301
8.3.2 Alkylations 303
8.3.3 Oxidations 304
8.4 Modification of carbonyl functions 306
8.4.1 Esterifications 306
8.4.2 Amidations 307
8.4.3 Aminations 307
8.5 Modification of amines 308
8.5.1 *N*-acylations 308
8.5.2 *N*-alkylations 308
8.6 Polysaccharide degradations 309
References 313

9 Carbohydrate–protein interactions 319

LOUIS T.J. DELBAERE and LATA PRASAD 319

9.1 General features 319
9.2 Basis of structural specificity 322
9.3 Examples 324
9.3.1 Lectins 324
9.3.2 Antibodies 325
9.4 Enzymes 326
References 328

Index 330

Contributors

John Brady	Institute of Food Science, New York State College of Agriculture and Life Sciences, Cornell University, Stocking Hall, Ithaca, NY 14853-7201, USA
Louis T.J. Delbaere	Department of Biochemistry, University of Saskatchewan, 107 Wiggins Road, Saskatoon, SK, Canada S7N 5E5
Ruth Falshaw	Industrial Research Ltd, Gracefield Road, PO Box 31-310, Lower Hutt, New Zealand
Paul Finch	Centre for Chemical Sciences, The Bourne Laboratory, Royal Holloway, University of London, Egham Hill, Egham TW20 0EX, UK
Alfred D. French	Cotton Fiber Quality Research Unit, Southern Regional Research Center, 1100 Robert E. Lee Boulevard, New Orleans, LA 70124, USA
Richard H. Furneaux	Industrial Research Ltd, Gracefield Road, PO Box 31-310, Lower Hutt, New Zealand
Jörg Habermann	Institut für Organische Chemie, Johannes Gutenberg-Universität Mainz, Postfach 3980, D-55099 Mainz, Germany
Karl J. Hale	Department of Chemistry, University College London, Christopher Ingold Laboratories, 20 Gordon Street, London WC1H 0HJ, UK
Milou Kouwijzer	Centre National de la Recherche Scientifique, Centre de Recherches sur les Macromolecules Végétales, Domaine Universitaire, 601 rue de la Chimie, Addresse Postale BP 53, 38041 Grenoble cedex 09, France

Horst Kunz	Institut für Organische Chemie, Johannes Gutenberg-Universität Mainz, Postfach 3980, D-55099 Mainz, Germany
Birgit Löhr	Institut für Organische Chemie, Johannes Gutenberg-Universität Mainz, Postfach 3980, D-55099 Mainz, Germany
Stefan Oscarson	Department of Organic Chemistry, Arrhenius Laboratory, Stockholm University, S-10691 Stockholm, Sweden
Serge Pérez	Centre National de la Recherche Scientitique, Centre de Recherches sur les Macromolecules Végétales, Domaine Universitaire, 601 rue de la Chimie, Addresse Postale BP 53, 38041 Grenoble cedex 09, France
Lata Prasad	Department of Biochemistry, University of Saskatchewan, 107 Wiggins Road, Saskatoon, SK, Canada S7N 5E5
Anthony C. Richardson	Department of Chemistry, King's College London, Campden Hill Road, Kensington, London W8 7AH, UK
George C. Slim	Industrial Research Ltd, Gracefield Road, PO Box 31-310, Lower Hutt, New Zealand
Manssur Yalpani	BioPolymer Technologies International Inc., PO Box 84, Westborough, MA 01581, USA

Preface

Longer ago than I enjoy remembering I found an annual frustration with the task of producing a concise and accessible reading list for students with more than a passing interest in carbohydrate chemistry. On some aspects good texts had been produced but were unavailable, on others concepts and knowledge had advanced considerably; in general there were considerable voids between the standard introduction in organic texts, and research-orientated reviews. Thus an invitation from Jon Walmsley of Blackie & Co actually took root and a project was initiated. The aim was to present authoritative overviews of the current status of some particular areas of structural and synthetic carbohydrate chemistry and which were perceived to be central to the subject. It is hoped thereby to underpin the steadily increasing perception and understanding of the roles of carbohydrates in nature.

Central to any consideration of carbohydrates is the monosaccharide unit; one testament to its enduring challenge and interest is the latest literature prediction of the anomeric ratio of glucose. Alfred French explains how the geometries of monosaccharides can be described and studied experimentally and theoretically. He makes clear how different geometries can be separated by relatively small differences in energy, and that in the condensed phase the influence of intermolecular interactions can be crucial. Whilst there is a steadily increasing body of experimental data, precise prediction remains extremely difficult. The structural theme is extended through oligo- and polysaccharides in chapters by John Brady and Serge Pérez and Milou Kouwijzer. John Brady summarises the experimental approaches to the determination of conformation at glycosidic linkages, and reviews the quantum and molecular mechanical methods for the calculation of conformational energies. Although convincing agreement between NMR-derived and modelled conformations is now a routine objective, Brady's careful analysis of the various limitations of calculations should caution against comfortable acceptance of a 'preferred' conformation. Serge Pérez and Milou Kouwijzer describe the methods for study of the geometry of extended carbohydrate chains. Repeat structural features give rise to ordered structures, especially in the condensed phase. However, the precise secondary and quaternary structure is not predictable from the repeat structure and linkage geometry, even for simple homopolysaccharides. In the hydrated state repeat structural features, especially if subject to interruptions, can lead to temporary or permanent gels. Pérez and Kouwijzer describe how the properties of some hydrated systems can vary with such factors as concentration, salts and temperature. Although these are difficult systems to study and interpret, especially when mixtures of polysaccharides are taken, the incentive is the understanding of materials of real significance to nature and industry. Discerning intermolecular structural interactions is the central theme of the discussion by Louis Delbaere and Lata Prasad. The interactions between proteins and carbohydrates are the subject of intense contemporary study because of their key importance to the regulatory roles of carbohydrates. Despite their highly hydrophilic appearance, it seems that carbohydrates interact with proteins via non-polar as well as polar, including water-mediated, contacts. Ever since Fischer's proof of the structure of glucose, chemical synthesis has responded to the challenges posed by carbohydrate structure elucidation. Karl Hale and Dick Richardson present abundant evidence of the ingenuity and skill of modern synthetic organic chemistry applied to monosaccharides. Beyond the common examples, the variety of structures presented by nature is considerable, but it seems that none is beyond access. The construction of glycosidically-linked sugar units has been an enduring theme in synthetic carbohydrate chemistry; Stefan Oscarson, Horst Kunz, Birgit Löhr and Jörg Habermann set out the principles and procedures for these endeavours. Perhaps there will never be a universal solution to the problems of yields and anomeric selectivity – perhaps the seduction of libraries and 'combinatorial' mixtures will attenuate the incentive – but the structural specificity of nature and the value of a pure substance cannot he denied. As described by Manssur Yalpani, synthesis and degradation at the polymer level usually require an acceptance of 'fuzzy structures'; however, the scope and significance are great. The chemistry and consequences of one specific type of chemical modification of carbohydrates, namely the replacement of a hydroxyl substituent by a hemiester sulphate, are comprehensively described by Ruth Falshaw, Richard Furneaux and George Slim. Despite the notional chemical similarity of sulphate and phosphate, nature seems to have evolved generally different types of materials and roles for carbohydrates containing these anionic substituents. Although a mechanical/structural role for sulphate is preeminent, some more subtle functions are beginning to emerge.

I warmly thank the authors for their efforts in producing such informed and enlightening exposition on these topics in carbohydrate chemistry and I acknowledge the staff of Kluwer (née Blackie) for their patience and assistance.

Paul Finch
Egham
August 1998

1 Monosaccharides: geometry and dynamics

ALFRED D. FRENCH and PAUL FINCH

1.1 Introduction

The geometries of sugars have long been recognized as determinants of their chemical and physical properties. For example, the 16 D-aldohexapyranoses do not differ in their empirical formula, $C_6H_{12}O_6$. Each aldohexapyranose has four hydroxyl groups and a hydroxymethyl group, yet, because of configurational differences at successive stereogenic carbon centres, each has distinct chemical reactivity and chromatographic behavior. Even the extents of their existence might seem to depend on the exact stereochemistry or chirality at each optical center. In the biosphere, D-glucose, mannose and galactose are nearly ubiquitous, while much less is seen of allose or gulose.

In 1971 Stoddart wrote a seminal book, *Stereochemistry of Carbohydrates* [1], that summarized much of the available knowledge. In the intervening quarter century, however, many advances have occurred. At that time, diffraction crystallography had solved a few structures but the 1970s and 1980s were especially fertile periods. Also, computer modeling was truly in its infancy. This chapter presents some of the new information as well as some of the basic ideas of carbohydrate geometry and dynamics. The reader is also referred to Eliel and Wilen's book, *Stereochemistry of Organic Compounds* [2], that covers many of the topics below in greater depth.

Monosaccharides are themselves important molecules. Just as important, however, is the role of monosaccharide moieties in oligosaccharides and polysaccharides. Substantial experience shows that the structural features of the polymer depend on the structural features of the monomeric building blocks as well as the geometries of the intermonomer linkages of the polymer. Therefore, the principles of monomeric geometry are essential to all carbohydrate chemistry.

1.1.1 *How to describe preferred sugar structures*

The preferred chemical forms of most monosaccharides and the majority of their derivatives are based on cyclic tetrahydropyran or tetrahydrofuran structures. In the solid state, and in glycosides, oligosaccharides and polysaccharides and other glycoconjugates, these cyclic structures are 'permanent'. In solutions of the 'free' sugars (aldoses, ketoses, some glycosylamines), the ring forms are in chemical equilibrium with, and connected via, the acyclic

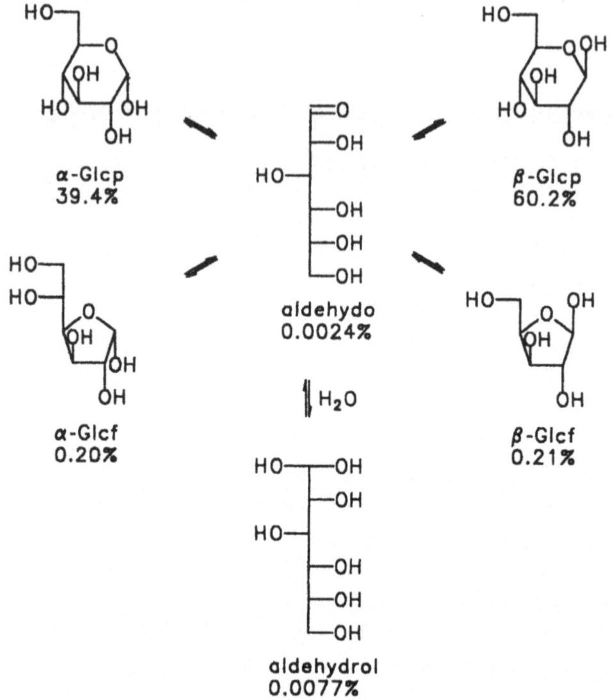

Figure 1.1 Structures of D-glucose forms at equilibrium in D_2O at 37°C, with percentages [113].

aldehydo or keto form which is structurally identical to the Fischer projection commonly used to present the stereochemical configurations of the monosaccharides. The proportion of the acyclic form, together with the hydrate, is generally small (less than 1%) for the common monosaccharides (Figure 1.1).

In some cases, the proportion of a carbonyl (e.g. pent-2-uloses, 15%–20%), or hydrate (e.g. dihydroxyacetone phosphate, 45%) may be much higher. Also, minor forms may be the relevant substrates for enzymes. In other cases, for example polyols and dithioacetals, the proportion of the acyclic form is necessarily 100%.

Within the approximation that variations of bond lengths and angles are unimportant, a complete description of a monosaccharide, for example in the crystalline, non-disordered solid state, requires choices among the following descriptors:

- number of carbon atoms (e.g. hexose, pentose);
- relative configuration at chiral centers, indicated by the sugar name prefix (e.g. *gluco-*, *ribo-*, *glycero-*);
- chirality (whether D or L);
- ring size (if any – furanose, pyranose, septanose);
- anomeric configuration (α or β);

- conformation (if a ring, envelope, chair, etc. or, if planar, zig-zag or sickle);
- orientations of exocyclic groups (described by torsion angles for each substituent).

Crystalline molecules are sometimes disordered, that is, one or more of the atoms have fractional occupancies of different positions. At room temperature this disorder can be dynamic, in which case any given example of the atom in question occupies two different positions. Alternatively, the disorder can be statistical. In that situation, some of the atoms of a given type are more or less permanently in one position and some in another. Disorder is not uncommon for the location of the glycosidic oxygen atoms of reducing sugars. In that case, the percentages of both anomeric forms must be specified.

In solution the dynamic nature of sugar structures requires, at least in principle, description of the relative populations of the various forms. In particular, descriptions in terms of averages can be misleading. Presentation as an average can mask the dynamic nature of the molecule and the resulting average structure may in fact be virtually impossible on relevant time scales. A trivial but emphatic example is provided by fructose, which forms both five-membered and six-membered rings. There are no rings of 5.5 atoms. Not so obviously erroneous are results from computationally augmented experiments that determine torsion angles. If the values of a torsion angle have a bimodal distribution, then reporting it as unimodal hides the true nature of the system. It is true that a single ring form predominates for many of the important sugars, but this cannot be assumed in general.

1.1.2 Flexibility

Even when there is only one important configuration and ring conformation the atoms in molecules still exhibit positional variation. Thus, only one chair-shaped ring is important for glucose, but that chair undergoes substantial variations in shape. At any temperature above absolute zero there is continuous variation ('breathing') from thermal motion. Even in crystals at room temperature there is still considerable thermal motion. According to one school of thought, crystallization selects a subset of the range of structures existing in solution. This subset of the structures is elegantly presented by crystallographers as the ellipsoids of thermal motion that correspond to the half-probability (in this example) limits of nuclear positions, as shown in Figure 1.2, a plot [3] of β-L-arabinose [4].

The largest relative motions are typically the rotations about bonds that are described by changes in torsion angles. Motions that change the bond angles are typically much smaller because the forces required to bend a bond angle are much larger. Changes in bond length require even stronger forces for a noticeable motion.

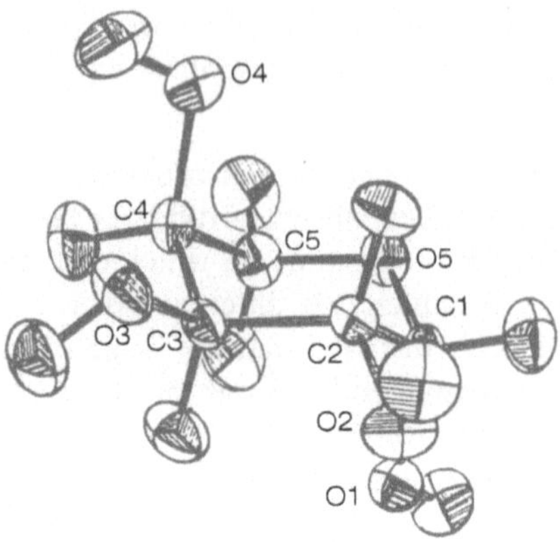

Figure 1.2 Oak Ridge thermal ellipsoid plot (ORTEP) [3] of β-L-arabinopyranose [4] at room temperature. The 50% probability ellipsoids are shown.

In solution, any bond for which the torsional barrier is 5 kcal/mol (*c.* 21 kJ/mol) or less, such as bonds to the exocyclic groups or within backbones of acyclic sugars, is considered to be 'freely rotating' at room temperature. The changes in the geometric details during such rotational motion are subtle but complex, because changes in these easily varied torsion angles lead to different bond lengths and bond angles. Bond angles can change by several degrees during changes in torsion angles while torsion-related bond-length stretching can amount to several picometres [5]. The changes in bond lengths and angles that are caused by changes in torsion angles result in part from changes in the atom–atom van der Waals repulsions. Another reason for the bond lengths and angles to change during torsional motions is that there are subtle shifts in the populations of the electrons as orbitals on adjacent atoms are given different relative orientations. Thus three types of flexibility of any molecule must be considered. The first is a degree of oscillation about a fixed ideal value, as in thermal motion, exemplified to a partial extent by the Oak Ridge thermal ellipsoid plots (ORTEPs). The second is a compensatory change in bond lengths and angles as a result of torsional movements. Third, there are many examples of well-populated multiple minima in torsion angle space, such as the two commonly found rotational orientations for O6 of glucopyranose (see section 1.6). For ring-forming molecules an additional type of flexibility is the concerted motion of all of the atoms, such as found in pseudorotation (section 1.4.4).

1.1.3 Prediction of structure

Because structure and behavior are so intertwined, much of carbohydrate chemistry has been devoted to the determination of geometries that characterize monosaccharides. If the observed molecular geometries of a range of

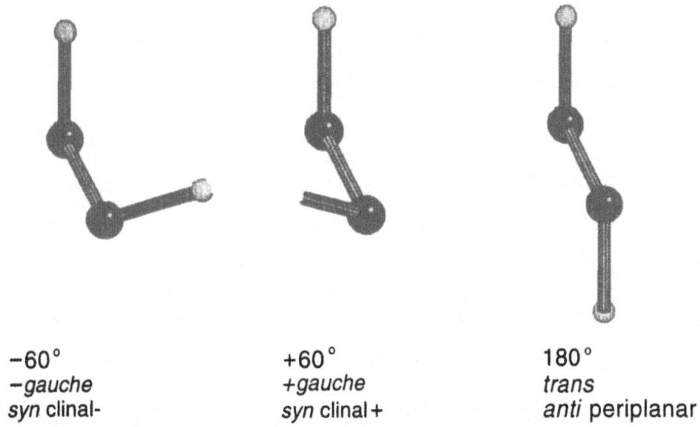

Figure 1.3 Various terms that define the torsion angle.

compounds can be predicted, then we can claim to understand the factors that govern their shapes. Thus, carbohydrate chemists have been driven to devise systems for predicting sugar ring shapes. These systems are based on important basic phenomena that must be understood.

The first principle governing sugar structures is that their sp^3 carbon atoms have tetrahedral bonding geometry, as proposed by van't Hoff [6] and Le Bel [7] in 1874, and later confirmed crystallographically in diamonds [8]. Second, it is well established that four-atom sequences of such carbon atoms, such as in *n*-butane, prefer to have a torsion angle (Figure 1.3) of about 180°, with secondary preferred values of ±60°. Torsion angles with 180° values do not lead to ring formation. However, alternating ±60° torsions for a six-atom sequence lead to ring closure by averaging to 0°. Alternation avoids the high-energy conformation having eclipsed, 0° torsion angles that could also lead to ring formation for five-membered furanoid rings. Because one or more butane-like sequences often appear in monosaccharides, this is an important factor in determining the ring shape. The sp^3 oxygen atoms also have a tendency for nearly tetrahedral bond angles, and O–C–C–C and C–O–C–C sequences are thought to have threefold torsional preferences somewhat similar to those of butane (Figure 1.4).

This secondary preference of saturated C–C bonds for ±60° torsion angles gives substantial stability to chair forms of cyclohexane, the tetrahydropyran ring and many of the pyranose sugars. In 1939 the crystal structures of α-D-glucosamine hydrochloride and hydrobromide were solved, revealing $^4C_1$ chairs (the nomenclature indicates a chair form, with carbon atom 4 high and carbon atom 1 low when viewed in the standard viewing orientation; section 1.1.4) [9]. Variation in placement of substitutents on these rings can either stabilize or destabilize that chair relative to the alternative chair or to the skew forms. Often, the preferred overall geometry of a

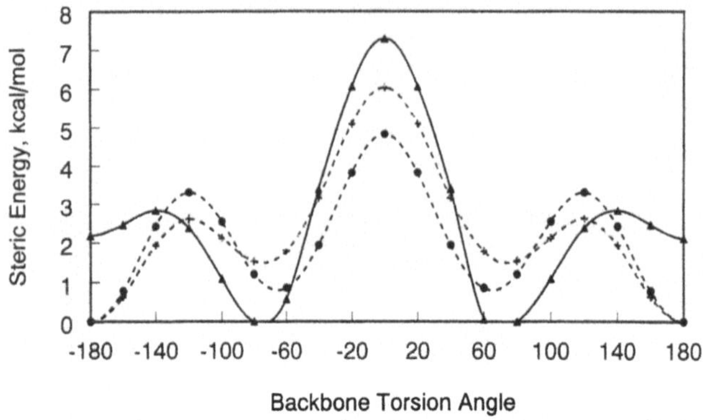

Figure 1.4 Energies, as calculated on the MM3(94) program, for methoxy methanol (▲), methoxy ethane (+) and *n*-butane (●). Note: 1 cal = 4.1868 J.

given monosaccharide can be deduced from this propensity for chair structures and the occurrence of a particular chemical derivatization. One of the earliest such studies was by Böseken [10], who proposed differentiation between the α and β anomers of glucose based on complexation by borate. His efforts were interpreted based on a planar furanose model of the α anomer (with *cis* hydroxyls on C1 and C2) that would complex borate, but which would not for the corresponding β form. In R. E. Reeves's studies of glucose [11, 12], the cuprammonium reaction occurred for *gauche* hydroxyl groups but not for other arrangements. Only one of the two possible chair forms of the glucopyranose ring causes a *syn gauche* disposition of the O2 and O3 hydroxyls. The other chair places them in an *anti* arrangement. The occurrence of the cuprammonium reaction suggested that the hydroxyls are almost always *gauche* and that the correct structure for the ring is $^4C_1$ (C1 in Reeves's nomenclature). Reeves's work in 1949 is now cited as the first proof of ring conformation in solution.

Angyal's nuclear magnetic resonance (NMR) studies [13] confirmed Reeves's work on solution conformation and determined the dominant shapes of many sugars. Angyal then further developed simple rules to aid in the prediction of the more likely chair forms of pyranoid sugars [14]. These rules or instability factors are penalties (in kcal/mol of free energy) for axial substituents. Thus, the favored shape of the pyranoid chair in water would be the one having the lowest penalty, or, alternatively, the largest number of equatorial substituents. (Similar work for cyclohexane rings has been reported by a number of authors. Large tabulations of these penalties for different groups are available; see Winstein and Holness [15], Bushweller [16] or Eliel and Wilen [2], and references therein.) This simple model suggests that there are repulsive interactions between axial hydrogen atoms or among the substituents when they are in axial positions, and these interactions lead to a higher free energy.

1.1.3.1 The anomeric effect

The exception to this rule occurs at anomeric centers, where electronegative substituents in axial positions tend to be much more likely for sugars than they are for pure carbocyclic rings. Thus, an axial methoxy group at C1 of methylglucopyranoside has a higher occurrence than does an equatorial methoxy group. This phenomenon is called the **anomeric effect** [17, 18]. In one explanation, the unexpectedly high population of the axial form is attributed to repulsive overlap of lone electron pairs (dipole–dipole repulsion) from the ring and equatorial glycosidic oxygen atoms in the β-glucoside [19]. Another theory of the anomeric effect calls for interaction of lone pair electrons on the ring oxygen with the C1–O1 antibonding σ^* orbital, creating partial double bond character there. At the same time, the population of electrons in the bond to the glycosidic oxygen is diminished, making the C1–O1 bond weaker. This double-bond–no-bond resonance model [20] has a lower energy for axially orientated glycosidic oxygens because the favorable overlap does not occur to such an extent when the glycosidic oxygen is in the equatorial disposition, and accounts for observed changes in bond length. Recent research suggests that for glucose or simpler models such as 2-hydroxytetrahydropyran, that delocalization, or hyperconjugation, accounts for the majority of the free energy differences between the two forms [21].

Other proposals for the origin of the anomeric effect include Walkinshaw's study, which suggested that hydrophilicity alone was sufficient to account for the concentrations of the anomers in aqueous solution equilibrium [22]. Booth *et al.* [23] proposed that entropy differences were the driving forces for axial group preferences. Attribution of the anomeric effect to any single source is almost certain to be incorrect, and explanation of the basis for the anomeric effect will continue to be sought.

Whatever the causes, aqueous solutions of glucose in equilibrium contain 36% α-glucopyranose, with its axial O1H group, compared with 64% of the all-equatorial β-form. Comparable substitution of cyclohexane gives almost exclusively equatorial substituent orientation. Similarly, α-mannose is more prevalent (75:25) than the β-form. The effect is enhanced in solvents with low dielectric constants. In the early stages of knowledge of a subject, deviations from expected behavior are considered to be special effects. A second level of understanding of the anomeric effect comes from analysis of the torsional energy for the sequence C5–O5–C1–O1 in aldose rings [24] (see Figure 1.5). This torsion angle takes values of 60° for the axial and 180° for the equatorial form. Despite the shared threefold character, the relative energies of the minima are rather different from the curve for butane with its C–C–C–C sequence (Figure 1.4). Instead of 180°, the ±60° values are preferred, leading to the relatively high amounts of the axial form. According to molecular mechanics calculations (MM3 computer program), the height of the eclipsed barrier is about 6 kcal (25 kJ) above the *gauche* minimum,

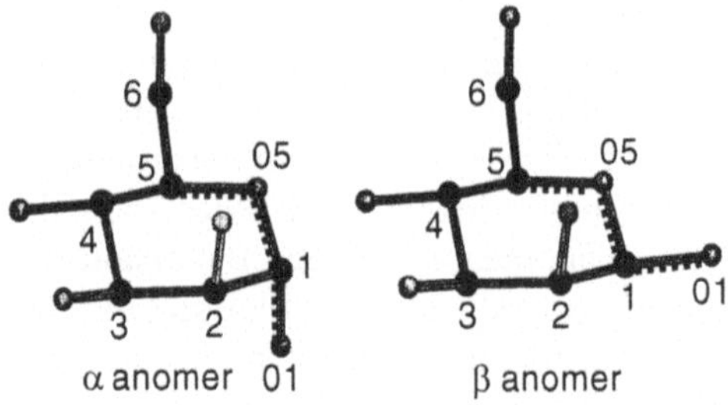

Figure 1.5 Diagram of the α and β anomers of mannose, showing the difference in the C5–O5–C1–O1 torsion angle.

and the O–C–O–C torsional energy itself has a rotational barrier of about 4 kcal (17 kJ) above the *gauche* minimum. In either case, the *gauche* minimum is about 2 kcal (8 kJ) lower than the *anti* minimum. Recent work with better *ab initio* and density functional theory quantum calculations shows that the constants used in the MM3 program for these torsional energies are approximately correct [25].

Attempts have also been made to understand the basis for the observed conformations of sugar structures that contain furanoid rings [26]. Here however, substituents have the less-differentiated, pseudo-axial and pseudo-equatorial dispositions. Furanoid rings are generally much more flexible than are pyranoid rings and it was felt that the more significant determinants of ring shape are the anomeric effect and a more controversial factor, the *gauche* effect. This latter effect is the observed propensity for hydroxyl groups on adjacent carbon atoms to favor *gauche* somewhat more than *anti* dispositions. In other words, the O–C–C–O atomic sequence has lower energy when its torsion angle is $\pm 60^\circ$ instead of 180°.

These two different sets of factors (the simple steric repulsions of axial groups, and the *gauche* and anomeric preferences for $\pm 60^\circ$ torsion angles) must be carefully balanced so as to yield a correct summation for a given sugar. The difficulty of balancing these forces underscores the likely utility of sophisticated, computerized computations in prediction of monosaccharide geometry, unavailable when Angyal's rules were proposed. Other determinants of monosaccharide geometry, besides their free energies as isolated entities, are likely to include general and specific solvent effects. Hydrogen bonds can form with solvent molecules, with other solute molecules or they can be intramolecular [27]. Although far weaker than covalent bonds, hydrogen bonds can be significant in comparison with other conformation interactions.

1.1.4 Conventional orientation of sugar ring drawings

Sugar structures are most easily recognized if they are drawn the same way every time. Over the years, several conventions have prevailed as knowledge

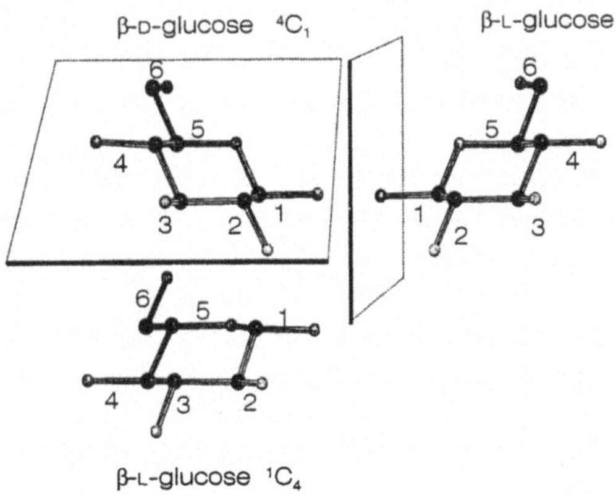

Figure 1.6 Conventional drawings of β-glucose, in the D and L forms, separated by horizontal and vertical mirror planes.

of structure has improved. For either Haworth or stereochemically correct drawings, convention [28] dictates a standard orientation with the ring oxygen at the rear and the lowest-numbered ring atom (C1 of glucose) to its right. The bond from the ring oxygen to the highest-numbered ring atom (C5–O5 in glucose) is parallel to the plane of the paper (Figure 1.6). The coplanar atoms (C2, C3, C5 and O5) are usually drawn in a plane that makes an angle of about 75° to the plane of the paper, so that the ring atoms closer to the viewer (C2 and C3) are below the more distant ring atoms (C5 and O5). This conventional orientation simplifies the determination of various stereochemical issues. The enantiomeric forms of a sugar can both be depicted according to the convention by drawing them one above the other, with the mirror plane of symmetry drawn in between. In such a drawing, the mirror image of the $^4C_1$ D form is seen to be the $^1C_4$ L form.

1.2 Methods for determination of monosaccharide geometry

1.2.1 Diffraction

If a crystal can be grown, it is often possible to determine the molecular geometry to a high degree of accuracy and simultaneously to confirm the chemical structure [29]. The most accurate studies, especially in terms of locating the hydrogen atoms, are done with neutron diffraction at low temperature, while most routine studies are done at room temperature with X-rays. Although such X-ray studies often suffice for understanding the structure, they may be no longer precise enough to aid the development of contemporary sophisticated modeling methods. Studies at low temperature

(123 K, the temperature of boiling liquid nitrogen, or even lower, cooled by liquid helium), give improved accuracy over those at room temperatures. They are often required to determine the hydrogen bonding systems and should be used whenever possible. One indicator of the overall quality of a crystallographic determination is the *R* factor, the sum of the discrepancies between the observed and calculated structure factors, divided by the sum of the observed structure factors. The very best determinations of carbohydrate crystal structures have yielded *R* factors of about 2.5%. Higher values indicate lower accuracy. Still, the heavy (i.e. non-hydrogen) atoms can be resolved at considerably higher (e.g. 10%) *R* values, and the structure is portrayed in a more general sense. At such a high error, however, the bond lengths, for example, could be less accurate than from a good computer model, and the hydrogen bonding can be uncertain at *R* values of 5% or less. Although low-temperature studies are beneficial in terms of precision, it should be remembered that some changes can occur when cooling.

Depending on crystal symmetry, individual crystallographic studies often produce only one structure for a molecule. Because of flexibility there may be a wide range of possible forms and a single structure from a diffraction study may correspond well to only one of many components of the solution equilibrium. For example, the inter-residue linkage conformation of methylmaltoside in aqueous solution is understood to be different from its crystal structure [30]. However, the crystal structure of methylmaltotrioside [31] has linkage conformations closer to the one thought best to represent the shape of maltose in aqueous solution. The crystallographic form may or may not be physiologically active. If a picture of the variability of the molecule is desired from the precise and definitive crystallographic studies, a variety of suitably related molecules must be studied. A potential problem, so far unaddressed, would be a propensity for crystal lattice packing forces to narrow the range of shapes observed in crystal structures. Thus, the molecule might be restricted to a subset of the full range experienced by a single tautomer in solution, despite consideration of a large range of relevant crystal structures. For example, furanosyl rings may be flattened by long-range attractive (and thus compressive) forces in crystal structures. Implicit in conclusions based on the variety of crystal structures of analogous molecules is the assumption that deviations from the minimum-energy form are random, with the extent determined by a Boltzmann distribution.

Although many of the fundamental aspects of carbohydrate structures were known from chemical analyses before their structures were determined by diffraction studies, the results of diffraction studies have advantages such as being highly quantified, providing precise distances, angles and torsion angles. This is especially important for carbohydrates because it helps establish the ranges of their molecular flexibilities. One other advantage is that the absolute configuration of molecules can be determined using Bijvoet's method [32]. Earlier descriptions of chiral structure had been fortuitously correct so the literature did not have to be rewritten. Still, some crystal structure reports inadvertently contain coordinates for the wrong enantiomer, giving opposite signs for torsion angles.

So that others can take full advantage of this specialized work, the results, in terms of fractional coordinates for the atoms, for almost all diffraction studies of small and medium-sized molecules are deposited (and checked) in the *Cambridge Crystal Structure Database* [33]. In the 1970s and 1980s, Jeffrey and Sundaralingam coauthored reviews in *Advances in Carbohydrate Chemistry and Biochemistry* [34]. *The Protein Data Bank* contains crystal structures of a few carbohydrates, complexed by proteins [35]. Those structures, however, are often of lower accuracy, and it is not always completely clear whether the distortions in the monosaccharide ring geometries are artifacts of those relatively low-resolution diffraction studies or are due to protein complexation.

1.2.2 Vibrational spectroscopy

The adoption of fast Fourier transform infra-red spectroscopy has revitalized this field, while the development of inexpensive lasers has forwarded Raman spectroscopy. A review of vibrational spectroscopy of carbohydrates is available [36].

1.2.3 Nuclear magnetic resonance spectroscopy

Although studies of the solid state by NMR can be useful, NMR is especially important for studying solutions and carbohydrates that will not crystallize. In particular the work by Angyal has been vitally important in determining the composition of equilibria of reducing saccharide solutions [37]. One book on carbon-13 (^{13}C) NMR spectroscopy contains a number of chemical shifts for monosaccharides and their derivatives [38]. Further information can be taken from coupling constants and nuclear Overhauser effects (nOes).

A frequent use of coupling constants is the determination of torsion angles through a Karplus equation [39]. Various types of coupling constants can be obtained (e.g. $^3J_{HH}$, $^3J_{CH}$, $^1J_{CH}$), and a variety of Karplus equations exists for their interpretation [40–42]. If enough coupling constants can be determined then the geometry of the molecule can be determined to a level sufficient for understanding, because bond lengths and bond angles are fairly standard. Also, especially in small molecules, the overall geometry is not very sensitive to variations in bond lengths and angles. In a software system developed by Altona, PSEUROT, the equilibria of two forms of furanoid rings can be studied through their coupling constants [43].

Similarly, nOes can be calculated for model carbohydrates, giving interproton distances [44]. Partly because the magnitude of the effect depends on the 6th power of the interproton distance, however, there is no unique solution if the molecules in question are not rigid, and nOes are not currently as helpful as coupling constants for structural study. Recently, the relationship between chemical shifts and torsion angles has been refined [45] with the aid of quantum mechanics calculations. Constancy of results regardless of level of quantum mechanical theory suggests that this promising method may be developed further and used without costing too much computer time.

Even though NMR studies with natural abundances of ^{13}C have been very worthwhile, recent work by Duker and Serianni [46] shows that substantial advantage in NMR analyses is gained by substituting ordinary carbon atoms with ^{13}C. This enhances the signal strength and improves sensitivity in conformational analyses based on coupling constants.

1.2.4 Optical rotation

Another approach to determining molecular structure, advanced by Stevens [47], is the calculation of the optical rotation of model carbohydrate structures for comparison with the observed value. Currently, the method is limited by the small number of atom types that can be treated, but optical rotation is simple to measure and is an independent indicator of molecular geometry.

1.2.5 Boltzmann weighting

Given a rigid molecule, it is reasonable to convert an experimental measurement, such as an NMR nOe, to a conformational specification. However, most carbohydrate molecules in solution are not rigid. An observed value such as a coupling constant or an optical rotation value reflects the contribution from each molecule in the solution so a particular experimental value could come from many different distributions of conformations. Therefore, a determination of the molecular geometry should be based on the relative probability of each possible structure. Computer modeling plays a key role in such studies because the calculation of an observable, such as a coupling constant or an optical rotation value, involves calculation of the free energy for each possible conformation, as well as the theoretical value for the observable. Through the Boltzmann equation, the probability of each conformer is calculated from the free energy (see Table 1.1) and that probability is used to weight the contribution for that conformer to the observable. Alternatively, the observable can be computed at intervals during a molecular dynamics study and averaged. It is possible only to check whether the computed prediction corresponds to the experimentally observed value, and even then correspondence may be fortuitous.

Calculations that are based on enthalpy (or potential energy) instead of free energy (for reasons of convenience and smaller computational expense) will have a small error. The sums of rotational and vibrational contributions

Table 1.1 Energy difference, ΔE, and resulting relative probability, P_i, at 303 K, calculated from $P_i = \exp(-\Delta E/RT)$

ΔE		
kcal	kJ	P_i
0.0	0.00	1.0000
0.1	0.42	0.8465
0.2	0.84	0.7165
0.3	1.25	0.6065
0.5	2.09	0.4346
1.0	4.18	0.1889
2.0	8.36	0.0357
3.0	12.54	0.0067
5.0	20.90	0.0002

to the free energies for the two chairs of β-glucopyranose differ by about 1.6 kcal (6.7 kJ) [48] at 298 K. These all-axial and all-equatorial structures are as different as any conformational variations of an aldohexapyranose, so this may correspond to the maximum range for variations in rovibrational free energy that might be expected in studies of pyranoid rings.

1.2.6 Computer modeling methods

As suggested in the preceding section, computerized molecular modeling is a useful adjunct to experiment. It is arguably even more useful when experimental data are unavailable. In its most elementary form, a computer model is just a list of atomic coordinates with information about the interatom connectivity. Beyond that, the phrase 'computer model' includes the software that is acting on the coordinate list. Computer modeling software falls into several categories: display, structure editing and energy studies such as molecular mechanics (MM) and quantum mechanics (QM) or molecular dynamics (MD). Often, commercial modeling systems incorporate, with varying emphases, all the above capabilities and many more. For example, databases of potential candidates for pharmacological testing can be searched, based on the similarities of various calculated properties. It is important for a program to have a range of structure input methods. Structures can be input by codes for substructures, by sketching (two-dimensional or three-dimensional) or from a variety of file formats.

It is often useful to learn how the energy varies as a result of changes in the molecular conformation. The energy curves in Figure 1.4 are one-dimensional 'energy surfaces'; the use of more variables requires a multidimensional hyperspace. The primary variables for structures that retain their chemical integrity are the torsional angles. Modeling software often provides a means systematically to vary the structure and compute the energy through use of 'dihedral (torsional angle) drivers'. When looking for structures with minimal energy, it is important to realize that, unless the molecule is driven through all possible arrangements, it is not certain that the overall minimum will be found. Although exhaustive searching is prohibitively time-consuming for larger molecules, it is feasible for monosaccharides.

Calculated energies depend to a large extent on the orientations of the hydroxyl groups because of the possibility of different hydrogen bonding systems. For isolated molecules, studied in the gas phase, intramolecular hydrogen bonding will occur. Therefore, it is necessary to consider a variety of arrangements to learn which of the combinations of substituent orientations has the lowest energy. In the case of glucopyranose, there are six rotating exocyclic substituents. If threefold barriers to rotation are assumed, there are 3^6 (729) combinations of exocyclic group orientations that must be examined, a task beyond current high-level *ab initio* calculations. Energy minimization studies based on molecular mechanics suggest that only about 100 of the 729 correspond to local minima on the potential surface.

Even after the question to be answered with a modeling study is formulated, the best way to carry out the study is often not established. As indicated above, simplifying assumptions are often needed. One major consideration is whether to include solvent, and, if so, whether to include it explicitly, through MD, or by a continuum method, which could be used in conjunction with any method of energy calculation.

1.2.6.1 Quantum mechanics

Ab initio QM calculations are based on solving the Schrödinger equation, with approximations, for the molecule. Because of these approximations, the energy values and structural details from affordable QM calculations may not compare as well with experimental results as do MM methods that are based on extrapolations of known behavior. On the other hand, if the theory required to describe the behavior is not incorporated in the MM program there is no hope for consistently adequate results from MM calculations. For example, hydrogen bonds are strengthened and shortened when they participate in networks of hydrogen bonds, compared with isolated examples. Most popular MM systems do not reproduce this effect but it is easily reproduced with QM. On the other hand, a study of the two chair structures of glucose with QM showed that misleading results were obtained when inadequate levels of theory were used. It was only with the most time-consuming levels of theory that could be afforded that QM gave a better correspondence with observed molecular geometry than did MM. The QM calculations that gave an apparently converged energy difference for the two chairs were so large that only 'single-point' calculations could be performed with a supercomputer, and the molecular geometry could not be optimized [48].

Perhaps the most frequent application for QM calculations in carbohydrate chemistry has been for the study of anomeric effects (including the exo-anomeric and reverse-anomeric effects). *Ab initio* quantum mechanics calculations by Salzner and von Ragué Schleyer [49], as well as natural bond orbital studies [50], confirm the importance of delocalization of electrons from the lone pairs of electronegative substituents (hyperconjugation). Further confirmation is given by calculations with even greater levels of electron correlation [21].

Studies of the variation in electronic aspects of molecular structure, such as chemical reactions, are done with QM. Studies of hydrogen bonding are another important use of QM [51]. Such detailed studies reveal both electrostatic and significant electron-sharing attractions in hydrogen bonds. Another major use for QM is the parameterization of MM potential energy functions (see below).

Density functional theory (DFT) QM methods have recently been applied to carbohydrates [52]. Compared with conventional *ab initio* methods with electron correlation, DFT methods offer considerable time savings and reduced memory requirements. Although semi-empirical QM calculations such as MNDO (modified neglect of differential overlap) and its derivatives (e.g. AM1 and PM3) consume much less time than full *ab initio* methods, they currently have too many limitations to be definitive for many of the important points of molecular geometry and energy. On the other hand, problems too big for *ab initio* calculations can be addressed when the limitations are tolerable.

1.2.6.2 Molecular mechanics

Although the basic input and output information for analysis of ground-state molecules by MM and QM programs are apparently similar, there are major differences between the two methods. MM programs incorporate large tables of data including, for example, ideal bond lengths and their stretching (or compression) constants, bond angles and bending constants, etc. for each type of bonded and non-bonded interaction. These data, and the equations that use these constants, constitute the potential energy function, or forcefield. These data come from experimental observations, as well as from QM calculations on small molecular fragments. Thus the results of a properly executed MM calculation should provide a convenient summary of available knowledge for a small molecule and an easy way to extrapolate from existing knowledge to conjecture about larger molecules. When the results of a molecular mechanics calculation are inadequate, this indicates that new theory is needed before understanding of the problem can be claimed. Another major difference between MM and QM is the huge disparity in the amount of time required. MM calculations for monosaccharides can now be done in a few seconds on personal computers, but adequate QM calculations take hours or days of supercomputer time.

Unless given further development, MM potential energy functions developed initially for proteins do not usually work very well for modeling carbohydrates. Additions to the AMBER forcefield have been proposed [53, 54], and the carbohydrate parameters developed in Brady's group [55] have been employed often in calculations with CHARMM [56]. Imberty *et al.* have proposed modifications to the force field in SYBYL [57]. These changes usually include anomeric torsion and ether oxygen-bending functions. Rasmussen and co-workers continue to develop special forcefields for carbohydrates [58, 59], and Vergoten and co-workers [60] have specialized in spectroscopic forcefields for carbohydrates. GROMOS [61] has also been used for carbohydrates, as have forcefields from Darmstadt (PIMM88 [62] and PIMM91 [63]). GLYCAM_93 has also been developed for carbohydrates [64]. More complex MM functions, such as MM3 [65] cover a larger range of structures at the cost of slightly increased computing time and larger computer memory requirements for a given sized molecule.

From the above partial listing of forcefields chosen to model carbohydrates it could be concluded that there is little agreement as to the necessary elements of a modeling system. Still, if the desired goal of a modeling calculation is merely to locate minimum-energy structures then the above programs will often give fairly comparable results. However, if the comparison of energies of several structures with similar energies is the goal then there is still room for improvement in all of the programs, and the results from two different programs will not be the same. From a scientific standpoint, use of the most recent software version is important. If erroneous results are obtained with new software, then the underlying theory or parameterization can be improved. Difficulties with older software may have already been resolved in the newer version. If not, there will be little interest in improving the older programs. In particular, versions of Allinger's MM2 program modified for carbohydrates were effectively made obsolete by later standard versions of MM2 and by MM3.

One reason that simple, protein forcefields are often inadequate for modeling carbohydrates is the necessity of defining torsional energies in terms of four atoms, not just the two atoms that define the bond subjected to torsional change. Forcefields designed for proteins or wide general applicability often depend only on the two central atoms of the torsional sequence. If the potential function for torsional change depends on the types of all four atoms then the anomeric energy differences can be reproduced by the standard energy functions of the program (see Figure 1.4). In MM3 a term is added, however, to attempt to account for the bond-length distance changes arising from torsional changes in anomeric (O–C–O–C) sequences. Because of stereoelectronic effects at the anomeric center, the normal bond stretching arising from torsional change is not adequate to represent all of the variation that occurs during rotation about the glycosidic bond. In some other programs the torsional energies mimic those of butane, regardless of the presence of two oxygen atoms in the sequence. Other differences between protein forcefields and complex general forcefields include stretch–bending and torsion–stretch terms which, in most cases, give small contributions to the total energy compared with other terms. Still, fractions of a kcal are important in many contexts, such as prediction of equilibria.

1.2.6.3 Molecular dynamics

MD software integrates Newton's equations of motion to depict the dynamic behavior of a molecule at a given temperature. These calculations permit information to be calculated beyond that from a simple energy minimization, such as the free energy. (Full matrix energy minimizations also permit calculations of free energy contributions for individual conformers.) In most MD studies of carbohydrates and other molecules the forces on the molecule are based on MM potential energy functions [66, 67]. Still, QM methods can be used; MD–QM is in the early stages of development. One of the chief advantages of MD studies is that the average properties of solvated molecules can be determined [68]. Because solvation is inherently a dynamic property, a full treatment requires explicit inclusion of the solvent molecules. Besides details of the carbohydrate structure, which are perturbed only a little by the presence of solvent in calculations to date, there is also information on the arrangement of water molecules. Typically, several hundred water molecules and a carbohydrate molecule are enclosed in a box with periodic boundaries so that long-range effects can be included without even more computational expense. Still, solvated dynamics calculations require much longer times than do calculations done without solvent. As computers and software evolve, longer and more comprehensive simulations will be practical. Simulations based on MM energy functions are currently carried out for a few hundred picoseconds with solvent, or for a nanosecond or more for vacuum studies. Although some molecular transformations occur on this time scale, others occur much more slowly and the simulations may not be completely representative. Improved software to accommodate the need for including many transformations in a shorter time is being developed [69].

1.3 Shapes of acyclic carbohydrates

The shapes of acyclic aldehydo and keto carbohydrates are described in terms of linear (planar zig-zag) or sickle shapes. Important considerations include whether intramolecular hydrogen bonding or particularly bulky groups in a 1,3-*syn* relationship can overcome the 180° torsion angle favored for the backbone atoms. A recent crystal structure of an unusual Amadori compound has one of its two fructose moieties in a planar, sickle shape, while the other fructose moiety makes a chair that has the normal $^{2}C_{5}$ conformation [70].

1.4 Shapes of carbohydrate rings

1.4.1 Size and energy

The most common rings are pyranoid or furanoid, with 6 and 5 members, respectively. Seven-membered rings also occur but are thought to have higher heats of formation, greater than the acyclic versions of the same compound. Larger rings for monosaccharide are less common. Approximate indicators for the propensity to form different ring shapes are given by heats of formation for seven-carbon analogs of the different sugar rings, as shown in Table 1.2 (all the energies therein are molar) [71, 72]. The energies for cyclohexane and its linear fragment are nearly identical, showing that there is no apparent penalty for formation of six-membered rings. In a similar comparison of cycloheptane and its linear fragment there is a cost of about 6 kcal (25 kJ) for ring closure.

Examination of the components of the calculated values for the various structures in Table 1.2 shows that the major differences arise from the torsional energies and bending energies (see Tables 1.3–1.7 for analogous information for a glyceraldehyde model). For sugar rings, with their heteroatoms and substituents, the furanoid and pyranoid forms can be equienergetic in selected environments, as observed for the ketosugars fructose and psicose. The acyclic tautomer usually has a higher energy [e.g. 2 kcal (*c.* 8 kJ)], and only a few per cent of it are found in the most favorable instances [35]. The shapes are described qualitatively in terms of where the rings are non-planar (planar rings have endocyclic torsion angles of 0°, a high-energy condition).

1.4.2 Characteristic furanose shapes

A plane can always be found that intersects any three points (or atoms). Of the five atoms in saturated furanoid rings having minimal energy, either one

Table 1.2 Heats of formation for various arrangements of C_6H_{12} and C_7H_{14} with single bonds

	Experimental value [71]	MM3 calculation [72]
n-hexane fragment of long chain (C_6H_{12})	−29.58[a]	−29.52[a]
cyclohexane (C_6H_{12})	−29.43	−29.69
n-heptane fragment of long chain (C_7H_{14})	−34.51[a]	−34.44[a]
cycloheptane (C_7H_{14})	−28.22	−28.03
n-heptane (C_7H_{16})	−44.89	−44.89
methyl cyclohexane (C_7H_{14})	−36.99	−37.18
cis 1,3 dimethyl cyclopentane (C_7H_{14})	–	−33.93[b]
ethyl cyclopentane (C_7H_{14})	−30.34	−30.80

[a] Calculated from the incremental differences between ΔH_f values for *n*-pentane, hexane, heptane, octane and nonane [−4.92 kcal × (6 or 7) $-CH_2-$ units] [72].
[b] Calculated by Alfred D. French for the present chapter.

Table 1.3 Bond lengths and stretching energy in glyceraldehyde

Bond	Current bond length	Standard bond length	Stretching constant for changes in bond length	Energy caused by stretching	Electronegativity correction of bond length	Miscellaneous bond-length correction[a]	Stretching constant correction of bond length
C1–C2	1.5202	1.5090	4.800	0.0421			
C1–O4	1.2094	1.2080	10.100	0.0014			
C1–H8	1.1191	1.1180	4.370	0.0004			
C2–C3	1.5382	1.5231	4.490	0.0708	−0.0140	0.0124 HY	
C2–O5	1.4331	1.4280	5.700	0.0105	0.0150		
C2–H9	1.1116	1.1082	4.931	0.0041	−0.0079	0.0041 HB	0.121
C3–O6	1.4386	1.4280	5.700	0.0446	0.0150		
C3–H7	1.1132	1.1106	4.721	0.0023	−0.0055	0.0041 BH	0.041
C3–H10	1.1122	1.1106	4.723	0.0008	−0.0055	0.0041 BH	0.043
O5–H11	0.9489	0.9470	7.630	0.0020			
O6–H12	0.9480	0.9470	7.630	0.0005			

[a] Depending on whether HY, HB or BH conditions apply.
Note: HY = hyperconjugation effect; BH = Bohlmann effect; HB = hyperconjugation plus Bohlmann effects.

Table 1.4 Non-bonded energies in glyceraldehyde greater than 0.1 kcal (0.4 J)

Atom pair	Current distance[a]	Distance after relocation[b]	Radii[c]	K^d	Energy (kcal/mol)	
C1, O6	2.9027		3.760	0.0575	0.3916	1, 4
C1, H7	2.8301	2.7926	3.560	0.0335	0.1795	1, 4
C1, H11	2.4861	2.4695	3.540	0.0299	0.6904	1, 4
C3, H11	2.6819	2.6494	3.640	0.0208	0.3011	1, 4
O4, O5	2.7448		3.640	0.0590	0.5538	1, 4
O4, H11	2.4612	2.4713	2.140	3.9750	−0.3410	H
O5, O6	2.8566		3.640	0.0590	0.3144	1, 4
O5, H10	2.6307	2.5969	3.440	0.0344	0.3177	1, 4
O6, H11	2.6328	2.6386	2.110	3.0000	−0.1756	H
H7, H9	2.4910	2.4087	3.240	0.0200	0.2249	1, 4
H7, H12	2.1458	2.0908	3.220	0.0179	0.8228	1, 4
H8, H9	2.5713	2.4816	3.240	0.0200	0.1523	1, 4
H9, H10	2.5366	2.4488	3.240	0.0200	0.1822	1, 4

Note: 1, 4 = interaction between atoms bonded to each other; H = hydrogen bonding interaction.
[a] Current distance between atoms of the pair.
[b] Distance between atoms after relocation of the hydrogen atom(s) to account for the movement of electron density for hydrogen atom(s).
[c] Sum of van der Waals radii.
[d] Constant for the given atom pair.

Table 1.5 Bending energies in glyceraldehyde

Atoms	Current bond angle θ (°)	Standard bond angle T_0(°)	Bending constant	Bending energy (kcal/mol)	Stretch bending constant	Stretch bending energy (kcal/mol)
C2–C1–O4	124.446	123.500			0.13	0.0039
in-plane 2–1–4	124.446	123.500	0.850	0.0164		
out-of-plane 1–8–1	0.082		1.180	0.0002		
C2–C1–H8	116.650	116.100			0.08	0.0014
in-plane 2–1–8	116.650	116.100	0.464	0.0031		
out-of-plane 1–4–1	0.074		0.650	0.0001		
O4–C1–H8	118.904	119.200			0.08	−0.0001
in-plane 4–1–8	118.904	119.200	0.850	0.0016		
out-of-plane 1–2–1	0.063		0.590	0.0001		
C1–C2–C3	111.776	110.600	0.800	0.0238	0.13	0.0101
C1–C2–O5	111.241	109.500	0.700	0.0454	0.13	0.0093(C)
C1–C2–H9	108.202	109.490	0.540	0.0200	0.08	−0.0038
C3–C2–O5	107.868	107.000	0.830	0.0135	0.13	0.0057
C3–C2–H9	109.066	109.800	0.590	0.0070	0.08	−0.0027
O5–C2–H9	108.626	110.000	0.820	0.0346	0.08	−0.0023
C2–C3–O6	109.308	107.900	0.830	0.0354	0.13	0.0118
C2–C3–H7	110.568	109.310	0.590	0.0201	0.08	0.0045
C2–C3–H10	110.486	109.310	0.590	0.0176	0.08	0.0039
O6–C3–H7	110.337	108.900	0.820	0.0363	0.08	0.0038
O6–C3–H10	109.094	108.900	0.820	0.0007	0.08	0.0005
H7–C3–H10	107.017	107.600	0.550	0.0041		
C2–O5–H11	107.578	106.800	0.750	0.0098	0.09	0.0012
C3–O6–H12	108.970	106.800	0.750	0.0751	0.09	0.0057

Note: C = values of 'trial' status were used for the bending constant.

Table 1.6 Torsional energies in glyceraldehyde

Atoms	Current torsion angle ω (°)	Torsional constants			Torsion energy (kcal/mol)	Torsion stretching constant	Torsion stretch energy (kcal/mol)
		V_1	V_2	V_3			
C1–C2–C3–O6	57.6	0.000	0.000	0.180	0.001	0.059	0.000(C)
C1–C2–C3–H7	−64.1	0.000	0.000	0.180	0.002	0.059	0.000
C1–C2–C3–H10	177.6	0.000	0.000	0.180	0.001	0.059	0.000
C1–C2–O5–H11	−45.4	0.000	0.000	0.090	0.013	0.100	0.001(C)
C2–C3–O6–H12	−111.4	0.400	0.000	0.000	0.222	0.100	−0.012
C3–C2–C1–O4	−118.2	−0.457	0.106	−0.160	0.579		
C3–C2–C1–H8	62.1	0.655	0.266	0.474	0.690	0.100	0.000
C3–C2–O5–H11	77.5	0.400	0.000	0.100	0.263	0.100	−0.001
O4–C1–C2–O5	2.5	−0.420	2.330	0.640	0.222		(C)
O4–C1–C2–H9	121.7	−0.154	0.044	−0.086	−0.091		
O5–C2–C1–H8	−177.3	0.000	0.000	0.000	0.000	0.100	0.000(C)
O5–C2–C3–O6	−65.0	0.500	−2.000	1.900	−1.255	0.059	0.000(B)
O5–C2–C3–H7	173.3	0.000	0.000	0.300	0.009	0.059	0.000
O5–C2–C3–H10	55.0	0.000	0.000	0.300	0.005	0.059	0.000
O6–C3–C2–H9	177.2	0.000	0.000	0.300	0.002	0.059	0.000
H7–C3–C2–H9	55.5	0.000	0.000	0.238	0.003	0.059	0.000
H7–C3–O6–H12	10.4	0.000	0.000	0.200	0.186	0.100	−0.012
H8–C1–C2–H9	−58.0	0.115	0.027	0.285	0.108	0.100	0.000
H9–C2–C3–H10	−62.8	0.000	0.000	0.238	0.001	0.059	0.000
H9–C2–O5–H11	−164.4	0.000	0.000	0.200	0.032	0.100	−0.001
H10–C3–O6–H12	127.6	0.000	0.000	0.200	0.192	0.100	−0.012

Note: C = V_1, V_2 and V_3 had trial status; B = V_1, V_2 and V_3 were of 'fair' status in terms of the quality of the estimates.

Table 1.7 Dipole–dipole interaction energies in glyceraldehyde

Bond 1	Dipole magnitude μ_1	Bond 2	Dipole magnitude μ_2	R_{12}[a]	Energy[b] (kcal/mol)
C1–C2	−1.010	+C3–O6	1.170	2.1634	0.5682
C1–C2	−1.010	+O5–H11	−1.670	1.9197	−1.0856
C1–C2	−1.010	+O6–H12	−1.670	2.7266	−0.3470
C1–O4	1.860	+C2–O5	1.170	2.1287	1.3137
C1–O4	1.860	+C3–O6	1.170	2.9772	0.3200
C1–O4	1.860	+O5–H11	−1.670	2.4162	−0.7348
C1–H8	−0.600	+C2–O5	1.170	2.3425	−0.2498
C2–O5	1.170	+C3–O6	1.170	2.1257	0.6515
C3–O6	1.170	+O5–H11	−1.670	2.5043	−0.5842

[a] Distance between centers of dipoles.
[b] Calculated energy for the dipole–dipole interaction.

or two atoms will be out-of-plane. Furanoid rings with four coplanar ring atoms are termed envelopes (E) because the bonds to the single out-of-plane atom resemble the partly folded flap of an envelope (Figure 1.7). There are 10 characteristic E shapes because each ring atom can be above or below the plane of the other atoms. The out-of-plane atom is indicated by a superscript or subscript character to represent the above-plane or below-plane atom, for example ^{3}E. There are also 10 characteristic rings with three coplanar atoms that are called twists (T). T forms have two adjacent out-of-plane atoms. In the characteristic T conformers the two out-of-plane atoms are displaced in opposite directions by equal but unspecified distances. The symbol for the above-plane atom is superscripted to the left of the T, and the below-plane atom is subscripted to the right. A refinement, seldom used, shows the atom that deviates the most from the mean plane on the left and the atom deviating the least on the right, subscripted or superscripted as appropriate. Equal deviations are designated with both atoms on the left side of the T.

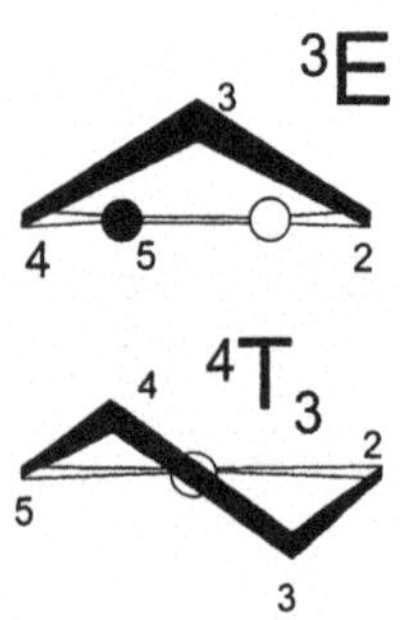

Figure 1.7 Envelope and twist forms for furanoid rings.

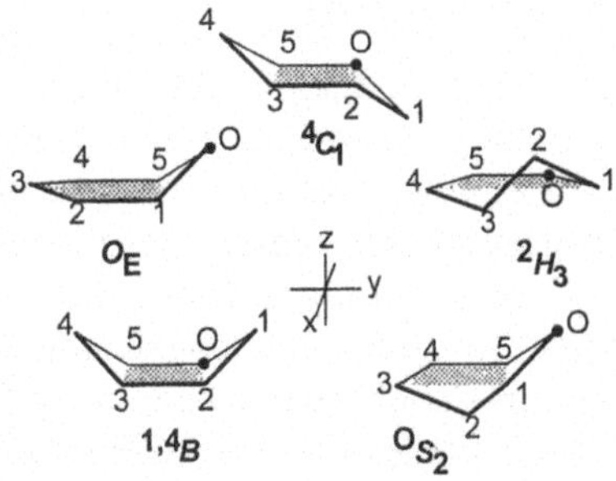

Figure 1.8 Examples of chair, boat, skew, envelope and half-chairs for pyranoid rings.

1.4.3 Characteristic pyranose shapes

The defined, characteristic shapes of six-membered rings have a minimum of four atoms in a plane. Pyranoid rings with endocyclic or exocyclic double bonds can also have stable structures with five atoms in a plane. Altogether there are 38 unique characteristic forms of pyranoid rings. Again, the out-of-plane atom positions are indicated by superscripts and subscripts (Figure 1.8). Unlike the furanoid rings, not all of the 38 characteristic forms are potentially stable if all bonds are single. Only the chairs, boats and skew forms are likely for saturated structures.

For a given chair or skew form it is not obvious which set of four atoms should be used to define the plane. For example, perfect glucose chairs with $^4C_1$, $^2C_5$ and $^oC_3$ shapes are identical. The other glucose chairs, $^1C_4$, $^5C_2$ and ^3C_o, also comprise a set of identical structures. Convention [28] dictates that the out-of-plane ring atom with the lowest number be part of the name. Thus, only the two unique chairs, one from each set, are mentioned. The characteristic boat conformers have each of the out-of-plane ring atoms displaced equally in the same direction from the plane. Each of the six boats (B) is unique, with the two out-of-plane atoms separated by two sets of atom pairs that form the plane. The skews (S) have out-of-plane atoms separated by one in-plane ring atom; only six are unique. Twelve half-chairs (H) for six-membered rings can be formed when four connected atoms are coplanar, with the two remaining adjacent atoms above and below the plane of the other four. The 12 structures with a single out-of-plane atom are 'envelopes'. This term, although official [28], is not universally used, with 'sofa' or 'half-chair' also being descriptive names. However, the S descriptor is used for the skew forms, and 'half-chair' describes a different condition. E and H structures are not stable for rings with saturated ring and exocyclic bonds. Another descriptor, the 'screw' shape, is used for rings similar to a boat and to a half-chair, with the two out-of-plane atoms on the same side of the plane but elevated by different amounts [73]. Such shapes are not stable for isolated rings but can occur in fused ring systems.

1.4.4 Pseudorotation in furanose rings

Flexibility and interchange among the possible shapes of furanoid rings are understood initially by analogy to cyclopentane. These structures undergo pseudorotation, the process that results in interconversion among the possible forms without passing through an intermediate structure with all ring atoms coplanar. Given a tetrahydrofuran ring with only the oxygen atom above a plane (the ^{O}E structure), conversion to the $^{O}T_1$ structure (carbohydrate nomenclature) is effected by lowering O while keeping C2, C3 and C4 in the original plane. C1 will be displaced downwards until both O and C1 are displaced equal (but opposite) distances from the plane of the three remaining ring atoms. At that point, the characteristic $^{O}T_1$ shape has been generated. Continuing, eventually C1 will have a full downward displacement and O will lie in the plane of the C2, C3 and C4 atoms, creating the E_1 structure. The movement of C1 is then reversed, and the held atoms are changed to C3, C4 and O. Upward movement of C1 then causes upward movement of C2, and the characteristic shape $^{2}T_1$ is generated when C1 and C2 are equidistant from the plane. This process can continue until all 20 characteristic furanoid conformers have been generated, at which time further pseudorotation generates the original ^{O}E structure. Thus, it is advantageous to represent the list of structures obtained by successive pseudorotational operations on a circle that is called a conformational (or pseudorotational) wheel (Figure 1.9).

The free energies of the envelope and twist forms of cyclopentane are equal, and there are no energy bariers to free pseudorotation. Inclusion of a ring oxygen and decoration of the ring with hydroxylic and hydroxymethyl substituents typically adds barriers of a few kcal to facile pseudorotation of furanoid sugars. This is usually not enough to prevent substantial pseudorotation in solution, but does cause some regions of pseudorotational space to be favored.

1.4.5 Puckering of furanose rings

Ring forms intermediate to the characteristic conformers exist (they are the rule rather than the exception) so it is useful to have a descriptive system that avoids reliance on phases such as 'a ^{3}E, distorted slightly towards $^{3}T_4$'. Because each of the 20 characteristic conformers is spaced 18° apart on the conformational wheel, any intermediate can be described in terms of a phase angle, Φ. The extent of out-of-plane deviation is called the amplitude, q. These puckering parameters may be plotted on a plane polar coordinate system, with the amplitude being the radius and the phase angle corresponding to the position on the pseudorotational wheel. Two conventions are in wide use for five-membered rings. One is the Altona–Sundaralingam (A–S) system [74], consisting of a phase angle Φ, and an amplitude angle, X.

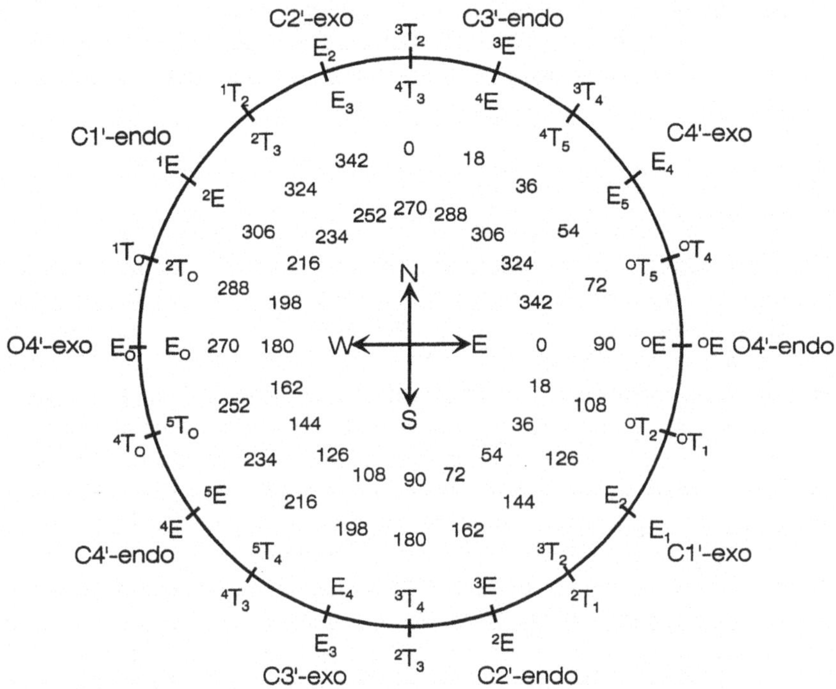

Figure 1.9 Conformational wheel for furanose rings. The inner ring is for the Cremer–Pople [75] parameters and the next ring is for the numeric values of Altona and Sundaralingam [74]. The next ring gives the letter designators for ketose conformations and the next for aldoses. The endo and exo notations are used mostly for ribose and deoxyribose Φ, θ locations for the 38 characteristic conformers of pyranoid aldoses (upper) and ketoses (lower). The interconversions between the numeric values of the parameters and the characteristic conformations depend on numbering the ring oxygen atom as atom number 1 in the input to puckering programs. Note: Φ = 'longitude'; θ = 'latitude'.

The other is the Cremer–Pople (C–P) system [75], with a different phase angle, Φ, and an amplitude scalar, q. In the A–S system, $\Phi = 0°$ for a T form with the two carbon atoms opposite the ring oxygen atom displaced below and above the plane of the other three (e.g. $^3T_2$ for an aldofuranose). In the C–P description, $\Phi = 0°$ for the ^OE form. In both systems, Φ increases with clockwise rotation, a reflection of the clockwise numbering of rings when they are viewed in the standard position. The degree of puckering, X, or the amplitude, q, is 0 for an all-planar structure that would be depicted at the center of the conformational wheel. Off the center, q is the mean deviation of the ring atoms from the mean plane. A useful coincidence is that $X \simeq 100q$, and $\Phi_{A-S} = \Phi_{C-P} + 90°$. The A–S system is used extensively for nucleotides and nucleosides, while the C–P system is favored by other crystallographers because the same system can be applied to rings of other sizes. Given the coordinates of the ring atoms, computer programs can calculate the puckering parameters [76]. Besides the quantitative puckering parameters and the qualitative letter notation, there is an even more informal notation that divides the conformational wheel into four quadrants, labeled north, south, east and west (Figure 1.9).

1.4.6 Puckering of pyranose rings

Pyranoid ring geometries are nominally similar to those of cyclohexane. Facile pseudorotation in cyclohexane is only possible among the high-energy B and S forms, in between which there are 0.5 kcal (2 kJ) barriers. For saturated pyranoid sugars, the barriers along the B–S pseudorotational itinerary are typically much higher. Interconversions between the C forms typically proceed through the 'symmetrical intermediate' E or H barriers, with a temporary (in most cases) visit to the B–S pseudorotational itinerary. Alternatively, ring opening and reclosing is involved. Because the pyranoid rings have a sixth atom, a third puckering parameter is needed. Φ depicts where the ring is puckered; θ indicates the type of ring puckering (C, B, S, E or H), and, again, q depicts the extent of deviation from the mean plane of the ring atoms. These three parameters constitute a spherical polar coordinate system, with analogy to values for a globe of the Earth [77]. Φ corresponds to the longitude, θ to the latitude and q to the altitude (measured from the center of the Earth). Figure 1.10 is a *platte carée* representation of the globe, similar to a Mercator projection but without the variations in latitudinal spacing to maintain proportionality of the land masses near the poles. Each of the chairs resides at one of the poles, with the favored $^4C_1$ form of D-glucose having, by convention, $\theta = 0°$. Although the $^4C_1$ form of D-glucose is said to be at the 'north' pole, most plotting software places the 0° values of θ at the bottom of the graph. The favored form of L-glucopyranose would be a $^1C_4$, and its θ puckering parameter would be 180°. The B and S forms are at $\theta = 90°$. Six of the E forms have θ values of 50° and six are at 130°. H forms are found at 54° and 126°. Refinements to the puckering parameter calculation have been presented [78].

1.4.7 Seven-membered rings

Seven-membered rings can take either chair (C) or twisted-chair (TC) structures [1]. Boat (B) and twisted-boat (TB) forms are also characteristic forms, but have systematically higher energies, at least for the cycloheptane hydrocarbon. The fourteen C and fourteen TC forms of septanoses constitute a pseudorotational itinerary, with the TC forms of cycloheptane having about 1 kcal (4 kJ) lower energy than the interspersed C forms that constitute small barriers to facile pseudorotation. The reason that the C forms are barriers is that the four-atom 'footrest' or 'backrest' of the chair requires a 0° torsion angle. The all-planar form has a very high MM3

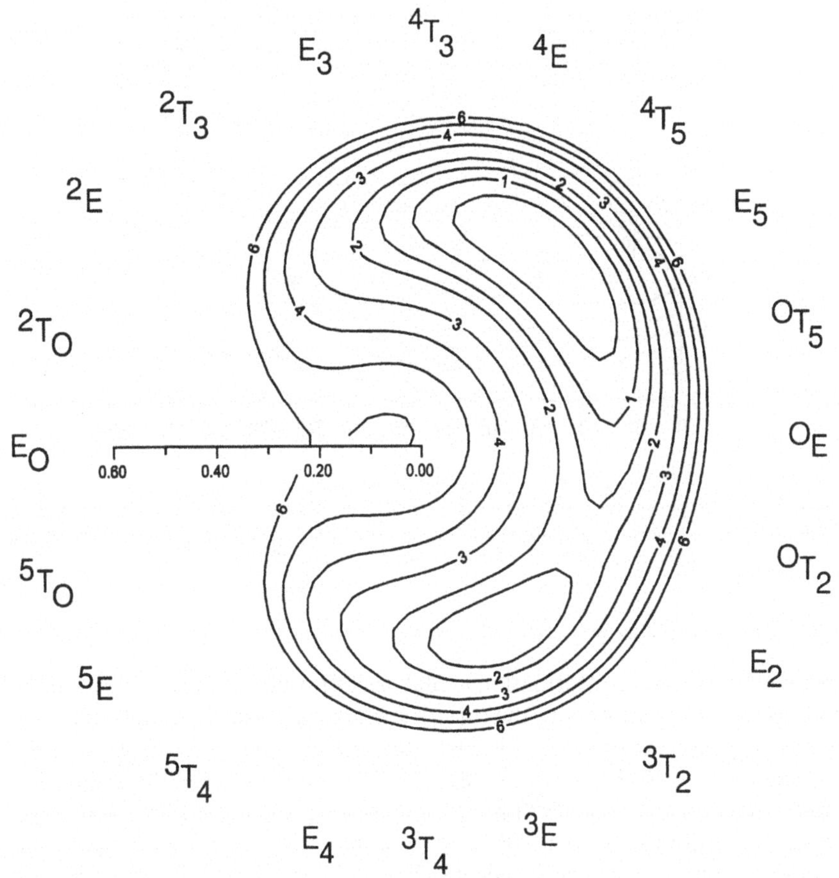

Figure 1.10 α-Fructofuranose energy surface calculated with MM3(92) and a dielectric constant of 4.0 [81]. [North–south interconversion takes an eastern path. The western path causes a conflict of the two primary alcohol groups and therefore has high energy [greater than 6.0 kcal/mol (25 kJ)]. The puckering amplitude is indicated by the horizontal line; the structures at the $^{4}T_{5}$ minimum have a value of q of about 0.42 Å. At the position of the southeastern minimum (*c.* $^{3}T_{2}$) q is slightly smaller, at about 0.38 Å.

energy [about 50 kcal (209 kJ) above the TC form] because of the energy of the 0° torsions and the endocyclic bond-angle strain (each C–C–C angle is about 130°). The fourteen B and TB septanoid ring shapes constitute a second pseudorotational itinerary. Transitions between the two itineraries are accomplished by high-energy 'symmetrical transitions' where the footrest of the chair, for example, participates in an intermediate with six coplanar atoms, when it is flipping up to make the boat shape. Because there are four major puckering parameters for septanoid rings, unambiguous two-dimensional representation is not possible. However, a puckering surface has been presented, based on two endocyclic torsion angles [79].

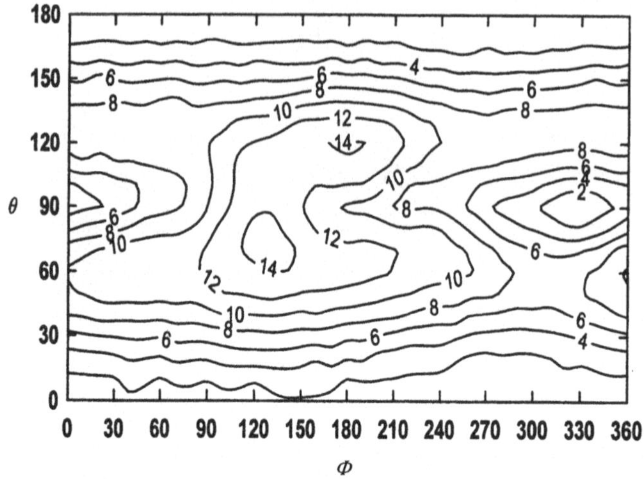

Figure 1.11 α-Idopyranose energy surface calculated with MM3(92) and a dielectric constant of 3.0 [85]. This plot shows that both chairs and a skew are predicted (see Figure 1.9 for the conversion). The path of interconversion is near Φ angles of 300°.

1.4.8 Energy surfaces

In a computational conformational analysis of furanoid rings, the computed energy can be plotted on the (Φ, q) surface (Figure 1.10) [80, 90]. For pyranoid rings, the contours of calculated energy have been displayed on a surface of θ vs. Φ (Figure 1.11) [82, 83]. The amplitude for the minimum-energy structure at each (Φ, θ) point can also be plotted on a θ vs. Φ surface. It varies substantially (0.4–0.8 Å) for different ring forms with optimized geometry, and q is highest for skews. Typically, the E and H structures are significant [e.g. 10–20 kcal (42–84 kJ)] barriers to interconversions between the chairs at the poles and the S and B structures at the equator. MM3 studies of all cyclic tautomers of fructose [81] and psicose have been published [84], as well as comprehensive MM3 treatments of the aldopyranoses [85]. For unsaturated sugars, another representation [86] may be more useful because the range of likely conformations (mostly B, E and H forms) is considerably restricted by the resistance to torsional change about the double bond. On the energy surfaces for such molecules the axes are the displacements (in Ångstrom) of the two atoms of the ring opposite the ends of the double bond.

1.5 Detailed factors that determine the shapes of sugar rings

It might be argued that the Schrödinger equation governs the molecular shape and that it is improper to attempt further analysis. The more pragmatic theoretician might apply a natural bond orbital analysis [21] to quantify

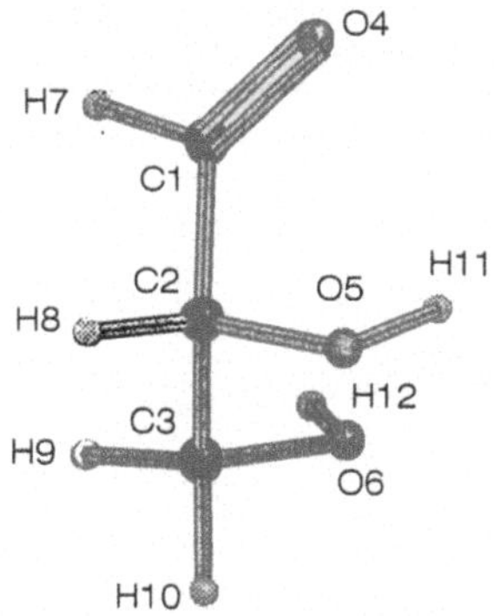

Figure 1.12 D-Glyceraldehyde in a low-energy form as computed by MM3(94). The atomic numbering is consistent with Tables 1.3–1.8 in text.

different contributions to the energies of various conformers. The distillation of the various types of contributions to the molecular potential energy gives rise to the possibility of predicting the most stable forms of unstudied molecules. Unfortunately, there are so many different contributions to the energy of even a monosaccharide that it is often necessary accurately to sum a large number of small contributions, both positive and negative (shown in Tables 1.3–1.7 for glyceraldehyde) (Figure 1.12). Because of the difference between the lengths of C–C and C–O bonds, formation of rings with an oxygen atom results in some bond-length distortion potential energy compared with cyclic hydrocarbons. Altogether there are 24 bonds to be considered in glucopyranose. There are also 42 bond angles, some of which are significantly strained, and 66 torsion angles. Intramolecular van der Waals interactions, and, finally, the electrostatic contributions must be considered. In computerized molecular models, the partial charges on the individual atoms of neutral molecules contribute to the calculated energy. The electrostatic energy is inversely proportional to the dielectric constant that is assumed for the model. MM3 calculations over a range in the dielectric constant of 1 to 80 have from 44 to 0 dipole–dipole interactions greater in energy than 0.2 kcal (0.8 kJ). Table 1.8 shows the sums of the component energies reported by the 1994 version MM3 at a dielectric constant of 4.0 for β-glucose. Although stereoelectronic energy values are not explicitly mentioned, they are incorporated in the MM3 forcefield in an elementary way, at least, and are reflected in the above molecular description.

The classical work of Angyal and later work by Franck [87] was based on a summation of terms (the so-called *A* values) arising from axial vs. equatorial substituents on chair-form pyranoid rings. It gives a good start to understanding conformational preferences. In a way, that method could be regarded as 'mental molecular mechanics', simple enough that a computer is not needed to determine relative potential energies. Still, there are several

Table 1.8 MM3 energy terms in β-glucose at a dielectric constant of 4.0

Term	Energy (kcal)
Compression	0.5489
Bending	1.1863
Bend–bend	−0.0931
Stretch–bend	0.1655
Van der Waals energy	
1,4	13.0980
Other	−1.5211
Torsional	−1.9616
Torsion–stretch	−0.0110
Dipole–dipole	−2.0715
Charge–dipole	0.0000
Charge–charge	0.0000
Total energy	9.3404[a]

[a] 39 kJ.
Note: Figures are subject to rounding errors.

reasons to look for a more advanced system for predicting preferred structures. Such systems are not general, being inapplicable to other types of structures, such as skew forms of pyranoid rings, unsaturated rings or five-membered rings. They do not apply to other structural features. Also, they are not able to take into account different molecular environments. Another difficulty is that there is only one energy term in this simple form of molecular mechanics, attributed to van der Waals repulsion, that logic suggests would not be additive. For example, the presence of two bulky axial groups on one side of the ring should increase the energy more than having one on each side of the ring. On the other hand, the contributions from torsional energy should be additive.

The final reason to look beyond 'mental molecular mechanics' is that reasonably good (and rapidly improving) computer software exists for performing just this type of task. For predicting ground-state structures there is little point to avoiding such a study. The MM formalism is expressed in terms familiar to chemists, so the dissection of molecular energy should be easily understood. The reader is cautioned that MM software in general aspires to give a total result that is approximately correct but usually is not robust enough to provide for precise delineation of the individual contributions to the energy of an observed structure. Still, such a decomposition with the best available software is likely to be more accurate than a simple summation of axial vs. equatorial substituent energies, or energies predicted by summing the *gauche* and anomeric effects.

In the *A* value systems no account was made for intramolecular hydrogen bonding or torsional values intermediate to favorable and unfavorable torsion angles. Although broadly understood, the finer points of these issues are current topics of modeling system development. In the case of torsion angles, accurate experimental distortion energies are difficult to obtain, and it is difficult to know the proper degree of hydrogen bonding to include in a calculation. The net favorable electrostatic interactions (hydrogen bonding and other dipole–dipole interactions) amount to as much as 8 kcal or 10 kcal (33 kJ or 42 kJ) in a monosaccharide model such as glucose which can form a peripheral ring of rather weak hydrogen bonds when in its usual $^4C_1$ conformation [88]. In a vacuum these hydrogen bonds would be likely, but in aqueous solution intermolecular hydrogen bonds with better geometry can be formed and intramolecular hydrogen bonds will be less of a factor in determining the structure. On the other hand, the never-observed $^1C_4$ β-D-glucose form could make very strong hydrogen bonds that might be more likely to be retained in solution. Such a strong intramolecular hydrogen bond is found in crystalline methyl-β-ribopyranoside [89].

A comparison of the components of the MM3 total energies for the two chair forms of β-D-glucose shows that the $^1C_4$ shape has 3.2 kcal (13.4 kJ) more torsional energy [48]. The net contribution from non-bonded interactions only adds 1.0 kcal (4.2 kJ) to the difference in favor of the $^1C_4$ shape. This latter difference is deceptive, however, because it balances the stronger hydrogen bonding of the $^1C_4$ shape against the lower repulsion in the $^4C_1$ structure. The 1,3 diaxial repulsions of the $^1C_4$ form are partly absorbed by bending of the bond angles and by small torsional distortions. Also, the endocyclic bonds are stretched somewhat in the $^1C_4$ shape. This energy decomposition contrasts sharply to the axial-group repulsions implied by Angyal's analysis even though the net energy differences for the two chairs by the two methods are similar.

1.6 Likely shapes for monosaccharides

The β-D-aldopyranoses tend to take the $^4C_1$ shape. Three of the α anomers have observable amounts of the $^1C_4$ form in equilibrium [85]. Some α-allose is present in the $^1C_4$ shape, and even more α-altrose in $^1C_4$. α-Idose is the most interesting example, with both chairs present in nearly equal amounts. In addition, one of its skew forms is also equi-energetic, as shown both by NMR and by modeling. A conformational energy map for the ring puckering of α-idose (α-idopyranose) is shown in Figure 1.11.

The β-ketoses are somewhat more ambiguous in their conformational preferences than the β-aldoses because of the presence of two substituents on the anomeric carbon, one of which will be axial regardless of which chair is preferred. Further, ketosugars are more likely than aldosugars to exist as furanoses, as indicated by the analogy to the lower heat of formation of *cis* 1,3 dimethyl cyclopentane (as in the ketohexafuranoses) compared with that of ethyl cyclopentane (analogous to the aldehexafuranoses) (see Table 1.2).

The furanose sugars typically have two low-energy forms (e.g. N and S), and relatively low [e.g. 2–4 kcal (8–17 kJ)] barriers to their interconversion (Figure 1.10). Although an all-planar ring shape is strained, such shapes have energies about 5–7 kcal (21–29 kJ) above the minimum. The energies of the eclipsed endocyclic torsion angles are partially offset by the reduced energies of nearly tetrahedral endocyclic bond angles enjoyed by planar five-membered rings. Typically, the contour 1 kcal above the minimum will extend to include several characteristic conformers, and the low-energy regions will be described in terms of which quadrant of the conformational wheel is preferred, such as the northern shape preferred by β-fructofuranose. The low-energy region in Figure 1.10 is northeastern. Because the furanose rings, as relatives of cyclopentane, have relatively low barriers to pseudorotation, the ring shape is easily deformed by factors such as crystal field and, presumably, specific solvent effects. Successful modeling also depends on the software, perhaps more so than for pyranose rings. It has been proposed that calculations that show an observed conformational feature to correspond to an energy more than 3.0 kcal/mol (12.6 kJ/mol) above the minimum for an isolated molecule indicate an erroneous energy calculation (or a defective experimental result) [90].

The exocyclic orientations have been obtained most reliably from experiment. Typically, the oxygen atoms of primary alcohol groups are orientated so as to avoid close 1–5 interactions with other oxygen atoms. The examples are divided according to whether the sugars have the *gluco* configuration at C4 (with equatorial O4) or the *galacto* configuration (with axial O4) (Figure 1.13) [91]. In both *gauche* + (*gt*) is one of two favored orientations. The other favored orientation is *gauche* – (*gg*) for glucose, and *trans* (*tg*) for galactose. The *trans* structures for glucose, almost never observed, and the *gauche* – form for galactose, seldom observed, have close interactions between O4 and O6. Modeling studies of isolated molecules have often favored these close interactions because the geometry is correct for the

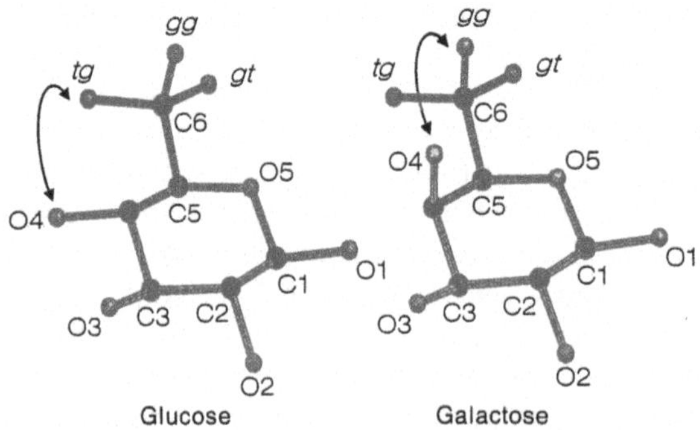

Figure 1.13 Glucose and galactose, showing the staggered conformations of the O6 primary alcohol group.

formation of strong hydrogen bonds, and the additional stabilization from these intramolecular hydrogen bonds is often enough to overcome factors working against the experimentally 'forbidden' orientation. In experiments involving condensed phases, other hydrogen bonding systems are even more attractive, and the systems with intramolecular O6–O4 hydrogen bonds are not found.

Despite the paucity of examples of 'forbidden' conformations from precise and accurate single-crystal diffraction studies or from NMR experiment, it has long been thought that O6 takes a *trans* (*tg*) conformation in glucose residues in native cellulose. It is an interesting dilemma because the experimental studies are based on the fiber diffraction technique that is much less definitive. However, there have been studies by different research groups, and they have reached the same conclusion regarding the O6 orientation despite different outcomes on other aspects of cellulose structure. A recent single-crystal study of cellotetraose, a mimic of cellulose II (the mercerized or regenerated form) conclusively showed all O6 atoms to be *gt* (*gauche* +), unlike the fiber diffraction studies of cellulose II, which found O6 in *tg* (*trans*) as well as *gt* positions [92].

Correct prediction of primary alcohol orientations requires very good balance between the O–C–C–O torsional energies and the hydrogen bonding energies, but since the isolated molecule is never studied experimentally the relevant experimental information may not be available. Molecular dynamics studies in water and molecular mechanics studies of isolated molecules with increased dielectric constants have been fairly successful in predicting that the *trans* structures of glucose would have higher energy than would the other forms. Recent MM3 calculations at the standard dielectric constant for isolated molecules ($\varepsilon = 1.5$) and *ab initio* calculations at the highest currently possible levels show that the *trans* position for O6 of glucose is higher in energy but not by an amount sufficient to account for the experimental observations. However, that situation is further improved by contributions from vibrational free energy [48].

1.7 Effects of condensed phases (solvents and crystals)

To date, the effects of solvent on monosaccharides are not well understood, although some substantial results are available. A pertinent example is the work by Dais and Perlin [93], who showed that in water fructose prefers the β-pyranose form but that in DMSO it prefers the furanose form, with substantial α-furanose present. Other work shows that the temperature of the solvent can also affect the fractions of various tautomers, and changes in solvent and temperature also affect anomeric ratios.

One of the difficulties in gaining knowledge of structures of molecules that are not distorted by the environment is that almost all experimental data for carbohydrates are from condensed phases, either solids or solutions. It would be very useful to have experimental data on molecular structure prior to solvation so that the effect of solvent could be determined by difference. Instead, we have results such as the above that show the changes in molecular structure resulting from changed solvent. Based on the changes in the ratios of tautomers in the work by Dais and Perlin, the net effects of changes in solvent on such equilibria correspond to a change in free energy of roughly 1 kcal (4 kJ).

Differences between the crystalline and solution states are substantial. In crystals, atomic movements resulting from thermal motion are much smaller than in solutions, and solutions lack the symmetry of crystals. Still, in crystalline solids the neighbors surrounding a central molecule could be considered to be frozen 'solvent' molecules. While the 'solvent' is not intuitively similar to solvents of more general interest, such as water, the study of crystals has the advantage that the 'solvent' can be located precisely by diffraction studies.

Surveys of monosaccharide conformations in crystal structures reveal significant ranges. The variations could be thought of as arising from 'specific solvent effects'. If the θ puckering parameter for crystalline α-glucosyl residues having a nomimal $^4C_1$ shape is allowed to take negative values for one of the two hemispheres defined by $\phi = 0°$ to $180°$, or $180°$ to $360°$, θ has an observed range of more than 25° in crystal structures. A given endocyclic torsion angle may range over 20° in the same moiety in different crystal structures. An even simpler indicator of the variation in ring shape is provided by the range of distances between O1 and O4 of α-glucosyl residues: 4.02–4.78 Å. One consequence of this flexibility is that computer models of amylosic polymers that were propagated from residues with shapes at these extremes have very large differences in their conformations. Energies calculated for the deformed glucose residues at the ends of this range are approximately 2–3 kcal (8–13 kJ) above the minimum [94]. 'Breathing' over a similar range is observed in molecular dynamics studies [67].

The cycloamyloses illustrate distortion of monosaccharide residues by the environment. Within a given molecule in a crystal structure of these cyclic oligomers of α-D-glucose, the lengths of the glucose residues (the distances between O1 and O4) vary in one ring from 4.02 Å to 4.50 Å [95], despite the fact that all glucose residues are chemically identical! In this example, the lengths vary because of the accommodation of a guest *para*-nitrophenol molecule that has a very planar shape. To form a complex with this guest, the cyclohexaamylose takes an oval shape instead of the circular form suggested by its hexagonal chemical symmetry. Intramolecular variations arising from simple geometric consequences of ring closure also exist in the cycloamyloses. The average O1–O4 distances in crystalline cyclohexa-, hepta- and octaamyloses are 4.25 Å, 4.40 Å and 4.48 Å, respectively [96]. More extreme distortion can occur. Permethylation (2,3,6 tri-*O*-methyl) of cycloheptaamylose results in a macro ring with six glucose rings in the normal $^4C_1$ conformation and one with the $^1C_4$ shape [97]. Another distorted glucose ring in a cycloheptaamylose structure has the $^oS_2$ shape [98], again attributed to the steric crowding caused by the methyl groups on O2 and O3, as well as the lack of hydrogen bonding. The conversion of glucose residues in cyclohexaamylose to altrose results in a cyclic molecule in which the altrose residues have alternating $^1C_4$ and $^4C_1$ shapes [99].

Another good example of environmental effects on molecular conformation of an unsubstituted monosaccharide is given by fructopyranose. It has a nearly perfect $^2C_5$ chair ($\theta = 180°$) in the neat crystal [100], and in the $CaCl_2$ complex [101]. On the other hand, it is quite distorted in the $CaBr_2$ complex, with $\theta = 163°$ [102]. Again this suggests that ring shapes are affected by the exigencies of crystal packing. Because the fructose ring in the $CaCl_2$ complex and the uncomplexed monosaccharide have similar ring shapes, it is not clear that anything other than simple packing forces are causing the distortion in the bromide complex. In crystalline cellotetraose, half of the glucose residues have θ values of about 10°, attributed to packing effects [92].

There may be a general effect on molecular properties arising from condensed phases: the diminution of the importance of intramolecular hydrogen bonding. Intermolecular hydrogen bonds are likely to have geometries that are more linear (and thus stronger), as well as shorter lengths (also stronger). Thus, intramolecular hydrogen bonds that are found in modeling studies of the isolated molecule, while they may stabilize the model by several kcal, are likely to be replaced by intermolecular hydrogen bonding systems when possible because even greater stabilization of the system could result. This is especially true for equatorial hydroxyl groups that make peripheral hydrogen bonds. Strong, *trans*-annular intramolecular hydrogen bonds might form among axial hydroxyl groups, even in aqueous solution. Thus, when modeling carbohydrates in condensed phases with a single, isolated molecule it is useful to diminish the strength of hydrogen bonding to compensate for the reduced probability of existence of the intramolecular hydrogen bonds. This reduction (but not elimination) in hydrogen bonding strength can be effected in many MM modeling systems by increasing the dielectric constant.

A somewhat parallel consideration is the effect of the existence of a surrounding field on the molecular properties. Again, an increased dielectric constant, compared with the vacuum value of 1.0, is used (MM3 uses a value of 1.5 for isolated molecules). Because the strength of the electrostatic forces is proportional to the reciprocal of the dielectric constant, the hydrogen bonding is reduced to unimportance as a determinant of conformation at values of the dielectric constant greater than 10.0, where the strength of a bond with ideal geometry is only about 0.5 kcal (2 kJ).

Solvation models are the subject of intense development. Continuum theory has been extensively applied to carbohydrates for more than a decade [103, 104] and has been developed further by other workers [105]. In this method, the energy of solvation is computed from the energy required to create a cavity in the solvent of the size needed to insert the solute as well as the van der Waals and electrostatic energies of interaction with a general field surrounding the solute. A recent study showed large [e.g. 5 kcal (21 kJ)] effects on the calculated equilibrium between the $^4C_1$ and $^1C_4$ forms of β-glucose in water [48]. Even more intuitively satisfying are molecular dynamics studies that include specific solvent molecules, but they tend to be impractically time-consuming.

1.8 Dynamic behavior of monosaccharides

1.8.1 Molecular dynamics studies

A recent molecular dynamics study of cyclopentane [106] showed how facile pseudorotation for an isolated molecule, interrupted by intermolecular collisions in the vapor phase, resumes by concerted thermal motion of the individual atoms. A cycle of pseudorotation required only 0.25 ps. Based on the E–H barrier heights found for cyclohexane and calculated for pyranosic sugars of about 12 kcal (50 kJ), chair-to-chair conversions could take place almost equally well by either ring breakage and subsequent reclosure or by the symmetrical interconversion without ring breakage.

An extensive dynamics study of a monosaccharide was reported in 1989 [107]. In this work, the single molecule of glucose was surrounded by 207 water molecules in a box with periodic boundaries. Despite the presence of solvent, the propensity for intramolecular hydrogen bonding was preserved. As a result, the O6 made a transition from the *gt* crystallographic position and was retained in the forbidden *tg* conformation. Other work, by Kroon-Batenburg and Kroon, showed that a simulation with a different potential function more closely reproduced the experimental observations of primary alcohol conformations [108]. Neither of these simulations showed conversion to alternate ring conformations. Other simulations, in vacuum with other potential functions, have shown larger amounts of intramolecular motion. Chair-to-boat-to-alternate chair and reverse transitions were noted during a relatively brief (75 ps) simulation of β-glucose, and the primary alcohol group visited the two experimentally observed positions.

Molecular dynamics perturbation methods were used to study the free energy difference between α-glucose and β-glucose [109]. The result was in fairly close agreement with the observed result of 64% β-glucose but actually favored the α form. (Because of the relationship between the Boltzmann equation and population, only a small difference in energy can make a large difference in the population ratio.) However, a large component [3 kcal (*c.* 13 kJ)] of the differential stabilization of the β-glucose form resulted from solvation, balanced by more than 3 kcal (*c.* 13 kJ) of electrostatic energy favoring α-glucose. With the modified potential for CHARMM, the hydroxymethyl group underwent three transitions between the *gt* and *gg* positions in 10 ps of simulation; the forbidden *tg* conformation did not occur.

1.8.2 Mutarotation

As first noted for glucose by Dubrunfaut in 1846, the change in optical rotation with time is a characteristic property of reducing sugars. The interpretation in terms of the establishment of an equilibrium mixture of structural isomers – through the process of mutarotation – of a single monosaccharide was first confirmed by the separate isolation of α-glucose and β-glucose and has continued to attract attention because of its technological and biochemical implications as well as its intrinsic interest. Although the phenomenon is inevitably associated with optical rotation it can be monitored by any property which depends on the composition of a sugar or solution such as solubility, liquid volume, refractive index, infra-red and NMR spectra, or enzyme activity.

Sugars can be classified into two groups according to their mutarotation data; those such as glucose, lactose, mannose and xylose which gave 'simple' first-order kinetics corresponding to the presence of just two interconverting forms, and those such as galactose, ribose and arabinose which give 'complex' two-phase kinetics [110]. The sugars which show complex mutarotations are those which give equilibrium solutions containing measurable amounts (greater than 2%) of furanose forms. Measurement of the single parameter optical rotation could give only a single overall kinetic constant, $k_{\alpha\beta} + k_{\beta\alpha}$, but this enabled the comparison of different sugars and the effects of acid and base catalysts. Simple mutarotation half-lives in water at 20°C range from about 5 min to 60 min (Table 1.9).

Table 1.9 Mutarotation half-lives of some 'simple' sugars

Sugar	Medium	Temperature (°C)	Half-life (min)
glucose	water	20	48
xylose	water	20	15
lactose	water	20	64
mannose	water	20	19
rhamnose	water	20	6.5
2-amino-2-deoxy-glucose	water	25	24
glucose	water	37	9.8
glucose	blood, serum	37	2.4
glucose	saliva, urine	37	4.4
fructose-6-P	pH 7	25	0.0053
fructose-1,6-bisP	pH 7	25	0.0013
glucose-6-P	pH 7	25	0.055

Although the likely intermediacy of the acyclic form of a sugar was clear by about 1900, and substantial amounts are present in solutions of trioses, δ-hydroxyketones and pentuloses, experimental evidence of its involvement was limited to the observation first made in 1941 that exchange of oxygen from the C1–OH of glucose into water is much slower than mutarotation. This implied also that routes via the hydrate or direct exchange of OH on cyclic forms are not likely. More direct evidence came with the detection of aldehydo forms of pentose and hexose sugars by ^{1}H (e.g. 2-deoxy-D-*erythro*-pentose [111]) or ^{13}C (e.g. ^{13}C-enriched glucose [112, 113]; ribose [114]; fructose [115]) NMR spectroscopy. In particular, Serianni and co-workers [116] confirmed the direct involvement of the aldehyde form of erythrose in mutarotation by single and double resonance saturation experiments. Thus the kinetic description of the mutarotation of a pentose or hexose according to the scheme in Figure 1.1 requires the specification of up to ten rate constants.

Determination of the individual rate constants can be approached in three ways:

- the change in composition of a sugar mixture starting from a pure anomer can be followed by gas–liquid chromatography of derivatives [117, 118] or directly by NMR spectroscopy;
- the rates of chemical exchange processes between particular nuclei can be measured by NMR line broadening, saturation transfer or two-dimensional exchange spectroscopy;
- the rate of formation of a particular form of a sugar can be obtained from the rate of a reaction catalyzed by an excess of an enzyme specific for that form.

The most generally applicable methods are those based on NMR spectroscopy, and they have been used with considerable success by Barker and Serianni and others to obtain partial or full sets of rate constants for erythrose and threose [119], pentose and hexose phosphates [116], fructose [115] and talose [120]. Application of these methods may be restricted to particular conditions, for example of temperature or presence of a catalyst in order to give measurable values, and the estimation of ring closing rates is especially prone to error because it depends on knowing the concentration of the acyclic form. Computer programs based on the reaction scheme have been used to obtain rate constants which are not measurable directly or accurately [115, 117]. The enzyme catalysis method has been applied to D-glucose (aldose reductase, specific for the aldehyde form) [121] and D-glucose-6-phosphate (glucose-6-phosphate dehydrogenase, specific for β-pyranose) [122]. Apart from the expected fact that ring opening rates are kinetically limiting, it is also the case that transformations of furanose forms are much faster than those of pyranoses.

Mutarotation rates are increased in the presence of acidic or basic catalysts [A or B, respectively]. As is apparent from the values for sugar phosphates in

Table 1.10 Unidirectional rate constants for mutarotation

	Conditions	Rate constant (per minute)				Ref
		$k_{\alpha x \rightarrow ac}$	$k_{ac \rightarrow \alpha x}$	$k_{\beta x \rightarrow ac}$	$k_{ac \rightarrow \beta x}$	
Pyranose:						
glucose	50 mM MOPS, pH 7.25	0.028	400	0.023	570	121
galactose	1 mM phthalate, pH 4.34	0.032	50	0.034	100	117
fructose	water, 30°C, pH 8.4			<6	<1200	115
talose	50 mM HOAc, pH 4, 28°C	0.19		0.11		120
Furanose:						
galactose	1 mM phthalate, pH 4.34	0.42	36	0.32	58	117
fructose-6-P	15% D_2O, pH 7.5, 40°C	1080		1260		116
fructose-1,6-bP	" " "	1680		8400		116
talose	50 mM HOAc, pH 4, 28°C	2.8		2.2		120
fructose	water, 30°C, pH 8.4	30	4800	210	30000	115

Note: x = p for pyranose and f for furanose; ac = acyclic.

Tables 1.9 and 1.10 [115–117, 120, 121], catalysis can be intramolecular as well as intermolecular. Classic studies of the catalysis of mutarotation by Brönsted and Guggenheim played an important role in the development of theories of acid–base catalysis and established that the observed rate arises from contributions from all the catalytic species in the solution (including the sugar anion):

$$k_{obs} = k_{H_2O}[H_2O] + k_{H_3O^+}[H_3O^+] + k_{OH^-}[OH^-] + k_A[A] + k_B[B]$$

The values of overall (i.e. the sum for forward and reverse directions) catalytic constants (Table 1.11) [123–127] show that bases are more effective catalysts of mutarotation than are acids. Thus the rate of mutarotation of glucose in water is at a minimum at pH 4, the value of k_{H_2O} is 4.3×10^{-4} l/mol/min at 25°C [123].

Unidirectional second-order catalytic constants have been determined for the OH^--catalyzed and H_3O^+-catalyzed ring opening and closing reactions of fructose [115] and 2-deoxy-erythropentose [111], and again show that

Table 1.11 Catalytic constants for catalysis of mutarotation of glucose

Temperature (°C)	Acid	k_A (l/mol/min)	Base	k_B (l/mol/min)	Refs
25	H_3O^+	6.7×10^{-1}	OH^-	$4.3 \times 10^{+3}$	123, 124
25			tris	8.7×10^{-1}	123
18	CH_3CO_2H	2.4×10^{-3}	$CH_3CO_2^-$	2.7×10^{-2}	125
25			PO_4^{3-}	5.9×10^{-1}	126
			HPO_4^{2-}	7.1×10^{-2}	
	H_3PO_4	6×10^{-3}	$H_2PO_4^-$	3×10^{-3}	
20	mutarotase	$9.8 \times 10^{+4}$ per min	(kcat)		127

the values for furanose interconversions are 1–2 orders of magnitude greater than those for pyranoses.

Although the nature of the individual reaction steps in mutarotation (i.e. ring opening and closing, proton donation and removal) is known, their timing (concerted or stepwise) and the involvement of water solvent remain the subject of debate. Capon [123] argued convincingly for mechanisms involving rapid formation of the conjugate acid or base form of the sugar followed by a rate-determining ring opening concerted with a second proton transfer.

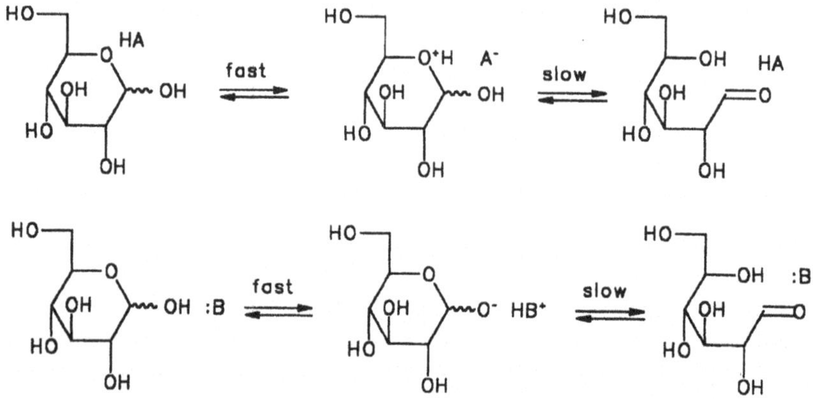

The data for reaction in water alone support a mechanism in which both proton transfers are rate determining [123].

Kinetic data for the mutarotation of carbohydrate metabolites are of central importance to considering whether chemical interconversion might limit or regulate the rates of metabolic reactions catalyzed by anomerically selective enzymes. Thus, for example, α-glucose-6-phosphate is the specific anomeric product of glycogen catabolism, but in the pentose phosphate pathway glucose-6-phosphate dehydrogenase is specific for the β anomer; the α → β anomerization might be rate limiting in conditions of starvation [128]. Such questions can be approached by comparing the unidirectional chemical anomerization rate (at *in vivo* concentration) with the metabolic flux either measured or calculated from enzyme activity in tissue. In a number of cases the two rates are of similar magnitudes suggesting that chemical anomerization could be significant, for example in the cycling of fructose-6-phosphate via fructose-1,6-bis-phosphate [129] or photosynthetic carbon reduction [116]. Since chemical mutarotation rates are higher for furanoses and for sugar phosphates these could be thought of as evolutionary devices for avoiding kinetic limitations of anomerization; the alternative is the provision of enzymes with anomerase activity, for example glucose-1-epimerase (mutarotase) [127], glucose-6-phosphate isomerase [130], glucose-6-phosphate-1-epimerase [128] and aldolase.

References

1. Stoddart, J.F. (1971) *Stereochemistry of Carbohydrates*, Wiley-Interscience, New York.
2. Eliel, E.L. and Wilen, S.H. (1994) *Stereochemistry of Organic Compounds*, John Wiley, New York, pp. 686–720.
3. Johnson, C.K. (1965) ORTEP: A FORTRAN Thermal-ellipsoid plot program for crystal structure illustration. Report ORNL-3794. Oak Ridge National Laboratory: Oak Ridge, TN.
4. Takagi, S. and Jeffrey, G.A. (1980) A neutron diffraction refinement of the crystal structures of β-L-arabinopyranose and methyl β-D-xylopyranoside. *Acta Crystallogr. Sect. B*, **33**, 3033–3040.
5. Tvaroška, I. and Carver, J.P. (1997) *Ab initio* molecular orbital calculation of carbohydrate model compounds. 4. The flexibility of ψ-type glycosidic bonds in carbohydrates. *J. Mol. Struct.* (*Theochem*), **395–396**, 1–14.
6. van't Hoff, J.H. (1874) *Archives Neérlandaises des Sciences Exactes et Naturelles*, **9**, 445–454; also in H.M. Leicester and H.S. Klickstein (eds) (1952) *A Source Book in Chemistry 1400–1900*. McGraw-Hill, New York, pp. 445–458.
7. Le Bel, J.A. (1874) *Société chimique de France Bulletin*, **22**, 337–347. Also in H.M. Leicester and H.S. Klickstein (eds) (1952) *A Source Book in Chemistry 1400–1900*. McGraw-Hill, New York, pp. 459–462.
8. Bragg, W.H. and Bragg, W.L. (1913) The structure of the diamond. *Nature* (*London*), **91**, 557.
9. Cox, E.G. and Jeffrey, G.A. (1939) Crystal structure of glucosamine hydrobromide. *Nature* (*London*), **143**, 894–895.
10. Böseken, J. (1913) Position in space of the hydroxyl groups of polyhydroxy compounds. *Chem. Ber.*, **46**, 2612–2628.
11. Reeves, R.E. (1949) Cuprammonium–glycoside complexes. III. The conformation of the D-glycopyranoside ring in solution. *J. Am. Chem. Soc.*, **71**, 215–217.
12. Reeves, R.E. (1951) Cuprammonium–glycoside complexes. *Adv. Carb. Chem.*, **6**, 107–134.
13. Angyal, S.J. (1969) The composition and conformation of sugars in solution. *Angewandte Chemie, Int. Ed.*, **8**, 157–226.
14. Angyal, S.J. (1968) Conformational analysis in carbohydrate chemistry. I. Conformational free energies. The conformations and α–β ratios of aldopyranoses in aqueous solution. *Aust. J. Chem.*, **21**, 2737–2746.
15. Winstein, S. and Holness, N.J. (1955) Neighboring carbon and hydrogen. XIX. *t*-Butylcyclohexyl derivatives. Quantitative conformational analysis. *J. Am. Chem. Soc.*, **77**, 5562–5578.
16. Bushweller, C.H. (1995) Stereodynamics of cyclohexane and substituted cyclohexanes. Substituent A-values. In *Conformational Studies of Six-membered Ring Carbocycles and Heterocycles* (ed. E. Juaristi), VCH, New York, pp. 25–58.
17. Tvaroška, I. and Bleha, T. (1989) Anomeric and exo-anomeric effects in carbohydrate chemistry. *Advances in Carbohydrate Chemistry and Biochemistry*, **47**, 45–123.
18. Thatcher, G.R.J. (ed.) (1995) The anomeric effect. *ACS Symposium Series*, **539**.
19. Lemieux, R.U. (1990) Explorations with sugars: how sweet it was. *Profiles, Pathways and Dreams*, American Chemical Society, Washington, DC.
20. Altona, C., Romers, C. and Havinga, E. (1959) Molecular structure and conformation of dihalodioxanes. *Tetrahedron Lett.*, **8**, 16–20.
21. Petillo, P.A. and Lerner, L.E. (1995) Origin and quantitative modeling of anomeric effect. *ACS Symposium Series*, **539**, 156–175.
22. Walkinshaw, M.D. (1987) Variation in the hydrophilicity of hexapyranose sugars explains features of the anomeric effect. *J. Chem. Soc. Perkin Trans.* **II**, 1903–1906.
23. Booth, H., Grindley, T.B. and Khedhair, A.K. (1982) The anomeric and exo-anomeric effects in 2-methoxytetrahydropyran. *J. Chem. Soc., Chem. Commun.*, 1047–1048.
24. Jeffrey, G.A., Pople, J.A., Binkley, J.S. and Vishveshwara, S. (1978) Application of *ab initio* molecular orbital calculations to structural moieties of carbohydrates. *J. Amer. Chem. Soc.*, **100**, 373–379.
25. Kneisler, J.R. and Allinger, N.L. (1996) *Ab initio* and density functional theory study of structures and energies for dimethoxymethane as a model for the anomeric effect. *J. Comput. Chem.*, **17**, 757–766.

26. Plavec, J., Tong, W. and Chattopadhyaya, J. (1993) How do the gauche and anomeric effects drive the pseudorotational equilibrium of the pentofuranose moiety of nucleosides? *J. Am. Chem. Soc.*, **115**, 9734–9746.
27. Jeffrey, G.A. and Saenger, W. (1991) *Hydrogen Bonding in Biological Structures.* Springer, Berlin.
28. IUPAC–IUBMB Joint Commission on Biochemical Nomenclature (JCBN) (1997) *Nomenclature of Carbohydrates*, Document 24.10. Prepared for publication by Alan D. McNaught, The Royal Society of Chemistry, Cambridge. (1996) *Pure and Appl. Chem.*, **68**, 1919–2008; (1997) *Carbohydr. Res.*, **297**, 1–92; *J. Carb. Chem.*, **16**, 1191–1280; *Advan. Carb. Chem. Biochem.*, **52**, 43–177.
29. Glusker, J.P., Lewis, M. and Rossi, M. (1994) *Crystal Structure for Chemists and Biologists*, VCH, New York.
30. Gress, M.E. and Jeffrey, G.A. (1977) A neutron diffraction study of the crystal structure of β-maltose monohydrate. *Acta Crystallogr. Sect B*, **33**, 2490–2495.
31. Pangborn, W., Langs, D. and Pérez, S. (1985) Regular left-handed fragment of amylose: crystal and molecular structure of methyl-α-maltrotrioside tetrahydrate. *Int. J. Biol. Macromol.*, **7**, 363–369.
32. Bijvoet, J.M., Peerdeman, A.F. and Van Bommel, A.J. (1951) Determination of absolute configuration of optically active compounds by means of x-rays. *Nature*, **168**, 271–272.
33. Allen, F.H., Bellard, S., Brice, M.D. *et al.* (1979) The Cambridge Crystallographic Data Centre: computer-based search, retrieval, analysis and display of information. *Acta Crystallogr. Sect. B*, **35**, 2331–2339.
34. Jeffrey, G.A. and Sundaralingam, M. (various years) Bibliography of crystal structures of carbohydrates, nucleosides and nucleotides. *Advances in Carbohydrate Chemistry and Biochemistry*: (1974) **30**, 445–466; (1975) **31**, 347–384; (1976) **32**, 353–384; (1977) **34**, 345–378; (1980) **37**, 373–436; (1981) **38**, 417–529; (1985) **43**, 203–421.
35. Bernstein, F.C., Koetzle, T.F., Wiliams, G.J.B. *et al.* (1977) The Protein Data Bank: a computer-assisted archival file for macromolecular structures. *J. Mol. Biol.*, **112**, 535–542.
36. Mathlouthi, M. and Koenig, J.L. (1986) Vibrational spectra of carbohydrates. *Advances in Carbohydrate Chemistry and Biochemistry*, **44**, 7–89.
37. Angyal, S.J. (1984) The composition of reducing sugars in solution. *Advances in Carbohydrate Chemistry and Biochemistry*, **42**, 15–68.
38. Breitmaier, E. and Voelter, W. (1990) *Carbon-13 NMR spectroscopy. High-resolution Methods and Applications in Organic Chemistry and Biochemistry.* VCH, Weinheim.
39. Haasnoot, C.A.G., De Leeuw, F.A.A.M. and Altona, C. (1980) The relation between proton–proton NMR coupling constants and substituent electronegativities. I. An empirical generalization of the Karplus equation. *Tetrahedron*, **36**, 2783–2792.
40. Haasnoot, C.A.G., De Leeuw, F.A.A.M. and Altona, C. (1980) Prediction of anti and gauche vicinal proton–proton coupling constants for hexapyranose rings using a generalized Karplus equation. *Bull. Soc. Chim. Belg.*, **89**, 125–131.
41. Tvaroška, I., Kožár, T. and Hricovíni, M. (1990) Oligosaccharides in solution. *ACS Symposium Series*, **430**, 162–176.
42. Tvaroška, I. and Taravel, F.R. (1995) Carbon–proton coupling constants in the conformational analysis of sugar molecules. *Advances in Carbohydrate Chemistry and Biochemistry*, **51**, 15–61.
43. Diez, E., Fabian, J.S., Guileme, J. *et al.* (1989) Vicinal proton–proton coupling constants. I. Formulation of an equation including interactions between substituents. *Mol. Phys.*, **68**, 49–63.
44. Neuhaus, D. (1989) *The Nuclear Overhauser Effect in Structural Conformational Analysis*, VCH, New York.
45. Durran, D.M., Howlin, B.J. and Gidley, M.J. (1995) *Ab initio* nuclear shielding calculations of a model α-(1-4)-glucan. *Carbohydr. Res.*, **271**, C1–C5.
46. Duker, J.M. and Serianni, A.S. (1993) (^{13}C)-Substituted sucrose: ^{13}C–^{1}H and ^{13}C–^{13}C spin coupling constants to assess furanose ring and glycosidic bond conformations in aqueous solution. *Carbohydr. Res.*, **249**, 281–303.
47. Stevans, E. and Duda, C.A. (1991) Solution conformation of sucrose from optical rotation. *J. Am. Chem. Soc.*, **113**, 8622–8627.

48. Barrows, S.E., Dulles, F.J., Cramer, C.J. *et al.* (1995) Relative stability of alternative chair forms and hydroxymethyl conformations of β-D-glucopyranose. *Carbohydr. Res.*, **276**, 219–251.
49. Salzner, U. and von Ragué Schleyer, P. (1994) *Ab initio* examination of anomeric effects in tetrahydropyrans, 1,3-dioxanes, and glucose. *J. Org. Chem.*, **59**, 2138–2155.
50. Glendening, E.D., Curtiss, L.A. and Weinhold, F. (1988) Natural resonance theory and studies of electronic delocalization in molecules. *Chem. Rev.*, **88**, 899–926.
51. Lii, J.-H. and Allinger, N.L. (1994) Directional hydrogen bonding in the MM3 force field. I. *J. Phys. Org. Chem.*, **7**, 591–609.
52. Csonka, G.I., Éliás, K. and Csizmadia, I.G. (1996) Relative stability of $^1C_4$ and $^4C_1$ chair forms of β-D-glucose: a density functional study. *Chem. Phys. Lett.* **257**, 49–60.
53. Glennon, T.M., Zheng, Y.-J., LeGrand, S.M. *et al.* (1994) A force field for monosaccharides and (1–4) linked polysaccharides. *J. Comput. Chem.*, **15**, 1019–1040.
54. Homans, S.W. (1990) A molecular mechanics force field for the conformational analysis of oligosaccharides: comparison of theoretical and crystal structures of Man α-(1-3) Man β-(1-4) GlcNAc. *Biochemistry*, **29**, 9110–9118.
55. Ha, S.N., Giammona, A., Field, M. and Brady, J.W. (1988) A revised potential energy surface for molecular mechanics studies of carbohydrates. *Carbohydr. Res.*, **180**, 207–221.
56. Brooks, B.R., Bruccoleri, R.E., Olafson, B.D. *et al.* (1983) CHARMM: a program for macromolecular energy, minimization and dynamics calculations. *J. Comput. Chem.*, **4**, 187–217.
57. Imberty, A., Hardman, K.D., Carver, J.P. and Pérez, S. (1991) Molecular modelling of protein–carbohydrate interactions. Docking of monosaccharides in the binding site of concanavalin A. *Glycobiology*, **1**, 631–642.
58. Engelsen, S.B., Fabricius, J. and Rasmussen, Kj. (1994) The consistent force field. 1. Methods and strategies for optimization of empirical potential energy functions. *Acta Chem. Scand.*, **48**, 548–552.
59. Engelsen, S.B., Fabricius, J. and Rasmussen, Kj. (1994) The consistent force field. 2. An optimized set of potential energy functions for the alkanes. *Acta Chem. Scand.*, **48**, 553–565.
60. Dauchez, M., Derreumaux, P., Lagant, P. and Vergoten, G. (1995) A vibrational molecular force field of model compounds with biological interest. IV. Parameters for the different glycosidic linkages of oligosaccharides. *J. Comput. Chem.*, **16**, 188–199.
61. Van Gunsteren, W.F. and Berendsen, H.J.C. (1990) Molecular dynamics computer simulation. Method, application and perspectives in chemistry. *Angew. Chem.*, **102**, 1020–1055.
62. Smith, A.E. and Lindner, H.J. (1991) π-SCF molecular mechanics PIMM: formulation, parameters, applications. *J. Comput. Aided Mol. Des.*, **5**, 235–262.
63. Kroeker, M. and Lindner, H.J. (1996) The PIMM force field – recent developments. *J. Mol. Model.*, **2**, 376–378.
64. Woods, R.J., Dwek, R.A., Edge, C.J. and Fraser-Reid, B. (1995) Molecular mechanical and molecular dynamical simulations of glycoproteins and oligosaccharides. 1. GLYCAM_93 parameter development. *J. Phys. Chem.*, **99**, 3832–3846.
65. Allinger, N.L., Rahman, M. and Lii, J.H. (1990) A molecular mechanics force field (MM3) for alcohols and ethers. *J. Am. Chem. Soc.*, **112**, 8293–8307.
66. Brady, J.W. (1986) Molecular dynamics of α-D-glucose. *J. Am. Chem. Soc.*, **108**, 8153–8160.
67. Brady, J.W. (1988) Molecular dynamics simulations of β-D-glucopyranose. *Carbohydr. Res.*, **165**, 306–312.
68. Brady, J.W. (1989) Molecular dynamics of α-D-glucose in aqueous solution. *J. Am. Chem. Soc.*, **111**, 5155–5168.
69. Simmerling, C., Fox, T. and Kollman, P.A. (1998) Use of locally enhanced sampling in Free Energy Calculations: testing and application to the α → β anomerization of glucose. *J. Am. Chem. Soc.*, **120**, 5771–5782.
70. Mossine, V.V., Barnes, C.L., Glinsky, G.V. and Feather, M.S. (1995) Interaction between two Cl–X–Cl′ branched ketoses: observation of an unprecedented crystalline *spiro*-bicyclic hemiketal tautomer of N,N-di(1-deoxy-D-fructos-1-yl)-glycine ('difructose glycine') having open chain carbohydrate. *Carbohydr. Letters*, **1**, 355–362.
71. Cox, J.D. and Pilcher, G. (1970) in *Thermochemistry of Organic and Organometallic Compounds*, Academic Press, New York.

72. Allinger, N.L., Yuh, Y.H. and Lii, J.-H. (1989) Molecular mechanics. The MM3 force field for hydrocarbons. *J. Am. Chem. Soc.*, **111**, 8551–8566.
73. Köll, P., Saak, W., Pohl, S. *et al.* (1994) Preparation and crystal and molecular structure of 6-*O*-[(2*S*)-2,3-epoxypropyl]-1,2:3,4-di-*O*-isopropylidene-α-D-galactopyranose. Pyranoid ring conformation in 1,2:3,4-di-*O*-isopropylidene-galactopyranose and related systems. *Carbohydr. Res.*, **265**, 237–248.
74. Altona, C. and Sundaralingam, M. (1972) Conformational analysis of the sugar ring in nucleosides and nucleotides: a new description using the concept of pseudorotation. *J. Am. Chem. Soc.*, **94**, 8205–8212.
75. Cremer, D. and Pople, J.A. (1975) A general definition of ring puckering coordinates. *J. Am. Chem. Soc.*, **97**, 1354–1358.
76. Cremer, D. RING88, Quantum Chemistry Program Exchange Program 288, QCPE, Creative Arts Building 181, Indiana University, Bloomington, IN 47405, USA.
77. Jeffrey, G.A. and Yates, J.H. (1979) Stereographic representation of the Cremer–Pople ring-puckering parameters for pyranoid rings. *Carbohydr. Res.*, **74**, 319–322.
78. Haasnoot, C.A.G. (1992) The conformation of six-membered rings described by puckering coordinates derived from endocyclic torsion angles. *J. Am. Chem. Soc.*, **114**, 882–887.
79. Entrena, A., Jaime, C., Espinosa, A. and Gallo, M.A. (1989) Theoretical (MM2) conformational analysis of 1,4-dioxepane. *J. Org. Chem.*, **54**, 1745–1748.
80. Levitt, M. and Warshel, A. (1978) Extreme conformational flexibility of the furanose ring in DNA and RNA. *J. Am. Chem. Soc.*, **100**, 2607–2613.
81. French, A.D., Dowd, M.K. and Reilly, P.J. (1997) MM3 modeling of fructose ring shapes and hydrogen bonding. *J. Mol. Struct.* (*Theochem*), **395–396**, 271–287.
82. Pickett, H.M. and Strauss, H.L. (1971) Symmetry and conformation of the cycloalkanes. *J. Chem. Phys.*, **55**, 324–334.
83. Pickett, H.M. and Strauss, H.L. (1970) Conformational structure, energy and inversion rates of cyclohexane and some related oxanes. *J. Am. Chem. Soc.*, **92**, 7281–7290.
84. French, A.D. and Dowd, M.K. (1994) Analysis of the ring-form tautomers of psicose with MM3(92). *J. Comput. Chem.*, **15**, 561–570.
85. Dowd, M.K., French, A.D. and Reilly, P.J. (1994) Modeling of aldopyranosyl ring puckering with MM3(92). *Carbohydr. Res.*, **264**, 1–19.
86. French, A.D. and Brady, J.W. (1990) Computer modeling of carbohydrate molecules. *ACS Symposium Series*, **430**, 1–19.
87. Franck, R.W. (1983) A revision of the value for the anomeric effect. *Tetrahedron*, **39**, 3251–3252.
88. Csonka, G.I., Kolossváry, I., Császár, P. *et al.* (1997) The conformational space of selected aldo-pyrano-hexoses. *J. Mol. Structure* (*Theochem*), **395–396**, 29–40.
89. James, V.J., Stevens, J.D. and Moore, F.H. (1978) Precision x-ray and neutron diffraction studies of methyl β-D-ribopyranoside, $C_6H_{12}O_5$. *Acta Crystallogr. Sect. B*, **34**, 188–193.
90. French, A.D. and Dowd, M.K. (1993) Exploration of disaccharide conformations by molecular mechanics. *J. Mol. Struct.* (*Theochem*), **286**, 183–210.
91. Kouwizjer, M.L.C.E. and Grootenhuis, P.D.J. (1995) A CHARMM-based force field for carbohydrates using the CHEAT approach: carbohydrate hydroxyl groups represented by extended atoms. *J. Phys. Chem.*, **99**, 13426–13436.
92. Gessler, K., Krauss, N., Steiner, T. *et al.* (1995) β-D-cellotetraose hemihydrate as a structural model for cellulose II. An x-ray diffraction study. *J. Am. Chem. Soc.*, **117**, 11397–11406.
93. Dais, P. and Perlin, A.S. (1987) Intramolecular hydrogen bonding and solvation contribution to the relative stability of the β-furanose forms of D-fructose in dimethyl sulfoxide. *Carbohydr. Res.*, **169**, 159–169.
94. French, A.D., Rowland, R.S. and Allinger, N.L. (1990) Modeling of glucose, the flexible monomer of amylose. *ACS Symposium Series*, **430**, 120–140.
95. Harata, K. (1977) The structure of the cyclodextrin complex. V. Crystal structures of α-cyclodextrin complexes of *p*-nitrophenol and *p*-hydroxybenzoic acid. *Bull. Chem. Soc. Jpn.*, **50**, 1416–1424.
96. Saenger, W. (1980) Crystal structures of the γ-cyclodextrin *n*-propanol inclusion complex: correlation of the α, β and γ-cyclodextrin geometries. *Biochem. and Biophys. Res. Comm.*, **92**, 933–938.

97. Caira, M.R., Griffith, V.J., Nassimbeni, L.R. and van Oudtshoorn, B. (1994) Unusual $^1C_4$ conformation of a methylglucose residue in crystalline permethyl-β-cyclodextrin monohydrate. *J. Chem. Soc. Perkin Trans.*, **2**, 2071–2072.
98. Harata, K. (1988) X-ray crystal structure of the permethylated β-cyclodextrin complex with *m*-iodophenol. A stabilized skew-boat pyranose conformation in the distorted macrocyclic ring. *J. Chem. Soc., Chem. Commun.*, 928–929.
99. Nogami, Y., Nasu, K., Koga, T. *et al.* (1997) Synthesis, structure, and conformational features of α-cycloaltrin: a cyclooligosaccharide with alternating $^4C_1/^1C_4$ pyranoid chairs. *Ange. Chem. Int. Ed. Engl.*, **36**, 1899–1902.
100. Kanters, J.A., Roelofsen, G., Alblas, B.P. and Meinders, I. (1977) The crystal and molecular structure of β-D-fructose, with emphasis on anomeric effect and hydrogen-bond interactions. *Acta Crystallogr. Sect. B*, **33**, 665–672.
101. Craig, D.C., Stephenson, N.C. and Stevens, J.E. (1974) Bis-(β-D-fructopyranose) calcium-chloride trihydrate. *Cryst. Struct. Comm.*, **3**, 195–199.
102. Cook, W.J. and Bugg, C.E. (1976) Effects of calcium interactions on sugar conformation: crystal structure of β-D-fructose-calcium bromide dihydrate. *Acta Crystallogr. Sect. B*, **32**, 656–659.
103. Tvaroška, I. and Kožár, T. (1980) Theoretical studies on the conformation of saccharides. 3. Conformational properties of the glycosidic linkage in solution and their relation to the anomeric and exo-anomeric effects. *J. Am. Chem. Soc.*, **102**, 6929–6936.
104. Tvaroška, I., Imberty, A. and Pérez, S. (1990) Solvent effect on the stability of isomaltose conformers. *Biopolymers*, **30**, 369–379.
105. AMSOL, Quantum Chemistry Program Exchange Program 606, by Hawkins, G.D., Lynch, G.C., Giesen, D.J., Rossi, I. *et al.* (1996) *QCPE Bulletin*, **14**, 11-13, QCPE, Creative Arts Building 181, Indiana University, Bloomington, IN 47405, USA.
106. Cui, W., Li, F. and Allinger, N.L. (1993) Simulation of conformational dynamics with the MM3 force field: the pseudorotation of cyclopentane. *J. Am. Chem. Soc.*, **115**, 2943–2951.
107. Brady, J.W. (1989) Molecular dynamics simulations of α-D-glucose in aqueous solution. *J. Am. Chem. Soc.*, **111**, 5155–5165.
108. Kroon-Batenburg, L.M.J. and Kroon, J. (1990) Solvent effect on the conformation of the hydroxymethyl group established by molecular dynamics simulation of methyl-β-D-glucoside in water. *Biopolymers*, **29**, 1243–1248.
109. Ha, S., Gao, J., Tidor, B. *et al.* (1991) Solvent effect on the anomeric equilibrium in D-glucose: a free energy simulation analysis. *J. Am. Chem. Soc.*, **113**, 1553–1557.
110. Pigman, W. and Isbell, H.S. (1968) Mutarotation of sugars in solution: part I History, basic kinetics, and composition of sugar solutions. *Advan. Carbohydr. Chem.*, **23**, 11–57; Isbell, H.S. and Pigman, W. (1969) Mutarotation of sugars in solution: part II Catalytic processes, isotope effects, reaction mechanisms, and biochemical aspects. *Advan. Carbohydr. Chem. Biochem.*, **24**, 13–65.
111. Bauer, H., Brinkmeier, A., Buddrus, J. and Jablonski, M. (1986) Ring-Ketten-Tautomerie von 2-Desoxy-D-Ribose: Vollständige Analyse der Dynamic durch NMR-Linienformsimulation und Spinnsättiggungsübertragung. *Liebigs Ann. Chem.*, 1804–1807.
112. Williams, C. and Allerhand, A. (1977) Detection of β-D-glucofuranose in aqueous solutions of D-glucose. Application of carbon-13 fourier-transform NMR spectroscopy. *Carbohydr. Res.*, **56**, 173–179.
113. Maple, S.R. and Allerhand, A. (1987) Detailed tautomeric equilibrium of aqueous D-glucose. Observation of six tautomers by ultrahigh resolution carbon-13 NMR. *J. Amer. Chem. Soc.*, **109**, 3168–3171.
114. King-Morris, M.J. and Serianni, A.S. (1987) ^{13}C NMR studies of [1-^{13}C] aldoses: empirical rules correlating pyranose ring configuration and conformation with ^{13}C chemical shifts and ^{13}C–^{13}C spin couplings. *J. Amer. Chem. Soc.*, **109**, 3501–3508.
115. Goux, W.J. (1985) Complex isomerization of ketoses: a ^{13}C NMR study of the base-catalysed ring-opening and ring-closing rates of D-fructose isomers in aqueous solution. *J. Amer. Chem. Soc.*, **107**, 4320–4327.
116. Pierce, J., Serianni, A.S. and Barker, R. (1985) Anomerisation of furanose sugars and sugar phosphates. *J. Amer. Chem. Soc.*, **107**, 2448–2456.

117. Wertz, P.W., Garver, J.C. and Anderson, L. (1981) Anatomy of a complex mutarotation. Kinetics of tautomerisation of α-D-galactopyranose and β-D-galactopyranose in water. *J. Amer. Chem. Soc.*, **103**, 3916–3922.
118. Cockman, M., Kubler, D.G., Oswald, A.S. and Wilson, L. (1987) The mutarotation of fructose and the invertase hydrolysis of sucrose. *J. Carbohydrate Chem.*, **6**, 181–201.
119. Serianni, A.S., Pierce, J., Huang, S.G. and Barker, R. (1982) Anomerisation of furanose sugars: kinetics of ring-opening reactions by ^{1}H and ^{13}C saturation-transfer NMR spectroscopy. *J. Amer. Chem. Soc.*, **104**, 4037–4044.
120. Snyder, J.R., Johnston, E.R. and Serianni, A.S. (1989) D-Talose anomerisation: NMR methods to evaluate the reaction kinetics. *J. Amer. Chem. Soc.*, **111**, 2681–2687.
121. Grimshaw, C.E. (1986) Direct measurement of the rate of ring opening of D-glucose by enzyme-catalysed reduction. *Carbohydrate Res.*, **148**, 345–348.
122. Bailey, J.M., Fishman, P.H. and Pentchev, P.G. (1968) Studies on mutarotases II. Investigations of possible rate-limiting anomerisation in glucose metabolism. *J. Biol. Chem.*, **243**, 4827–4831.
123. Capon, B. (1974) Kinetics and mechanism of mutarotation of aldoses. *J.C.S. Perkin Trans. II*, 1600–1610.
124. Bell, R.P. and Prue, J.E. (1950) Kinetic studies in heterogeneous buffer systems part II. The system quinine, quinine sulphate and potassium sulphate. *Trans. Faraday Soc.*, **46**, 14–21.
125. Brönsted, J.H. and Guggenheim, E.A. (1927) Contribution to the theory of acid and base catalysis. The mutarotation of glucose. *J. Amer. Chem. Soc.*, **49**, 2554–2584.
126. Bailey, J.M., Fishman, P.N. and Pentchev, P.G. (1970) Anomalous mutarotation of glucose-6-phosphate. An example of intramolecular catalysis. *Biochemistry*, **9**, 1189–1194.
127. Bentley, R. and Bhate, D.S. (1960) Mutarotase from *Penicillium notatum* II. The mechanism of the mutarotation reaction. *J. Biol. Chem.*, **235**, 1225–1233.
128. Wurster, B. and Hess, B. (1974) Glucose-6-phosphate 1-epimerase from baker's yeast: purification, properties and possible biological function. *Hoppe-Seyler's Z. Physiol. Chem.*, **354**, 255–265.
129. Schray, K.J. and Benkovic, S.J. (1978) Anomerisation rates and enzyme specificity for biologically important sugars and sugar phosphates. *Acc. Chem. Res.*, **11**, 136–141.
130. Wurster, B. and Hess, B. (1973) Anomeric specificity and anomerase activity of glucose-phosphate isomerase (EC 5.3.1.9) from baker's yeast. *Hoppe-Seyler's Z. Physiol. Chem.*, **354**, 407–420.

Software

AMBER: Oxford Molecular Group, The Medawar Centre, Oxford Science Park, Oxford OX4 4GA, UK.

CHARMM: Prof. Martin Kaplus, Department of Chemistry, Harvard University, 12 Oxford Street, Cambridge, MA 02138, USA.

GROMOS: BIOMAS b.v., Laboratory of Physical Chemistry, ETH Zentrum, Universitätstrasse 6, CH-8092 Zurich, Switzerland.

MM2, MM3, and the parameterizations AMI, PM3, MNDO: Quantum Chemistry Program Exchange (QPE), Creative Arts Building 18, Indiana University, Bloomington, IN 47405, USA.

PIMM(88), PIMM(91): Prof. Hans Jörg Lindner, Department of Organic Chemistry, Technische Hochschule Darmstadt, Petersenstrasse 22, D-64287 Darmstadt, Germany.

SYBLY: Tripos Inc., 1699 South Hanley Road, St Louis, MO 63144-291, USA.

2 Chemical synthesis of monosaccharides

KARL J. HALE and ANTHONY C. RICHARDSON

In this chapter we present an overview of recent developments that have occurred in the chemical synthesis of monosaccharides and their derivatives.

2.1 The asymmetric synthesis of monosaccharides from achiral starting materials

Prior to 1980, the asymmetric synthesis of monosaccharide derivatives from simple achiral precursors was considered to be a near-impossible task owing to the paucity of asymmetric reagents available for performing important C–C and C–O bond-forming reactions enantioselectively. This situation changed quite dramatically, however, soon after Katsuki and Sharpless made their landmark discovery of the asymmetric epoxidation (AE) reaction of allylic alcohols [1]. Suddenly, organic chemists had at their disposal an extremely powerful and versatile tool for the creation of synthetically useful 2,3-epoxy alcohols with high levels of enantiospecificity, and they readily embraced it for the construction of many types of monosaccharide.

In the Sharpless AE route to the L-hexoses [2], a reiterative two-carbon homologation/asymmetric functionalisation cycle was applied to 4-benzhydryloxy-(*E*)-2-buten-1-ol **1**. Scheme 2.1 illustrates the successful implementation of this strategy for the synthesis of L-glucose. The functionalisation sequence began with the asymmetric epoxidation of **1** using *tert*-butyl hydroperoxide, titanium tetraisopropoxide and (+)-diisopropyl tartrate (DIPT). This furnished the chiral 2,3-epoxy alcohol **2** which was then subjected to base-mediated epoxide migration to give the terminal epoxide and *in situ* ring-opening with thiophenoxide to produce diol **3**. Isopropylidenation of **3** followed by oxidation afforded a mixture of diastereomeric sulfoxides that were subjected to Pummerer rearrangement to give **4**. The latter were converted to aldehyde **5** without epimerisation at the α centre by low-temperature reduction with one equivalent to diisobutylaluminium hydride and aqueous work-up. At this point, another two-carbon homologation/asymmetric functionalisation cycle was executed, but this time the mixed thioacetals **7** were hydrolysed under mildly basic conditions to bring about efficient epimerisation at the chiral centre adjacent to the aldehyde. This quite general reaction established the desired *syn*-relationship between the C(2) and C(3) stereocentres in **8**. By further protecting group adjustments, L-glucose **9** was obtained in high

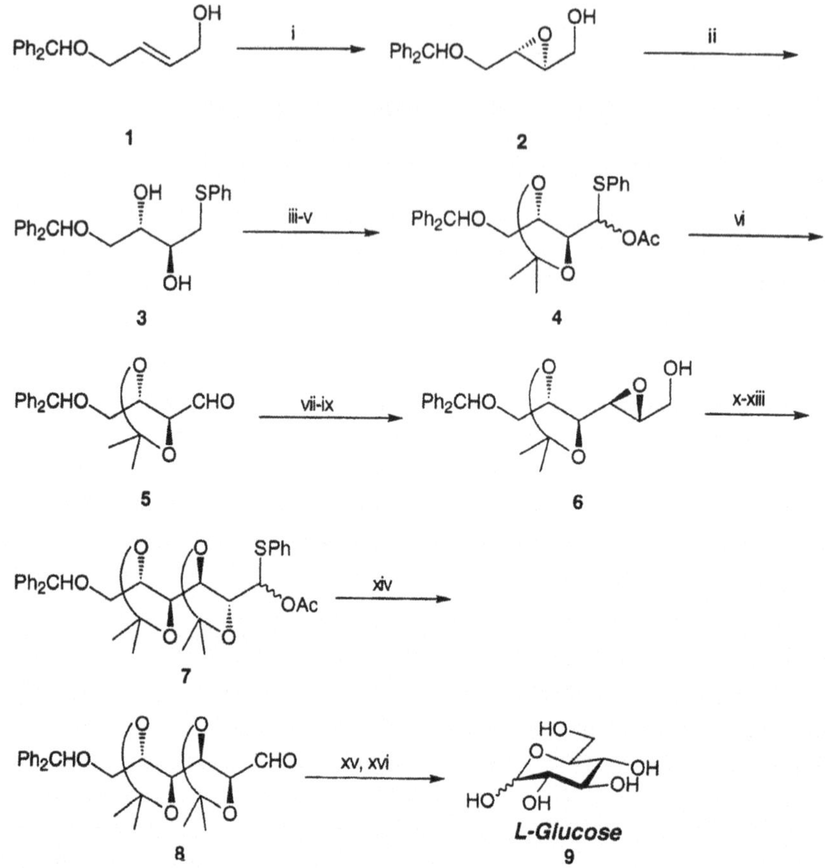

Scheme 2.1 i, (+)-Diisopropyl tartrate, $Ti(OPr\text{-}i)_4$, t-BuO_2H, CH_2Cl_2, −20°C, overnight, 92%; ii, 0.5 N aq. NaOH, t-BuOH, Δ, add PhSH over 3 h, 71%; iii, $Me_2C(OMe)_2$, $POCl_3$, CH_2Cl_2, room temperature, 100%; iv, m-CPBA, CH_2Cl_2, −78°C; v, NaOAc, Ac_2O, Δ, 8 h, 93%; vi, i-Bu_2AlH, CH_2Cl_2, −78°C, 91%; vii, $Ph_3P{=}CHCHO$, C_6H_6, 88%; viii, $NaBH_4$, MeOH, −40°C, 91%; ix, (−)-DIPT, asymmetric epoxidation, 84%; x, 0.5 N aq. NaOH, t-BuOH, Δ, add PhSH over 3 h, 63%; xi, CSA, $Me_2C(OMe)_2$, CH_2Cl_2, 100%; xii, m-CPBA, CH_2Cl_2, −78°C; xiii, Ac_2O, NaOAc, Δ, 8 h, 87%; xiv, K_2CO_3, MeOH, overnight, 65%; xv, 90% aq. CF_3CO_2H, 10 min; xvi, H_2, Pd-C, MeOH, 38%, 2 steps.

optical purity. Through simple variants of these two cycles all eight hexoses were prepared with complete stereocontrol.

A powerful sequence of reactions for 2-deoxy-hexose synthesis is the combination of Sharpless AE and asymmetric allylboration [3]. In this method, illustrated in Scheme 2.2 for the synthesis of 2-deoxy-D-galactose **16**, a partially-protected (*Z*)- or (*E*)-bisallylic alcohol is converted to a 2,3-epoxy alcohol via the AE, and the remaining primary hydroxyl is oxidised to an aldehyde. An asymmetric allylboration is then carried out either with

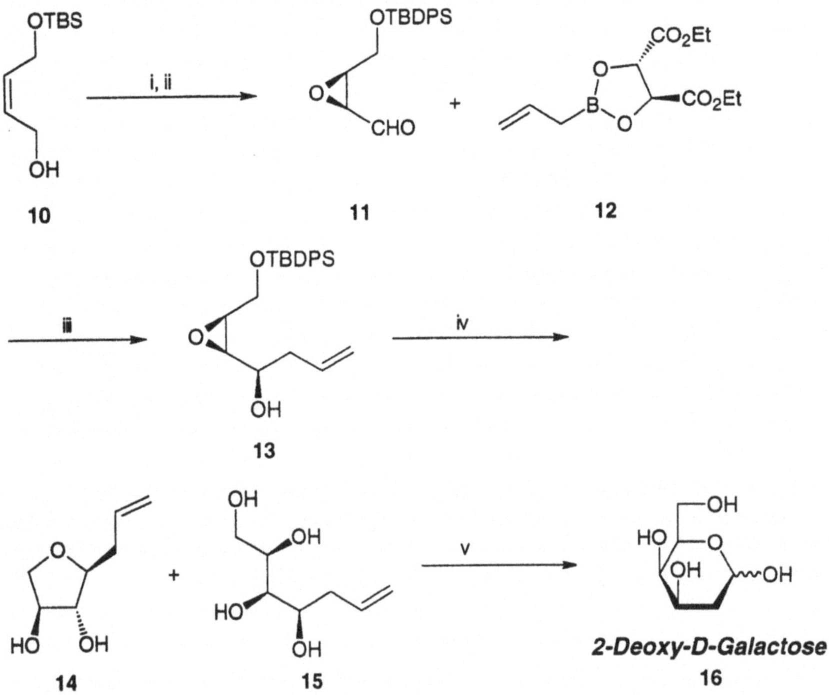

Scheme 2.2 i, (−)-Diethyl tartrate, Ti(OPr-*i*)$_4$, *t*-BuO_2H, CH_2Cl_2, 4 A sieves, −20°C, 57% yield, 90% enantiomeric excess (ee); ii, PCC, NaOAc, CH_2Cl_2, 100%; iii, PhMe, −78°C, 65%, 99:1 selectivity, 97% ee; iv, NaOH, H_2O, *t*-BuOH, Δ (ratio of **14**:**15** = 2:1); v, O_3, MeOH, −20°C, Me_2S work-up.

the (*RR*)- or the (*SS*)-tartrate allylboronate at low temperature. This furnishes a homoallylic alcohol with excellent levels of stereocontrol, provided that both reagents constitute a 'matched' pair of reactants. However, if stereochemical 'mismatching' occurs between the chiral coupling partners then much poorer levels of stereoselection result. Hydrolysis of the allylated epoxy alcohol with base and ozonolytic degradation of the alkene complete this rather concise entry into 2-deoxy-hexoses.

Roush and coworkers [4] have devised an effective pathway to 2,6-dideoxy hexoses that capitalises on the regioselective opening of epoxy alcohols obtained from Sharpless kinetic resolution (Scheme 2.3). These workers found that epoxy alcohols such as **20** react with aqueous acid to give 1,2-diols in which the water molecule has attacked at C(6) with high regioselectivity. In contrast, when a phenylurethane is used for intramolecular delivery of the oxygen nucleophile under anhydrous Lewis acid conditions, as is the case with **18**, the product of C(5) opening is preferentially formed. Scheme 2.3 illustrates the operation of this approach for the synthesis of D-olivose **19** and D-digitoxose **21**.

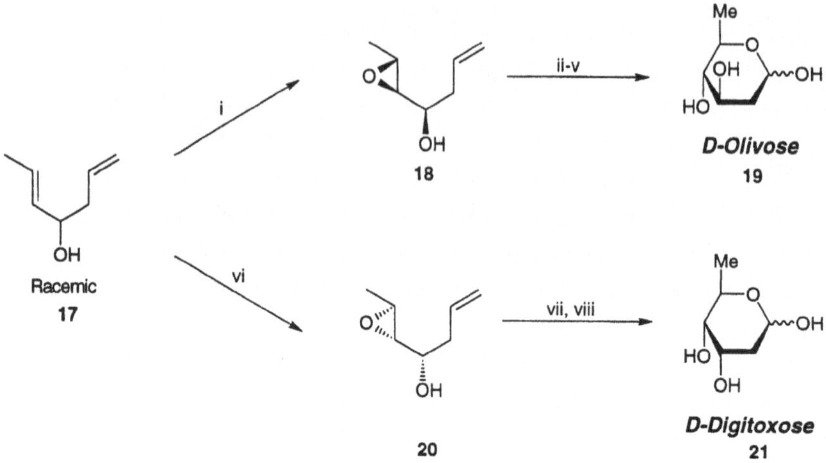

Scheme 2.3 i, (−)-Diisopropyl tartrate, Ti(OPr-*i*)$_4$, *t*-BuO$_2$H (0.4 equiv.), 33% yield, >95% enantiomeric excess (ee); ii, PhNCO, Py; iii, Et_2AlCl, Et_2O, −20°C then H_3O^+; iv, NaOMe, MeOH; v, O_3, MeOH, −20°C, Me_2S work-up, 55% yield for 4 steps; vi, (+)-diisopropyl tartrate, Ti(OPr-*i*)$_4$, *t*-BuO$_2$H (0.4 equiv.), 37% yield, >95% ee; vii, H_3O^+, Me_2SO; viii, O_3, MeOH, −20°C, Me_2S work-up, 70% yield for 2 steps.

A particularly useful intermediate for the preparation of 2,3,6-trideoxy-3-amino-L-hexoses is the heterocyclic alcohol (*S*)-1-(2-furyl)ethanol **23**, available in 98% enantiomeric excess (ee) via the low-temperature asymmetric reduction of 2-acetylfuran with the chiral reducing agent obtained from lithium borohydride and (*S,S*)-*N,N*-dibenzoylcystine. Alcohol **23** has served as a chiral starting material for a number of very elegant 2,3,6-trideoxy-3-amino-L-hexose syntheses that have included the antibiotic sugar L-daunosamine hydrochloride **30** (Scheme 2.4). In this route to **30**, developed by Sammes [5], **23** was initially converted to a 3:1 mixture of enones **24** and **25** that were separated by chromatography. The major isomer **25** was then reduced with sodium borohydride to a 13:1 mixture of allylic alcohols that was enriched in the desired *erythro* isomer **26**. A modified Mitsunobu reaction with dibenzamide was employed for introducing the imidate moiety in **27** with inversion of configuration. This was the prelude to the stereocontrolled installation of the nitrogen functionality at C(3) in **30** via a bromonium-ion-assisted cyclisation. This method of ring closure smoothly delivered **28** in 91% yield. Compound **28** was then reduced with lithium aluminium hydride to obtain **29**, which was converted to L-daunosamine hydrochloride **30** by deblocking with acid.

An alternative pathway to L-daunosamine has been described by Uskokovic and co-workers at Roche (Scheme 2.5) [6]. Their approach to **39** relied on the highly enantioselective hydroboration of **31** with (−)-diisopinylcampheylborane. After oxidative rearrangement of the resulting trialkylborane with basic

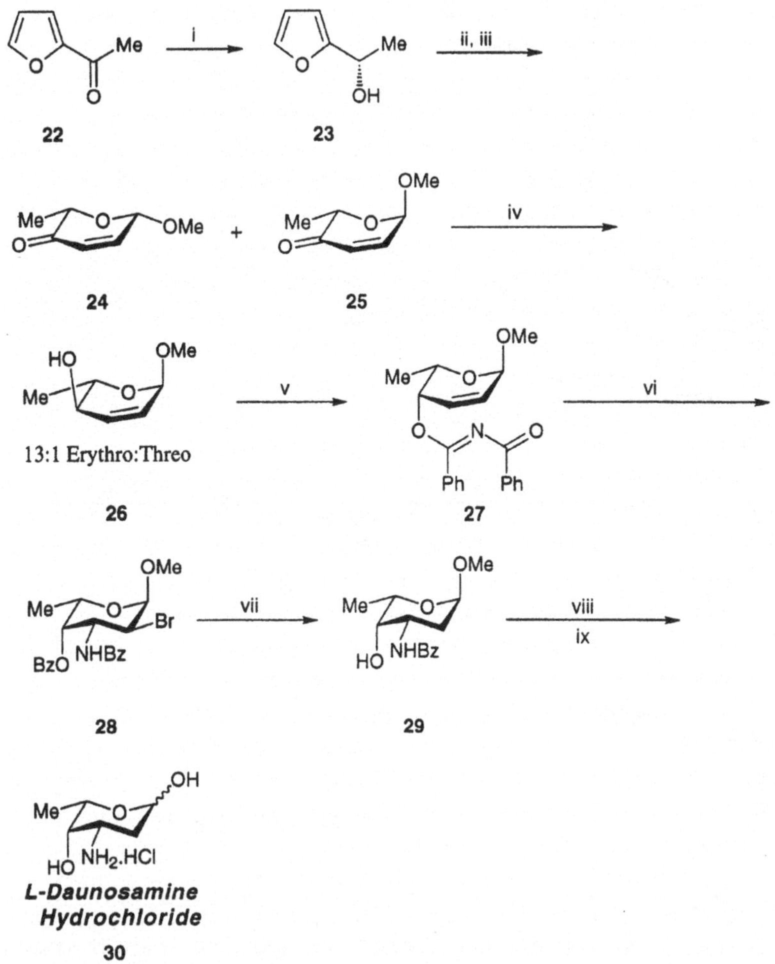

Scheme 2.4 i, (*S*,*S*)-*N*,*N*-Dibenzoylcystine, $LiBH_4$, *t*-BuOH, THF, Δ, 12 h, then cool to −78°C and add 2-acetylfuran, 82%, 89% ee; ii, Br_2, MeOH, −35°C, 93%; iii, HCO_2H, MeOH (ratio of **24**:**25** = 1:3), 83%; iv, $NaBH_4$, THF, H_2O, 92%; v, $NaBH_4$, THF, H_2O, 92%; vi, Ph_3P, DEAD, PhCONHCOPh, THF, 0°C to room temperature, 86%; vii, *N*-bromosuccinimide (3 equiv.), EtOH-$CHCl_3$ (25:1), room temperature, 6 h, 91%; viii, $LiAlH_4$, THF, room temperature, 95%; ix, HCl, MeOH, H_2O, Δ, 24 h, 45%.

hydrogen peroxide, alcohol **33** was isolated in 95% ee and 33% yield. Hydroxyl-directed epoxidation of **33** with *m*-chloroperbenzoic acid (MCPBA) enabled the *syn*-stereochemistry to be established between the hydroxyl and epoxide in **34**. After oxidation to the cyclopentanone, the carbohydrate skeleton was fashioned by Baeyer–Villiger ring expansion. Reduction of lactone **36** to hemi-acetal **37** followed by Fischer glycosidation under carefully defined

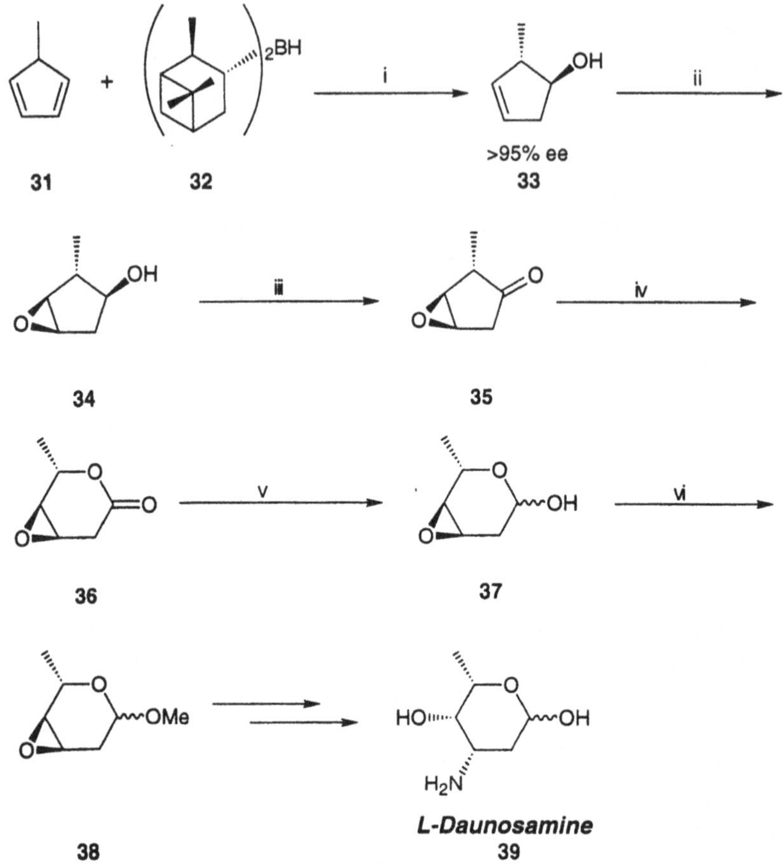

Scheme 2.5 i, THF, −78°C for 5 h, then 0°C for 16 h; 3 N NaOH, 20% aq. H_2O_2, 33%; ii, *m*-CPBA, $NaHCO_3$, CH_2Cl_2, 69%; iii, CrO_3, Py, Me_2CO, 0°C, 65%; iv, *m*-CPBA, $NaHCO_3$, CH_2Cl_2, 98%; v, *i*-Bu_2AlH, PhMe, −75°C, 2 h, 77%; vi, catalyst BF_3 in MeOH, 85%. Note: ee = enantiomeric excess.

conditions afforded methyl 3,4-anhydro-2,6-dideoxy-α- and β-L-ribopyranosides **38** as a separable mixture. The α anomer had previously been converted to L-daunosamine **39** by Goodman and co-workers [7]. Besides offering a convenient avenue to 2,3,6-trideoxy-3-amino hexoses of the L-daunosamine class, this route appears capable of being modified to produce 2,6-dideoxy-hexoses from both the D and L series.

Llera and co-workers [8] have cleverly exploited the condensation reaction between the carbanion of (*R*)-methyl *p*-tolylsulfoxide and ethyl 2-furancarboxylate for monosaccharide assembly (Scheme 2.6). This yielded a β-ketosulfoxide **42** that was reduced either with DIBAL to give alcohol **43**, or with DIBAL/$ZnCl_2$ to give **46**, both with greater than 95% diastereomeric

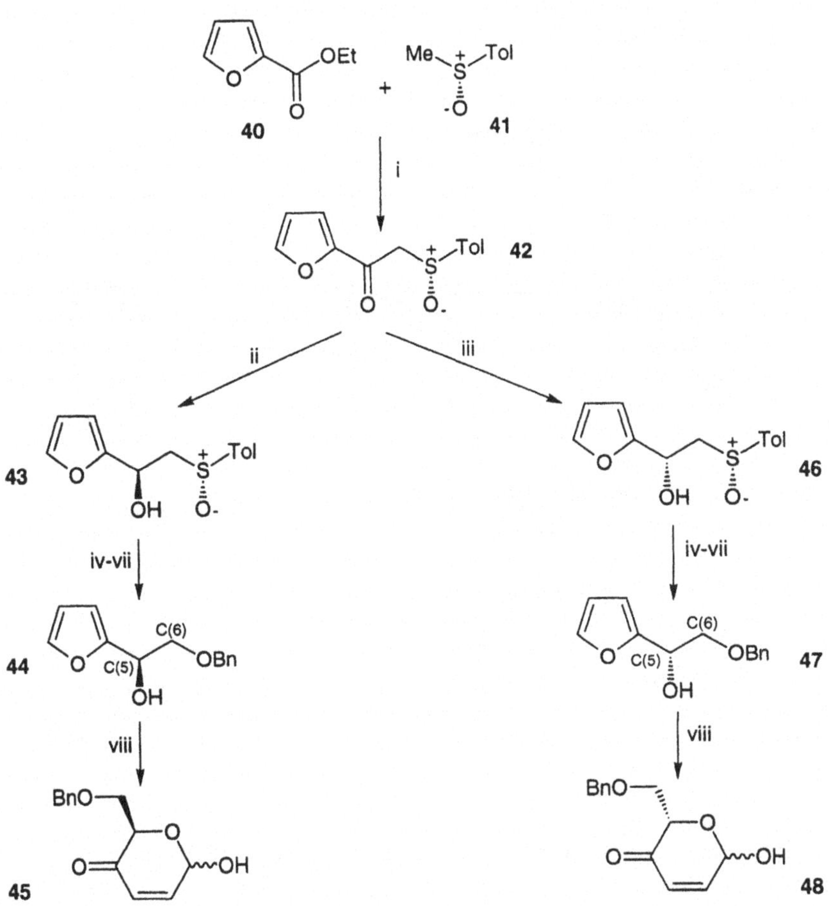

Scheme 2.6 i, *i*-Pr_2NLi, THF, −78°C, 75%; ii, *i*-Bu_2AlH, THF, −78°C, 5 h, 85%; iii, *i*-Bu_2AlH, $ZnCl_2$, THF, −78°C, 5 h, 80%; iv, SEM-Cl, *i*-Pr_2NEt, CH_2Cl_2, 40°C, 95%; v, $(CF_3CO)_2O$, MeCN, collidine, 0°C, then aq. $NaBH_4$, 88%; vi, aq. NAOH, BnBr, CH_2Cl_2, *n*-Bu_4NBr, 94%–95%; vii, *n*-Bu_4NF, DMPU, 80%–93%; viii, Br_2, MeCN, H_2O, −5°C to room temperature, 80%–89%.

excess (d.e.). After protection of the newly introduced alcohol in **43** and **46**, Pummerer rearrangement was used to oxygenate at C(6). Reduction of the resulting trifluoroacetoxysulfide mixtures led to the enantiomeric alcohols **44** and **47** after protecting group manipulation. Both furyl alcohols **44** and **47** underwent rearrangement to **45** and **48**, respectively, after oxidation with bromine in acetonitrile. Either molecule can be considered a viable hexose precursor since Achmatowicz has developed methodology for the manipulation of such systems into monosaccharides with the *allo*, *altro*, *manno*, *gluco*, *galacto* and *ido* configurations [9].

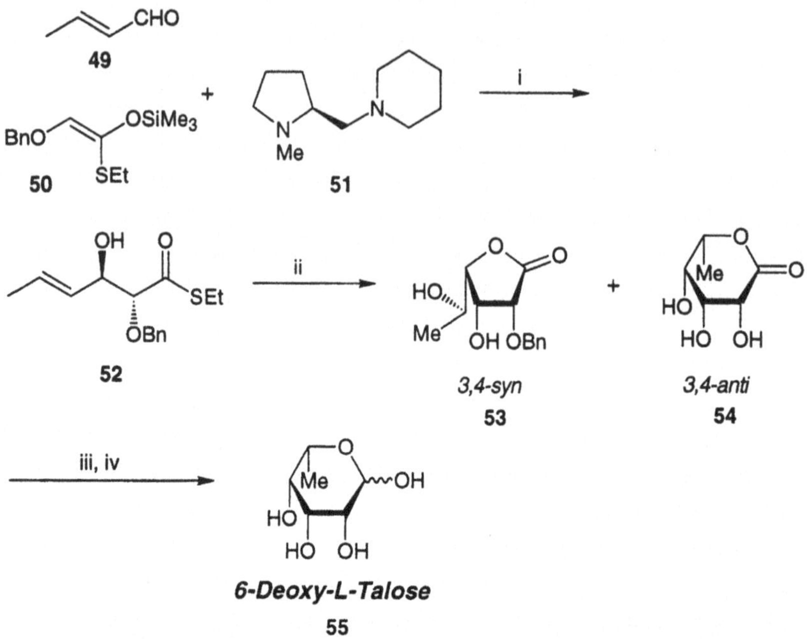

Scheme 2.7 i, $Sn(OTf)_2$, n-$Bu_2Sn(OAc)_2$, CH_2Cl_2, $-78°C$, 85% (*anti*:*syn* $\geq$ 98:2, >97% enantiomeric excess); ii, catalyst OsO_4, NMO, Me_2CO-H_2O (8:1), room temperature, then H_2S, 72% (ratio of **53**:**54** = 28:72); iii, i-Bu_2AlH, CH_2Cl_2, $-78°C$, 71%; iv, H_2, Pd-C, EtOH, 100%.

Mukaiyama and co-workers have observed that dihydroxylation of the aldol adducts obtained from chiral Lewis acid-promoted reactions of α,β-unsaturated aldehydes with α-benzyloxy thioketene acetals offers an economical way of synthesising monosaccharides in high ee. By capitalising on this approach, 6-deoxy-L-talose **55** was synthesised by Mukaiyama in only four steps from simple achiral precursors (Scheme 2.7) [10]. In this route to **55**, the key asymmetric aldol reaction between **49** and **50** was executed at −78°C in dichloromethane in the presence of the chiral Lewis acid generated from tin (II) triflate, [(*S*)-1-methyl-(2-piperidin-1-yl)methyl]pyrollidine **51**, and dibutyltin acetate. This afforded the *anti* aldol **52** with very high levels of diastereoselection and enantiomeric excess. Dihydroxylation of **52** with osmium tetraoxide resulted in a 72:28 mixture of lactones **54** and **53**. These were separated and the *O*-benzyl protecting group removed from **54** by catalytic hydrogenation in methanol over a palladium catalyst. The fact that 6-deoxy-L-talose was prepared in a noteworthy 31% overall yield serves as a testament to the synthetic efficiency of the method. Mukaiyama and co-workers have utilised this strategy for the production of 6-deoxy-D-allose [11] (Scheme 2.8) and the branched sugar 4-*C*-methyl-D-ribose [10]

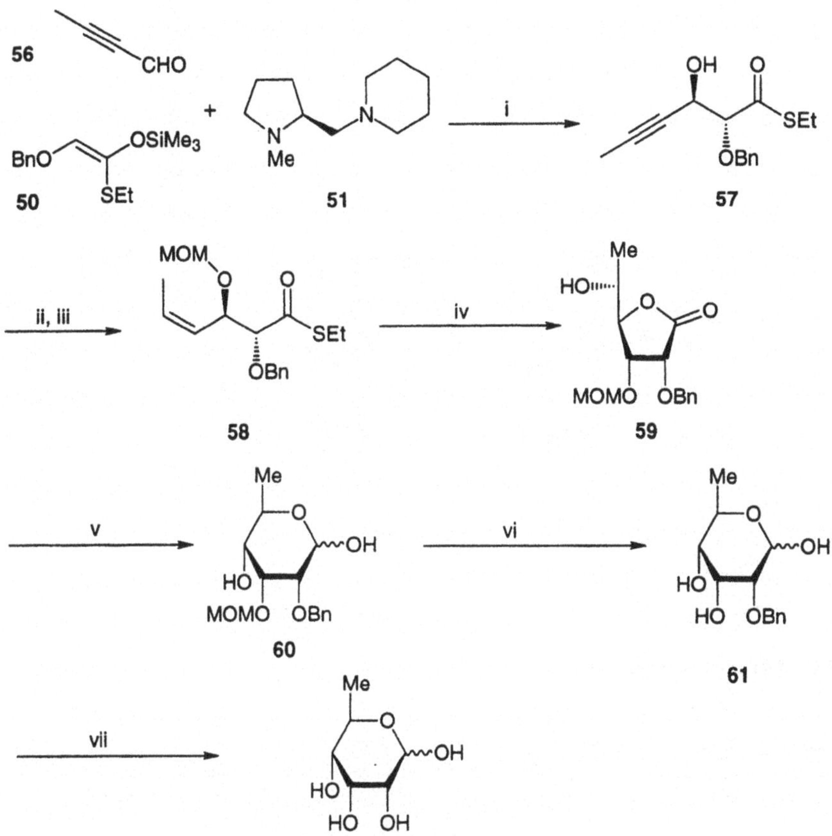

6-Deoxy-D-Allose
62

Scheme 2.8 i, $Sn(OTf)_2$, n-$Bu_2Sn(OAc)_2$, CH_2Cl_2, −78°C, 82% (*anti*:*syn* ≥ 94:6, >92% enantiomeric excess); ii, MOM-Cl, i-Pr_2NEt; iii, H_2, Pd-C, 100%; iv, catalyst OsO_4, NMO, 87%, 3,4-*anti*:*syn* = 4:1; v, i-Bu_2AlH, CH_2Cl_2, −78°C, 75%; vi, 6 N HCl, THF, 94%; vii, H_2, Pd-C, MeOH, 95%.

(Scheme 2.9). Clearly, an even greater range of monosaccharides should be accessible if the Sharpless asymmetric dihydroxylation [12, 13] reaction is incorporated into the planning.

Enders and Jegelka [14] have demonstrated the worth of 1,3-dioxan-5-one as a three-carbon synthon for the asymmetric synthesis of unusual ketoses via the SAMP/RAMP hydrazone alkylation method. In this procedure, a SAMP [(*S*)-1-amino-2-methoxymethyl-pyrollidine] or RAMP [the corresponding (*R*)-enantiomer] chiral auxiliary is temporarily attached to 1,3-dioxan-5-one **68** and a chiral aza enolate generated from the hydrazone by treatment with strong base; an asymmetric alkylation is then performed at −78°C with an

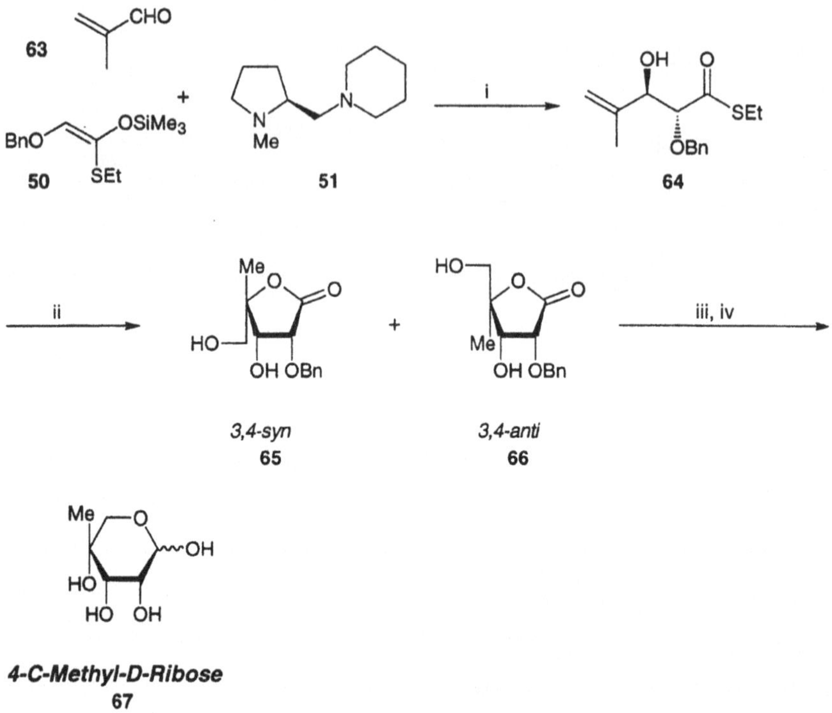

Scheme 2.9 i, $Sn(OTf)_2$, $n\text{-}Bu_2Sn(OAc)_2$, CH_2Cl_2, −78°C, 71% (*anti*:*syn* ≥ 91:9, 96% enantiomeric excess); ii, catalyst, OsO_4, NMO, $Me_2CO\text{-}H_2O$ (8:1), room temperature, then H_2S (ratio of **65**:**66** = 83:17); iii, $i\text{-}Bu_2AlH$, CH_2Cl_2, −78°C, 63%; iv, H_2, Pd-C, EtOH, 100%.

iodoalkane or an activated alkyl chloride. A second asymmetric alkylation is then carried out under kinetically controlled conditions. Ozonolysis detaches the hydrazone auxiliary from the dialkylated product to finalise the synthesis of the protected ketose. The operational workings of this approach are revealed more fully in Scheme 2.10 for the preparation of 5-deoxy-L-(−)-*threo*-3-pentulose.

In recent years, asymmetric cycloaddition has become a popular vehicle for *ab initio* monosaccharide construction, with the groups of Danishefsky and Vogel making particularly eminent contributions in this area. In Danishefsky's synthesis of L-glucose, the key asymmetry-inducing step was a highly diastereoselective hetero Diels–Alder cycloaddition between benzaldehyde and the chiral diene **75** catalysed by chiral Lewis acid (+)-$Eu(hfc)_3$ {a tris[3-(heptafluoropropylhydroxymethylene)-(+)-camphorato]europium (III) derivative} (Scheme 2.11) [15]. At −78°C this reaction furnished a 25:1 mixture of diastereoisomers that was enriched in the desired β-L-pyranoside; the latter was isolated pure by fractional crystallisation. It underwent transformation into α-acetoxy enone **77** by sequential reaction with trifluoroacetic

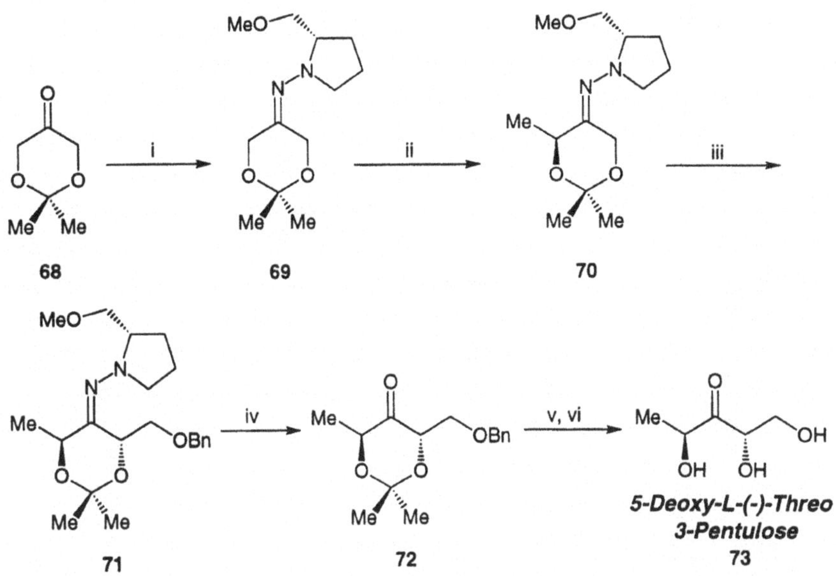

Scheme 2.10 i, SAMP, PhMe, Δ, 20 h, 95%; ii, *t*-BuLi, THF, −78°C, 1.5 h, then cool to −95°C, add MeI; iii, *t*-BuLi, THF, −78°C, 1.5 h, then cool to −95°C, add $BnOCH_2Cl$; iv, O_3, CH_2Cl_2, −78°C, 62% yield, 3 steps; v, 3 N HCl, MeOH, 0.5 h, 95%; vi, H_2, Pd-C, MeOH, 95%.

acid and manganese triacetate. Reduction of the carbonyl group in **77** with sodium borohydride and cerium trichloride followed by acetylation then afforded the L-glucal derivative **78**. Oxygenation of the enol ether in **78** was accomplished by dihydroxylation with osmium tetraoxide; acetylation of the resulting diol produced **79**. At this juncture, the phenyl ring at C(5) was degraded to a hydroxymethyl group by ozonolysis and diborane reduction. Sequential acetylation and deacetylation furnished L-glucose **9** in optically pure condition.

Vogel and associates have reported routes to a number of hexoses and pentoses that proceed from optically pure 7-oxabicyclo[2,2,1]hept-5-en-2-yl derivatives [16]. Such intermediates are accessible from the zinc iodide catalysed [4 + 2]-cycloaddition of furan to the respective enantiomers of 1-cyanovinyl-camphanate. While these cycloadditions display little intrinsic diastereofacial bias [diastereoselectivity (ds) approximately 65:35], the high crystallinity of the adducts allows the major component to be isolated in high optical purity after three successive recrystallisations. The utility of these intermediates as chiral templates for monosaccharide synthesis is illustrated by the route developed to the partially protected D-ribose derivative **87** (Scheme 2.12) [17]. It commenced from the major diastereoisomer **82**, obtained from the Diels–Alder reaction between furan and the (*1S*)-camphanate **80**. Stereospecific dihydroxylation of **82** proceeded exclusively from the less hindered *exo* face to give acetal **83** after acetonation. The camphanate

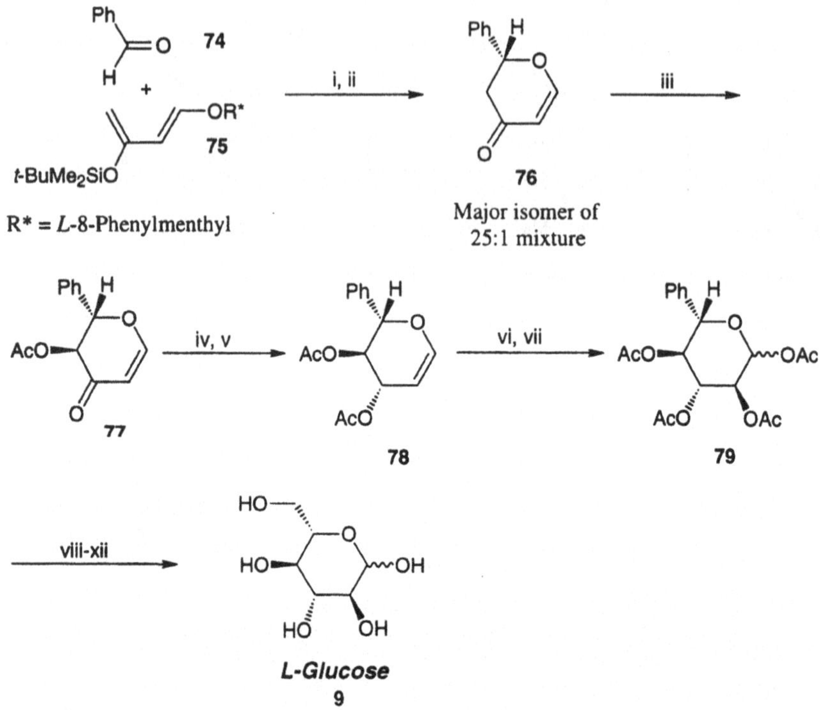

Scheme 2.11 i, (+)-Eu(hfc)$_3$, hexane, −78°C, 72 h, then room temperature, 24 h, 81%; ii, CF_3CO_2H, CH_2Cl_2, 0°C, 87%; iii, Mn(OAc)$_3$, C_6H_6, 90°C, 12 h, 52%; iv, $NaBH_4$, $CeCl_3$, MeOH; v, Ac_2O, CH_2Cl_2, Et_3N, DMAP, 80%, 2 steps; vi, catalyst OsO_4, THF, NMO, t-BuOH, H_2O; vii, Ac_2O, CH_2Cl_2, Et_3N, DMAP, 80%, 2 steps; viii, O_3 in O_2, AcOH, 16 h, room temperature; ix, H_2O_2, H_2O, 24 h; x, BH_3-THF, room temperature, 12 h; xi, Ac_2O, Et_3N, DMAP, CH_2Cl_2, 42%, 4 steps; xii, NaOMe, MeOH, 86%.

auxiliary was detached from **83** by basic hydrolysis in the presence of formaldehyde, liberating ketone **84** in 92% yield. Regiospecific Baeyer–Villiger reaction of **84** elaborated the hexofuranurono-6,1-lactone **85**, which was saponified and transformed into methyl glycoside **86** by Fischer glycosidation. In order to convert **86** into D-ribose derivative **87** a brominative decarboxylation was performed with bromine and red mercuric oxide in carbon tetrachloride at reflux. This afforded methyl 5-bromo-5-deoxy-2,3-isopropylidene-β-D-ribofuranoside in 60% yield. Nucleophilic displacement with acetate anion and mild alkaline hydrolysis then completed this elegant synthesis of **87**.

The versatility of ketone **84** was further exemplified by the synthesis of monosaccharides such as 3-amino-3-deoxy-D-altrose [18] and D-allose [19]. For the latter, **84** was converted to silyl enol ether **88** by reaction with *N*-(*tert*-butyldimethylsilyl)-*N*-methyltrifluoroacetamide and triethylamine in

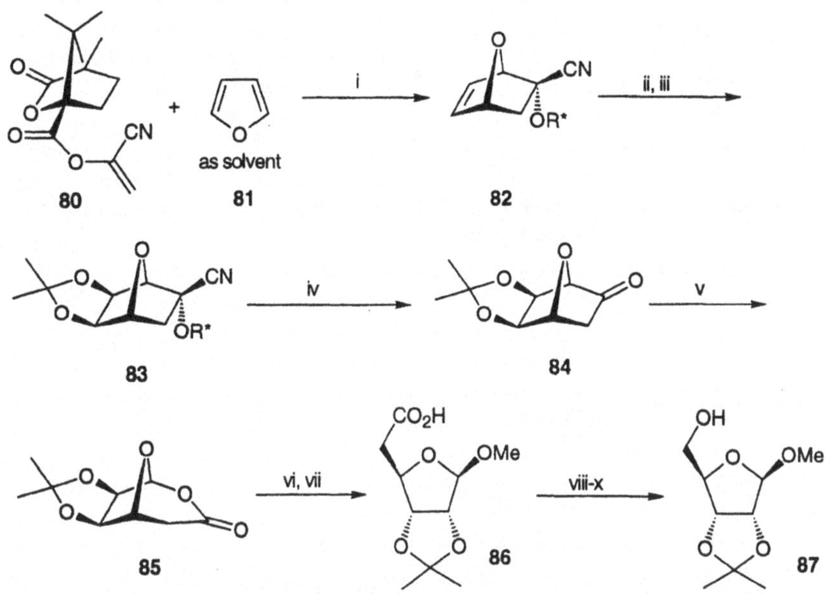

Scheme 2.12 i, catalyst ZnI_2, 60°C, sealed tube, 24 h, 29% yield, and 98% enantiomeric excess after recrystallisation; ii, catalyst OsO_4, *t*-BuOH, Me_2CO, 30% H_2O_2, 20°C, 15 h; iii, $Me_2C(OMe)_2$, Me_2CO, *p*-TsOH, 65%, 2 steps; iv, 3 N aq. KOH, THF, H_2O, HCHO, 92%; v, *m*-CPBA, $NaHCO_3$, $CHCl_3$, 20°C, 98%; vi, catalyst MsOH, $Me_2C(OMe)_2$, MeOH, 82%; vii, KOH, H_2O, THF, 98%; viii, Br_2, HgO (red), CCl_4, Δ, 80%; ix, KOAc, DMF; 75%; x, $NaHCO_3$, HMPA-H_2O (17:3), 130°C, 75%.

dimethylformamide (Scheme 2.13). Oxidation of **88** with MCPBA in tetrahydrofuran (THF) then created the α-*m*-chlorobenzoyloxy ketone by ring opening of the intermediary epoxide with *m*-chlorobenzoic acid and *in situ* acyl migration. Heating at 200°C effected de-esterification to deliver α-hydroxy ketone **89**. Baeyer–Villiger oxidation allowed rupture of the tricyclic ring system to expose the 1,6-hexuronolactone framework. A transesterification with potassium carbonate in methanol furnished lactol **90**, which was converted to D-allose **91** by transketalisation, reduction and glycoside hydrolysis.

Another homochiral building block, with a useful application profile for monosaccharide synthesis, is 7-oxabicyclo[2,2,1]hept-5-en-2-one **97**. It has been successfully utilised for the preparation of methyl 3-deoxy-D-arabino-hexofuranoside [20] and 4-deoxy-D-mannose [20], respectively. Corey and Loh [21] have described an efficient method for synthesising **97** through enantioselective Diels–Alder reaction between furan and 2-bromoacrolein, in the presence of catalytic quantities (10 mol%) of the oxazaborolidine derived from *N*-tosyl(α*S*,β*R*)-β-methyltryptophan **93** (Scheme 2.14) [22]. This furnished adduct **94** in 98% yield with 92% ee. After reduction to the bromohydrin **95**, and conversion to the peroxide, the latter was opened

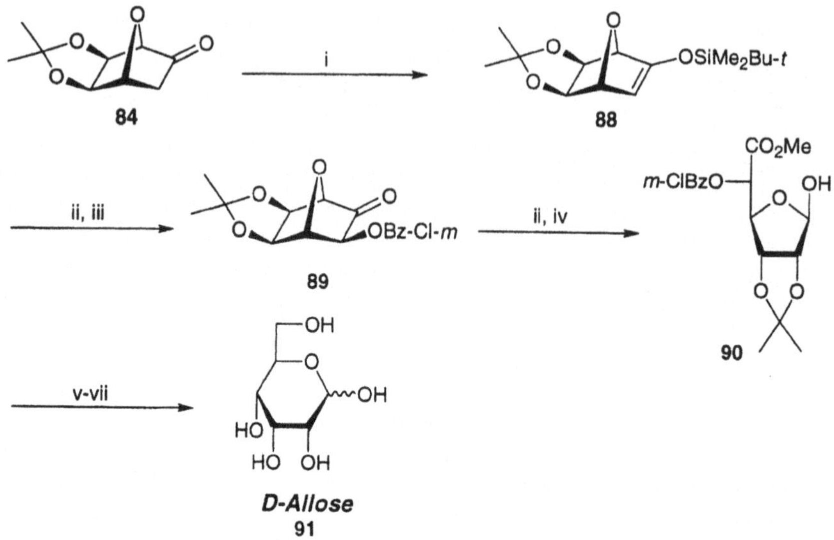

Scheme 2.13 i, TBSN(Me)C(O)CF_3, Et_3N, DMF, 85%; ii, *m*-CPBA, THF; iii, 200°C, 15 min; iv, K_2CO_3, MeOH; v, MeOH/H^+, 92%; vi, $LiAlH_4$, THF, 71%; vii, 2% aq. H_2SO_4.

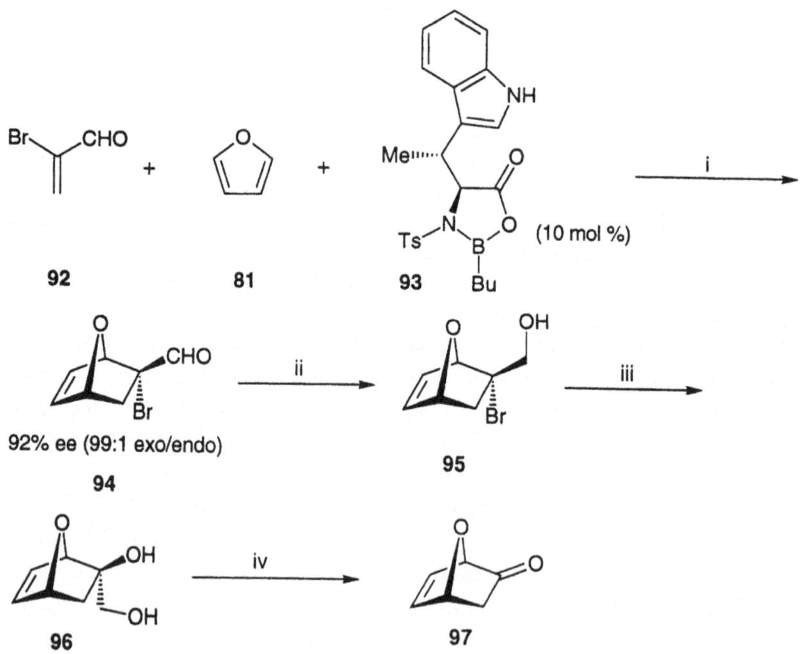

Scheme 2.14 i, CH_2Cl_2, −78°C, >98%; ii, $NaBH_4$, THF-H_2O (9:1), >98% yield; iii, 10% aq. K_2CO_3, dioxane, H_2O (3:1), Δ, 16 h, 92%; iv, $Pb(OAc)_4$, CH_2Cl_2, 0°C, 85%.

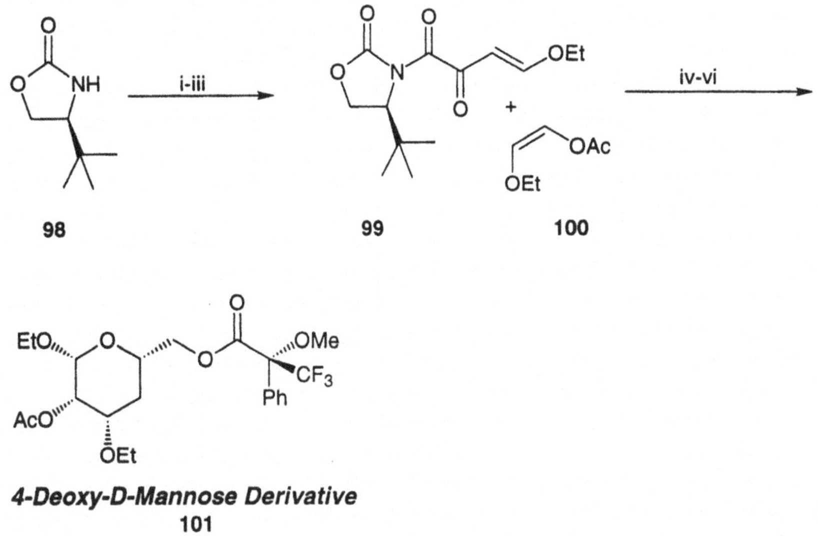

Scheme 2.15 i, *n*-BuLi, THF; ii, $(COCl_2)$; iii, ethyl vinyl ether; iv, Me_2AlCl, CH_2Cl_2, −40°C; v, H_2, Pd-C; vi, reduction/acylation.

in situ with hydroxide ion to provide diol **96**, which was oxidatively cleaved **97** with lead tetraacetate. The main advantage of this synthesis of ketone **97**, over the camphanate auxiliary-based approach, derives from its high stereoselectivity and overall synthetic efficiency, which clearly makes monosaccharide syntheses based on this starting material take on a much more practical flavour.

Tietz has described a synthesis of a 4-deoxy-α-D-mannoside which utilises an inverse-demand hetero Diels–Alder reaction between the chiral oxabutadiene **99** and enol ether **100** to construct the carbohydrate framework with excellent *endo* selectivity (>50:1) and high d.e. (97%) (Scheme 2.15) [23]. Quite clearly, the synthetic potential of this reaction for the rapid construction of monosaccharide derivatives is enormous, and we expect that further refinements of this tactic will take place in future years.

Matteson and Peterson [24] have discussed a conceptually new approach to the synthesis of L-ribose that is based on the addition of (dibromomethyl)-lithium to (*s*)-pinanediol [(benzyloxy)methyl]-boronate (**102**). This generated a boronate complex that underwent a zinc chloride catalysed stereospecific 1,2-rearrangement to the (*1S*)-[2-benzyloxy)-1-bromoethyl]boronate **103** (Scheme 2.16). The bromide in **103** was then displaced with lithium phenylmethoxide to yield (*s*)-pinanediol (*1R*)-[1,2-bis(benzyloxy)ethyl]boronate **104** with high diastereocontrol (35:1). Two reiterations of this homologation cycle led to (*s*)-pinanediol (*1R*, *2R*, *3S*)-[tetrakis(benzyloxy)butyl]boronic ester **105** uncontaminated by other diastereoisomers. Unfortunately,

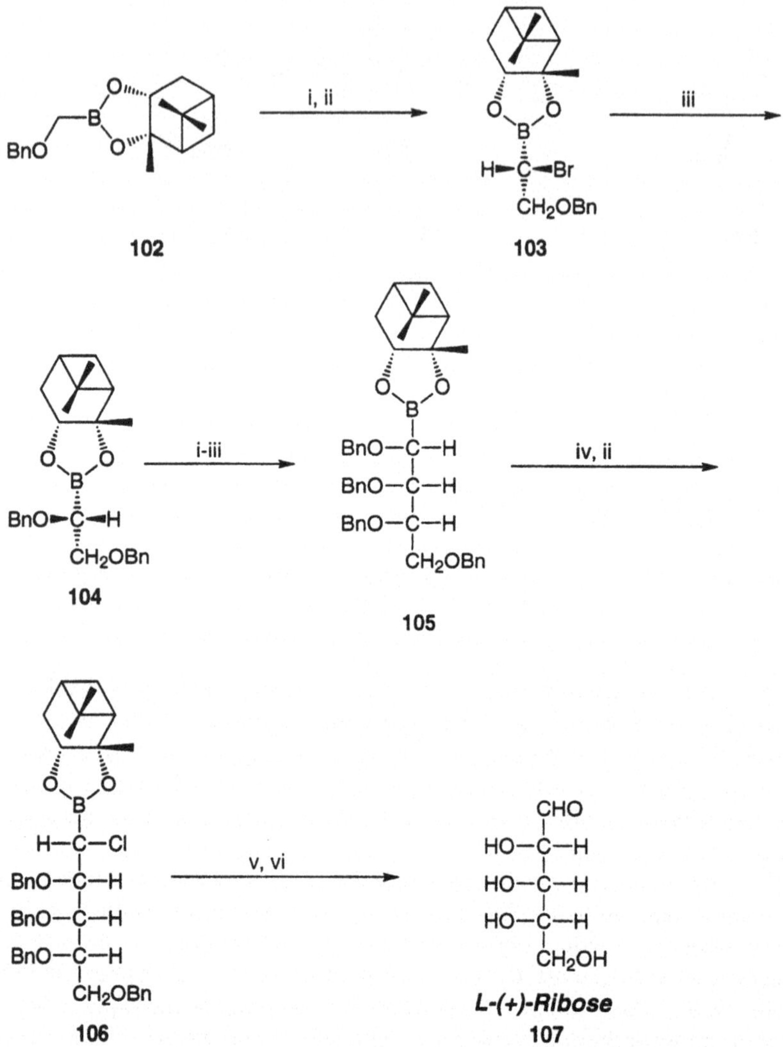

Scheme 2.16 i, Br_2CHLi, THF, −78°C, 1 h; ii, $ZnCl_2$, THF, −78°C to room temperature; iii, BnOLi, THF, −78°C; iv, Cl_2CHLi, THF, −78°C; v, borate buffered H_2O_2, THF; vi, H_2, Pd-C, 95% EtOH, 13% overall yield for the 10 steps.

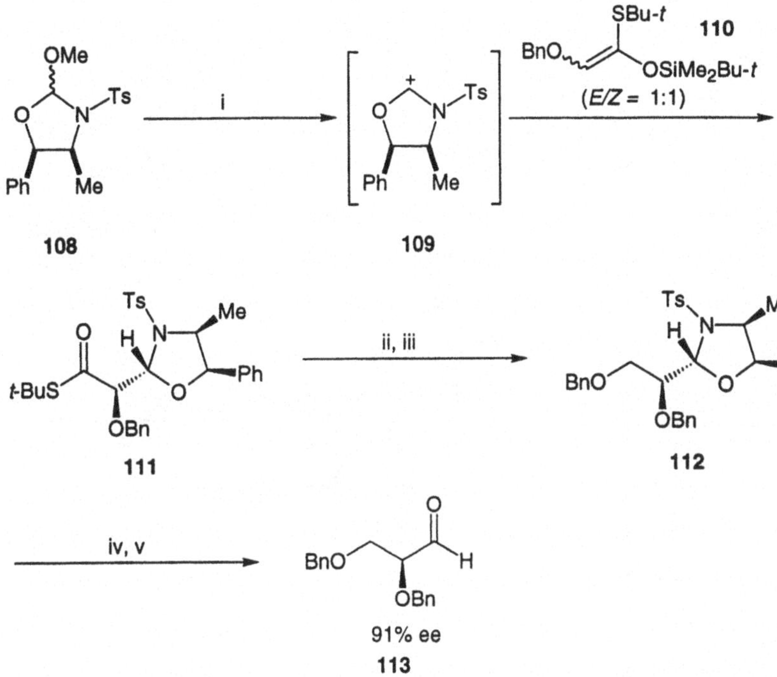

Scheme 2.17 i, BF_3-Et_2O, PhMe, −50°C, 3 h, 65%; ii, $LiAlH_4$, THF, 0°C, 93%; iii, 50% aq. NaOH, n-Bu_4NHSO_4, BnBr, CH_2Cl_2, 83%; iv, $(CH_2SH)_2$, CH_2Cl_2, BF_3-Et_2O, 63%; v, $CaCO_3$, MeI, Me_2CO, 60°C, 12 h, 50%.

addition of the fifth carbon required for completion of the total synthesis failed using the (dibromomethyl)lithium protocol. Instead, (dichloromethyl)-lithium was used as the one-carbon homologating agent. It produced L-ribose in 13% overall yield after oxidation of **106** with hydrogen peroxide and hydrogenation over palladium.

Scolastico and co-workers [25] have investigated the Lewis-acid-mediated addition of silyl ketene acetals to the chiral orthoamide **108** as a means of obtaining homochiral monosaccharide derivatives. In an initial demonstration of this approach (Scheme 2.17), silyl ketene acetal **110** was condensed with **108** in the presence of boron trifluoride etherate at low temperature to afford thioester **111** with 96:4 diastereoselectivity. Reduction of **111** with lithium aluminium hydride followed by *O*-benzylation was then used to obtain **112**. (*S*)-Di-*O*-benzyl glyceraldehyde was obtained in 91% ee by simple transacetalisation with 1,2-ethanedithiol and hydrolysis of the resulting dithioacetal. Extension of these ideas to trimethylsilyloxyfuran allowed a very efficient synthesis of D-ribose propane-1,3-diyl dithioacetal **118** to be achieved with excellent stereocontrol (Scheme 2.18) [26].

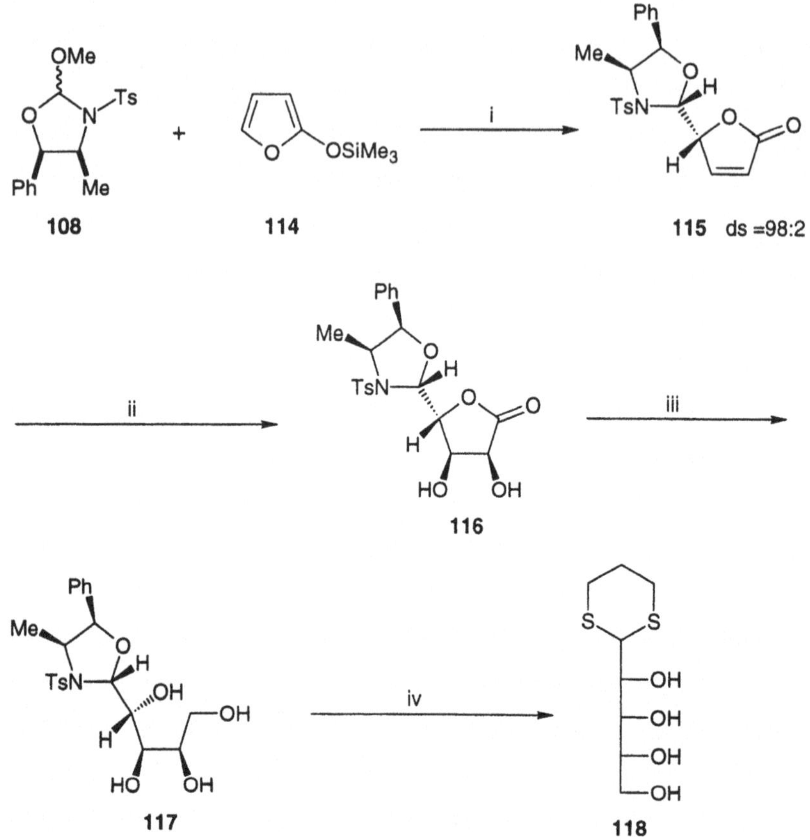

Scheme 2.18 i, *t*-$BuMe_2SiOTf$ (0.05 equiv.), CH_2Cl_2, −78°C, 96%; ii, catalyst dicyclohexyl-18-crown-6, $KMnO_4$, CH_2Cl_2, −20°C, 18 h, 60% yield; iii, $NaBH_4$, MeOH, room temperature, 24 h, 86%; iv, propane-1,3-dithiol, CH_2Cl_2, BF_3-Et_2O, 2.5 h, 43%. Note: ds = diastereoselectivity.

2.2 Asymmetric synthesis of monosaccharides via chemical homologation and degradation

By far the most popular pathway to monosaccharides has been through chain extension of suitably protected glyceraldehyde derivatives. Dondoni and co-workers have examined the utility of 2-trimethylsilylthiazole as a reagent for the stereocontrolled chain elongation of 2,3-*O*-isopropylidene-D-glyceraldehyde **119** [27]. In this procedure (Scheme 2.19), the sugar aldehyde is mixed with 2-trimethylsilylthiazole **120** in dichloromethane at room temperature or below for several hours. Without any added catalyst, a highly stereocontrolled reaction ensues between both species which results in a silylated D-erythrose derivative being formed in good yield. Mechanistically,

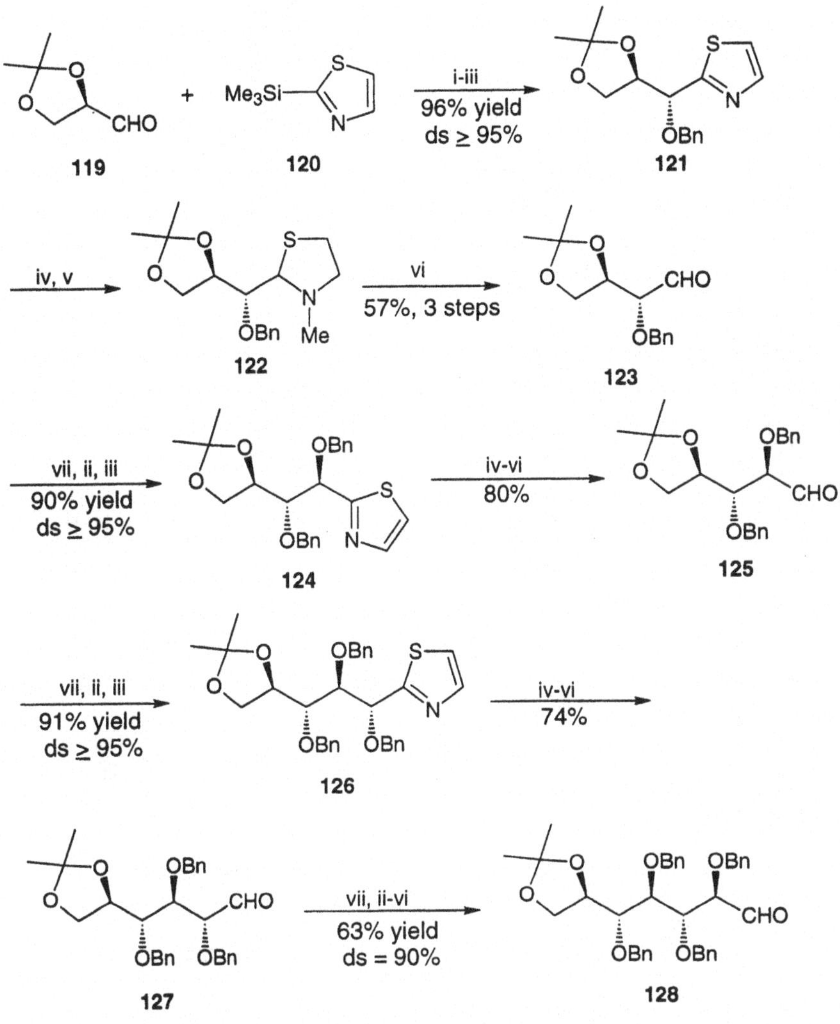

Scheme 2.19 i, CH_2Cl_2, 0°C to room temperature, 12 h; ii, *n*-Bu_4NF, THF, room temperature; iii, NaH, BnBr, THF; iv, MeI, MeCN, Δ; v, $NaBH_4$, MeOH, −10°C; vi, $HgCl_2$, MeCN, H_2O; vii, 2-trimethylsilylthiazole, CH_2Cl_2, 0°C to room temperature, 12 h. Note: ds = diastereoselectivity.

this reaction is believed to proceed by addition of the Si–C bond in **120** to the aldehyde carbonyl to give a thiazolium-2-ylide that combines with a second molecule of the aldehyde to produce a 2:1 adduct that breaks down to the product. After *O*-desilylation, *O*-benzylation and a three-step aldehyde unmasking operation the D-erythrose derivative **123** is obtained in 53% overall yield from the starting sugar **119**. Repetition of this six-step protocol over three more cycles produces the D-allose derivative **127** with very high levels of

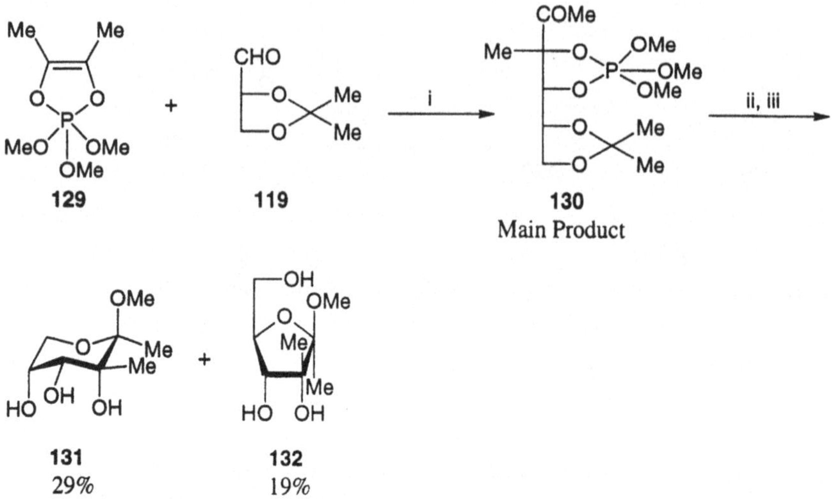

Scheme 2.20 i, Mix neat compounds, room temperature, 48 h; ii, H_2O; iii, MeOH, H^+.

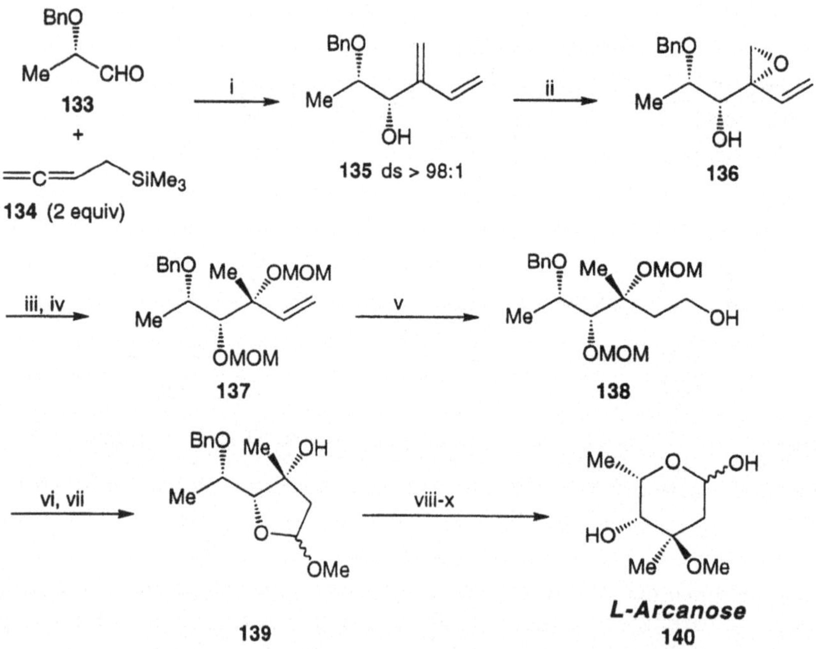

Scheme 2.21 i, $TiCl_4$ (1.1 equiv.), CH_2Cl_2, −78°C, 12 h, 77%; ii, catalyst $VO(acac)_2$, *t*-BuO_2H, 0°C, 80%; iii, $LiAlH_4$, THF, 0°C; iv, MOM-Cl, *i*-Pr_2NEt, CH_2Cl_2, Δ, 80%; v, thexylborane, THF, 0°C, then 30% H_2O_2, NaOH, 0°C, 75%; vi, $(COCl)_2$, Me_2SO, CH_2Cl_2, −60°C, then Et_3N, 0°C; vii, catalyst *p*-TsOH, MeOH, 72%; viii. KH, MeI, THF, 92%; ix, H_2, Pd-C in 0.2% $HClO_4$-MeOH; x, 1 N H_2SO_4 : THF (1 : 1), 78%.

anti-diastereoselectivity [27]. The method has been applied with considerable success to the construction of certain higher carbon monosaccharides [27].

A reaction of outstanding simplicity for the construction of D-fructopyranoside derivatives branched at C(3) is the Ramirez dioxaphosphazole condensation; an example of this process can be found in Scheme 2.20 [28].

Several important 2,6-dideoxy-L-hexoses with branching at C(3) have been synthesised in totally stereocontrolled fashion from (*S*)-2-benzyloxy-propanal via a strategy involving Lewis-acid-mediated addition of 1-trimethylsilyl-2,3-butadiene as the initial key step (Scheme 2.21). With titanium tetrachloride as the Lewis acid additive, high stereoselectivity in favour of the *syn*-diene-ol **135** was observed, presumably as a result of the formation of a chelated transition state. Compound **135** was converted to L-arcanose in a further nine steps, as shown in Scheme 2.21 [29]. When boron trifluoride etherate was used, however, stereoselectivity was in favour of the *anti* product **141** (Scheme 2.22); the latter was converted to L-olivomycose by a route similar to that in Scheme 2.21 [29].

A highly original four-carbon homologation protocol has been introduced by Marshall and co-workers [30] that is based on the reaction of non-racemic γ-(methoxymethyloxy) allylic stannanes with monosaccharide aldehydes. Its application to the synthesis of the L-iduronate derivatives **150** and **151** is shown in Scheme 2.23. The requisite (*R*)-allylic OMOM stannane **144** was prepared by asymmetric 1,2-reduction of acyl stannane **143** with (*S*)-(+)-BINAL, protection of the resulting (*R*)-α-hydroxystannane with methoxymethyl chloride, and boron trifluoride etherate mediated rearrangement. Allylstannane **145** smoothly added to a mixture of aldehyde **146** and a slight excess of magnesium bromide etherate in dichloromethane at −78°C

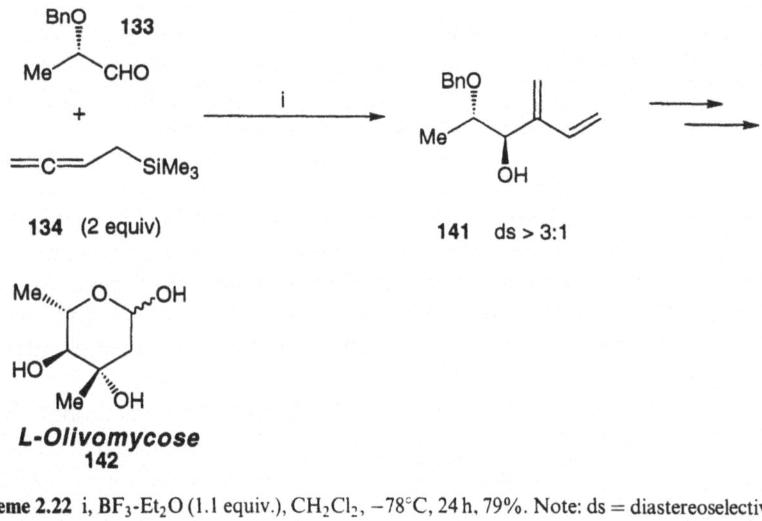

Scheme 2.22 i, BF_3-Et_2O (1.1 equiv.), CH_2Cl_2, −78°C, 24 h, 79%. Note: ds = diastereoselectivity.

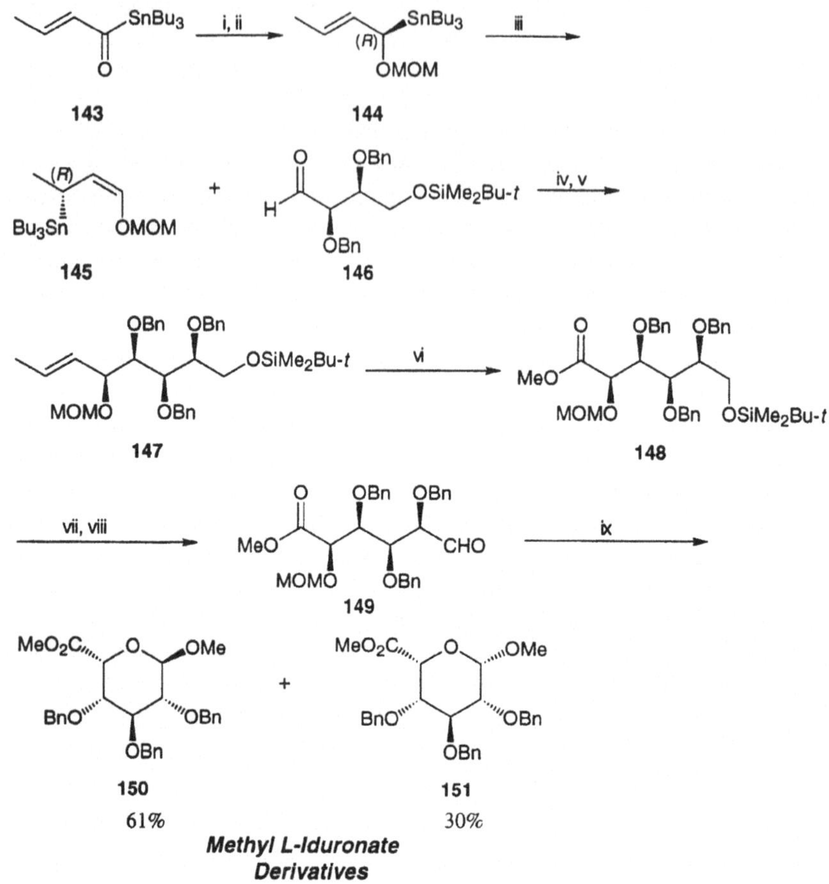

Scheme 2.23 i, (*S*)-(+)-BINAL-H, THF, −85°C; ii, *i*-Pr_2NEt, MOM-Cl, CH_2Cl_2, 40%–50%; iii, BF_3-Et_2O, −78°C, 80%–90%; iv, $MgBr_2$-Et_2O, CH_2Cl_2, −23°C to room temperature, 68%; v, NaH, DMF, BnBr, 0°C, 87%; vi, O_3, NaOH, MeOH, −78°C to room temperature, then $NaBH_4$, MeOH; vii, *n*-Bu_4NF, THF, AcOH, room temperature, 98%; viii, Dess–Martin reagent, CH_2Cl_2, 89%; ix, 1% HCl-MeOH, Δ, 1.5 h.

to give the *syn,syn* diol derivative **147** as the sole detectable product in 68% yield after benzylation. Ozonolysis in methanol-dichloromethane containing sodium hydroxide converted **147** into methyl ester **148**. Desilylation followed by Dess–Martin periodinane oxidation furnished aldehyde **149**, which underwent Fischer glycosidation to afford **150** and **151** as a separate mixture of anomers.

Marshall and co-workers [31] have further extended this methodology to monosaccharide-derived enals to provide a totally stereoselective route to higher carbon sugars. A spectacular demonstration of the power of their method is provided by their synthesis of L-*threo*-D-galactooctonic-1,4-lactone

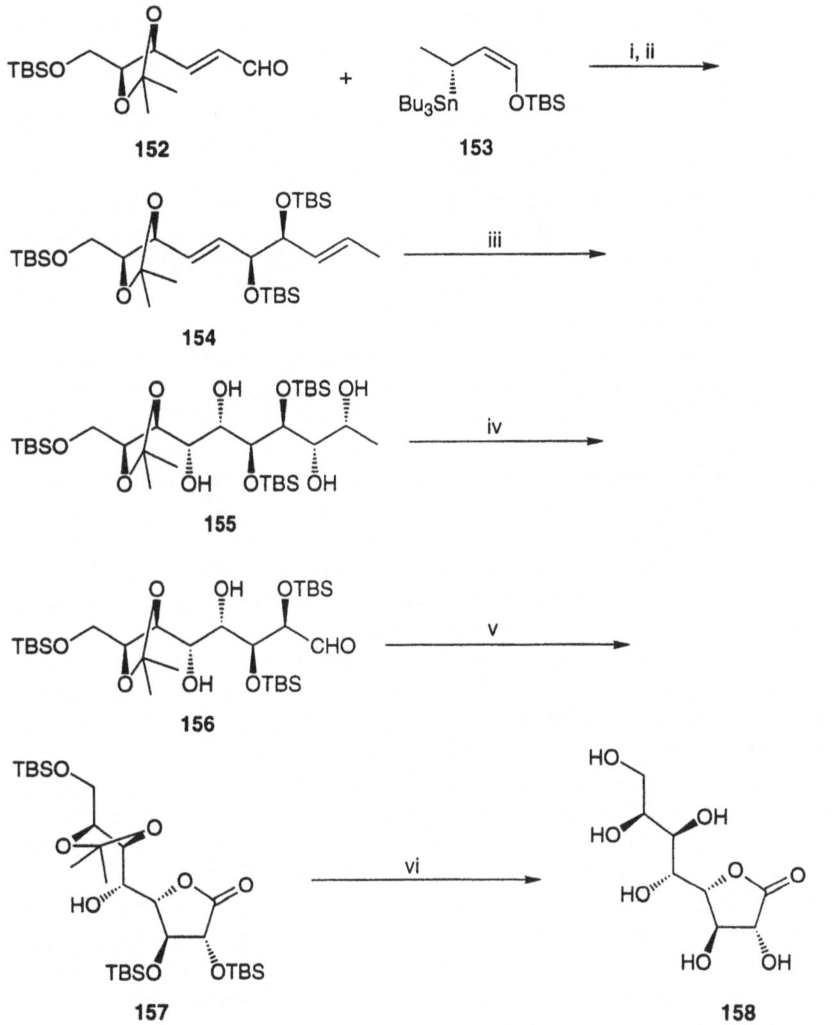

Scheme 2.24 i, BF_3-Et_2O, CH_2Cl_2, −78°C, 97%; ii, *t*-$BuMe_2SiOTf$; iii, OsO_4, NMO, 73%; iv, H_5IO_6, 86%; v, PCC, 92%; vi, *p*-TsOH, 88%.

158 (Scheme 2.24), where six contiguous hydroxy stereocentres were set in a mere three steps. Treatment of enal **152** with the (*R*)-allylic OTBS stannane **153** and boron trifluoride etherate in dichloromethane at −78°C led to a single protected *syn*-diol derivative **154** after silylation with *tert*-butyldimethylsilyl triflate. Dihydroxylation with osmium tetraoxide proceeded with very high *anti* diastereoselectivity to give **155** which underwent a remarkably chemoselective cleavage of the more exposed terminal diol with periodic acid to deliver an epimeric mixture of lactols **156**. Oxidation

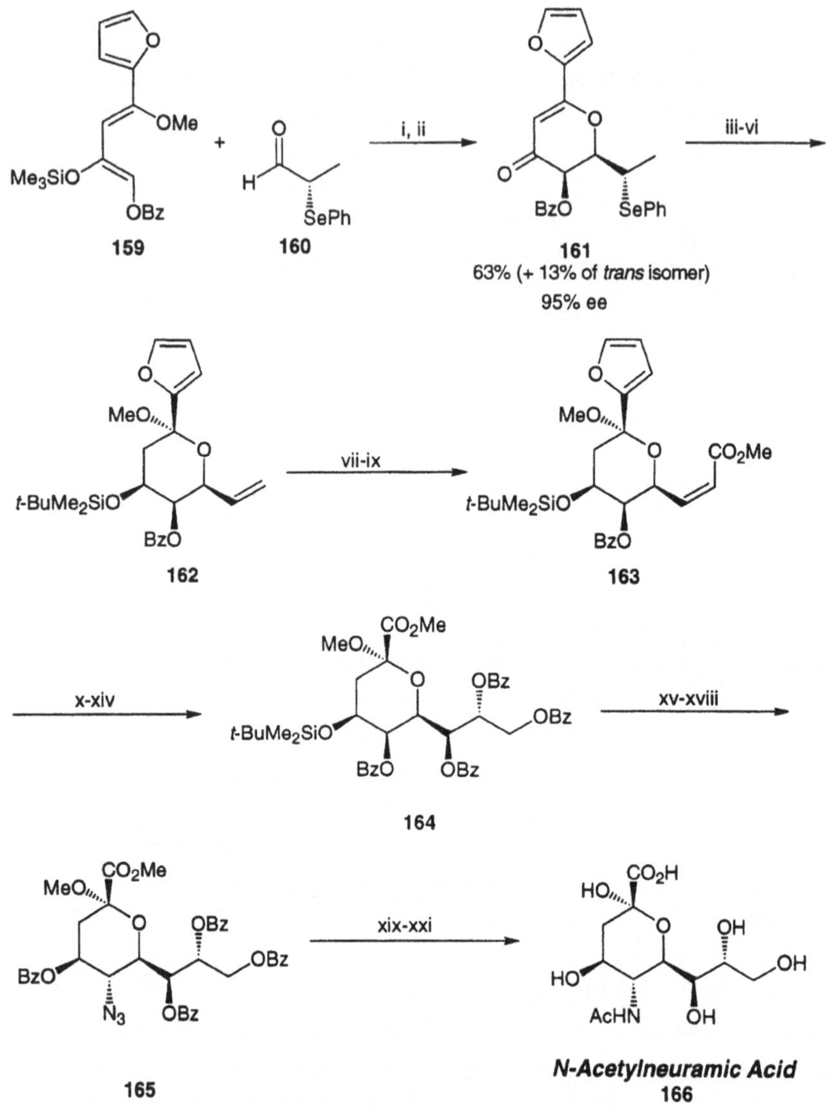

Scheme 2.25 BF_3-Et_2O, CH_2Cl_2, hexanes, DMF, −78°C; ii, CF_3CO_2H, CH_2Cl_2, room temperature; iii, $NaBH_4$, $CeCl_3$, CH_2Cl_2, EtOH, −78°C to −20°C; iv, CSA, MeOH; v, *t*-$BuMe_2SiOTf$, CH_2Cl_2, 2,6-lutidine; vi, 30% H_2O_2, Py, THF, 81% for 4 steps; vii, OsO_4, NMO, THF, H_2O, 94%; viii, $Pb(OAc)_4$, CH_2Cl_2; ix, $(CF_3CH_2O)_2P(O)CH_2CO_2Me$, $KN(SiMe_3)_2$, 18-crown-6, 71% for 3 steps; x, OsO_4, Py, 92%; xi, $LiEt_3BH$, THF, −78°C; xii, BzCl, CH_2Cl_2, DMAP, 80% for 2 steps; xiii, catalytic RuO_2, $NaIO_4$, $NaHCO_3$, $MeCN:CCl_4:H_2O$; xiv, CH_2N_2, Et_2O, 90% for steps; xv, aq. HF, MeOH, 45°C, 4 h; xvi, K_2CO_3, CH_2Cl_2, 61%, 2 steps; xvii, Tf_2O, Py, CH_2Cl_2, 0°C to room temperature; xviii, *n*-Bu_4NN_3, C_6H_6, 86%, 2 steps; xix, H_2, Pd/$CaCO_3$-Pb, EtOAc, MeOH, 5 h; xx, Ac_2O, DMAP, CH_2Cl_2, 94% for 2 steps; xxi, 1 N aq. NaOH, THF, 50°C, 2 h; then Dowex HCR-S resin, 50°C, 8 h.

to lactone **157** was achieved with pyridinium chlorochromate (PCC) in dichloromethane. After deprotection with 10 mol% of *p*-TsOH in methanol, L-*threo*-D-galacto-1,4-lactone **158** was obtained as a crystalline solid.

Danishefsky and co-workers [32] have shown that chiral α-phenylseleno aldehydes are useful starting materials for the construction of higher carbon monosaccharides. When condensed with the furyl diene **159** at −78°C under boron trifluoride etherate catalysis, aldehyde **160** yielded a 5:1 mixture of the *cis*:*trans* Cram–Felkin cycloaddition products (Scheme 2.25). This was subjected to β-elimination with trifluoroacetic acid to give **161** in 63% yield and 95% ee after separation. Stereospecific 1,2-reduction, followed by glycosidation, protection and selenoxide elimination enabled enone **161** to be converted to alkene **162** in 81% yield for the four steps.

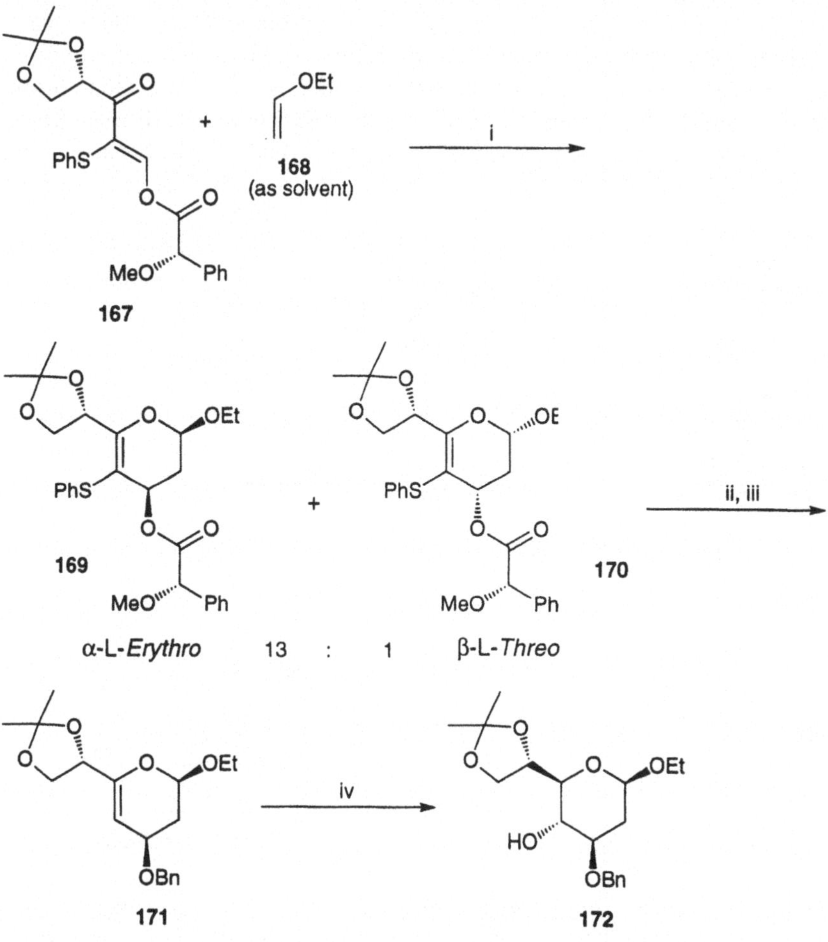

Scheme 2.26 i, 75°C, sealed tube, 6 h, 30% yield; ii, aq. NaOH, DMF, then BnBr, 83%; iii, Raney Ni, THF, 86%; iv, BH_3-THF, H_2O_2, NaOH, EtOH, room temperature, 1 h, 73%.

The olefin in **162** was then subjected to dihydroxylation and oxidative cleavage with lead tetraacetate to produce the aldehyde. Horner–Emmons reaction with the Still phosphonate afforded the *Z*-alkene **163** almost exclusively, to set the stage for a stereocontrolled osmylation reaction. At this point, the methyl ester was reduced to the primary alcohol, the three hydroxyl groups benzoylated, and the furan moiety oxidatively degraded to provide **164** after esterification. Fortunately, *O*-desilylation of **164** triggered off a partial *O*-benzoyl migration from O(5) to O(4), which was driven to completion by further treatment with base. The nitrogen functionality was installed at C(5) in **165** by triflate displacement with azide ion. This was then converted to *N*-acetylneuramic acid **166** by a series of standard manipulations.

Schmidt and co-workers have explored the utility of the chiral heterodiene **167** for heptose construction through inverse-demand Diels–Alder cycloaddition with ethyl vinyl ether (Scheme 2.26) [33]. High stereoselectivity in favour of the α-L-*erythro* product **169** was observed in this cycloaddition. After saponification and *O*-benzylation in the same pot, Raney nickel desulfurisation furnished glycal **171** in good yield. Hydroboration and

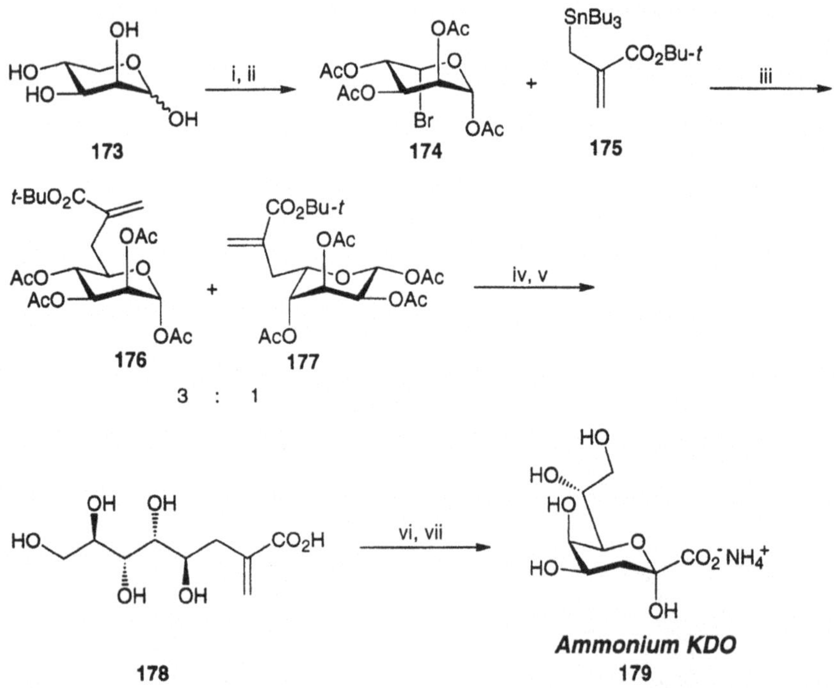

Scheme 2.27 i, Ac_2O, Py, 68%; ii, NBS, CCl_4, Δ, 1 h, $h\nu$, 35%; iii, AIBN, C_6H_6, 80°C; iv, NaOH, H_2O, room temperature, 0.5 h; v, $NaBH_4$, H_2O, room temperature, 0.5 h, 84%, 2 steps; vi, O_3, THF-H_2O (9:1), 0°C, 10 min, then Me_2S work-up, room temperature, 10 h; vii, NH_4OH, 92% yield for 2 steps.

oxidation completed this noteworthy synthesis of 2-deoxy-α-L-*galacto*-heptopyranoside **172**.

Free radical reactions are now beginning to show promise for the preparation of higher monosaccharides as illustrated by Giese and Linker's recent synthesis of ammonium 3-deoxy-D-*manno*-octulosonate (ammonium KDO) **179** from D-lyxose **173** (Scheme 2.27) [34]. This exploited the Ferrier photobromination procedure for functionalisation of C(5) to obtain **174**, and a

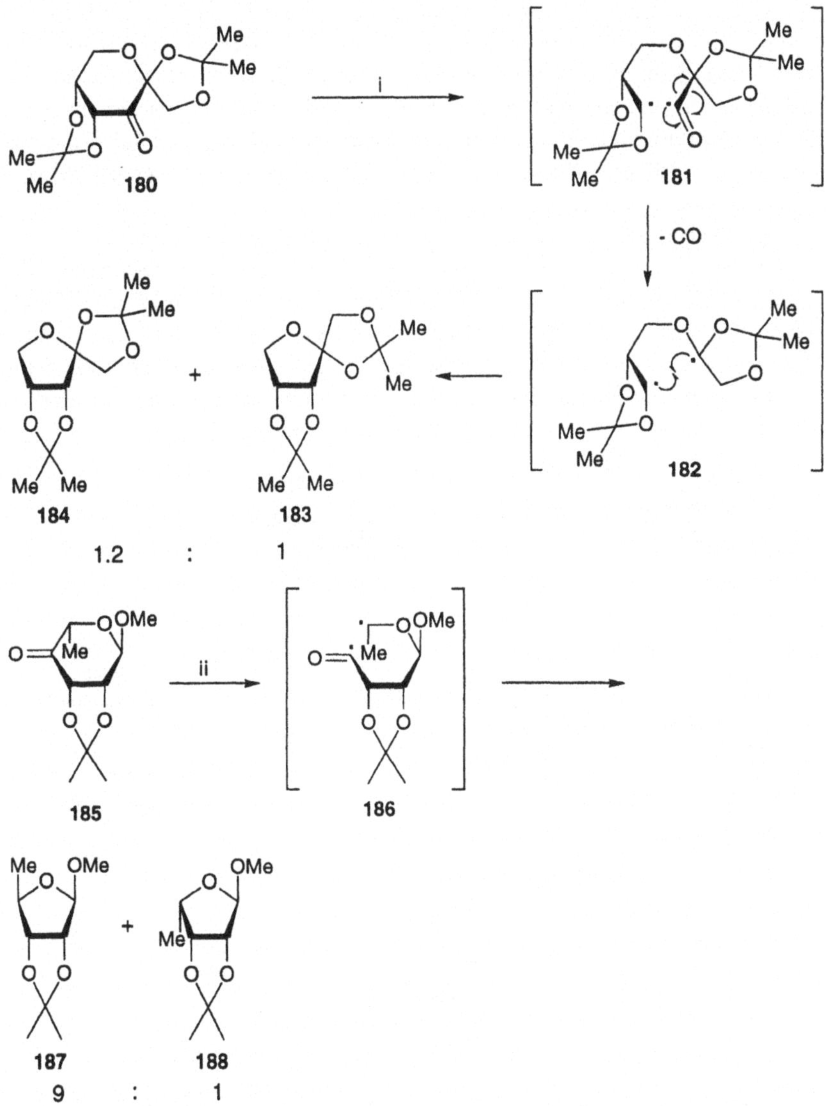

Scheme 2.28 i, $h\nu$, C_6H_{14} (or C_6H_6), 35%–40%; ii, $h\nu$, C_6H_6, 60%.

thermally mediated intermolecular free radical coupling between **174** and allylic stannane **175** in the presence of azobis-isobutyronitrile (AIBN) as initiator. The latter reaction furnished a 3:1 mixture of **176** and **177** which were separated by preparative high-pressure liquid chromatography (HPLC). The desired D-*manno* product **176** was saponified and the resulting hemiacetal reduced to primary alcohol **178** with sodium borohydride in the presence of sodium hydroxide. After ozonolysis and treatment with ammonium hydroxide, ammonium KDO **179** was obtained in 92% for the last two steps.

Considerably less synthetic effort has gone into developing methods for the efficient and controlled degradation of monosaccharide derivatives. One useful breakthrough has been Collins' photochemical degradation of monosaccharide uloses; two examples of this method are shown in Scheme 2.28 [35]. On ultraviolet (UV) irradiation in a range of solvents, acetal-protected uloses such as **180** and **185** undergo Norrish type I cleavage to give alkyl acyl biradicals that spontaneously decarbonylate to produce biradical intermediates that undergo ring closure to give chain-shortened sugars.

An alternative way of extruding one carbon atom and descending the monosaccharide series is through treatment of protected sugar hemiacetals such as **189** and **192** with diacetoxyiodobenzene and a catalytic quantity of iodine in dichloromethane for several hours (Scheme 2.29) [36]. This leads to the formation of mixed acetal formates which can be converted to

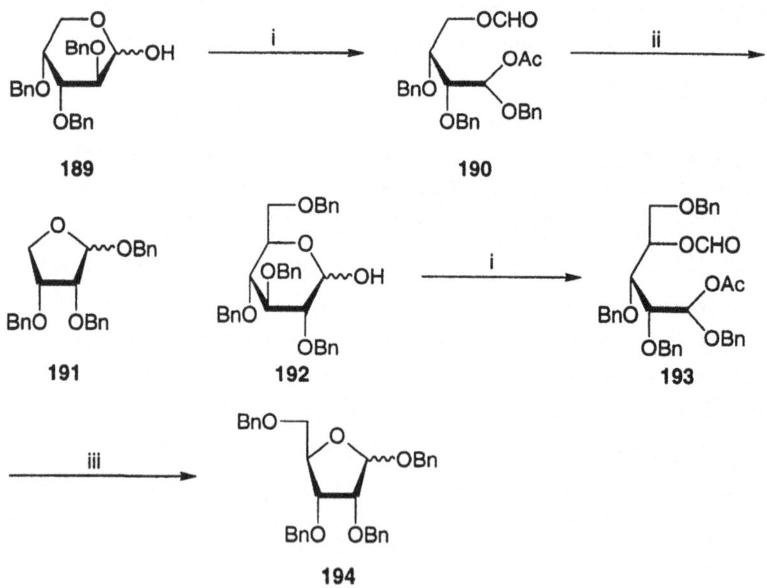

Scheme 2.29 i, $PhI(OAc)_2$, catalyst I_2, CH_2Cl_2, room temperature, 2 h, 87%–98%; ii, 15% $HClO_4$, MeOH (1 equiv.), EtOH, room temperature, 8 h, 82%; iii, 14% NH_4OH, MeOH, room temperature, 2 h, then 15% $HClO_4$, room temperature, 2 h, 77%.

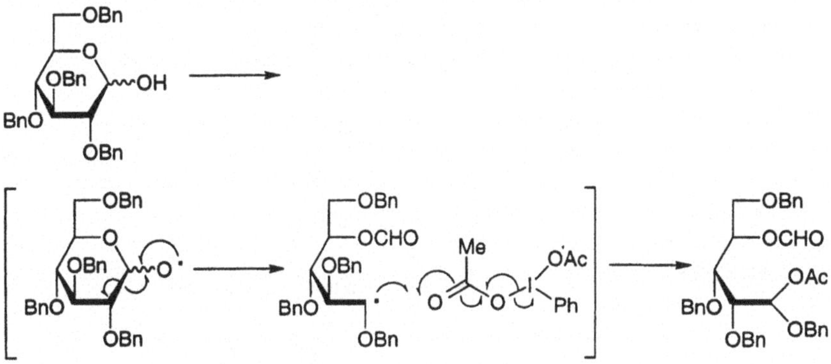

Scheme 2.30

furanosides by acid-catalysed transacetalisation. Mechanistically, these reactions are believed to proceed though C(1)-alkoxy radicals which decarbonylate and combine with the reagent to give the product (Scheme 2.30).

2.3 Monosaccharide mimetics and pseudo sugars

Recognition of the fact that cell–surface glycoconjugates can function as cell–cell recognition molecules and mediate the adhesion of cells within the immune system has stimulated interest in the synthesis of molecules that are recognised as particular monosaccharides by biological systems but which do not elicit the same biological response [37]. Such molecules are frequently termed monosaccharide mimetics, and one class of mimetic in which there has been particular interest for many years are molecules that are identical to the parent monosaccharide in every sense except for the ring oxygen atom. In this section, we will attempt to give the reader a flavour of the strategies used to build such compounds.

2.3.1 Preparative methods for 5-thioaldoses

A number of strategies have been followed to create 5-thio-D-glucose **198**; the one shown in Scheme 2.31 is due to Driguez and Henrissat [38]. It involves tosylation of 1,2-*O*-isopropylidene-3,6-glucuronolactone **195**, reduction of the lactone ring system to give the 3,6-diol and formation of the L-ido-epoxide **196** by treatment with methanolic sodium methoxide. Reaction of **196** with thiourea converted it to the D-gluco-5,6-episulfide which ring-opened to give **197** when subjected to mild acetolysis. Aqueous trifluoroacetic acid unmasked the 1,2-*O*-isopropylidene unit and transesterification brought about cyclisation to 5-thio-D-glucose **198**.

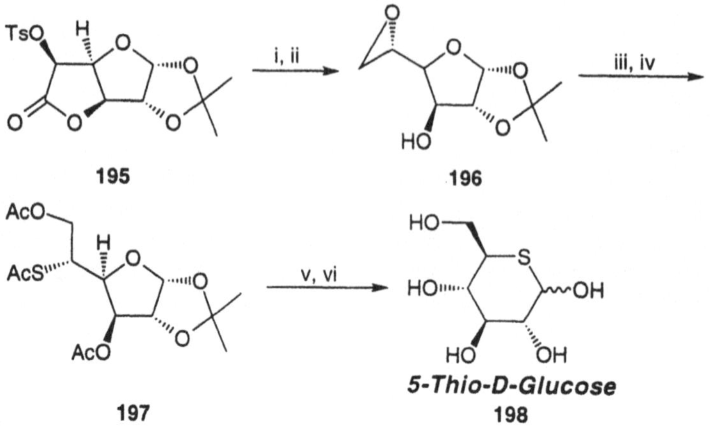

Scheme 2.31 i, $LiBH_4$, THF, 5 h, 81%; ii, NaOMe, MeOH, −15°C to room temperature; iii, thiourea, MeOH, 96 h, room temperature; iv, Ac_2O/AcOH (10:1), NaOAc, 36 h, 130°C, 77%, 3 steps; v, 90% aq. CF_3CO_2H, CH_2Cl_2; vi, NaOMe, MeOH, 71%, 2 steps.

5-Thio-D-glucose pentaacetate has served as a valuable precursor of tri-*O*-acetyl-5-thio-D-glucal **200** [39], which is itself a useful intermediate for the production of 5-thio-D-glucosamine **201** and 5-thio-D-mannosamine **202** via 1,2-azidonitration with ceric ammonium nitrate and sodium azide in acetonitrile at −20°C (Scheme 2.32).

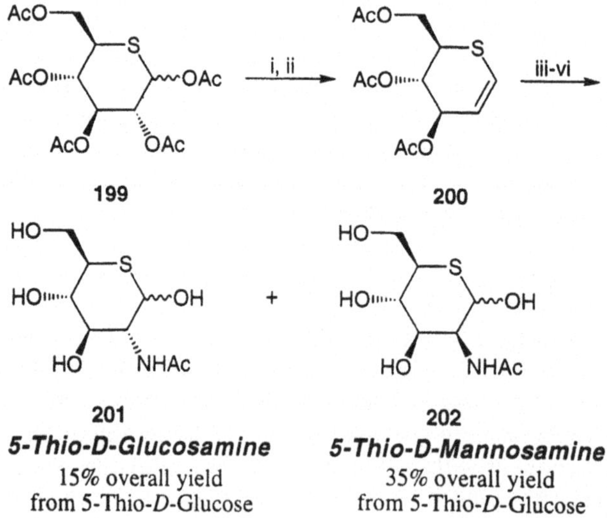

Scheme 2.32 i, HBr, AcOH; ii, Zn, AcOH; iii, $Ce(NH_4)_6NO_3$ (1.5 equiv.), NaN_3, (3.0 equiv.), MeCN, −20°C; iv, NaOAc, AcOH, 100°C, 1 h; v, H_2, PtO_2, MeOH, Ac_2O, 1 atm; vi, NaOMe, MeOH.

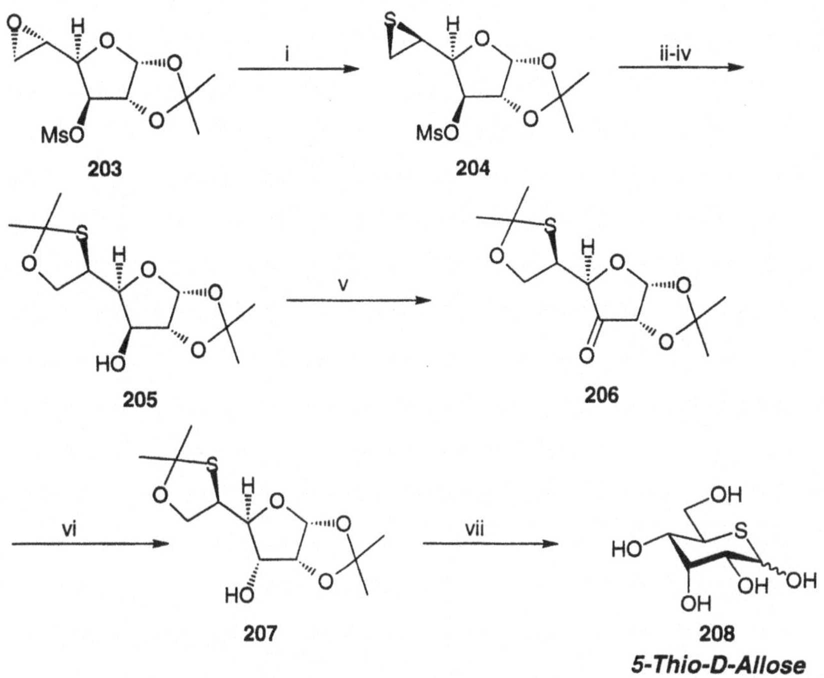

Scheme 2.33 i, Thiourea, MeOH, room temperature, overnight, 95%; ii, Ac_2O, NaOAc, AcOH, Δ, 15 h, then H_2O, 2 h, 65%; iii, *p*-TsOH, $Me_2C(OMe)_2$, MeOH, overnight, room temperature, 54%; iv, NaOMe, MeOH, Δ, 4 h, 80%; v, Me_2SO, Ac_2O, room temperature, 48 h, 44%; vi, $NaBH_4$, EtOH, H_2O, 88%; vii, 50% aq. AcOH, 90°C, 2 h, 85%.

For the preparation of 5-thio-D-allose **208** (Scheme 2.33), Al-Masoudi and Hughes [40] converted the 5,6-D-*ido*-epoxide **203** into the related 5,6-D-gluco-episulfide **204** and then performed a selective acetolysis, an acetal exchange and an *O*-desulfonylation with methanolic sodium methoxide to obtain **205**. The stereochemistry at C(3) was then inverted by an oxidation–reduction sequence, with 5-thio-D-allose **208** being obtained after acid hydrolysis. For 5-thio-D-altrose **214** [41], the same workers reacted **209** with methanolic HCl to obtain **210**. This was then isopropylidenated across O(4) and O(6) and **211** was treated with base to yield the D-*allo*-epoxide **212**. *Trans*-diaxial epoxide ring-opening with hydroxide ion and graded acid hydrolysis then furnished the target molecule **214** (Scheme 2.34).

Hashimoto and Izumi [42] have reported a synthesis of 5-thio-L-fucose **218** (Scheme 2.35) that sets off with an oxidation of **215** and a stereoselective addition of methyllithium to the resulting 3-*O*-allyl-1,2-*O*-isopropylidene-β-D-*arabino*-pentodialdo-1,4-furanose. This furnished a 10:1 mixture of alcohols in which the desired D-*altro* isomer was predominant. This mixture was then triflated at O(5), subjected to displacement with potassium thioacetate

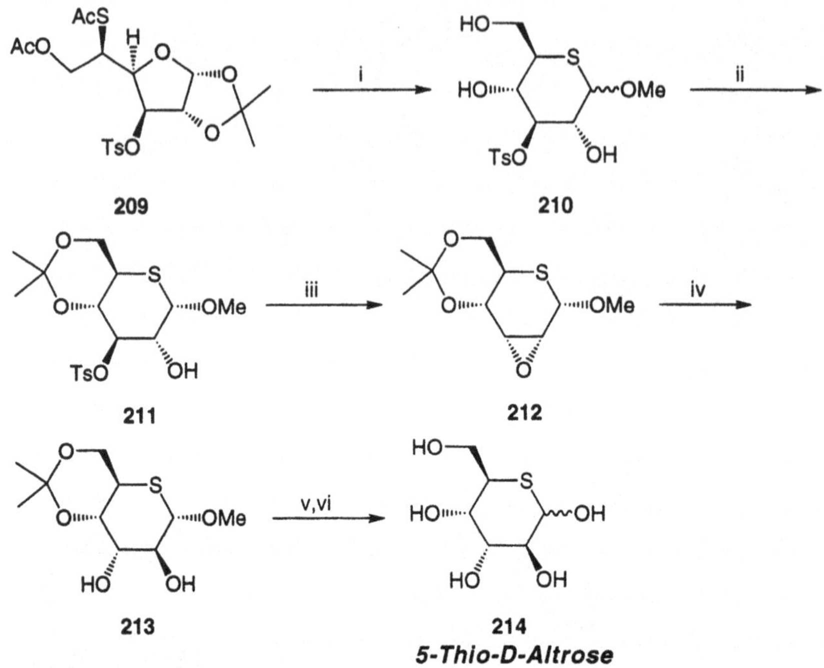

Scheme 2.34 i, Concentrated HCl, MeOH, Δ, 6 h; ii, $Me_2C(OMe)_2$, Me_2CO, *p*-TsOH; iii, NaOMe, MeOH, room temperature, overnight; iv, aq. NaOH, Δ, 3 h; v, 80% aq. AcOH, room temperature; vi, 0.05 M aq. H_2SO_4, 75°C, 2 h.

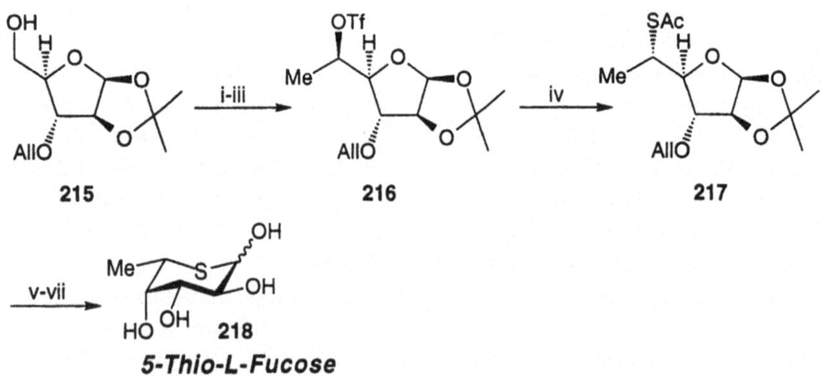

Scheme 2.35 i, Me_2SO, $(COCl)_2$, CH_2Cl_2, −78°C, 90%; ii, MeLi, Et_2O, −78°C, 0.5 h, 10:1 selectivity, 77%; iii, Tf_2O, Py, CH_2Cl_2; iv, KSAc, DMF, 48%, 2 steps; v, NaOMe, MeOH; vi, 70% aq. AcOH; vii; Pd-C, H_2O, *p*-TsOH.

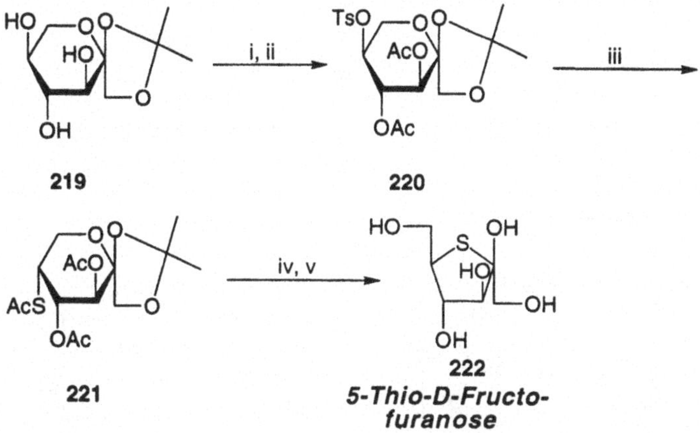

219 220 221 222

Scheme 2.36 i, TsCl, Py, 0°C, 48 h, 39%; ii, Ac_2O, Py; iii, KSAc, DMF, 75°C, 70 h, 82%; iv, 90% aq. CF_3CO_2H, 3 h, room temperature, 24%; v, NaOMe, MeOH. Note: DMF = dimethylformamide.

and carried forward through a series of standard deprotections to give 5-thio-L-fucose **218**.

Chmielewski and Whistler [43] have devised a strategy for obtaining 5-thio-D-fructofuranose **222** that is based upon thioacetate displacement of 4-*O*-(*p*-toluenesulphonyl)-1,2,3,4-tetra-*O*-acetyl-α-D-sorbopyranose **220** in dimethylformamide and subsequent transesterification with methanolic sodium methoxide (Scheme 2.36).

2.3.2 *Preparative methods for 5-amino-5-deoxy-aldoses*

Considerably less success has accrued in developing viable synthetic routes to 5-amino-5-deoxy-D-hexose derivatives, primarily because of the readiness with which these compounds undergo dehydration and aromatisation to the corresponding pyridines. However, there has been some progress in this area.

There has been particular interest in developing preparative routes to 5-amino-5-deoxy-D-glucose (nojirimycin) since it functions as an antibiotic against gram-positive bacteria. Vasella and Voeffray [44] have put together a particularly elegant sequence of reactions for obtaining **227**, in which the pivotal step was an asymmetric [3 + 2] cycloaddition between furan and a glyoxalate-derived D-*manno*-furanosylnitrone; it successfully installed three of the four asymmetric centres embedded within the target structure (Scheme 2.37). The final stereocentre was incorporated through a stereoselective dihydroxylation of **224** with osmium tetraoxide to give **225** after acetonation. After fission of the N–O bond and removal of the auxiliary to obtain **226**, a series of protecting group adjustments allowed nojirimycin **227** to be obtained in good yield.

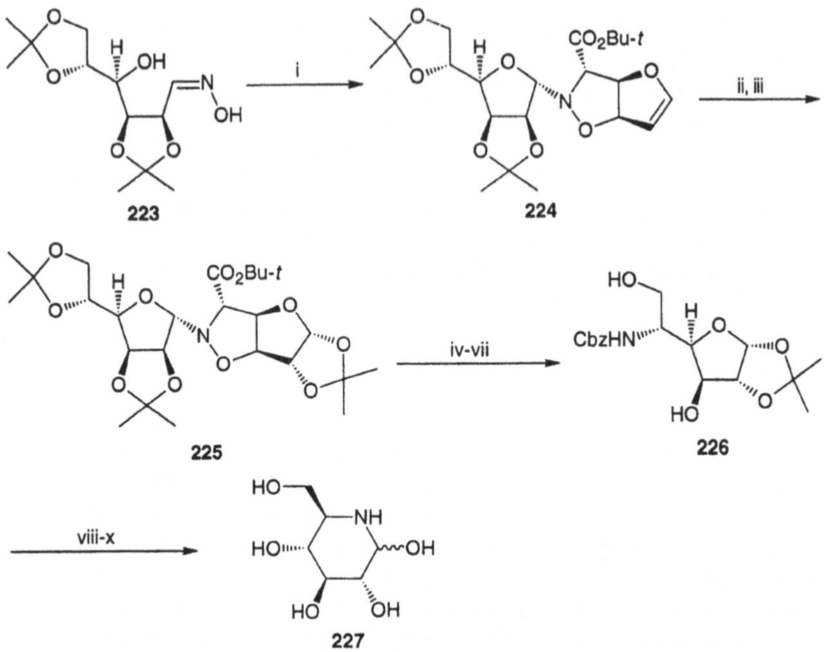

Scheme 2.37 i, t-BuO_2CCHO, furan, $CHCl_3$, 80°C; ii, catalyst OsO_4, NMO, Me_2CO, 52%, 2 steps; iii, $FeCl_3$, Me_2CO, 0°C, 77%; iv, H_2, Raney Ni, EtOH, 80°C, 50 bar; v, BnOC(O)Cl, aq. $NaHCO_3$, $CHCl_3$, 69%, 2 steps; vi, BuCl, CF_3CO_2H, 87%; vii, $LiEt_3BH$, THF, 97%; viii, H_2, Pd-C, EtOH, EtOAc; ix, SO_2, MeOH, 0°C, 91%, 2 steps; x, Dowex (OH^-), H_2O, 92%.

Auberson and Vogel [45] have described an asymmetric pathway to 5-amino-5-deoxy-D-allopyranose (allonojirimycin) that involves acid-catalysed ring opening/Fischer glycosidation of α-bromolactone **228** with allyl alcohol to give **229** (Scheme 2.38). The allyl ester was then selectively hydrolysed to the carboxylic acid with [$Rh(PPh_3)_3Cl$] and 1,4-diazabicyclo[2,2,2]octane (DABCO), and the caesium carboxylate reacted with caesium azide to give **230** after benzylation, with retention of configuration at C(5). Reduction of **230** with lithium aluminium hydride led to **231** which was transformed into its bisulfite **232** by treatment with aqueous SO_2.

Synthetic routes to 5-amino-5-deoxy-D-mannopyranose [46] and 5-amino-5-deoxy-D-galactopyranose [47] have also been described by Legler and co-workers.

2.3.3 *Preparative methods for monosaccharides containing a phosphorus atom in the ring*

A comprehensive review [48] appeared in 1984 on the preparation and chemistry of monosaccharides containing a phosphorus atom in the ring, and so

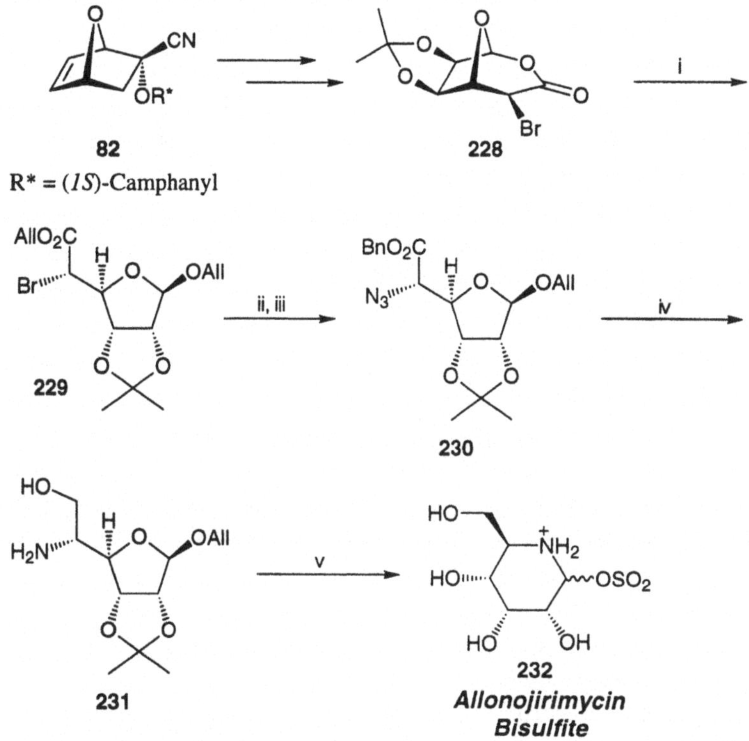

Scheme 2.38 i, Allyl alcohol, $MeSO_3H$, 20°C, 66%; ii, $[Rh(PPh_3)_3Cl]$ (0.1 equiv.), DABCO (0.1 equiv.), EtOH/H_2O (9:1), 87%; iii, Cs_2CO_3, DMF, then CsN_3, 60°C, 8 h, then BnBr added at 20°C, 96%; iv, $LiAlH_4$, THF, 20°C, 1 h, 61%; v, SO_2 bubbled into H_2O at 55°C for 72 h, 54%.

this topic will be dealt with only briefly here. Yamamoto and colleagues [49] have formulated a route to (*5RS*)-hydroxyphosphinyl-D-glucopyranose that attaches the phosphorus substituent to C(5) by base-mediated Michael addition of dimethylphosphonoacetate at 0°C (Scheme 2.39). After catalytic hydrogenation of the nitro group over platinum oxide, the ensuing amine was diazotised in the presence of water to give alcohol **234**. After *O*-tritylation of the primary hydroxyl in **234** the dimethoxyphosphinyl was reduced with REDAL, and acid treatment then removed the 1,2-acetal to bring about ring closure to **235**.

2.3.4 Preparative methods for carba-monosaccharides

Carba- or *pseudo*-monosaccharides are cyclic monosaccharide analogues in which the ring oxygen atom has been replaced by a methylene group. Their chemistry has been extensively reviewed by Suami and Ogawa [50],

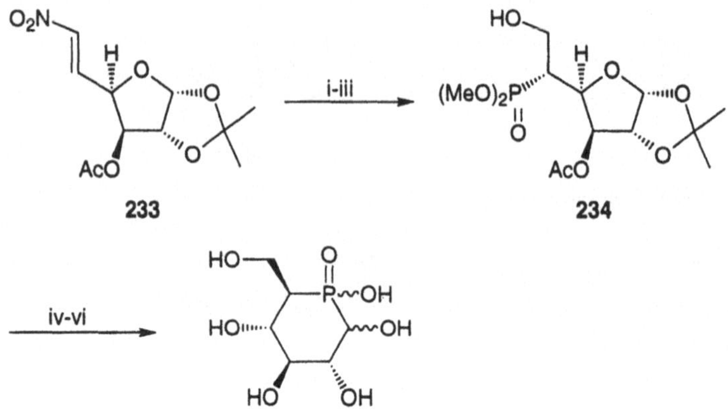

233 **234**

235
(5RS)-Hydroxyphosphinyl -D-Glucose

Scheme 2.39 i, Neat $(MeO)_2P(O)H$, Et_3N, 0°C, 57%; ii, H_2, PtO_2, HCl, 15 h, 91%; iii, $NaNO_2$, AcOH, H_2O, 2 h, 63%; iv, Ph_3CCl, Py, 60°C, 30 h, 45%; v, REDAL, PhMe, C_6H_6, 0°C–5°C; vi, propanol-0.25 M aq. H_2SO_4 (1 : 1), 90°C, 82%.

and so in this section we will attempt to illustrate the approaches used for accessing such compounds.

The synthesis of carba-β-D-glucopyranose pentaacetate and carba-β-L-altropyranose pentaacetate achieved by Suami and co-workers [51] (Scheme 2.40) featured a tandem aldol/alkylative cyclisation on 2,3,4-tri-*O*-benzyl-5-deoxy-5-iodo-L-arabinose **238** with dimethyl malonate in the presence of sodium hydride. After acetylation this led to **239** as a mixture of epimers. These were then subjected to thermal decarboxylation and simultaneous β elimination with sodium chloride in aqueous DMSO to give allylic alcohol **241** after reduction of the methyl ester with lithium aluminium hydride. Hydroboration of **241** turned out to be non-stereoselective, affording an equal mixture of the β-D-gluco and β-L-altro diol derivatives after oxidative work-up. After *O*-debenzylation with lithium in liquid ammonia and acetylation the target molecules **242** and **243** were obtained.

Paulsen and co-workers [52] have prepared enantiomerically pure carba-α-D-galactopyranose from the quebrachitol derivative **244** by the sequence of reactions, shown in Scheme 2.41. The first step involved conversion of **244** into L-*chiro*-inositol **245** by treatment with boron tribromide. This was then fully acetonated and the *trans*-fused isopropylidene group selectively hydrolysed with 95% aqueous acetic acid. A regioselective *O*-benzylation was then carried out to give **246**, which was oxidised and olefinated with methylenetriphenylphosphorane. Hydroboration proceeded from the less hindered alkene face of **247** to yield the primary alcohol after oxidative work-up. The latter was then protected as an MEM ether, the benzyl ether

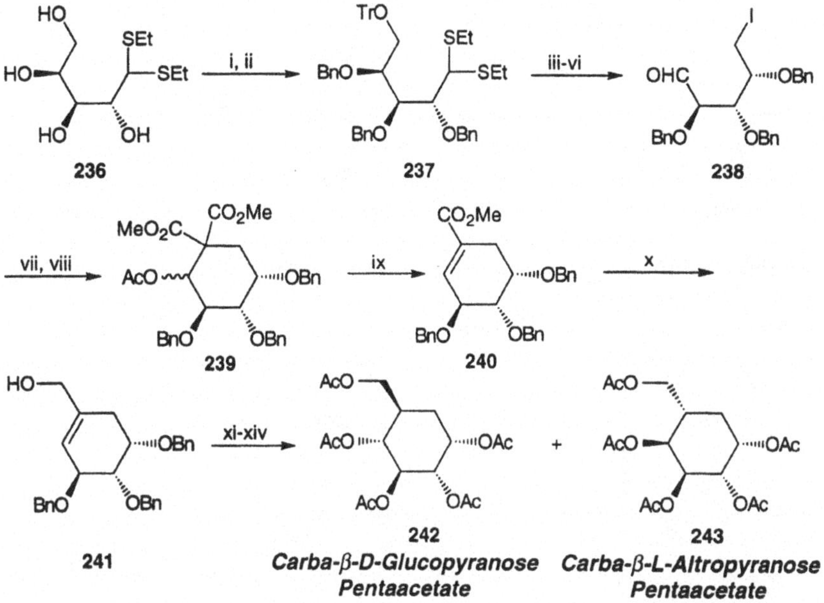

Scheme 2.40 i, TrCl, Py, DMAP, 85%; ii, NaH, BnBr; iii, *p*-TsOH, 79%, 2 steps; iv, TsCl, Py; v, $HgCl_2$, $CaCO_3$; vi, NaI, 54%, 3 steps; vii, dimethyl malonate, NaH (2 equiv.), DMF, 0°C; viii, Ac_2O, Py, 43% (+33% of a tetrahydropyran by-product); ix, aq. NaCl, DMSO, 110°C–170°C, 75%; x, $LiAlH_4$, 15°C, 83%; xi, BH_3-THF, THF, followed by 35% aq. H_2O_2, aq. NaOH; xii, Ac_2O, Py, 69%, 2 steps; xiii, Na, liq. NH_3; xiv, Ac_2O, Py, 80% yield, 2 steps.

was removed, and a Barton xanthate deoxygenation executed on **248** to give **249**. Several deblocking steps then completed the synthesis of **250**.

Starting from the aforementioned benzyl ether derivative **246**, Paulsen and associates [52] carried out a synthesis of carba-α-D-mannopyranose (Scheme 2.42). Iodination of **251** followed by elimination led to allylic alcohol **252** which underwent oxidation and conjugate addition to give ketone **254**. Thioketal hydrolysis and reduction then provided **255** which was chain-shortened and deprotected to deliver the target molecule **257**.

Suami and co-workers [53] have developed a unified synthetic strategy to carba-α-L-altro-, carba-β-D-gluco-, and 2-amino-2-deoxy-carba-α-D-glucopyranose peracetates **267**, **270** and **273**, respectively, which proceeded from the chiral epoxide **263**. This was available in eight steps from D-glucose by the sequence shown in Scheme 2.43 for the synthesis of **267**. This involved three-carbon Wittig homologation at C(3) and stereospecific hydrogenation to obtain **259**; oxidative cleavage across the C(5)–C(6) bond to create **260**, and intramolecular aldol reaction and dehydration of **260** to produce enone **261**. This was epoxidised stereospecifically with basic hydrogen peroxide, and the ketone reduced with sodium borohydride to give **263** as the major product of a 5:1 mixture of epimeric alcohols. In order to

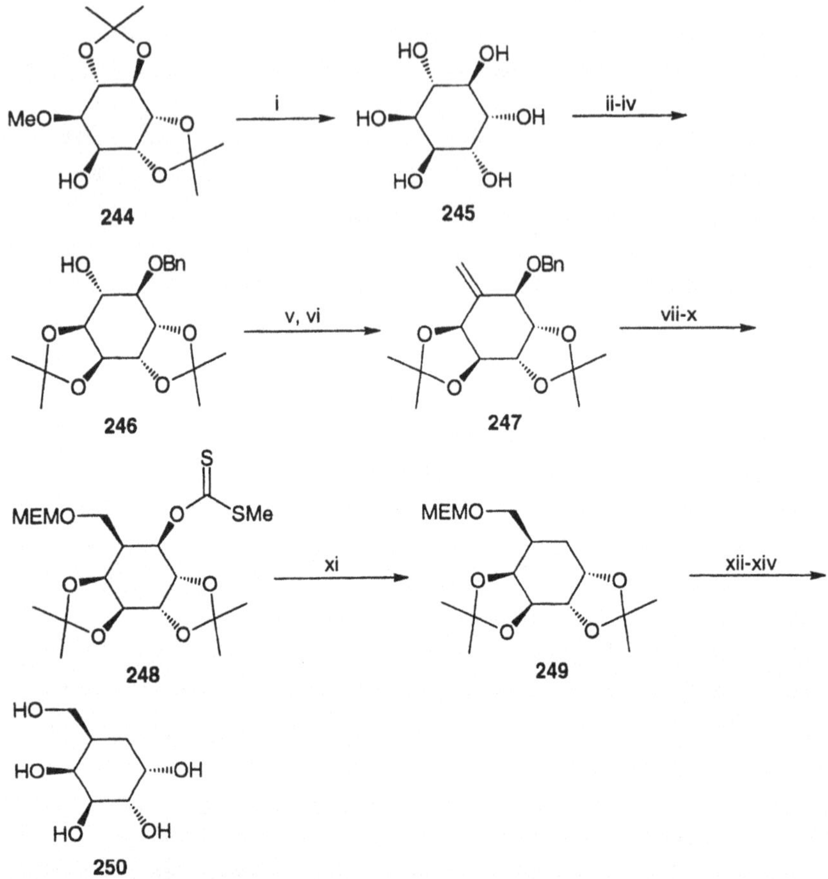

Carba-α-D-Galactopyranose

Scheme 2.41 i, BBr_3, CH_2Cl_2, 0°C, 10 min, then room temperature, 20 h; ii, $Me_2C(OMe)_2$, H^+, 77%; iii, 95% AcOH, 88%; iv, Et_4NI, BnBr, aq. NaOH, CH_2Cl_2, 87%; v, Me_2SO, $(COCl)_2$, CH_2Cl_2, −78°C, *i*-Pr_2NEt, 96%; vi, $Ph_3P^+CH_3Br^-$, *n*-BuLi, THF, hexanes, 85%; vii, BH_3-THF, THF, −50°C to 0°C, then 30% H_2O_2, aq. NaOH, 82%; viii, MEM-Cl, *i*-Pr_2NEt, CH_2Cl_2, 85%; ix, H_2, Pd-C, MeOH, 48 h, 73%; x, NaH, imidazole, THF, CS_2, MeI, 91%; xi, Bu_3SnH, AIBN, PhMe, Δ, 81%; xii, 1 M HCl in MeOH; xiii, Ac_2O, Py, 60%, 2 steps; xiv, NaOMe, MeOH, then Amberlite IR-120 (H^+) resin, 78%.

obtain **264** it was necessary to instigate a Payne rearrangement in **263**, open the newly formed epoxide with hydroxide ion, and *O*-benzylate. The isopropylidene acetal was then hydrolysed and the lactol reduced with sodium borohydride. The resulting triol **265** was degraded with aqueous sodium periodate, aldehyde **266** reduced with sodium borohydride, and the *O*-benzyl ethers removed by dissolving metal reduction. Acetylation of the resulting polyol afforded **267**.

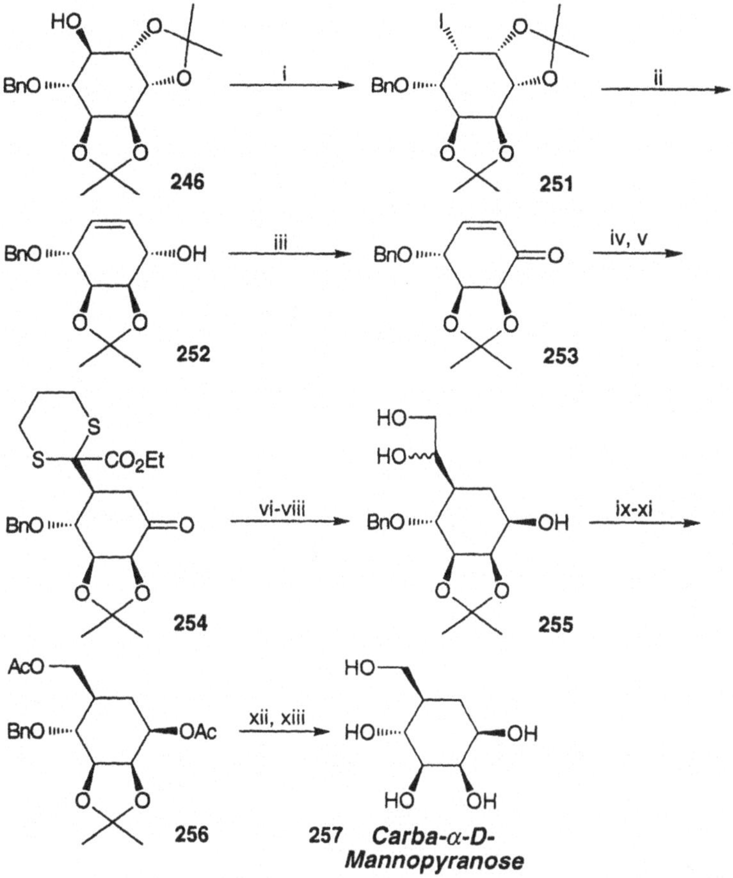

Scheme 2.42 i, Ph_3P, I_2, imidazole, PhMe, 90°C, 3 h, 92%; ii, $LiAlH_4$, Et_2O, 3 h, 65%; iii, Me_2SO, $(COCl)_2$, CH_2Cl_2, −78°C, *i*-Pr_2NEt, 97%; iv, 2-ethoxycarbonyl-1,3-dithiane, *i*-Pr_2NLi, THF, −78°C, 71%; v, $LiAlH_4$, THF, −40°C; vi, Ac_2O, Py, 69%, 2 steps; vii, HgO, $HgCl_2$, aq. MeCN, 2 h, then $NaBH_4$ followed by Ac_2O, Py, 92%; viii, NaOMe, MeOH, then IR-120 (H^+) resin; ix, aq. $NaIO_4$, CH_2Cl_2, Bu_4NBr; x, $NaBH_4$, EtOH; xi, Ac_2O, Py, 94%, 4 steps; xii, 80% AcOH, catalyst CF_3CO_2H, 2 h then Pd-C, H_2, 24 h; xii, Ac_2O, Py, 77%, 2 steps; xiii, NaOMe, MeOH, 86%.

For the preparation of **270** [53], aldehyde **266** was converted to allylic alcohol **268** by β elimination and reduction. Sequential hydroboration and oxidation were then used to gain access to **269** in 63% yield; approximately 11% of the α-L-*altro* isomer was also formed in this reaction. Deprotection of **269** and acetylation completed the synthesis of **270** (Scheme 2.44).

The carba-α-D-glucosamine derivative **273** was also available from epoxide **263** via the pathway outlined in Scheme 2.45 [53]. This entailed opening the epoxide with sodium azide and protecting the resultant hydroxyls as

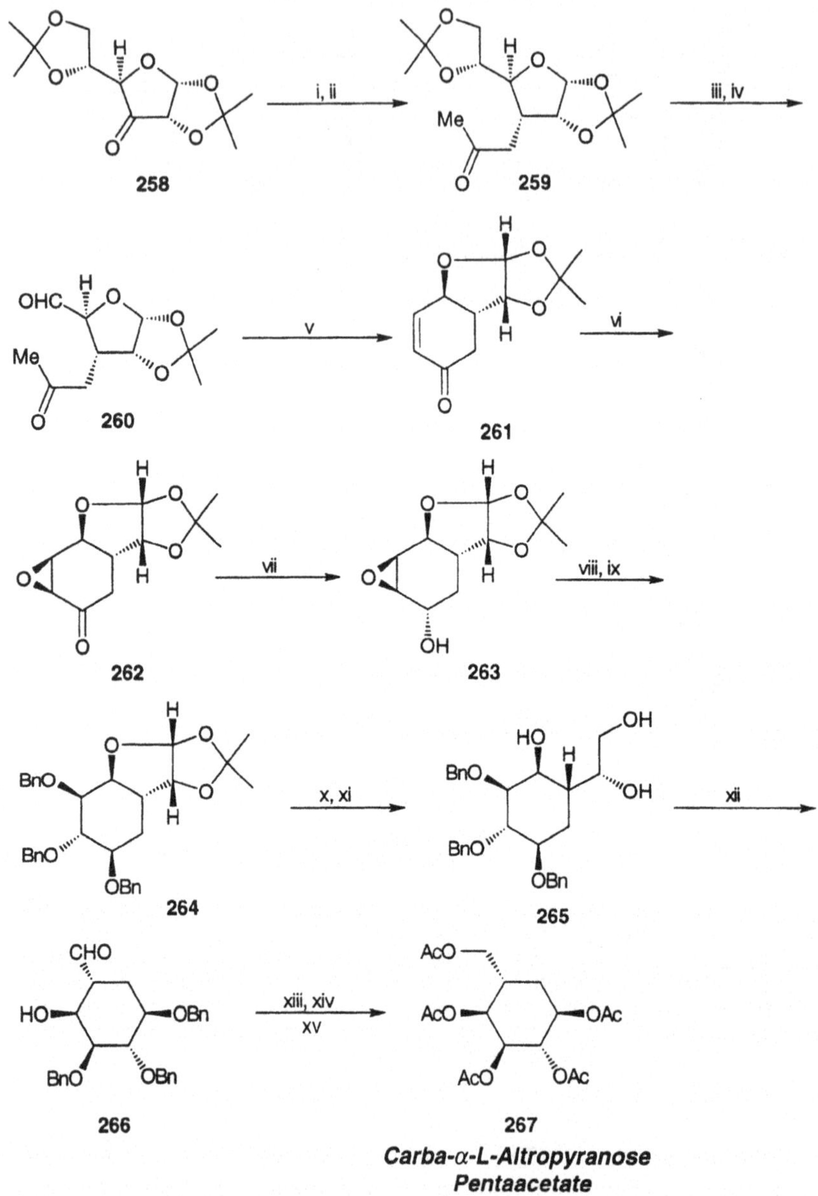

Scheme 2.43 i, $Ph_3P{=}CHCOMe$, C_6H_6, Δ, 96% ($Z/E = 3:1$); ii, H_2, T4 Raney-Ni, MeOH, 20 h, 88%; iii, 60% aq. AcOH, room temperature, 17 h, 98%; iv, $NaIO_4$, MeOH, H_2O, 100%; v, DBU, C_6H_6, Δ, 35 h, 45%; vi, 30% aq. H_2O_2, aq. NaOH, room temperature, 3 h, 96%; vii, $NaBH_4$, EtOH, 0°C, 70%; viii, methoxyethanol/H_2O (10:3), NaOAc, Δ, 9 h, 73%; ix, NaH, BnBr, DMF, 93%; x, 80% aq. AcOH, 1,4-dioxane, Δ, 2 h; xi, $NaBH_4$, MeOH, 0°C, followed by 35% aq. H_2O_2; xii, $NaIO_4$, H_2O, MeOH; xiii, $NaBH_4$, EtOH, then 35% aq. H_2O_2; xiv, dissolving metal reduction; xv, Ac_2O, Py, 52% for 4 steps.

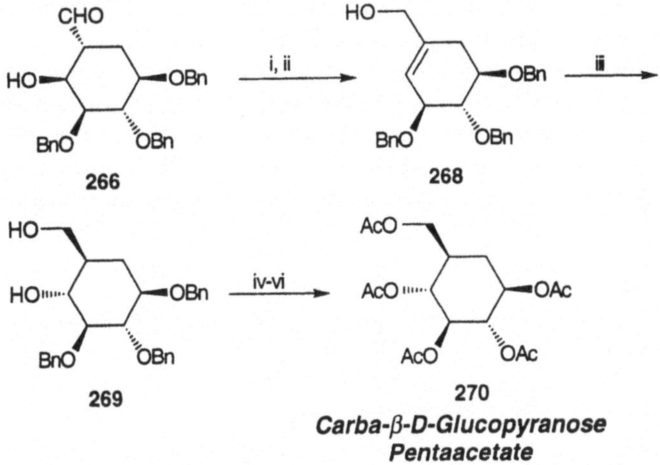

Scheme 2.44 i, MsCl, Py, room temperature, 24 h; ii, $NaBH_4$, EtOH, room temperature, 3.5 h, 22% from **264**; iii, BH_3-THF, THF, 0°C for 20 min, room temperature for 1 h, then 35% aq. H_2O_2, aq. NaOH; iv, Ac_2O, Py, 5 h; v, H_2, Pd-C, MeOH, 14 h; vi, Ac_2O, Py, 98%.

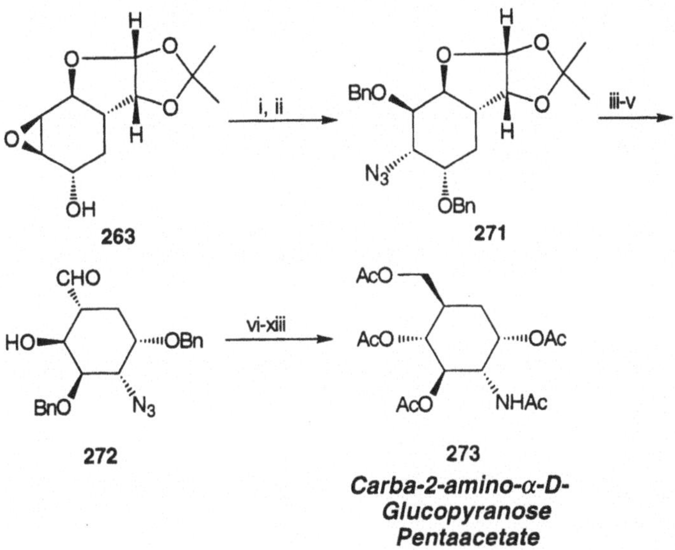

Scheme 2.45 i, 2-Methoxyethanol, H_2O, NaN_3, Δ, 4 h, 84%; ii, NaH, DMF, BnBr, 82%; iii, aq. HCl, 1,4-dioxane; iv, $NaBH_4$, EtOH, 0°C, 1 h; v, $NaIO_4$, H_2O, MeOH; vi, MsCl, Py, room temperature, 8 h; vii, *i*-Bu_2AlH, CH_2Cl_2, PhMe, −15°C, 27% yield for 5 steps; viii, BH_3-THF, THF, 0°C, aq. NaOH, 35% H_2O_2; ix, Ac_2O, Py, 21%, 2 steps; x, H_2, T4 Raney Ni, EtOH; xi, Ac_2O, Py, 66%, 2 steps; xii, H_2, Pd-C, EtOH; xiii, Ac_2O, Py, 92%, 2 steps.

O-benzyl ethers. Compound **271** was then converted to **272** using a strategy similar to that already discussed for **267** in Scheme 2.43.

Interest in *pseudo*-pentofuranoses has intensified in recent years since they are potentially valuable intermediates for the production of carba-nucleosides. Yoshikawa and co-workers [54] have conceived a particularly facile synthesis of carba-α-D-arabinofuranose **282** from D-arabinose (Scheme 2.46). The cornerstone of this was a stereoselective nitromethyl

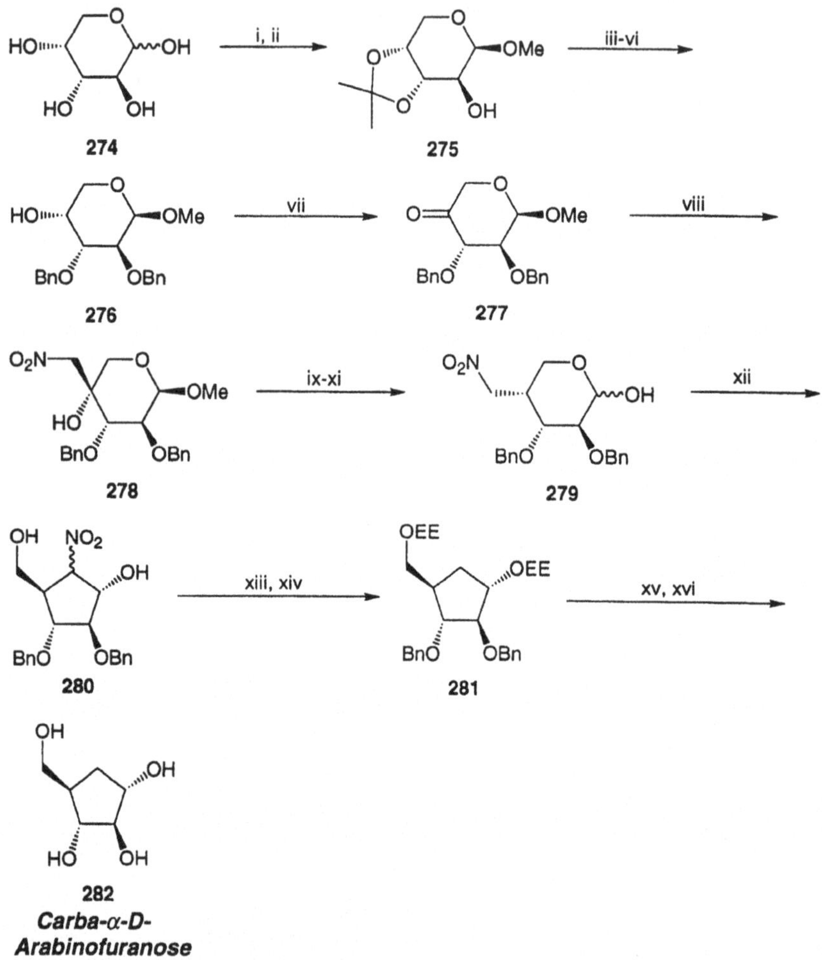

Scheme 2.46 i, HCl, MeOH, Δ; ii, $Me_2C(OMe)_2$, *p*-TsOH, DMF, 100%, 2 steps; iii, BnCl, NaH, DMF, 100%; iv, 80% aq. AcOH, 50°C; v, Bu_2SnO, PhMe, Δ; vi, CsF, BnBr, DMF, 100%, 3 steps; vii, $(COCl)_2$, Me_2SO, CH_2Cl_2, −40°C, Et_3N; viii, CH_3NO_2, KF, 18-crown-6, DMF, 72%, 2 steps; ix, *p*-TsOH, Ac_2O; x, $NaBH_4$, EtOH, 85%, 2 steps; xi, concentrated HCl, AcOH, 30°C; xii, CsF, DMF, 86%; xiii, ethyl vinyl ether, PPTS, CH_2Cl_2; xiv; Bu_3SnH, AIBN, PhMe, Δ, 52%, 2 steps; xv, H_2, Pd-black, EtOAc; xvi, PPTS, 80% aq. Me_2CO, 40°C, 90%.

anion addition to ketone **277** to furnish branched nitropyranose **278**. This was transformed into the nitroalkene by β elimination and this conjugatively reduced from the less hindered alkene face with sodium borohydride in ethanol. The methyl glycoside was then unmasked to provide **279** which underwent internal Henry reaction after treatment with caesium fluoride in DMF. The hydroxyl groups in **280** were briefly protected as ethoxyethyl ethers, and the nitro group then removed by free radical reduction with tri-*n*-butylstannane to give **282** after deblocking.

Several *pseudo*-ketoses have been prepared in optically active form from carbohydrate and non-carbohydrate precursors. A seven-step synthesis of carba-β-D-fructofuranose appeared from Wilcox's laboratory in 1986 (Scheme 2.47) [55]. Their initial subtarget was the olefinic ketose **284**; it was available from 2,3,5-tri-*O*-benzyl-D-arabinose by Wittig olefination, Swern oxidation and acidolysis. After being converted to its lithium carboxylate salt, **284** reacted with (dibromomethyl)lithium to afford a single-product diastereoisomer **285** after treatment with diazomethane. Subsequent radical cyclisation of **285** in the presence of tri-*n*-butylstannane led to **286** as sole product. A Barbier–Wieland one-carbon degradation was then carried

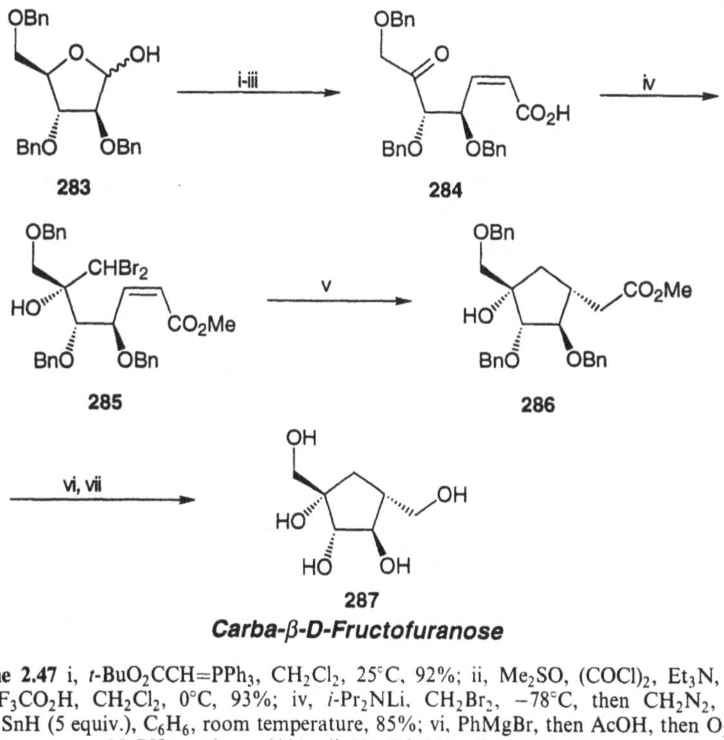

283 284 285 286 287

Carba-β-D-Fructofuranose

Scheme 2.47 i, *t*-$BuO_2CCH{=}PPh_3$, CH_2Cl_2, 25°C, 92%; ii, Me_2SO, $(COCl)_2$, Et_3N, 96%; iii, CF_3CO_2H, CH_2Cl_2, 0°C, 93%; iv, *i*-Pr_2NLi, CH_2Br_2, −78°C, then CH_2N_2, 78%; v, Bu_3SnH (5 equiv.), C_6H_6, room temperature, 85%; vi, PhMgBr, then AcOH, then O_3 with $NaBH_4$ work-up, 50%; vii, H_2, $Pd(OH)_2$-C, EtOH, 98%.

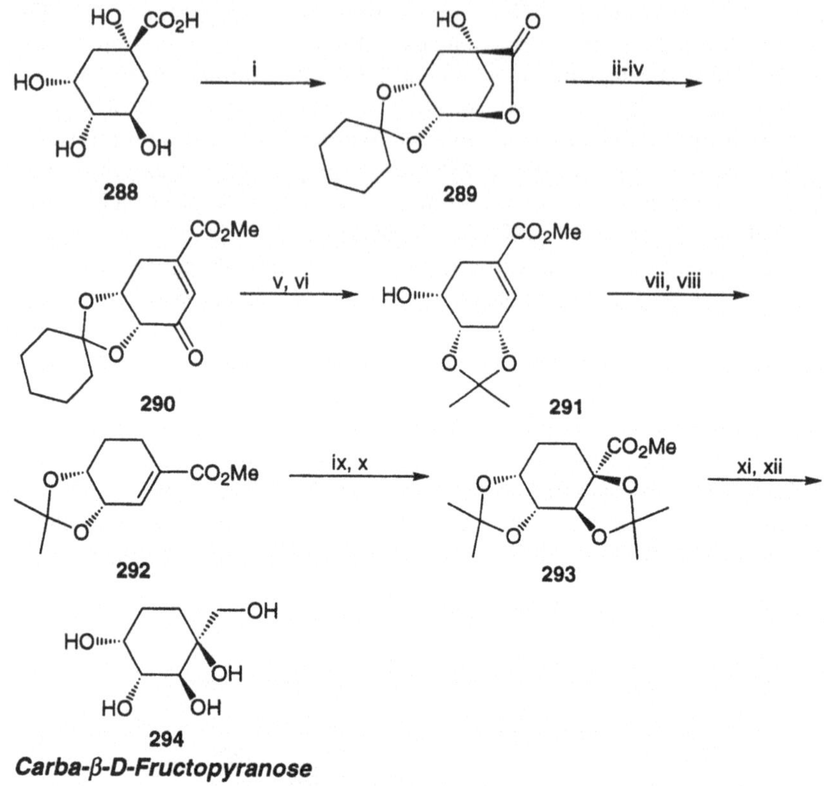

Scheme 2.48 i, Cyclohexanone, C_6H_6, DMF, Dowex 50WX8 resin, Δ, 79%; ii, NaOMe, MeOH, 0°C, 96%; iii, Me_2SO, $(COCl)_2$, Et_3N, −78°C to room temperature; iv, $POCl_3$, Py, room temperature, 76%, 2 steps; v, $NaBH_4$, MeOH, 0°C, 82%; vi, Me_2CO, H^+; vii, PhOC(S)Cl, Py, CH_2Cl_2, DMAP, room temperature, 2 h, 67%; viii, Bu_3SnH, $(t\text{-BuO})_2$, PhMe, Δ, 1.5 h, 76%; ix, OsO_4, NMO, *t*-BuOH, 81%; x, 2-methoxypropene, CSA, CH_2Cl_2, 80%; xi, *i*-Bu_2AlH, THF, −20°C to room temperature, 75%; xii, CF_3CO_2H, H_2O (1 : 1), 91%.

out on the methyl ester function in **286**. Hydrogenation over Pearlman's catalyst finally liberated the desired *pseudo*-sugar **287** in high yield.

By modifying an already good route to carba-β-D-fructopyranose originally devised by Shing and Tang [56], McComsey and Maryanoff [57] have developed an even better means of obtaining **294** in enantiomerically pure form (Scheme 2.48). In this hybrid route to **294**, (−)-quinic acid was first transformed into enone **290** and this reduced and transketalised to afford alcohol **291**. The hydroxyl group in **291** was then removed by tri-*n*-butyltin hydride reduction of the corresponding phenylthionocarbonate, and the resulting alkene **292** dihydroxylated and protected to give **293**. The methyl ester was then reduced to the primary alcohol with diisobutylaluminium hydride and the *O*-isopropylidene groups cleaved with acid to finalise the synthesis.

2.4 Recent methodological developments in the chemical modification and manipulation of monosaccharide derivatives

In this part of the chapter we highlight some important methodological developments that have occurred in synthetic carbohydrate chemistry over the past fifteen years.

2.4.1 Protecting groups

The utility of diphenylmethyl ethers as monosaccharide protecting groups has been firmly demonstrated by Webber and his colleagues [58]. These workers observed that the diphenylmethyl group can be readily introduced into monosaccharide derivatives by treatment with 2 mol equiv. of diazo(diphenyl)methane per hydroxyl in benzene or acetonitrile. In the case of methyl 4,6-*O*-benzylidene-α-D-mannopyranoside **301**, a high yielding O(3)-selective etherification proved possible when 1.25 mol equiv. of diazo(diphenyl)methane was used with catalytic tin (II) chloride, although little regioselectivity was observed when this reaction was applied on other diol substrates. A particularly noteworthy aspect of the uncatalysed diazo(diphenyl)methane etherification procedure resides in its compatibility with acid and base labile functionality within a sugar substrate (Scheme 2.49). This is beautifully illustrated by a consideration of the *O*-diphenylmethylation of **299** which afforded the 6-*O*-diphenylmethyl ether **300** in good yield without loss or migration of the *O*-acetate groups. In contrast, *O*-benzylation of **299** with silver oxide and benzyl bromide in dimethylformamide furnished a mixture of the 4- and 6-*O*-benzyl ethers as a result of acetate migration. Like *O*-benzyl ethers, *O*-diphenylmethyl ethers are stable towards moderately basic conditions. For instance, methyl 2,3,6-tri-*O*-acetyl-4-*O*-diphenylmethyl-α-D-glucopyranoside was transformed into methyl 2,3,6-tri-*O*-methyl-4-*O*-diphenylmethyl-α-D-glucopyranoside in high yield by sequential treatment with methanolic sodium methoxide and sodium hydride/methyl iodide. A further advantage of *O*-diphenylmethyl ethers resides in their reasonable acid stability, which often allows acetal-protecting groups to be selectively removed in their presence. The favoured method for deblocking monosaccharide diphenylmethyl ethers is through catalytic hydrogenation over palladium on charcoal. In our opinion, the diphenylmethyl ether protecting group is currently much underemployed in synthetic carbohydrate chemistry, and it is our hope that chemists will start to take more advantage of it in future years.

Another major breakthrough in protecting group design has been Ley *et al.*'s introduction of Dispoke acetals for the selective protection of *trans* diequatorial vicinal diols in carbohydrate derivatives [59, 60]. Such acetals are generally introduced by reaction with a *bis*-dihydropyran (*bis*-DHP) under mild acid conditions (Scheme 2.50). *Cis* vicinal diols or 1,3 diols

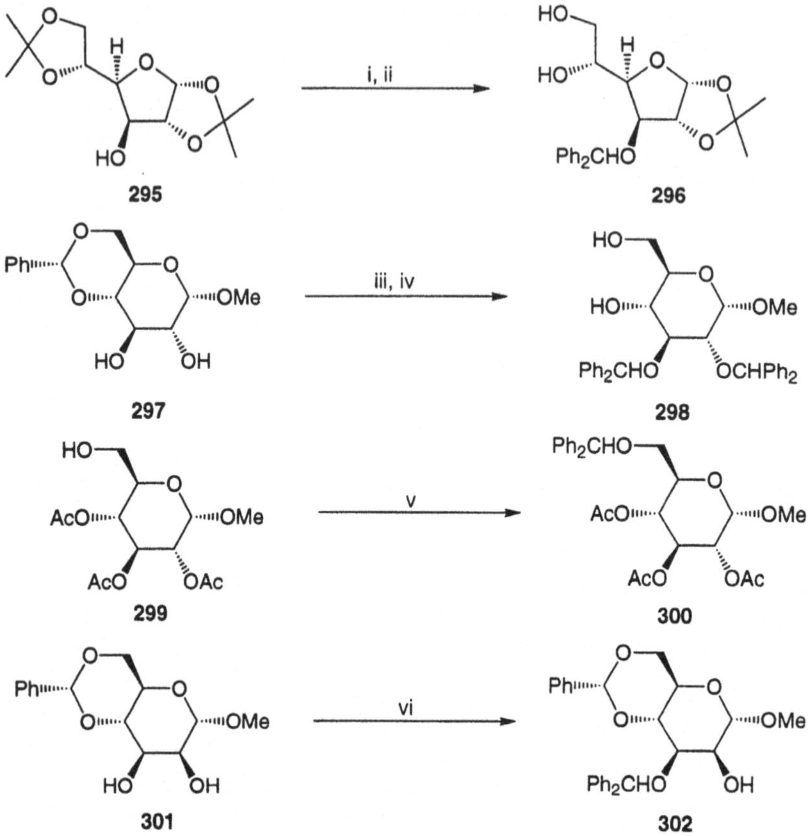

Scheme 2.49 i, Ph_2CHN_2, C_6H_6, Δ, 20 h, 78%; ii, 70% aq. AcOH, room temperature, 60 h, 95%; iii, Ph_2CHN_2, MeCN, room temperature, 71%; iv, concentrated HCl in Me_2CO (*c.* 0.2 M), 80%; v, Ph_2CHN_2, C_6H_6, 87%; vi, Ph_2CHN_2, DME, $SnCl_2$, 0°C, 1.5 h, 80%.

generally do not engage in acetalation with a *bis*-DHP reagent. For the first time ever, therefore, it is possible selectively to bridge, in high yield and in a single step, the 2- and 3-positions of methyl α-D-galactopyranoside with an acetal protecting group [59]. Similar selectivity was also observed with the 1-thio-β-D-ethyl galactoside **306** when treated with *bis*-DHP **304** (Scheme 2.51). Dispoke acetals are stable to the *N*-iodosuccinimide/triflic acid conditions used to promote glycosidation reactions of thioalkyl glycosides and can be cleaved without destroying the newly installed *O*-glycosidic linkage by simple treatment with aqueous trifluoroacetic acid (Scheme 2.51) [59]. When confronted with monosaccharides in which all the secondary hydroxyls are *trans* and equatorial, as is the case with methyl α-D-glucopyranoside, the *bis*-DHP reagent **304** gives rise to inseparable mixtures of the 2,3- and 3,4-*O*-dispoke acetals. In order to overcome this difficulty, Ley's group [60] have recently introduced several enantiomerically pure *bis*-dihydropyran reagents

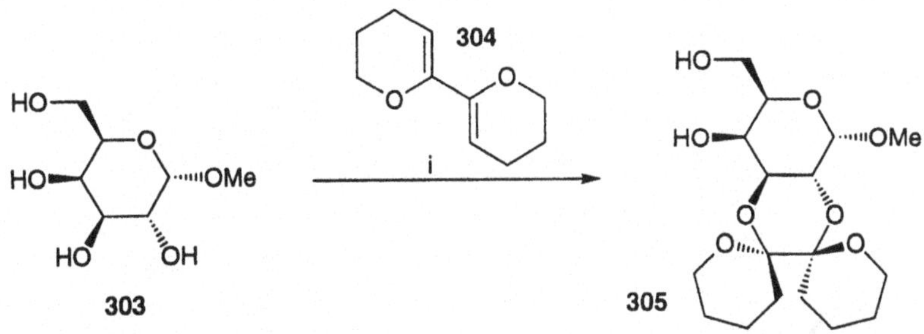

Scheme 2.50 i, CSA, $CHCl_3$, Δ, 1.5 h, then ethylene glycol, Δ, 0.5 h, 76%.

that are functionalised at the C(2) and C(2′) positions, whose chirality can be used to determine whether the 2,3- or the 3,4-acetal is formed. With the *R*,*R*-diene **312**, the D-gluco substrate **311** furnished the 2,3-acetal **313** exclusively in 88% yield, whereas with the *S*,*S*-diene **315**, the 3,4-acetal was the sole

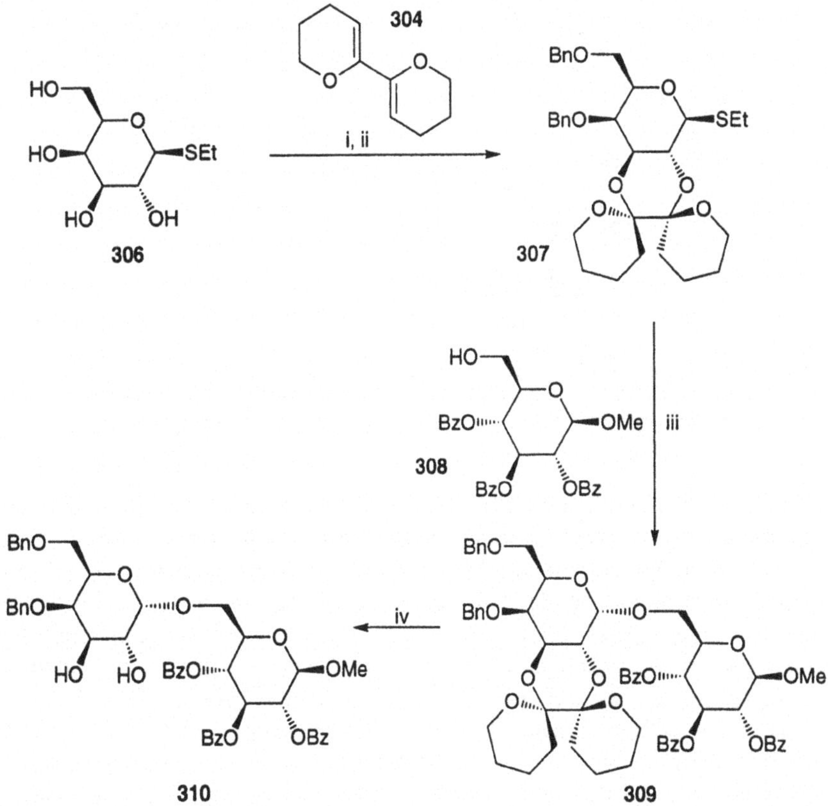

Scheme 2.51 i, CSA, $CHCl_3$, Δ, 1.5 h, then ethylene glycol, Δ, 0.5 h, 59%; ii, NaH, BnBr; iii, *N*-iodosuccinimide, CF_3SO_3H, $ClCH_2CH_2Cl$, Et_2O, 42%; iv, CF_3CO_2H-H_2O (19 : 1), 59%.

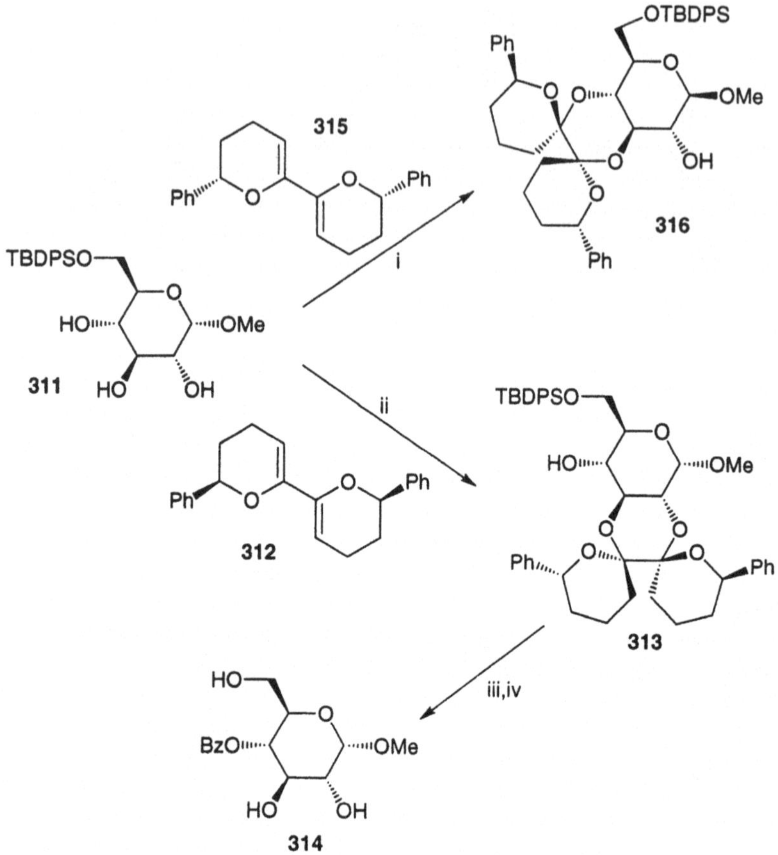

Scheme 2.52 i, Catalyst CSA, diene (1.5 equiv.), $CHCl_3$, Δ, overnight, 75%; ii, catalyst CSA, diene (1.5 equiv.), $CHCl_3$, Δ, overnight, 88%; iii, BzCl, Py, DMAP, room temperature, overnight, 96%; iv, $FeCl_3$ (5 equiv.) CH_2Cl_2, room temperature, overnight, 50%.

product obtained (Scheme 2.52). Somewhat surprisingly, both these phenyl-substituted dispoke acetals proved highly resistant to hydrogenolysis, and so Ley *et al.* [60] recommend their removal with ferric chloride in dichloromethane overnight. While *O*-benzoate esters maintain their positional integrity under these deprotection conditions, *O*-*tert*-butyldiphenylsilyl ethers tend to be unstable. In an effort to develop a dispoke acetal that retains the aforementioned regiodiscriminatory properties, but which could be removed without causing loss of *O*-*tert*-butyldiphenylsilyl ethers, Ley and co-workers investigated the utility of dispoke acetals with allyl groups located at C(2) and C(2′) (Scheme 2.53) [60]. They showed that such acetals could be deblocked by a two-step procedure involving ozonolysis and base-mediated β elimination of the resulting β-alkoxy aldehyde.

317

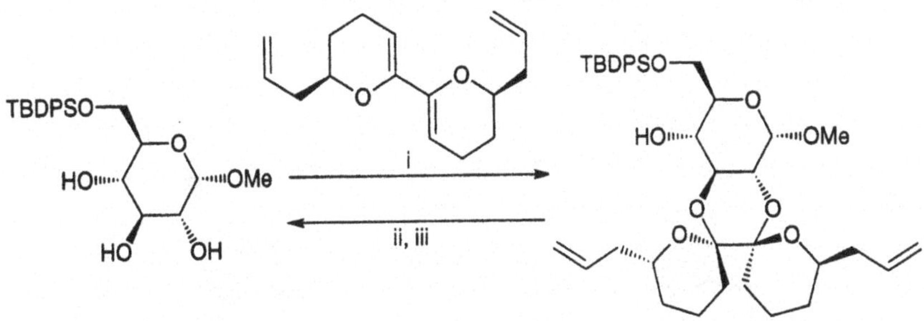

311 318

Scheme 2.53 i, Catalyst CSA, diene (1.16 equiv.), $CHCl_3$, Δ, 48 h, 76%; ii, O_3, CH_2Cl_2, −78°C, then Ph_3P (1.4 equiv.), 7 h, room temperature, 100%; iii, DBU (1.0 equiv.), PhMe, 80°C, 21 h, 54%.

Other potentially important developments that have occurred in protecting-group technology in recent years have included the introduction of a method for the oxidative cleavage of 4,6-*O*-benzylidene acetals of methyl hexopyranosides with *tert*-butyl hydroperoxide and cupric chloride [61]. While at present the regioselectivity of these acetal openings is at best modest (Scheme 2.54), further work in this area might allow improvements to be made.

While considerable success has accrued in the regioselective *O*-benzylation of monosaccharides via their *O*-stannylene acetals [62] it is only recently that highly regioselective de-*O*-benzylations of perbenzylated monosaccharides have been achieved [63–65]. So far, the best results have been obtained with Lewis acids such as tin tetrachloride and titanium tetrachloride (Scheme 2.55) [64, 65]. While, at present, it is difficult to predict *a priori* which substrates will undergo de-*O*-benzylation regioselectively, it is clear from all the substrates

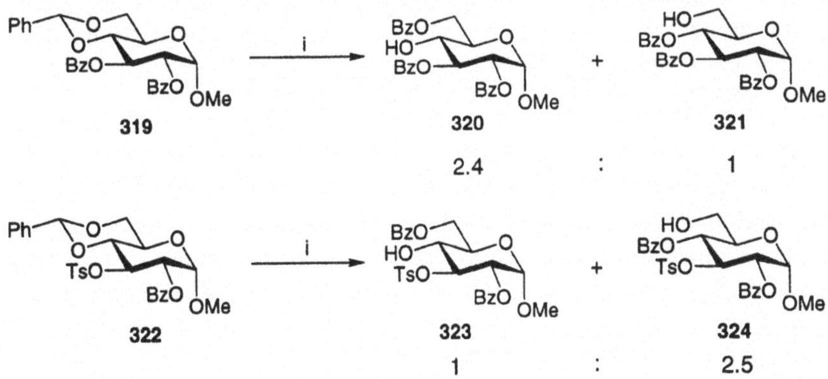

Scheme 2.54 i, Anhydrous *t*-BuO_2H in C_6H_6 (70% solution), $CuCl_2$ (10 mol%), 50°C, argon, 15 h, 88%–99% yield.

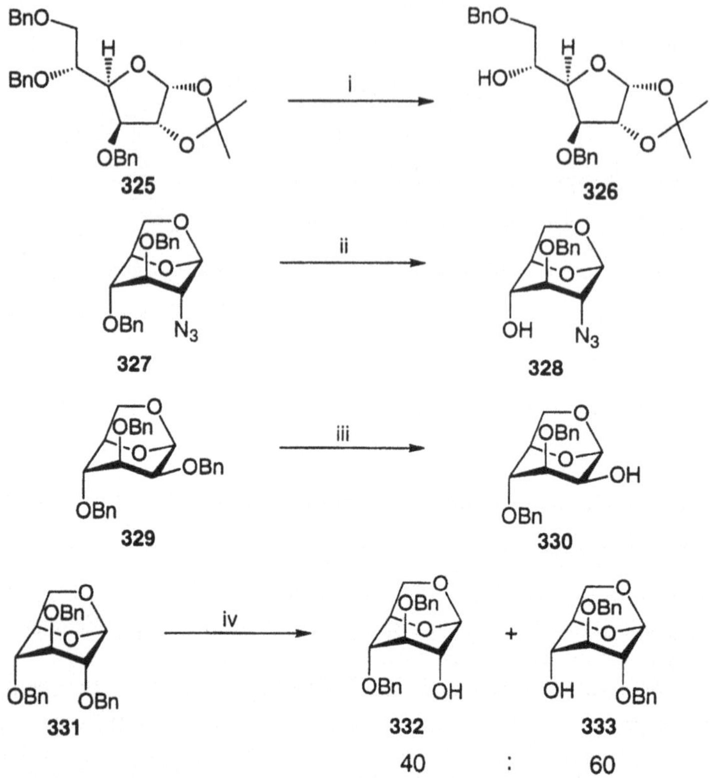

Scheme 2.55 i, $TiCl_4$, DMF, CH_2Cl_2, room temperature, 8 h, 54%; ii, $TiCl_4$, CH_2Cl_2, room temperature, 20 min., 88%; iii, $SnCl_4$, CH_2Cl_2, 0.5 h, 92%; iv, $SnCl_4$, CH_2Cl_2, room temperature, 8 h, 95%

so far examined that at least three properly arranged metal chelating groups need to be present, and that one of these must be the benzyloxy group undergoing deprotection. A six-coordinate octahedral tin or titanium complex has been suggested as the intermediate in these reactions [65].

2.4.2 New methodology for the synthetic manipulation of monosaccharides

Tremendous advances have occurred over the past fifteen years in the chemical modification of monosaccharide derivatives through nucleophilic displacement reactions. In particular, the introduction of the trifluoromethanesulfonate (triflate) ester group into carbohydrate chemistry has made possible many nucleophilic displacement reactions that were previously considered futile with the corresponding *O*-mesyl and *O*-tosyl ester monosaccharide derivatives. An illustration of the scope of *O*-triflate esters for the chemical modification of monosaccharides is given by the examples shown in Scheme 2.56 [66]. Two

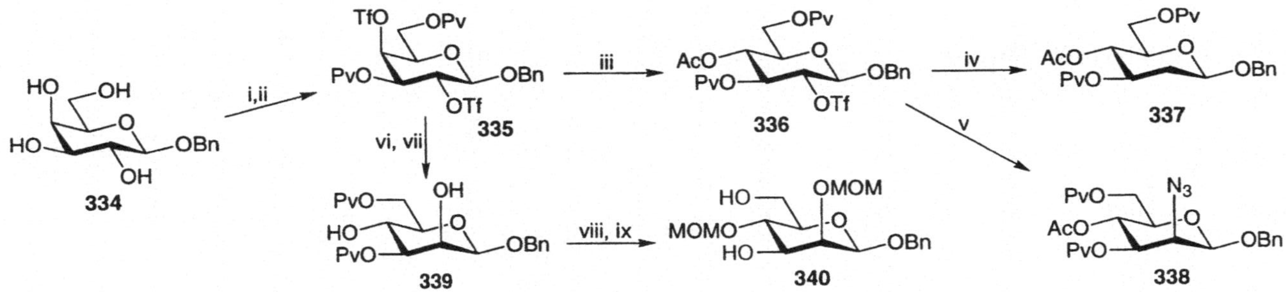

Scheme 2.56 i, $(Bu_3Sn)_2$, PhMe, Δ, then PvCl (3.0 equiv.), room temperature, 85%; ii, $(CF_3SO_2)_2O$ (3.0 equiv.), Py, CH_2Cl_2, 0°C to room temperature, 98%; iii, CsOAc (1.5 equiv.), 18-crown-6, PhMe, room temperature, 84%; iv, *n*-Bu_4NBH_4, C_6H_6, room temperature, ultrasound, 12 h, 82%; v, *n*-Bu_4NN_3 (2.0 equiv.), C_6H_6, room temperature, ultrasound, 12 h, 91%; vi, $CsOC(O)CF_3$ (2.0 equiv.), 18-crown-6, PhMe-DMF, 80°C; vii, aq. $NaHCO_3$, MeOH, 76%, 2 steps; viii, P_2O_5, $CH_2(OMe)_2$, $(ClCH_2)_2$, 40°C, 93%; ix, NaOMe, MeOH, ultrasound, room temperature, 100%.

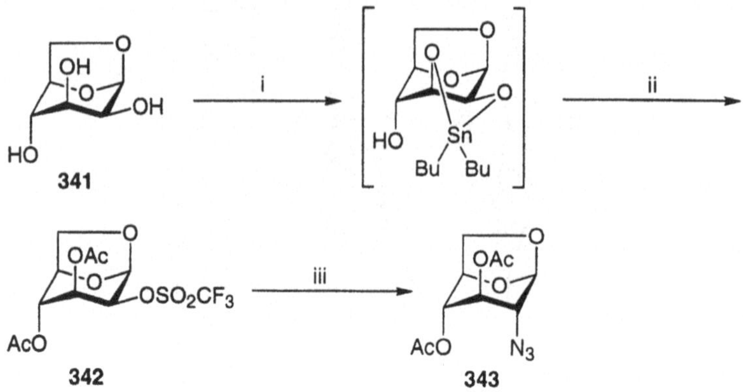

Scheme 2.57 i, Bu_2SnO, PhMe, DMF, Δ; ii, CF_3SO_2Cl, CH_2Cl_2, −10°C, 1.5 h, then Ac_2O, Py, 0°C; iii, NaN_3, DMF, 50% overall yield.

items of particular merit are the reductive deoxygenation at C(2) in **336** with tetra-*n*-butylammonium borohydride to give **337**, and the double displacement of **335** with the trifluoroacetate anion. These are especially noteworthy given the sterically hindered locations of both these *O*-triflate esters and the relatively poor nucleophilicity of both these reagents. Another reaction of note is the selective displacement of the C(4) triflate in **335** by the basic acetate anion which proceeded in high yield without a competing elimination reaction occurring. *O*-Triflate esters have traditionally been prepared by treatment of the parent monosaccharide alcohol with triflic anhydride and 2,6-lutidine in dry dichloromethane at 0°C. Several high-yielding procedures are also now available for regioselective *O*-triflation of certain monosaccharide polyols using *O*-stannylene acetal derivatives and trifluoromethanesulfonyl chloride in dichloromethane at low temperature; one of these is shown in Scheme 2.57 [67].

Nucleophilic ring opening of five-membered cyclic sulfates is becoming increasingly popular for the stereoselective introduction of new functionality into monosaccharide derivatives. Quite frequently, high levels of regioselectivity are displayed in these S_N2 displacements, but it should be emphasised that the regiochemical outcome is often very difficult to predict *a priori*, it depending quite dramatically on the nature of the nucleophile used and the exact substrate undergoing reaction (Scheme 2.58) [68, 69].

Considerable regioselectivity can often be observed in the halogenation of monosaccharides with the 'cocktail' formed from triphenylphosphine and a carbon tetrahalide in pyridine [70]. An example of high regiocontrol being observed in this type of halogenation reaction can be found in Scheme 2.65 [71]. Garegg and Samuelsson [72] have also reported high C(6) selectivity in the iodination of methyl α-D-gluco- and α-D-manno-pyranosides with the reagent prepared from triphenylphosphine, iodine and imidazole in toluene (Scheme 2.59). Another useful system for the conversion of a

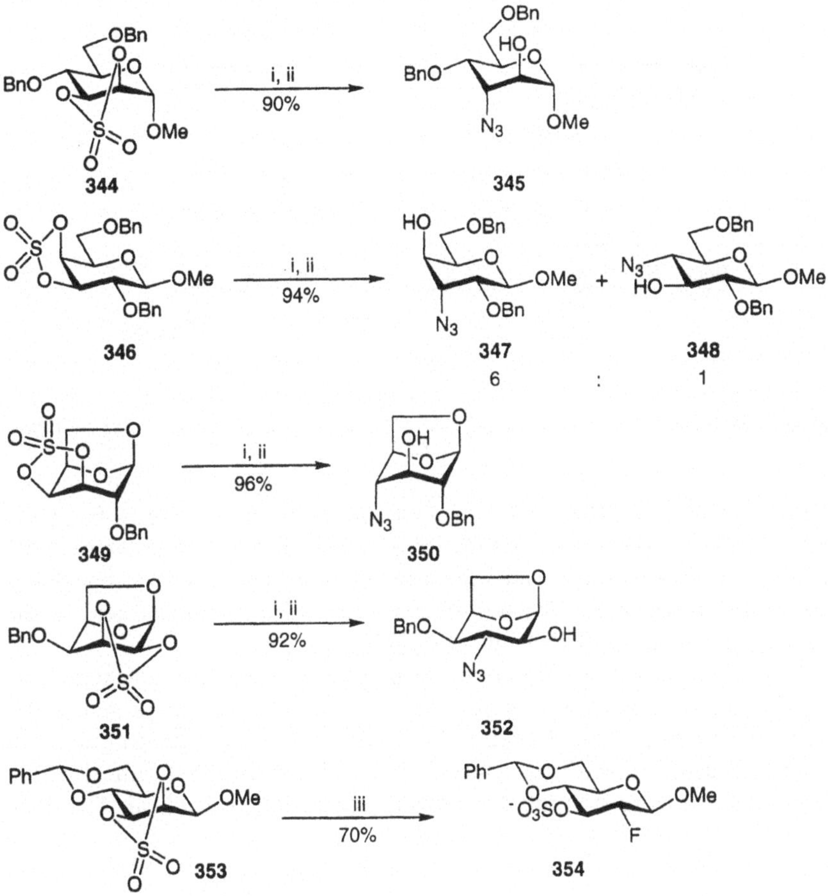

Scheme 2.58 i, LiN_3, DMF, 80°C; ii, aq. H_2SO_4, THF; iii, Me_4NF, MeCN, Δ, 10 min.

monosaccharide alcohol directly into an iodide is the combination of triphenylphosphine and 2,4,5-tri-iodoimidazole in toluene. It allows methyl 3,4,6-tri-*O*-benzyl-2-deoxy-2-iodo-α-D-glucopyranoside **358** to be synthesised in 82% yield directly from methyl 3,4,6-tri-*O*-benzyl-α-D-mannopyranoside **357** [72].

A particularly attractive reaction for the selective thiofunctionalisation of monosaccharides is the Mitsunobu reaction with triphenylphosphine, diethyl azodicarboxylate (DEAD) and mercaptobenzothiazole [73]. When **359** is allowed to react with these reagents at room temperature for 4 h and then at 110°C for 2 h, and the mixture *O*-acetylated, the 3,6-dithio-β-D-allopyranoside **360** is isolated along with an elimination product (Scheme 2.60). Free radical desulfurisation of **360** with tri-*n*-butylstannane then gives access to the corresponding 3,6-dideoxy sugar **361**. The Mitsunobu reaction

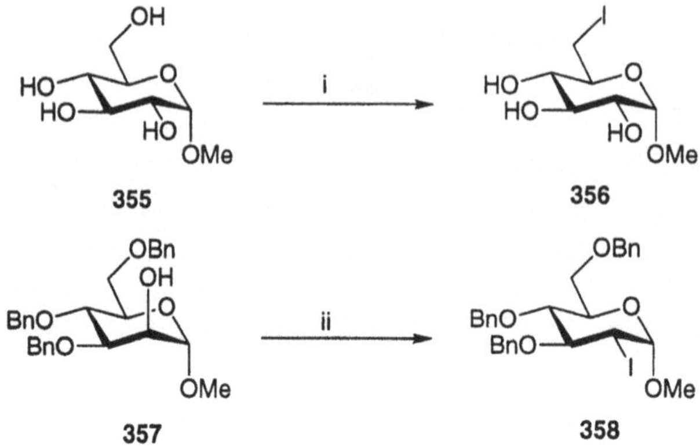

Scheme 2.59 i, Ph_3P, I_2; imidazole, PhMe, Δ, 80%; ii, Ph_3P, 2,4,5-tri-iodoimidazole, PhMe, Δ, 82%.

is also valuable for the preparation of anhydro-sugars directly from the monosaccharide precursors with a *trans* vicinal diol system (Scheme 2.61) [74, 75]. The regioselective introduction of azide into monosaccharides is also viable through the Mitsunobu reaction when hydrazoic acid is the nucleophilic component (Scheme 2.62) [75].

Barton and co-workers have recently announced a new free radical method for the conversion of a monosaccharide xanthate into an amino-sugar derivative in two steps (Scheme 2.63) [76]. It involves generation of a monosaccharide radical from the xanthate and trapping that radical with a diazirine to generate a cyclic three-membered hydrazinyl radical. This then

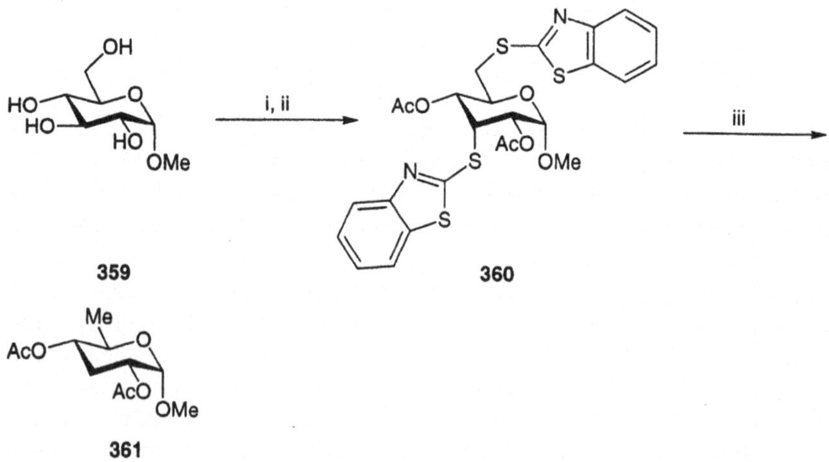

Scheme 2.60 i, Ph_3P, DEAD, mercaptobenzothiazole, Py, 4 h, room temperature, then 110°C for 3 h; ii, Ac_2O, 42%, 2 steps; iii, Bu_3SnH, AIBN, 82%.

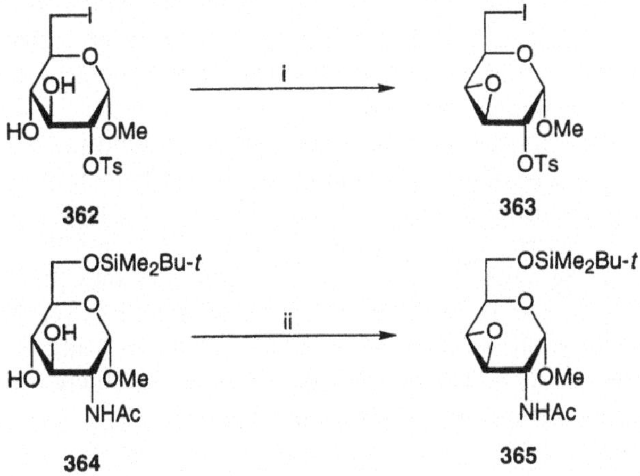

Scheme 2.61 i, Ph_3P, $i\text{-}PrO_2CN{=}NCO_2Pr\text{-}i$, PhMe, 90°C, 2 h, 56%; ii, Ph_3P, DEAD, neat, 3A MS, 60°C, 80%.

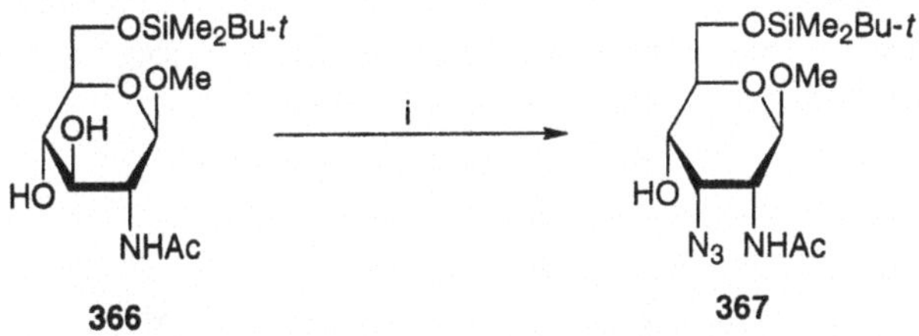

Scheme 2.62 i, Ph_3P, DEAD, HN_3, 62%.

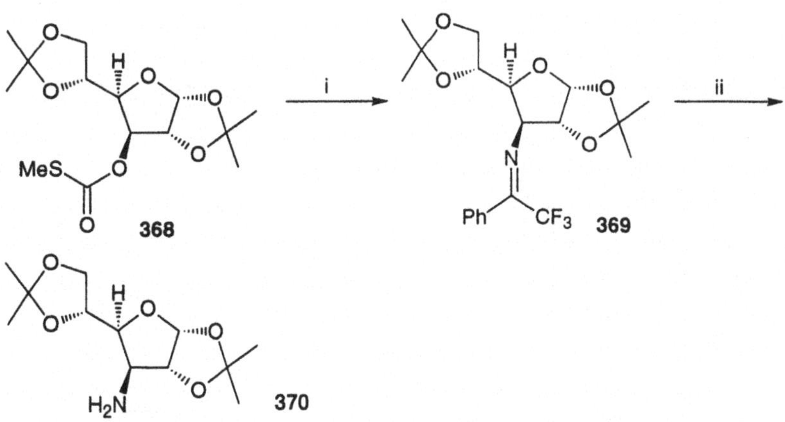

Scheme 2.63 i, 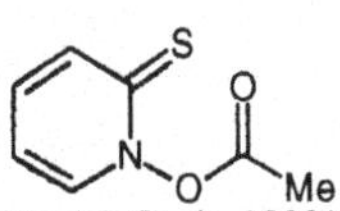 , 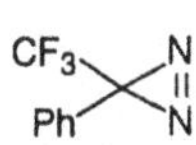 , CH_2Cl_2, $h\nu$, 60%; ii, $B(OH)_3$, EtOH, H_2O, Δ, 100%.

undergoes dimerisation to a tetrazene derivative which spontaneously rearranges to the imine with loss of nitrogen. In the case of **368**, the imine formed was **369**; it underwent hydrolysis to the free amine **370** by heating with boric acid in aqueous ethanol.

The discovery of the *Calicheamicin* family of antitumour antibiotics unearthed a new type of monosaccharide in which a hydroxylamine function is present [77]. Danishefsky and co-workers [78] have developed a high-yielding pathway to hydroxylamino sugars that is predicated on the oxidation of the corresponding amino sugars with 1 mol equiv. of dimethyldioxirane in acetone at low temperature (Scheme 2.64). Occasionally, overoxidation can be a problem with this method, with oximes sometimes being formed as the exclusive products (e.g. **377** to **378**), but, despite this, there are still many examples of the method being successfully applied.

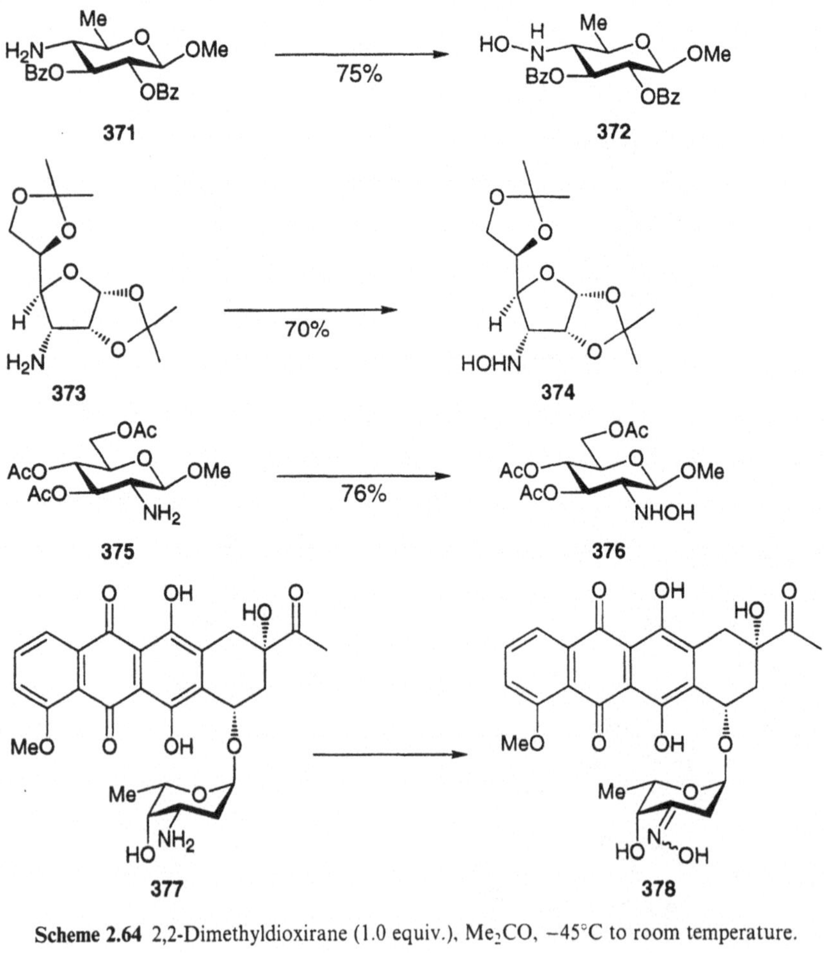

Scheme 2.64 2,2-Dimethyldioxirane (1.0 equiv.), Me_2CO, −45°C to room temperature.

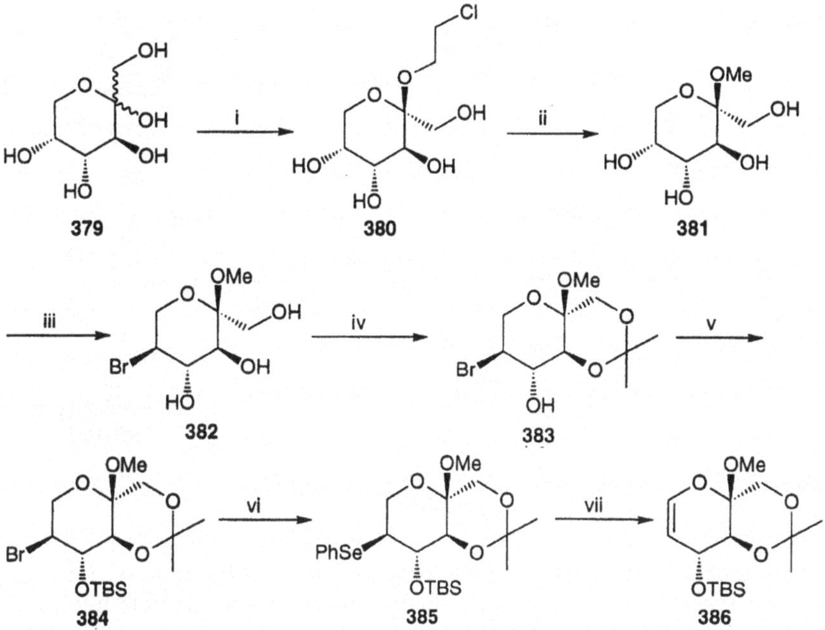

Scheme 2.65 i, 2-Chloroethanol, MeCOCl, room temperature, 3 h, 97%; ii, MeCOCl, MeOH, room temperature, 9 h, 63%; iii, Ph_3P (3.6 equiv.), CBr_4 (3.8 equiv.), Py, 80°C, 4.5 h, 92%; iv, $Me_2C(OMe)_2$, *p*-TsOH, DMF, 80°C, 58%–64%; v, *t*-$BuMe_2SiCl$, imidazole, DMF, room temperature, 97%; vi, $(PhSe)_2$ (2 equiv.), $NaBH_4$ (4 equiv.), DMF, 70°C, 24 h, 88%; vii, $NaIO_4$ (2 equiv.), THF-H_2O (1 : 1), room temperature, 92%.

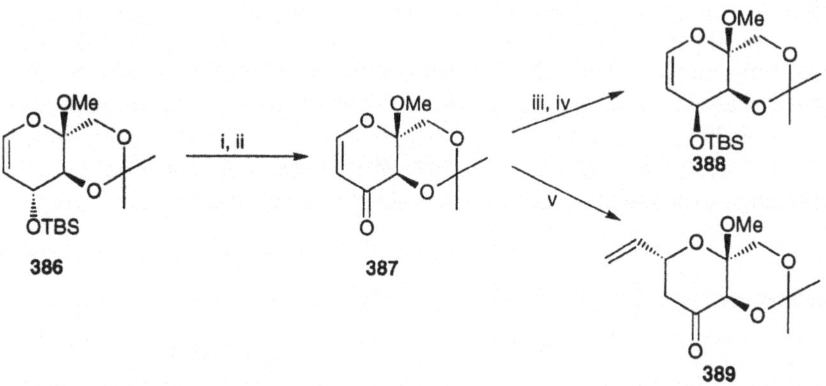

Scheme 2.66 i, *n*-Bu_4NF, THF, room temperature, 96%; ii, *n*-Pr_4NRuO_4 (0.05 equiv.), NMO (1.5 equiv.), 4A MS, CH_2Cl_2, 75%; iii, *i*-Bu_2AlH, PhMe, CH_2Cl_2, −78°C, 75%; iv, *t*-$BuMe_2SiCl$, imidazole, DMF, room temperature, 77%; v, vinylmagnesium bromide, CuI, Me_3SiCl, THF, −78°C, 84%.

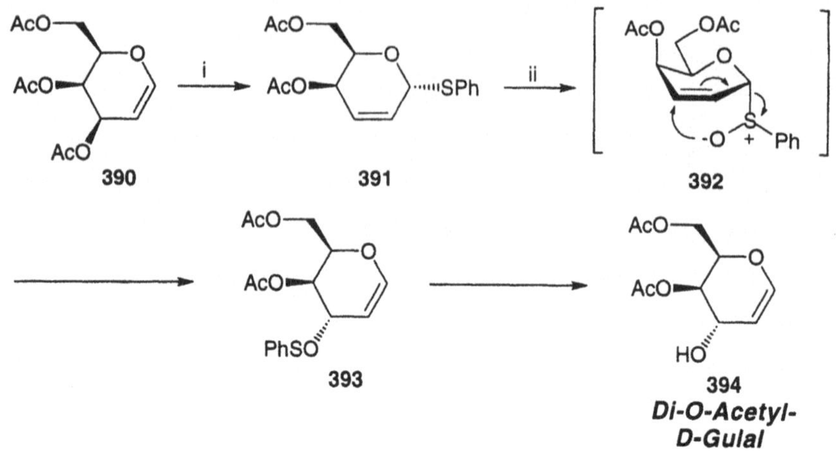

Scheme 2.67 i, PhSH, BF_3-Et_2O, CH_2Cl_2, −78°C; ii, *m*-CPBA, CH_2Cl_2, 0°C, followed by piperidine, 60%.

Development of novel methodology for the construction of rare and unusual types of glycal continues to be a topic of considerable interest to synthetic carbohydrate chemists. One very interesting pathway to differentially protected 5,6-glycals of ketoses involves selective bromination at C(5) in methyl β-D-fructopyranoside, protection and nucleophilic displacement of **384** with sodium phenylselenide to deliver **385**; *syn* elimination of the selenoxide then gives **386** in good overall yield [71]. Compound **386** could be converted to its 4-epimer **388** by a stereospecific oxidation–reduction sequence (Schemes 2.65 and 2.66) [71].

Danishefsky and associates have reported a novel pathway for obtaining hitherto rare glycals of the allal and gulal series [79]. In an illustration of their approach to the latter series of compounds (Scheme 2.67) D-galactal was subjected to allylic displacement with thiophenol under Lewis acid conditions. The resulting thioglycoside **391** was then oxidised to the phenylsulfoxide **392** and this was treated with base. This triggered a highly stereoselective [2,3]-rearrangement to give **393** which was converted to the differently protected D-gulal derivative **394** by simple hydrolysis.

References

1. Katsuki, T. and Sharpless, K.B. (1980) *J. Amer. Chem. Soc.*, **102**, 5974.
2. Ko, S.Y., Lee, A.W.M., Masamune, S., Reed, L.A. III, Sharpless, K.B. and Walker, F.J. (1990) *Tetrahedron*, **46**, 245.
3. Roush, W.R., Straub, J.A. and VanNieuwenhze, M.S. (1991) *J. Org. Chem.*, **56**, 1636.
4. Roush, W.R., Brown, R.J. and DiMare, M. (1983) *J. Org. Chem.*, **48**, 5083; Roush, W.R. and Brown, R.J. (1993) *J. Org. Chem.*, **48**, 5093; Roush, W.R. and Hagadorn, S.M. (1985) *Carbohydr. Res.*, **136**, 187.

5. Sammes, P.G. and Thetford, D. (1988) *J. Chem. Soc. Perkin Trans.*, **1**, 111.
6. Grethe, G., Sereno, J., Williams, T.H. and Uskokovic, M.R. (1983) *J. Org. Chem.*, **48**, 5313.
7. Marsh, J.P., Mosher, C.W., Acton, E.M. and Goodman, L. (1967) *Chem. Commun.*, 973.
8. Llera, J.-M., Trujillo, M., Blanco, M.-E. and Alcudia, F. (1994) *Tetrahedron: Asymmetry*, **5**, 709.
9. Achmatowicz, O., Bukowski, P., Szechner, B. *et al.* (1977) *Tetrahedron*, **27**, 1973.
10. Mukaiyama, T., Shiina, I. and Kobayashi, S. (1990) *Chem. Lett.*, 2201.
11. Kobayashi, S., Onozawa, S. and Mukaiyama, T. (1992) *Chem. Lett.*, 2419.
12. Sharpless, K.B., Amberg, W., Bennani, Y.L. *et al.* (1992) *J. Org. Chem.*, **57**, 2768.
13. Kolb, H., VanNieuwenhze, M.S. and Sharpless, K.B. (1994) *Chem. Rev.*, **94**, 2483.
14. Enders, D. and Jegelka, U. (1993) *Tetrahedron Lett.*, **34**, 2453.
15. Bednarski, M. and Danishefsky, S.J. (1986) *J. Am. Chem. Soc.*, **108**, 7060.
16. Vogel, P., Fattori, D., Gasparini, F. and Le Drian, C. (1990) *SYNLETT*, 173.
17. Vieira, E. and Vogel, P. (1983) *Helv. Chim. Acta*, **66**, 1865; Wagner, J., Vieira, E. and Vogel, P. (1988) *Helv. Chim. Acta*, **71**, 624.
18. Nativi, C., Reymond, J.-L. and Vogel, P. (1989) *Helv. Chim. Acta*, **72**, 882.
19. Auberson, Y. and Vogel, P. (1989) *Helv. Chim. Acta*, **72**, 278.
20. Fattori, D., de Guchteneere, E. and Vogel, P. (1989) *Tetrahedron Lett.*, **30**, 7415.
21. Corey, E.J. and Loh, T.-P. (1993) *Tetrahedron Lett.*, **34**, 3979.
22. Corey, E.J. and Loh, T.-P. (1991) *J. Amer. Chem. Soc.*, **113**, 8966; Corey, E.J., Loh, T.-P., Roper, T.D. *et al.* (1992) *J. Amer. Chem. Soc.*, **114**, 8290.
23. Tietze, L., Schneider, C. and Montenbruck, A. (1994) *Angew. Chem. Int. Ed. Engl.*, **33**, 980.
24. Matteson, D.S. and Petersen, M.L. (1987) *J. Org. Chem.*, **52**, 5116.
25. Bernadi, A., Piarulli, U., Poli, G. *et al.* (1990) *Bull. Soc. Chim. Fr.*, **127**, 751.
26. Casiraghi, G., Colombo, L., Rassu, G. and Spanu, P. (1990) *J. Org. Chem.*, **55**, 2565.
27. Dondoni, A. (1992) *Modern Synthetic Methods 1992, Vol. 6* (ed. R. Scheffold), VHCA, Basel, 377; Dondoni, A., Fantin, G., Fogagnolo, M. *et al.* (1989) *J. Org. Chem.*, **54**, 693.
28. David, S. and Lépine, M.C. (1985) *J. Chem. Soc. Perkin Trans.*, **1**, 1262.
29. Hatakeyama, S., Suguwara, K. and Takano, S. (1991) *Tetrahedron Lett.*, **32**, 4513.
30. Marshall, J.A., Seletsky, B.M. and Luke, G.P. (1994) *J. Org. Chem.*, **59**, 3413.
31. Marshall, J.A., Beaudoin, S. and Lewinski, K. (1993) *J. Org. Chem.*, **58**, 5876; Marshall, J.A. and Beaudoin, S. (1994) *J. Org. Chem.*, **59**, 6614.
32. Danishefsky, S.J., DeNinno, M.P. and Chen, S.-h. (1988) *J. Amer. Chem. Soc.*, **110**, 3929.
33. De Gaudenzi, L., Apparao, S. and Schmidt, R.R. (1990) *Tetrahedron*, **46**, 277.
34. Giese, B. and Linker, T. (1992) *Synthesis*, 46.
35. Collins, P.M. (1971) *J. Chem. Soc., C* (1960); Collins, P.M. and Gupta, P. (1971) *J. Chem. Soc., C* (1965); Collins, P.M. and Travis, A.S. (1980) *J. Chem. Soc. Perkin Trans.*, **1**, 779; Collins, P.M., Farnia, F. and Travis, A.S. (1979) *J. Chem. Res. (S)*, 266.
36. Inanaga, J., Sugimoto, Y., Yokoyama, Y. and Hanamoto, T. (1992) *Tetrahedron Lett.*, **33**, 8109.
37. Musser, J.H. (1992) *Ann. Rep. Med. Chem.*, **27**, 301.
38. Driguez, H. and Henrissat, B. (1981) *Tetrahedron Lett.*, **22**, 5061.
39. Csuk, R. and Glanzer, B.I. (1986) *J. Chem. Soc. Chem. Comm.*, 343.
40. Al-Masoudi, N.A.L. and Hughes, N.A. (1986) *Carbohydr. Res.*, **148**, 25.
41. Al-Masoudi, N.A.L. and Hughes, N.A. (1986) *Carbohydr. Res.*, **148**, 39.
42. Hashimoto, H. and Izumi, M. (1992) *Chem. Lett.*, 25; see also Takahashi, S. and Kuzuhara, H. (1992) *Chem. Lett.*, 21.
43. Chmielewski, M. and Whistler, R.L. (1975) *J. Org. Chem.*, **40**, 639.
44. Vasella, A. and Voeffray, R. (1982) *Helv. Chim. Acta.*, **65**, 1134.
45. Auberson, Y. and Vogel, P. (1989) *Angew. Chem. Int. Ed. Eng.*, **28**, 1498.
46. Legler, G. and Julich, E. (1984) *Carbohydr. Res.*, **128**, 61.
47. Legler, G. and Pohl, S. (1986) *Carbohydr. Res.*, **155**, 119.
48. Yamamoto, H. and Inokawa, S. (1984) *Adv. Carbohydr. Chem. and Biochem.*, **42**, 135.
49. Yamamoto, H., Hanaya, T., Kawamoto, H. *et al.* (1985) *J. Org. Chem.*, **50**, 3516.
50. Suami, T. and Ogawa, S. (1990) *Adv. Carbohydr. Chem. and Biochem.*, **48**, 22.
51. Suami, T., Tadano, K., Kameda, Y. and Iimura, Y. (1984) *Chem. Lett.*, 1919.
52. Paulsen, H., von Deyn, W. and Roben, W. (1984) *Liebigs Ann. Chem.*, 433.
53. Tadano, K., Ueno, Y., Fukabori, C. *et al.* (1987) *Bull. Chem. Soc. Jpn.*, **60**, 1727.

54. Yoshikawa, M., Yokokawa, Y., Inoue, Y. *et al.* (1994) *Tetrahedron*, **50**, 9961.
55. Wilcox, C.S. and Gaudino, J.J. (1986) *J. Amer. Chem. Soc.*, **108**, 3102.
56. Shing, T.K.M. and Tang, Y. (1990) *J. Chem. Soc. Chem. Comm.*, 748; (1990) *Tetrahedron*, **46**, 6575; (1991) *Tetrahedron*, **47**, 4571.
57. McComsey, D.F. and Maryanoff, B.E. (1994) *J. Org. Chem.*, **59**, 2652.
58. Jackson, G., Jones, H.F., Petursson, S. and Webber, J.M. (1982) *Carbohydr. Res.*, **102**, 147; Petursson, S. and Webber, J.M. (1982) *Carbohydr. Res.*, **103**, 41.
59. Ley, S.V., Leslie, R., Tiffin, P.D. and Woods, M. (1992) *Tetrahedron Lett.*, **33**, 4767.
60. Entwistle, D.A., Hughes, A.B., Ley, S.V. and Visentin, G. (1994) *Tetrahedron Lett.*, **35**, 777.
61. Sato, K., Igarishi, T., Yanagisawa, Y. *et al.* (1988) *Chem. Lett.*, 1699.
62. David, S. and Hanessian, S. (1985) *Tetrahedron*, **41**, 643.
63. Cruzado, M. del C. and Martin-Lomas, M. (1986) *Tetrahedron Lett.*, **27**, 2497.
64. Martin, O.R., Kurz, K.G. and Rao, S.P. (1987) *J. Org. Chem.*, **52**, 2922.
65. Hori, H., Nishida, Y., Ohrui, H. and Meguro, H. (1989) *J. Org. Chem.*, **54**, 1346.
66. Sato, K. and Yoshimoto, A. (1995) *Chem. Lett.*, 39.
67. Dasgupta, F. and Garegg, P.J. (1988) *Synthesis*, 626.
68. van der Klein, P.A.M., Filemon, W., Veeneman, G.H. *et al.* (1992) *J. Carbohydr. Chem.*, **11**, 837.
69. Tewson, T.J. (1982) *J. Label. Comp. Radiopharm.*, **19**, 1629; (1983) *J. Org. Chem.*, **48**, 3507.
70. Anisuzzaman, A.K.M.S. and Whistler, R.L. (1978) *Carbohydr. Res.*, **61**, 511.
71. Hale, K.J. and Manaviazar, S. (1994) *Tetrahedron Lett.*, **35**, 8873; see also: Chan, J.Y.C., Cheong, P.P.L, Hough, L. and Richardson, A.C. (1985) *J. Chem. Soc. Perkin Trans.*, **1**, 1447.
72. Garegg, P.J. and Samuelsson, B. (1979) *J. Chem. Soc. Chem. Comm.*, 978.
73. Dancy, I., Laupichler, L., Rollin, P. and Thiem, J. (1992) *SYNLETT*, 283.
74. Ferrier, R.J., Schmidt, P. and Tyler, P.C. (1985) *J. Chem. Soc. Perkin Trans.*, **1**, 301.
75. Brandstetter, H. and Zbiral, E. (1980) *Helv. Chim. Acta*, **63**, 327.
76. Barton, D.H.R., Jaszberenyi, J.C. and Theodorakis, E.A. (1992) *J. Amer. Chem. Soc.*, **114**, 5904.
77. Lee, M.D., Dunne, T.S., Siegel, M.M. *et al.* (1987) *J. Amer. Chem. Soc.*, **109**, 3464; Lee, M.D., Dunne, T.S., Chang, C.C. *et al.* (1987) *J. Amer. Chem. Soc.*, **109**, 3466.
78. Wittman, M.D., Halcomb, R.L. and Danishefsky, S.J. (1990) *J. Org. Chem.*, **55**, 1981.
79. Wittman, M.D., Halcomb, R.L. and Danishefsky, S.J. (1990) *J. Org. Chem.*, **55**, 1979.

3 Carbohydrate sulphates

RUTH FALSHAW, RICHARD H. FURNEAUX and
GEORGE C. SLIM

3.1 Introduction

The presence of negatively charged sulphate groups on a wide variety of carbohydrates from simple sugars to complex polysaccharides provides numerous interesting and useful materials. The position(s) of sulphate ester(s) on the sugar ring, the nature of the carbohydrate and the presence of other functional groups, such as amines or uronic acids, all play an integral role in the diversity of carbohydrate sulphates. The commercial significance of sulphated carbohydrates is substantial. For example, carrageenans used in the food industry were worth US$200 million per annum in 1994 (Bixler, 1995), while 33 tonnes of heparin and related anticoagulant products were sold in 1991 (Linhardt, 1991). The extent and content of recent published work indicates the commercial potential for many other natural materials or synthetic analogues. Here we summarise the state of knowledge in the structure, function, reactions and utility of naturally occurring carbohydrate sulphates and their synthetic analogues, and the studies of mono- and oligosaccharide sulphate derivatives.

3.2 Natural occurrence and biological role

The classification we have used here of polysaccharides as homopolymers or heteropolymers is somewhat arbitrary as small amounts of other sugars are commonly found even in materials where one type of sugar is predominant.

3.2.1 Sulphated galactans

A range of economically important sulphated galactans composed of alternating 3-linked β-D-galactopyranosyl and 4-linked α-D- or L-galactopyranosyl residues are the major structural cell wall components of most red seaweeds. Various hydroxyl groups in these polymeric chains may be sulphated, methylated or glycosylated by simple monosaccharide units. The 3-linked β-D-galactopyranosyl residue may be substituted (at the 4- and 6-positions) with a pyruvate acetal while the 4-linked residues are frequently in the form of the 3,6-anhydride. Polymers where the 4-linked

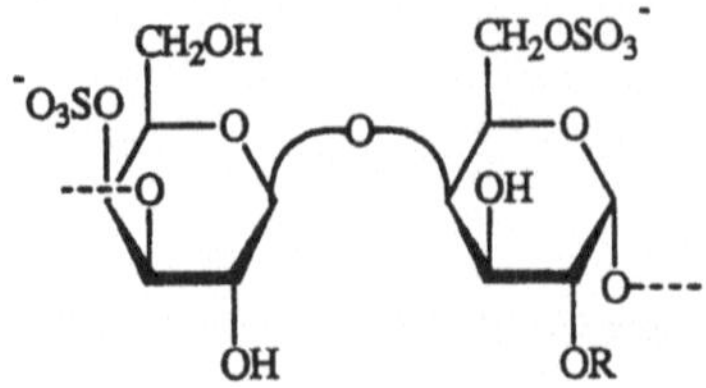

1 R = H
2 R = SO_3^-

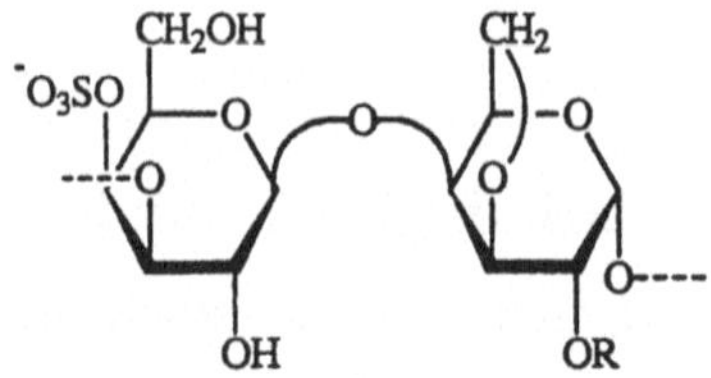

3 R = H
4 R = SO_3^-

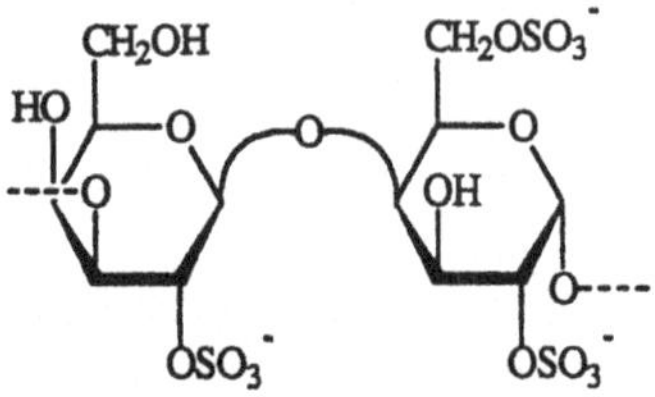

5

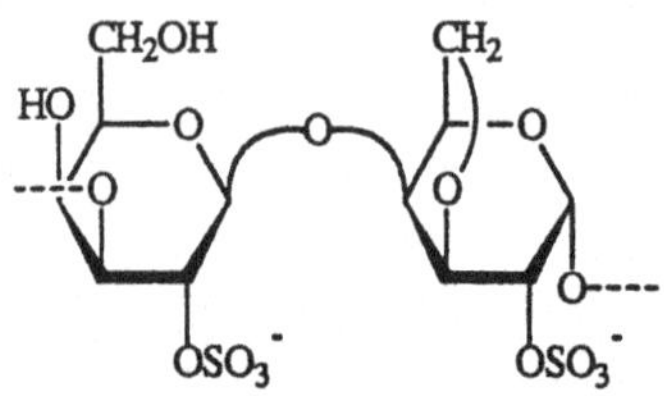

6

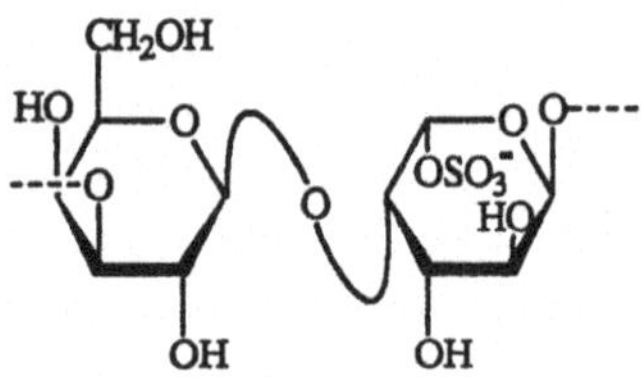

7

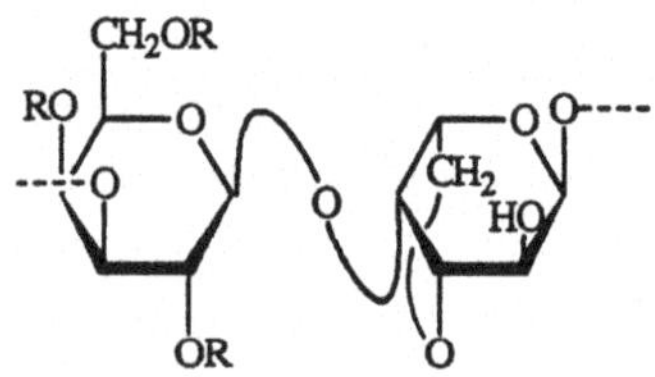

8 1xR = SO_3^-, 2xR = H

Figure 3.1 Disaccharide repeat units of some red algal galactan sulphates: **1** = μ-carrageenan (G4S-D6S); **2** = ν-carrageenan (G4S-D2S,6S); **3** = κ-carrageenan (G4S-DA); **4** = ι-carrageenan (G4S-DA2S); **5** = λ-carrageenan (G2S-D2S,6S); **6** = θ-carrageenan (G2S-DA2S); **7** = agarose precursor (G-L6S); **8** = 2-, 4- or 6-sulphated agarose (G2S-LA, G4S-LA, G6S-LA). Notation in parentheses is defined by Knutsen *et al.* (1994).

residues are in the L configuration are known as agars and those having these residues in the D configuration are carrageenans. Both groups of polysaccharides have been studied widely and reviewed recently (Armisen and Galatas, 1987; Craigie, 1990; Matsuhashi, 1990; Selby and Whistler, 1993; Stanley, 1990; Therkelsen, 1993). Some common disaccharide repeating units are shown in Figure 3.1 with the traditional Greek letter classification of carrageenans and the new structural notation defined by Knutsen *et al.* (1994). Frequently, a given seaweed polysaccharide contains more than one type of structural unit.

The presence of agar or carrageenan in a red seaweed is taxonomically important with agars found in the Gelidiaceae and Gracilariaceae, for example, and carrageenans found in the Gigartinaceae and Solieriaceae. The type of carrageenan also varies between the life stages of algae in the Gigartinaceae (McCandless *et al.*, 1983). Other red algal families such as the Cryptonemiaceae contain sulphated polysaccharides with significant amounts of both 4-linked D-galactose and L-galactose (Miller *et al.*, 1995). Seasonal and environmental factors affecting the growth of seaweeds may also influence the structure of the polysaccharides. Variables such as light, temperature, salinity, nitrogen and carbon have been manipulated to gain a better understanding of polysaccharide biosynthesis but a better understanding is still being sought (Hemmingson *et al.*, 1995). The distribution of polysaccharides within the cell walls has been studied using antibodies and fluorescent probes (Vreeland *et al.*, 1992).

The commercially valuable property of agar, its thermoreversible aqueous gelation, results from the dominant component, agarose, a neutral linear polymer of alternating 3-linked β-D-galactopyranosyl and 4-linked 3,6-anhydro-α-L-galactopyranosyl residues. The gel strength of agars, however, can be weakened by the presence of naturally occurring sulphate substituents. This is particularly so for the most common sulphation in natural agar-type polysaccharides, which is at O-6 of the 4-linked L-galactose units (L6S). Such units are considered the biological precursors of the 3,6-anhydride moiety, which is associated with agar's gel-forming ability.

Sulphate groups can also occur at other positions such as O-2, O-4 and/or O-6 of the 3-linked β-D-galactose residue. Agar from *Gracilaria dominguensis*, containing significant sulphation at the 6-position of the 3-linked galactose units, has been found to be anti-tumour active (Fernández *et al.*, 1989). Agar 4-sulphate (G4S-LA), which is diastereomeric to κ-carrageenan (G4S-DA), has been found in *Dasyclonium incisum* (Miller *et al.*, 1993) and *Gracilaria bursapastoris* (Murano *et al.*, 1995). The latter exhibits ion-dependent gelation similar to but weaker than that of κ-carrageenan. Agar 2-sulphate (G2S-LA) disaccharide units have been characterised in a complex galactan (containing both 4-linked D-galactose and L-galactose) from *Pachymenia lusoria* (Miller *et al.*, 1995). The polysaccharide from *Champia novae-zealandiae* also contains both 4-linked D-galactose and L-galactose but these are unusually sulphated at both the 2 and 3 positions (Miller *et al.*, 1996).

Carrageenans are generally more heavily sulphated than agars and their major commercial uses derive from their ion-dependent aqueous gelation. The number and position of the sulphate substituent determine the ions required and the texture of the gel. κ-Carrageenan (G4S-DA) forms firm gels with potassium ions, while ι-carrageenan (G4S-DA2S) forms softer gels with calcium ions. The presence of significant amounts of 6-sulphated 4-linked galactose 'precursor' residues [G4S-D6S, (μ-carrageenan) and G4S-D2S,6S, (ν-carrageenan)] weaken the gelling properties of κ- and ι-carrageenans (as with agars) by decreasing the structural regularity required to produce strong interactions and hence gels. Conversion of precursor units into the corresponding 3,6-anhydrides with hot alkali (Section 3.4.1) improves gelling properties. Alkali treatment can also produce 3,6-anhydride 2-sulphate units from the 4-linked 2,6-disulphated D-galactose residues of λ-carrageenan (G2S-D2S,6S) but only low levels of the anhydride have been reported to occur naturally in λ-type carrageenan extracts (Falshaw and Furneaux, 1995a).

The anionic sulphate groups of carrageenans enable them to interact with proteins. This has led to significant use in meat products (e.g. poultry and pet food) and the creation of a wide range of dairy foods that utilise the ability of carrageenan to gel or thicken milk products (Bixler, 1995). Non-food applications include a hair-conditioning effect due to carrageenan interaction with keratin (Witt, 1984), whilst carrageenan–ovalbumin adducts have useful antiallergenic properties (Hioki *et al.*, 1991).

Food safety issues have been studied in response to the widespread use of carrageenans in food. Low molecular weight carrageenans, whilst not genotoxic carcinogens, produce a carcinogenic effect in rats by an epigenetic mechanism (Mori *et al.*, 1984), can induce colonic ulcers (Marcus *et al.*, 1992) and have been shown to affect human skin fibroblasts (Tveter-Gallagher *et al.*, 1982). Carrageenans have useful antiviral properties (e.g. Witvrouw *et al.*, 1994; García-Villalón and Gil-Fernández, 1991) and inhibit non-immune lymphocyte–macrophage interactions (Chong and Parish, 1985).

While galactan sulphates are largely associated with red algae, a β-D-galactan sulphate with a backbone of alternating 3-linked and 6-linked β-D-galactopyranosyl units with most sulphate groups at the O-6 of the 3-linked units has been identified from the brown seaweed *Laminaria angustata* var. *longissima* (Nishino *et al.*, 1994) and a range of sulphated α-L-galactans has been characterised from different species of ascidians (Pavão *et al.*, 1989).

3.2.2 Sulphated fucans

Numerous naturally occurring polysaccharides have been found to contain fucose but the terms fucan and fucoidan are generally applied to sulphated polymers with a high fucose content or where there is a fucosyl backbone. Studies on the growth of fertilised zygotes from the brown seaweed genus *Fucus* have shown that fucans when sulphated are localised in the rhizoid cell and are required for adhesion of the embryo to the substratum. Results from early studies on the chemical structure of fucoidan from *Fucus vesiculosus* were interpreted in terms of a predominantly (1 → 2)-linked polymer of 4-sulphated fucopyranosyl residues with some (1 → 3)- and (1 → 4)-linkages. This work was reviewed by Medcalf (1978) and Painter (1983). Recent work by Patankar *et al.* (1993) on a commercially available extract from *Fucus vesiculosus* and by Nishino *et al.* (1991b; 1991c) on purified high fucose content fractions from *Ecklonia kurome*, however, led to a revised view of these polymers as having predominantly a (1 → 3)-linked fucopyranosyl backbone with some sulphate groups at O-4 and single fucosyl branches at O-2 and O-4 (Structure **1** in Figure 3.2). Similar, but more regular structures are found in echinoderms (Mulloy *et al.*, 1994) (structures **2** and **3** in Figure 3.2). These polymers are found in the body walls of sea cucumber and the jelly coat of sea urchin eggs. The presence and spatial orientation of sulphate groups of the sea urchin fucan and a molecular weight $\geq$15 000 are required for binding to sea urchin sperm protein, a necessary step in the fertilisation process. That polyvinyl sulphate also has a high affinity for this protein indicates that it is the spatial arrangement of sulphate groups rather than the nature of the backbone that is critical for binding (DeAngelis and Glabe, 1987). Similar results have been found for the interaction of porcine sperm protein and algal fucoidan (Jones, 1991; Urch and Patel, 1991). Sulphate content is also an important factor in the anticoagulant and antithrombotic activities of fucans from *E. kurome* (Nishino and Nagumo, 1991; 1992) and *F. vesiculosus* (Soeda *et al.*, 1992), and on the antiproliferative activity of *F. vesiculosus* fucan against arterial smooth muscle cells (Vischer and Buddecke, 1991). The antithrombin effect of *E. kurome* fucans has also been linked to molecular weight (Nishino *et al.*, 1991a).

The binding of fucoidan (from *F. vesiculosus*) to L-selectin has been utilised in the analysis of general liquid-phase binding of this lectin. The ability of glucan sulphates to inhibit fucoidan–L-selectin binding was dependent on molecular weight, sulphate density and the nature of the intraglucosyl linkages (Yoshida *et al.*, 1994). The same fucoidan does not bind to a mouse macrophage lectin, although certain sulphated galactans (carrageenans) do (Imai and Irimura, 1994). Fucoidans, and other sulphated polysaccharides, exhibit activity against a range of viruses (Baba *et al.*, 1988). Against African swine fever, however, the activity of fucoidan is less than that of carrageenans (García-Villalón and Gil-Fernández, 1991). The anti-HIV (anti-human-immunodeficiency-virus) properties of fucoidan (and other sulphated polysaccharides) results from their interaction with target cells to inhibit virus entry rather than through neutralisation of virons directly (McClure *et al.*, 1992).

3.2.3 Sulphated mannans

A sulphated 3-linked α-D-mannan has been isolated from the red seaweed *Nemalion vermiculare* (Usov *et al.*, 1973).

3.2.4 Sulphated heteropolysaccharides

Sulphated heteropolysaccharides have been found in all major divisions of seaweeds (Craigie, 1990; Medcalf, 1978; Painter, 1983). Examples from red seaweeds include a sulphated xylogalactan and a sulphated xylomannan

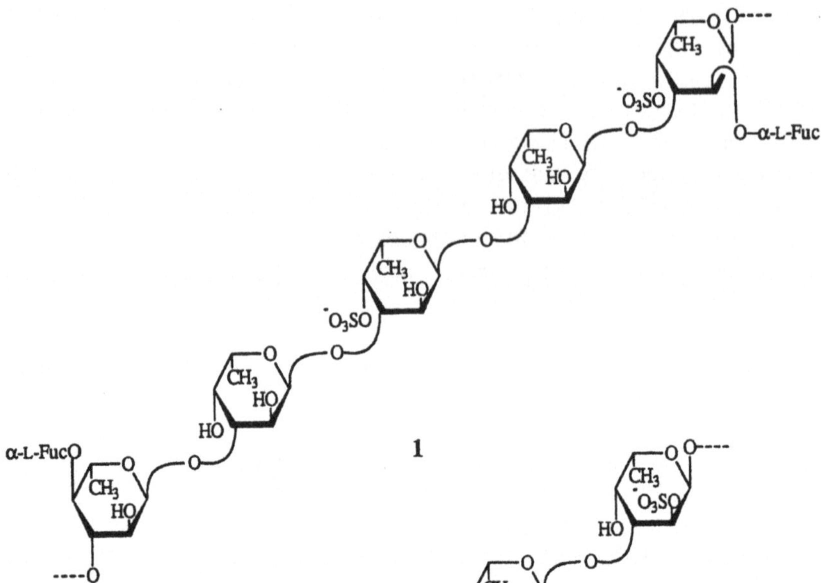

1

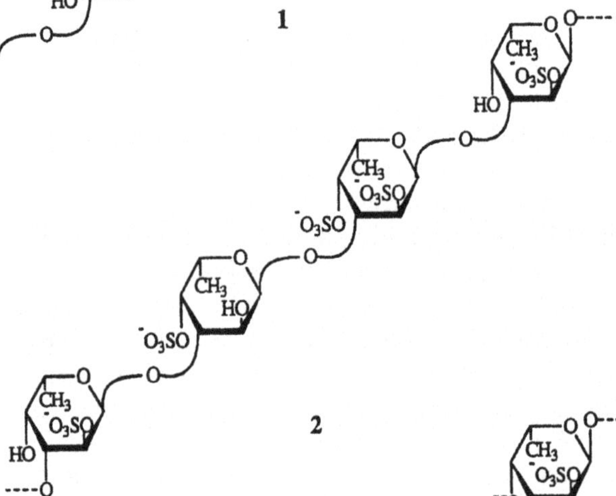

2

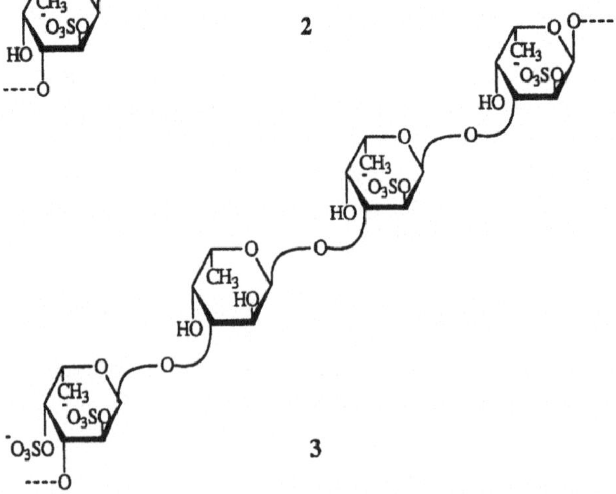

3

Figure 3.2 Structures of fucoidans from various sources: **1** = average structure of fucoidan from *Fucus vesiculosis* (Patankar *et al.*, 1993); **2** = repeating tetrasaccharide unit of fucoidan from sea urchin (Mulloy *et al.*, 1994); **3** = repeating tetrasaccharide unit of fucoidin from sea cucumber (Mulloy *et al.*, 1994).

from *Nothogenia fastigiata* (Matulewicz *et al.*, 1994). The latter inhibited replication of herpes simplex virus. Further examples include sulphated xylogalactans from *Chondria macrocarpa* (Furneaux and Stevenson, 1990) and *Corallina officinallis* (Cases *et al.*, 1992). A uronic acid-containing sulphated heteropolysaccharide from *Schizymenia dubyi* is active against several viruses and affects the growth of certain cells (Bourgougnon *et al.*, 1993; 1994). Craigie (1990) has reviewed the complex mucilages of red microalgae. Recent work has identified a 3-O-(α-D-glucopyranosyluronic acid)-L-galactose repeating unit that is common to three *Porphyridium* and one *Rhodella* species (Geresh *et al.*, 1990). Glucose 6-sulphate and galactose 3- and 6-sulphates were isolated from partial acid-hydrolysates of *Porphyridium* sp. (Lupescu *et al.*, 1991). Polysaccharide structure was found to have altered in a mutant strain resistant to the herbicide 2,6-dichlorobenzonitrile (Arad *et al.*, 1993). Pulse-chase experiments with $^{35}SO_4^-$ showed there was no sulphate turnover in the metabolism of the sulphated polysaccharide in *Rhodella maculata* (Millard and Evans, 1982).

The carbohydrate sulphates of the Chlorophyta have not been widely studied. The literature to 1983 was reviewed by Painter (1983). Recently, polysaccharides from 23 green seaweeds were partially characterised but no correlation between their anticoagulant activity and either their structure or the systematic position of the seaweed was found. A high rhamnose-containing sulphated polysaccharide from *Monostroma nitidum* was the most active (Maeda *et al.*, 1991). The sulphated rhamnogalactoarabinan from *Chaetomorpha anteninna* has a branched structure with a predominance of (1 → 4)-linkages and sulphate esters on O-2 of some arabinose units (Rao and Ramana, 1991). *Spongomorpha indica* contains a branched sulphated arabinogalactan with the arabinose present in both pyranose and furanose forms (Rao *et al.*, 1991). A sulphated xylorhamnoglucuronan from *Ulva lactuca* forms gels only with mixtures of borate and calcium ions (Haug, 1976).

Numerous heteropolysaccharides have been isolated from brown seaweeds. It is difficult to generalise about their structures as they are clearly complex mucilages. The composition of a given fraction is influenced by the extraction and purification conditions, as demonstrated by Mabeau *et al.* (1990) for *Pelvitia canaliculata*, *Fucus vesiculosus*, *Sargassum muticum* and *Laminaria digitata* subjected to four different extraction conditions. Frequently, a range of chemically related, but distinct, polymers exist within one species that can be separated using techniques such as selective precipitation, gel filtration, ion-exchange chromatography and gel electrophoresis. For example, three antitumour active uronic acid-containing fucan fractions prepared by an eight-step purification procedure from *Sargassum thunbergii* contained 10.4%–11.2% uronic acid, 28.4%–36.2% sulphate and 1.6%–2.0% protein. Their molecular weights ranged from 13 500 to 76 000. One fraction also contained some galactose, and another contained galactose and xylose (Itoh *et al.*, 1993). Electrophoretic subfractionation of one of four polysaccharide fractions from *Hizikia fusikome* revealed a 42 000 molecular weight sulphated galactofucan containing small amounts of uronic acid, xylose and mannose and a 95 000 molecular weight sulphated mannogalactofucan containing small amounts of uronic acid and xylose, both of which had anticoagulant properties (Dobashi *et al.*, 1989). Generic names such as 'sargassan' and 'ascophyllan' have been applied to certain polymer fractions obtained from, in these cases, *Sargassum linifolium* and *Ascophyllum nodium*, respectively (Abdel-Fattah *et al.*, 1974; Larsen and Haug, 1963). Whilst such fractions have been characterised using various techniques, the creation of generic names and their application to other extracts of potentially different structure may be misleading and is not to be encouraged.

The Bacillariophyceae class of algae, known as diatoms, produce a range of sulphated heteropolysaccharides which serve a variety of functions. Details of the diverse constituent sugar and sulphate contents of various diatom extracellular polymeric substances are included in a comprehensive review (Hoagland *et al.*, 1993).

A complex sulphated heteropolysaccharide fraction has been isolated from the extracellular polysaccharide of a *Cryptomonas* sp. soil alga (Paulsen *et al.*, 1992). A strain of *Pseudomonas* bacterium isolated from the brown seaweed *Undaria pinnatifida* produced a (1 → 4)-linked β-D-galactan with alternate residues substituted with either a sulphate group at the 2-position or a sulphate at the 3-position, and a β-D-glucose unit at the 6-position (Matsuda *et al.*, 1993). A different strain of the same bacterium isolated from the green seaweed *Codium fragile* produced a polysaccharide containing (1 → 3)-linked 2-*O*-acetyl-β-D-galactopyranosyl residues alternating with (1 → 2)-linked α-L-rhamnopyranosyl 2-sulphate residues (Worawattanamateekul *et al.*, 1993). Bacteria isolated from deep-sea hydrothermal vents produce complex exopolysaccharide sulphates containing a range of neutral sugars and hexuronic acids (Guezennec *et al.*, 1994).

3.2.5 Glycosaminoglycans

Glycosaminoglycans (GAGs, previously also known as mucopolysaccharides) are ubiquitous polysaccharide components of connective tissue (for an excellent review of mammalian GAGs see Fransson, 1985). They are classified into seven structural groups, heparin, heparan sulphate, keratan sulphate, dermatan sulphate, hyaluronate and chondroitin 4- and 6-sulphates. All are linear polymers of a repeating disaccharide unit composed of a 3- or 4-linked hexosamine and a 4-linked uronic acid moiety, except for keratan sulphate which has a simple hexose in place of the uronic acid. The classes are differentiated on the basis of the stereochemistry and sulphation pattern of the disaccharide units (Figure 3.3). This gives rise

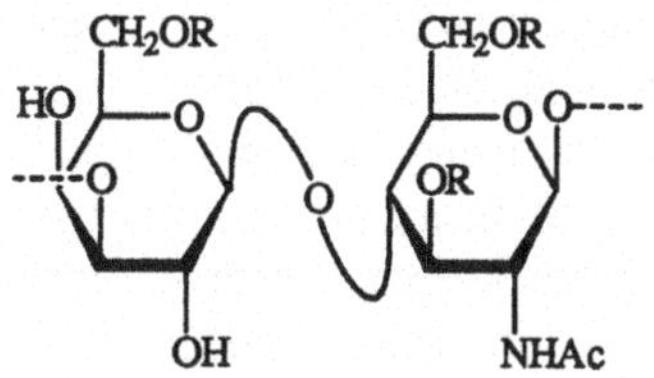

1 R=H, SO_3^-

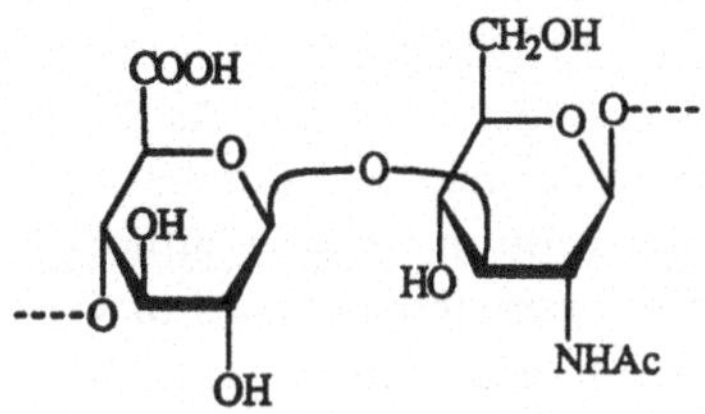

2

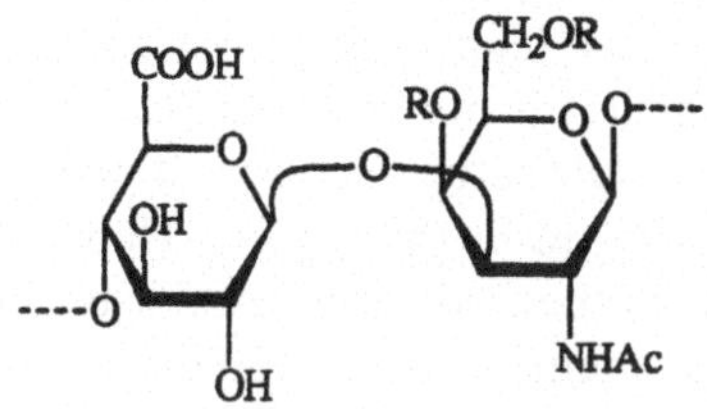

3 R=H, SO_3^-

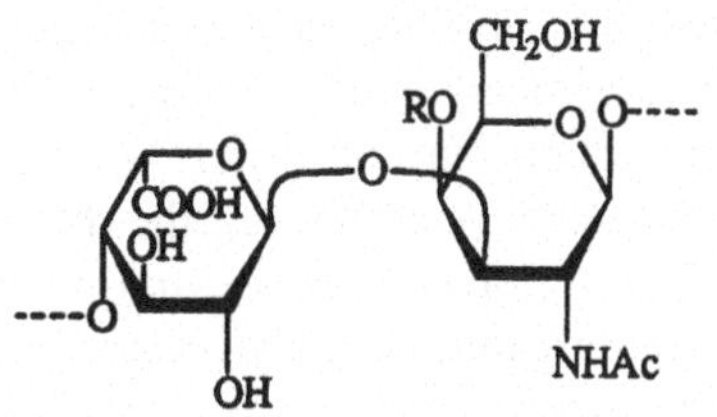

4 R=SO_3^-

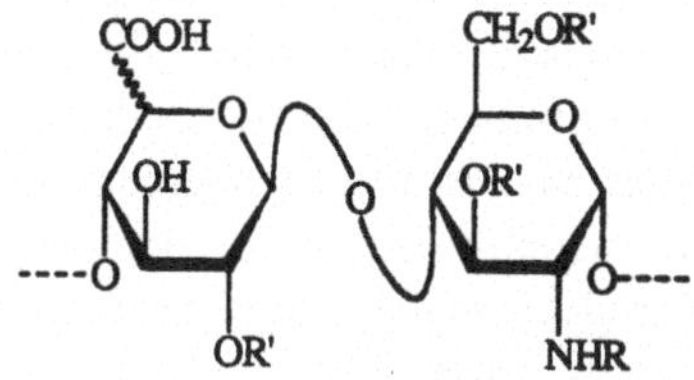

5 R=Ac, SO_3^-
R'=H, SO_3^-

Figure 3.3 Predominant disaccharide units in glycosaminoglycans (the predominant sulphated positions are given in parentheses): **1** = keratan sulphate (hexosamine C-6); **2** = hyaluronic acid (none); **3** = chondroitin 4-sulphate (hexosamine C-4); **3** = chondroitin 6-sulphate (hexosamine C-6); **4** = dermatan sulphate (hexosamine C-4); **5** = heparin and heparan sulphate (very varied).

to a group of compounds whose structures and functions are quite diverse and yet whose physical properties are broadly the same so that their isolation and characterisation is not a trivial task. GAG nomenclature is further complicated by the fact that a single chain may be made up of a number of different GAG classes (Scott, 1993).

GAGs are widely incorporated as active ingredients in a variety of health-care and pharmaceutical products such as cosmetics, dietary supplements, wound dressings, surgical fluid replacements (Van Brunt, 1986) and anti-coagulants (Linhardt, 1991). Despite this, the difficulties surrounding their isolation and characterisation, owing largely to their structural complexity, have meant that the biological roles of GAGs are only poorly understood and have only recently been studied in depth.

With the exception of hyaluronic acid and heparin, GAGs are found in nature as chains of around 40 disaccharide units covalently linked to a peptide backbone in aggregates known as proteoglycans. The name arose because the mass of the carbohydrate is generally far larger than the mass of protein, in contrast to the usual run of glycoproteins. However, with the discovery of proteins bearing a single GAG chain the term has come to be used for proteins bearing any GAG chains at all. Hyaluronic acid occurs as very high molecular weight chains, up to 12 000 disaccharide units to which it is not clear whether there is a protein attached or not. As the only non-sulphated member of the group, hyaluronic acid will not be further discussed in this review. Heparin is also an exception because, although it is synthesised on a protein backbone, like the remaining classes of GAG, the chains of up to 150 disaccharide units are subsequently cleaved off and cut down to between 10 and 40 disaccharide units by specific enzymes (Casu, 1985). The mature heparin is then stored in the mast cells from where it is released during inflammation.

In general the GAG chains are connected to the protein backbone of their proteoglycan via a single xylose residue *O*-glycosidically linked to serine. The most common exception is keratan sulphate which may be either *N*- or *O*-linked via *N*-acetylgalactosamine or mannose, depending on the tissue from which it is isolated (Fransson, 1985). There is also evidence that particular glycoproteins have heparin, heparan sulphate and possibly chondroitin sulphate chains attached to their *N*-linked complex-type oligosaccharides (Sundblad *et al.*, 1988a).

The structural complexity of GAG chains arises as a consequence of their biosynthesis (Linhardt, 1991). Once the linkage region has been synthesised, the alternating hexosamine and uronic acid residues are added sequentially by glycosyltransferases using the appropriate nucleotide activated sugars, in an analogous manner to normal oligosaccharide biosynthesis. The resulting regular precursor polymer is then modified to a varying extent by other enzymes which cause *N*-deacetylation, *N*- or *O*-sulphation or inversion of the uronic acid residue at C-5. Since none of these modifications is carried out to completion, and the extent of the modification carried out by one enzyme affects the extent of modification by the next, the result can be a very heterogenous, but non-random, collection of GAG chains on a single protein backbone. The signals which initiate the synthesis of a particular GAG-type at a particular attachment site on the protein and those for subsequent modification of the polymer are not at all understood.

Proteoglycans play a wide variety of roles in nature (for reviews see Kjellen and Lindahl, 1991; Poole, 1986) and it is not clear how much of their activity is due to the protein backbone and how much to the GAG chains. The GAGs have two major functions, to act as purely structural components of the extracellular matrix and to provide ligands for the binding of other molecules. As structural molecules they provide the organisation and tensile strength of tissues such as cartilage, the vitreous humour of the eye, synovial fluid, etc., acting generally as molecular shock absorbers. Structural properties such as hydrodynamic volume and viscosity are mediated both by charge density and by the stereochemistry of the disaccharide units, in particular the ratio of iduronic acid to glucuronic acid. The binding of other molecules can arise from specific interactions with defined species (as exemplified by the interaction of heparin with antithrombin III, see Section 3.2.5.3) or more general charge–charge type non-specific interactions, or, indeed, a combination of the two. In the case of the non-specific interactions, it is difficult to separate the structural role from the molecular recognition role. As proteoglycans are studied more intensively so the number of their known interactions with other molecules, both specific and non-specific, increases (Gallagher, 1989).

3.2.5.1 Keratan sulphate

Keratan sulphate is the only member of the group which does not contain a uronic acid residue. The repeating unit of keratan sulphate is *N*-acetyllactosamine which can be sulphated at O-6 of either the *N*-acetylglucosamine or the galactose unit (Figure 3.3). There are two major forms of keratan sulphate, structurally distinguished by their linkage regions. Type I keratan sulphate, found in the cornea and some other tissues, is *N*-linked to asparagine in the peptide backbone via a branched mannose rich core oligosaccharide. It generally has a very low level of sulphation. Type II keratan sulphate, found along with chondroitin sulphate on the proteoglycans of skeletal tissue, is *O*-linked via serine and is generally sulphated at least on O-6 of the *N*-acetylglucosamine. In shark cartilage both O-6 positions are usually sulphated.

3.2.5.2 Chondroitin sulphate and dermatan sulphate

Chondroitin sulphate and dermatan sulphate are sometimes referred to as the galactosaminoglycans because they consist of repeating units of a uronic acid linked to *N*-acetylgalactosamine (Figure 3.3). Chondroitin sulphate contains D-glucuronic acid and the galactosamine is sulphated at O-6 (chondroitin 6-sulphate) or O-4 (chondroitin 4-sulphate) and sometimes both. Dermatan sulphate contains L-iduronic acid (the C-5 epimer of D-glucuronic acid) and is completely sulphated at O-4 of the galactosamine, with occasional sulphates on O-2 of the uronic acid. In the past, chondroitin 4-sulphate was known as chondroitin sulphate A, dermatan sulphate as chondroitin sulphate B and chondroitin 6-sulphate as chondroitin sulphate C. This nomenclature persists in their degradative enzymes and commercial GAG products.

Dermatan sulphate chains invariably contain regions of chondroitin sulphate as a consequence of their biosynthesis (Figure 3.4). As mentioned

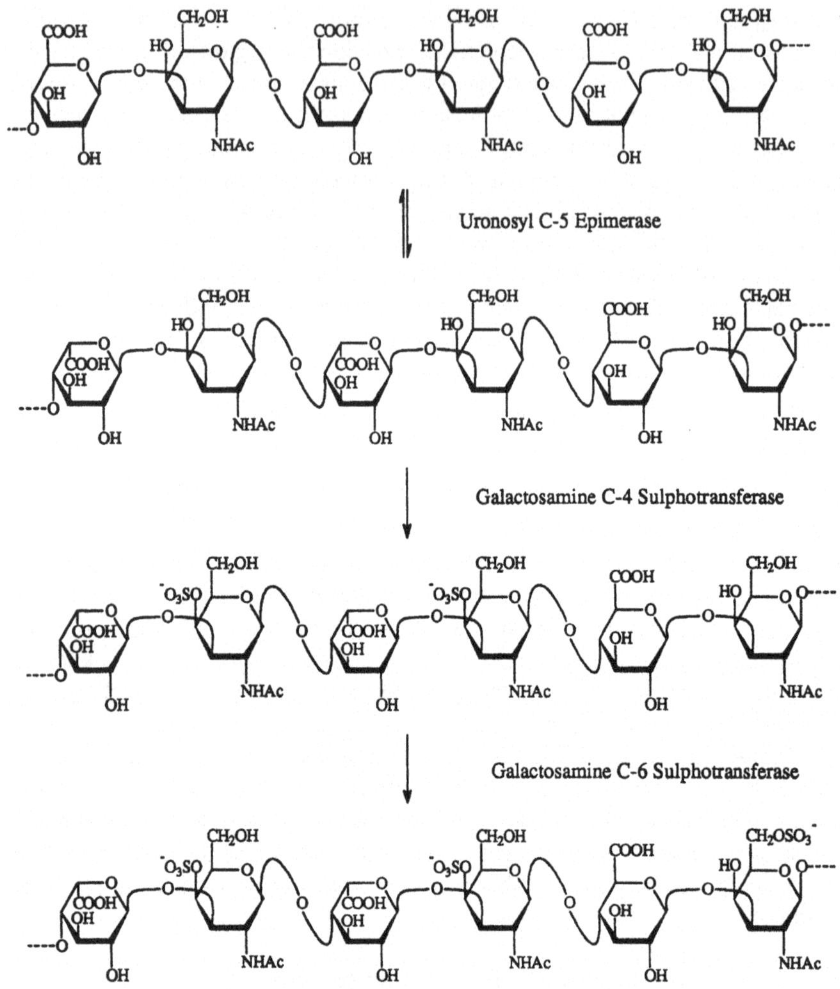

Figure 3.4 Biosynthetic scheme for dermatan sulphate.

above, GAG synthesis is initiated by polymerases that construct regular polymers on the linkage region. In the case of dermatan sulphate, the precursor polymer has the chondroitin stereochemistry. To form chondroitin sulphate, this is then sulphated at either the O-4 or O-6 position of the galactosamine. In tissues which produce dermatan sulphate, the non-sulphated polymer is the substrate for a uronosyl C-5 epimerase which sets up an equilibrium between the glucuronic- and iduronic-acid forms in which the glucuronic acid dominates. Sulphation at O-4 of the galactosamine attached to the iduronic acid pulls these regions out of the equilibrium because they are no longer recognised by the epimerase. This accounts for the almost complete sulphation of dermatan sulphate at O-4 of the galactosamine residue. The remaining glucuronic-acid-containing disaccharide units may then be sulphated at O-4 and O-6 of the galactosamine residues to varying degrees. The mechanism of control of this sequence is unknown but the dermatan segments generally occur in blocks, the location of which in the chain is roughly constant in a particular proteoglycan from a particular source (Cheng *et al.*, 1994).

Chondroitin sulphate chains are found on a wide variety of proteoglycans, especially in association with keratan sulphate in cartilage and with dermatan sulphate in the skin, where the chondroitin sulphate chains contribute to the organisation and resilience of the extracelluar matrix. The ratio of chondroitin 4-sulphate to chondroitin 6-sulphate in the GAG chains varies in a tissue-specific and species-specific manner. So, for example, in the cartilage of bovine trachea, the ratio of 4- to 6-sulphate is around 60 : 40, whereas in shark cartilage the ratio is 30 : 70. Shark cartilage chondroitin sulphate is also notable for its high level of disaccharide units sulphated at both O-4 and O-6 of the galactosamine.

Dermatan sulphate is largely found in the proteoglycans of skin and basement membrane. While chondroitin sulphate displays very weak protein binding properties, and has no proven biological activity beyond its structural role, dermatan sulphate acts as an antithrombotic agent and anticoagulant agent and binds strongly to plasma molecules such as platelet factor 4, heparin cofactor II, fibronectin, etc. These interactions are nonspecific charge–charge binding events and, since chondroitin sulphate and dermatan sulphate have a similar charge density, it is the configuration of the uronic acid which accounts for the difference (Casu *et al.*, 1988). Glucuronic acid is found in the stable $^4C_1$ chair conformation whereas iduronic acid exists in an equilibrium between the $^1C_4$ chair and $^2S_0$ skew forms with a small contribution from the $^4C_1$ chair (Figure 3.5). The position of the equilibrium is affected by any sulphates on the iduronic acid and flanking hexosamines because of electrostatic interactions between them and the carboxyl group. The net effect of this is that iduronic-acid-containing polymers are much more flexible than the equivalent glucuronic-acid-containing polymers, as evidenced by their better binding of divalent metal ions. Their

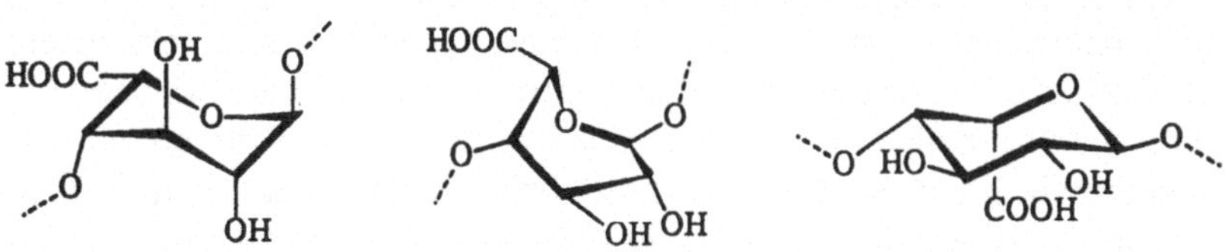

A **B** **C**

Figure 3.5 $^1C_4$ (**A**), $^2S_0$ (**B**) and $^4C_1$ (**C**) conformations of iduronic acid.

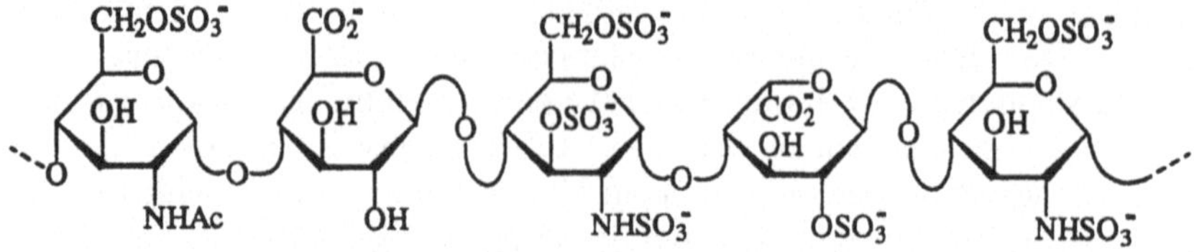

Figure 3.6 Antithrombin III recognition site in heparin.

greater flexibility allows them to interact better with the basic groups on tissue and plasma receptor molecules. The same argument is used for the greater biological activity of heparin over heparan sulphate.

3.2.5.3 Heparan sulphate and heparin

Heparan sulphate and heparin are glucosaminoglycans, like keratan sulphate. They consist of repeating units of uronic acid linked to *N*-acetylglucosamine (Figure 3.3). Heparin and heparan sulphate are made by essentially the same processes but whereas heparin is only synthesised in mast cells heparan sulphate is widely distributed through epithelial tissues. The structural distinctions between the two groups are essentially that heparin is more highly *N*-sulphated (Lindahl and Kjellen, 1991), has a higher iduronic acid to glucuronic acid ratio and has a significant number of 3-*O*-sulphated *N*-sulphated glucosamine residues (Gallagher and Walker, 1985). This last group is an essential part of the antithrombin III binding region of heparin (Figure 3.6) which accounts for the bulk of its anticoagulant properties. There is, however, some overlap between the properties of heparin and heparan sulphate. Again, the variation in the structures arises as a consequence of biosynthesis. The initial biosynthetic product for both GAGs is a regular polymer of glucuronic acid linked to *N*-acetylglucosamine, joined to a protein backbone by a specific sequence of sugars. The protoheparin chains have a molecular weight of around 60 000. The key step is the subsequent *N*-deacetylation/*N*-sulphation which occurs at more than 80% of the available sites in heparin and between 40% and 50% of the sites in heparan sulphate. The following C-5 epimerisation of D-glucuronic acid to L-iduronic acid, iduronic acid 2-*O*-sulphation, and glucosamine 6-*O*- and 3-*O*-sulphation are concentrated around the *N*-sulphated residues. As before, none of these reactions goes to completion and the extent of each is dependent on the previous reaction, giving rise to a very diverse final product. The mechanisms that control these reactions are not well understood (Campbell *et al.*, 1994; Wlad *et al.*, 1994). Heparan sulphate chains remain attached to their protein backbone but the heparin chains are removed by specific enzymes and cut down to around a third of their original length.

Heparin is the most widely known of the GAGs because of its general use since 1937 as an anticoagulant drug (Linhardt, 1991). This activity is largely due to the specific interaction of the pentasaccharide sequence, illustrated in Figure 3.6, with the protein antithrombin III. Antithrombin III is a regulatory protein which binds to the proteinase thrombin, preventing it from cleaving fibrinogen into fibrin and initiating the clotting process. Heparin strongly potentiates the activity of antithrombin in preventing blood clotting. The antithrombin III binding site makes up around 5% of the total weight of commercial heparin preparations and is present on approximately one chain in three. There has been considerable interest in the chemical synthesis of the antithrombin III binding pentasaccharide, both in the natural form and as modified derivatives. Simpler derivatives, that retain the binding activity but which are more readily synthesised than the natural binding site, show promise as anticoagulant drugs as they do not have the side effects of heparin's other biological activities (van Boeckel and Petitou, 1993).

Heparin also binds with a large number of other circulatory proteins, including lipoproteins, platelet factor 4 and lipoprotein lipase, which gives it a variety of biological effects (for a review, see Casu, 1985). The most important of these, besides anticoagulation, are antithrombotic properties and the lowering of circulating lipid levels. Considerable effort is being put into separating the pharmacological activities of heparin so that these other activities can be used without the complications of patient bleeding caused by its anticoagulant activity. Low molecular weight heparins, which have been chemically or enzymatically produced by the degradation of natural heparin, are finding increasing use as antithrombotic agents with lowered anticoagulant effects.

Since the level of heparin circulating in the plasma is very low and heparin is only released by mast cells under specific conditions, such as anaphylactic stress and inflammation, it is not clear how relevant its anticoagulant properties are to its physiological role. In fact, the role of heparin in inflammation has yet to be elucidated.

Heparan sulphate, in contrast, is widely distributed in the proteoglycans of plasma components and endothelial cell walls (Gallagher *et al.*, 1986). It also binds to a large number of other plasma components by both non-specific charge–charge and specific sequence recognition interactions. In particular, two fibroblast growth factors have been shown to recognise a hexasaccharide present in both heparin and heparan sulphate (Tyrrell *et al.*, 1993). It is likely that heparan sulphate proteoglycans have a major role in the control of blood-flow properties such as viscosity and prevention of clotting on vessel linings.

3.2.5.4 Analysis of glycosaminoglycans

The diversity of GAG species has made their analysis a complex task. Proteoglycans are generally isolated from tissue by extraction with guanidinium chloride or other chaotropic agents to destroy protein secondary structure and prevent proteolysis. They are separated by isopycnic centrifugation and may be identified by their chromatographic behaviour or immunochemical methods (Lyon, 1993). The GAGs are then freed by specific enzymes which cleave the glycosidic linkage to serine or a general protease digest. Alternatively, the GAGs may be released from the tissue directly by protease digestion if there is no desire to analyse their attached protein.

The molecular weight of intact GAG chains can be determined by standard means such as gel permeation chromatography, low-angle laser light scattering or viscometry (Ahsan *et al.*, 1994), although the standards used in this work must be chosen with care because of the highly charged and linear nature of GAGs. The charge density and carboxyl to sulphate ratio are determined by potentiometric titration or elemental analysis (Volpi, 1994). Proton (Ludwig-Baxter and Perlin, 1991) and particularly carbon-13 (Bociek *et al.*, 1980; Cockin *et al.*, 1986) nuclear magnetic resonance (NMR) spectroscopy can be used to characterise GAG components in favourable cases, but structurally diverse species such as heparin and heparan sulphate give very complex spectra.

The most complete assignment of GAG composition is carried out by degradation of the GAG chains by enzymic or chemical means and identification of the fragments. In the case of the galactosaminoglycans, chondroitin sulphate and dermatan sulphate, enzymes which will degrade the polymers into disaccharide units are available (Slim *et al.*, 1994). The enzymes are lyases which cleave the glycosidic bond between the hexosamine and the uronic acid by elimination to leave a 4,5-unsaturated uronic acid residue (Figure 3.7). This has the advantage that the unsaturated acid group provides

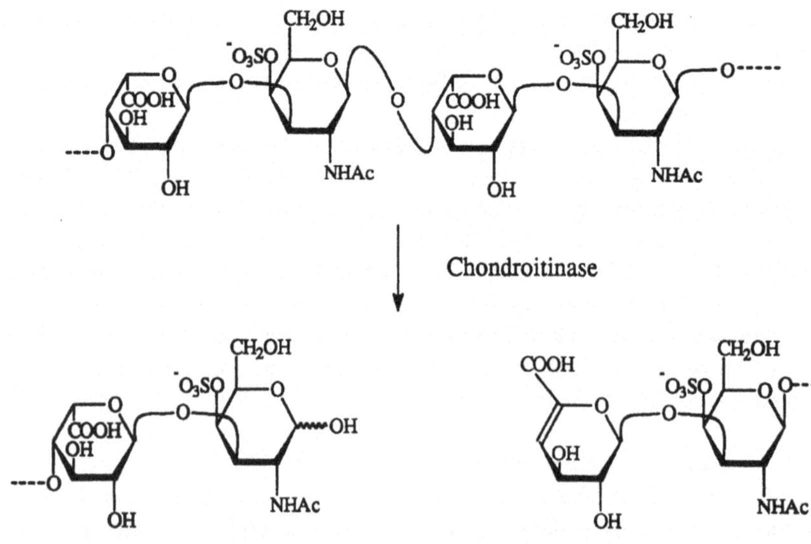

Figure 3.7 Enzyme cleavage of galactosaminoglycans.

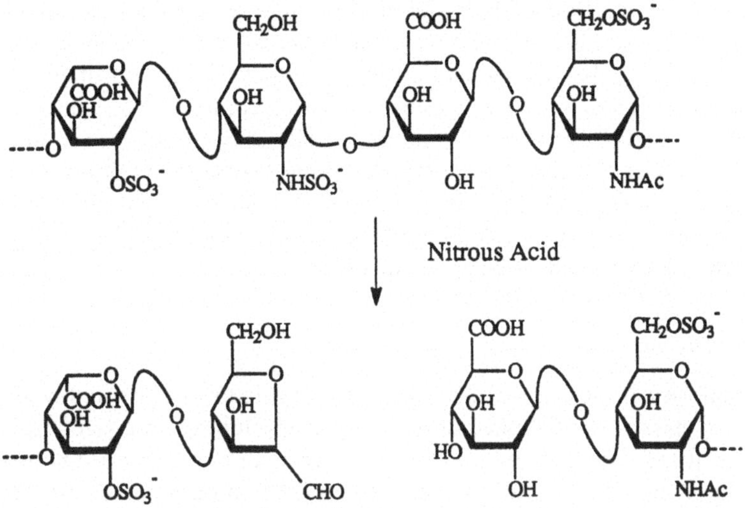

Figure 3.8 Nitrous acid cleavage of heparin.

a convenient ultraviolet (UV) chromophore. The resulting disaccharides can be separated by anion exchange high-performance liquid chromatography (HPLC) and identified by comparison with standards or, where standards are not available, by NMR, mass spectrometry and other standard techniques (Sugahara *et al.*, 1994).

For heparan sulphate and heparin the available enzymes are more specific for particular sequences than the chondroitinases. They generally leave large amounts of oligosaccharides of 10 or more units which are much harder to assign structurally (Jandik *et al.*, 1994) so that chemical degradation is often preferred. Treatment with nitrous acid causes ring contraction of *N*-sulphated and *N*-deacetylated glucosamine residues with concomitant chain cleavage to leave a terminal 2,5-anhydromannose unit (Figure 3.8) which can be conveniently radiolabelled and used as a marker for agarose gel electrophoresis or HPLC (Conrad, 1993). This method also depends on the ability to obtain suitable standards which are not readily available. Nitrous acid degradation is a stoichiometric process so that the size of the oligosaccharides produced can be controlled by the amount of reagent used. Capillary zone electrophoresis at low pH values is a particularly sensitive technique for the separation of natural and synthetic heparin oligosaccharides with molecular weights of 3500 or less (Damm *et al.*, 1992).

Recently, these methods have been elaborated into the technique of oligosaccharide mapping whereby the intact GAG chain is labelled on one end and treated with a specific depolymerisation agent. Size analysis of the resulting fragments allows analysis of the position of the specific feature recognised by the depolymerisation technique (Turnbull, 1993). So, for example, the location of iduronic acid residues in a galactosaminoglycan chain can be established by cleaving the GAG from the protein by alkaline hydrolysis and labelling the end group by reduction with borotritiide. The labelled polymer is partially digested with chondroitinase AC, which cleaves the glycosidic linkage to glucuronic acid residues but not those to iduronic acid, and the size of the fragments is determined by gradient polyacrylamide gel electrophoresis. Approaches such as this have shown that iduronic acid regions occur in specific patterns in the galactosaminoglycan chains of different proteoglycans (Cheng *et al.*, 1994).

3.2.6 Glycoproteins

The majority of the oligosaccharides of glycoproteins can be roughly divided into two classes: the *O*-linked or mucin type which have an *N*-acetylgalactosamine at the reducing end attached to a serine or threonine of the protein, and the *N*-linked type which have a common mannose rich core attached to asparagine via *N*-acetylglucosamine (Kornfeld and Kornfeld, 1985; see also Section 5.2). Very broadly speaking, the *N*-linked oligosaccharides have a complex branching structure and are involved in cell surface recognition events whereas the *O*-linked oligosaccharides consist of simpler chains or multiple repeats of simple sequences and confer particular physicochemical properties to the glycoprotein (Kobata, 1993). In general, the *O*-linked oligosaccharides are more likely to be sulphated, but many exceptions to these generalisations are being found as more oligosaccharides are identified and their biological roles elucidated.

Sialylation (substitution with neuraminic acid) and sulphation very often occur together in the same group of oligosaccharides. The biological roles of sulphate groups and sialic acids may be linked, presumably by virtue of the fact that they both display negatively charged binding sites for molecular recognition events. Oligosaccharides can be substituted at the same sites by either of these substituents and there may be competition between the two for these sites during the biosynthesis of mucins, which are heavily sialylated and sulphated, and glycoprotein hormones (Thotakura and Blithe, 1995). It has been shown that the cell adhesion molecule E-selectin, which binds to the tetrasaccharide sialyl Lewis X, binds equally strongly to the ligand in which the sialic acid has been replaced by a sulphate group (Yuen *et al.*, 1994).

3.2.6.1 O-*Linked sulphated oligosaccharides*

The archetypal *O*-linked oligosaccharide-substituted glycoproteins are the mucins, which make up the bulk of the mucus excreted by the cells of mucous membranes. Sulphated oligosaccharides have been isolated from gastrointestinal (Roberton *et al.*, 1993), respiratory (Sangadala *et al.*, 1993) and salival (Piotrowski *et al.*, 1994) mucus. The sulphate groups, along with other negatively charged substituents such as sialic acids and phosphates, serve to control the pH of the medium in contact with the cells making up the membrane and also to prevent the binding of pathogens to the cells by providing a simple physical barrier or alternative oligosaccharide binding sites (Piotrowski *et al.*, 1994). Despite the importance of sulphomucins as a first line of defence against pathogens very little is known about the structure of the sulphated oligosaccharides they contain.

Sulphated oligosaccharides have been isolated from the respiratory mucus of cystic fibrosis patients, who express elevated levels of sulphomucins. Cystic fibrosis is caused by a genetic defect in chloride and sodium ion transport across cell membranes. The relationship between the increase in mucus production and the genetic defect is not obvious but it has been speculated that the increased levels of sulphomucin give enhanced protection against pathogens as compared with normal mucin. In one study, sulphated oligosaccharides were isolated from a cystic fibrosis patient and assigned as containing galactose 4-sulphate and galactose 6-sulphate residues at the non-reducing termini with internal *N*-acetylglucosamine 6-sulphate and galactose 6-sulphate residues (Mawhinney and Chance, 1994) by fast atom bombardment mass spectroscopy and glycosidase degradation. In another study, the only sulphated residues found were terminal galactose 3-sulphate and internal *N*-acetylglucosamine 6-sulphate as determined by NMR spectroscopy (Lo-Guidice *et al.*, 1994). It is worth noting that of the 40 neutral and acidic oligosaccharides isolated by one group and the 24 isolated by the other none were found to be in common. The reasons for this apparent variation are not clear.

Gastrointestinal mucus glycoproteins have been shown to contain sulphated galactose, glucose and *N*-acetylglucosamine, but without the positions of the sulphate groups being clearly established. The ion exchange properties of mucus are particularly important in the gastrointestinal tract as they help prevent stomach acid damaging the gut lining. It has been demonstrated that bacteria such as *Helicobacter pylori*, which cause peptic ulcers, bind to acidic oligosaccharide moieties on the cell wall (Telford *et al.*, 1994) and that this can be prevented by sulphomucins (Piotrowski *et al.*, 1991).

3.2.6.2 N-*Linked sulphated oligosaccharides*

The structure and biosynthesis of *N*-linked sulphated oligosaccharides has received far less attention than have GAGs or *O*-linked oligosaccharides, presumably because of their comparatively low levels in tissue. More recently, however, sulphated asparagine-linked oligosaccharides have been isolated from a number of sources, including viral coat proteins, mammalian leutropins, cell adhesion molecules, pituitary hormones and lysosomal enzymes and other glycoproteins with important regulatory functions (Freeze and Wolgast, 1986). Few have been fully characterised; mannose

6-sulphate, mannose 4-sulphate, *N*-acetylgalactosamine 4-sulphate and sulphated *N*-acetylglucosamine residues have been identified in *N*-linked oligosaccharides from various tissues (Roux *et al.*, 1988). The range of structures is very varied, indicating that there are a number of sulphating enzymes in different tissues (Yamashita *et al.*, 1983). The extent and distribution of sulphation on asparagine-linked oligosaccharides from cultured cell lines also vary with the culture conditions and the phase of cell growth (Sundblad *et al.*, 1988b).

The biological significance of sulphated *N*-linked oligosaccharides is largely unknown. The fact that they are being found to be so widely spread, albeit at relatively low levels, and so heterogeneous, may indicate that they play a role in cell differentiation and development. The wide variety of structures available from the post-translational glycosylation and sulphation of proteins gives great scope for specific structure–function relationships.

3.2.7 Miscellaneous sulphated carbohydrates

3.2.7.1 Sulphated glycolipids

Sulphatides, cerebrosides bearing a sulphate ester, usually on O-3 of a terminal galactose moiety, are found in brain and nerve tissue. In the inherited disease metachromatic leukodystrophy they accumulate in neural and other tissue owing to a deficiency in a catabolic sulphatase enzyme. Seminolipids, which have a sulphated oligosaccharide moiety attached at O-3 of a 1-*O*-alkyl-2-*O*-acyl or 1,2-di-*O*-alkyl-glycerol moiety, are found in human gastric secretion, testes and spermatozoa (Gigg, 1988). Sulphated glycolipids isolated from the human peripheral nervous system (Ilyas *et al.*, 1984) have been shown to contain 3-*O*-sulphated glucuronic acid residues on a paragloboside core structure (Figure 3.9) (Nakano *et al.*, 1993). Sulphated glycolipids with oligosaccharide structures similar to the *O*-linked glycoproteins have also been shown to be ligands for cell–cell recognition events (Krivan *et al.*, 1989).

Saponins are glycosylated terpenes that form a stable soapy foam when shaken with water. They have been isolated from a wide variety of plants

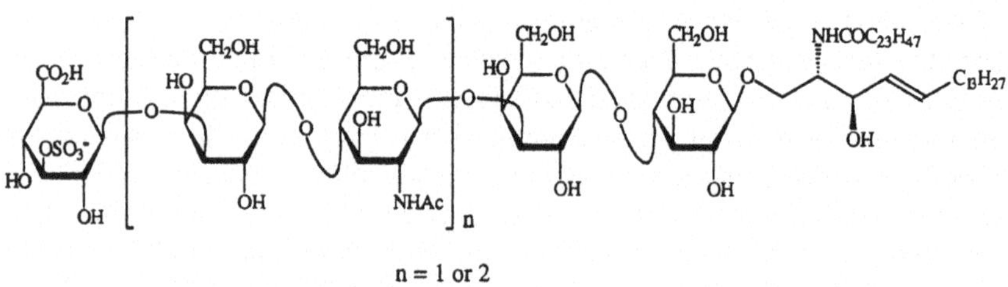

Figure 3.9 Sulphated glycosphingolipid from human peripheral nervous system.

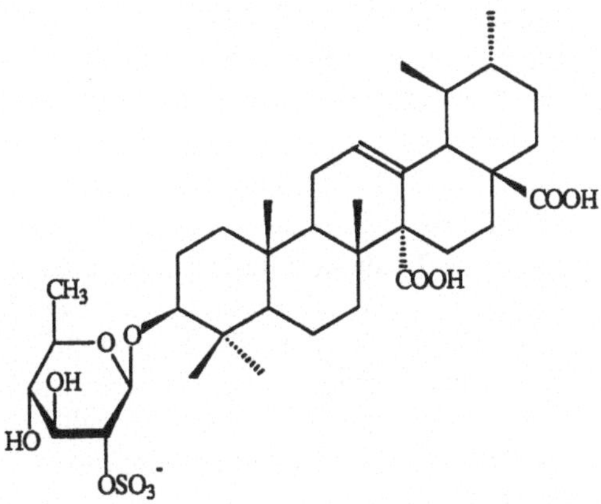

Figure 3.10 Sulphated triterpenoid saponin from *Zygophyllum propinquum*.

and from some marine organisms. Saponins are generally divided into two classes, triterpenoid (for a review, see Mahato *et al.*, 1988) and steroidal (for a review, see Mahato *et al.*, 1982) on the basis of their aglycone structure. Both the aglycone and the sugar moiety may be sulphated. The saponins from marine organisms such as starfish and sea cucumbers are most commonly sulphated on the glycoside, for example see Figure 3.10 (Ahmad *et al.*, 1993). The first isolation of a sulphated saponin from a plant source was reported in 1983 by Watanabe *et al.* The isolation and structural elucidation of saponins have received a great deal of attention because of their diverse pharmacological activity, including anti-tumour, antibiotic and cholesterol-lowering effects, but their physiological role in the organism remains unclear.

The signalling factors from *Rhizobium* bacteria which initiate root nodule formation in leguminous plants (Nod factors) are lipo-oligosaccharides such as that shown in Figure 3.11 (Lerouge *et al.*, 1990). The host specificity is determined, at least in part, by the sulphation pattern of the oligosaccharide.

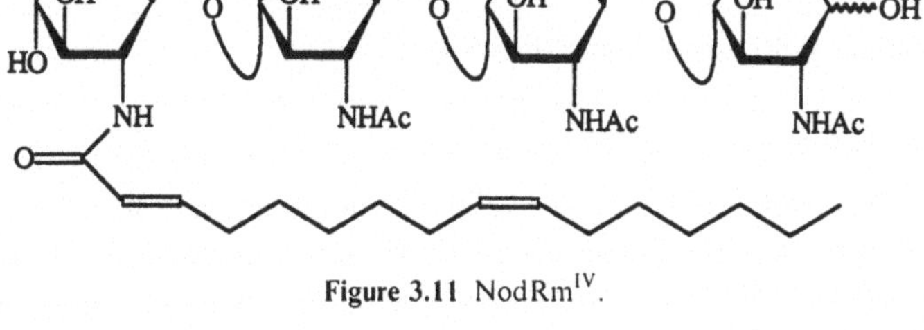

Figure 3.11 NodRm^{IV}.

Removal of the 6-sulphate from the reducing sugar of NodRm^{IV} stops it from nodulating alfalfa, the host plant of *Rhizobium meliloti* from which it derives, but makes it capable of nodulating vetch, a non-host plant (Tailler *et al.*, 1994).

3.2.7.2 Sulphated glycosides

Benzyl-β-D-glucopyranoside 2-sulphate has been isolated from the stem of the plant *Salvadora persica* L. (Salvadoraceae) (Kamel *et al.*, 1992) and the biologically active 4-sulphated 2-acetamido-2-deoxy-β-D-glucopyranoside, bulgecin A, has been isolated from a *Pseudomonas* species (Shinagawa *et al.*, 1984). 4-*O*-(-β-D-Glucopyranosyl 6-sulphate)-gallic acid is a periodic leaf movement factor (Neier-Augenstein and Schildknecht, 1990).

3.2.8 Differentiation of sulphate and phosphate groups in biological systems

Sulphate and phosphate groups are isosteric, but have significantly different pK_a values. Interchanging these groups can have a significant impact upon the biological properties of the carbohydrate derivative concerned. Replacement of the 6-sulphate group on the non-reducing terminal 2-*N*,6-*O*-disulphated D-glucosamine of an anti-thrombin III binding pentasaccharide (Section 3.2.5.3) with a 6-phosphate group, eliminated activity. Similarly the trisulphated analogue of the secondary messenger D-myo-inositol 1,4,5-trisphosphate is biologically inactive. The ability of protein receptors to differentiate sulphate and phosphate has been ascribed to the greater energy requirements for dehydration of phosphate and the different networks of hydrogen bonding (van Boeckel and Petitou, 1993; Vos *et al.*, 1991); in the case of the sulphate and phosphate binding proteins that strip away water of solvation and bind SO_4^{2-} and $HOPO_3^{2-}$ ions themselves, there is a hydrogen bond acceptor group in the binding site of the former but not the latter (Luecke and Quiocho, 1990). Glucose-6-phosphate dehydrogenase will accept glucose 6-sulphate as a substrate. Although the specific activity of the enzyme is much lower with this substrate, the combination is advantageous for NAD(P)H cofactor regeneration in enzyme-catalysed ketone-to-alcohol reductions because, unlike glucose 6-phosphate, glucose 6-sulphate does not catalyse the decomposition of NAD(P)H (Wong *et al.*, 1981).

3.3 Chemical synthesis and modification

3.3.1 Sulphation reagents

The strategies and reagents employed for the synthesis of carbohydrate sulphates have changed little since they were reviewed 30 years ago (Turvey, 1965). The reagents most frequently employed for sulphation are complexes of sulphur trioxide with a tertiary amine (e.g. trimethylamine or pyridine) or an amide (formamide or dimethylformamide), in pyridine, dimethyl sulphoxide or dimethylformamide as solvent. The vigour of such reagents is inversely proportional to the strength of the complexing Lewis base. The complex with trimethylamine is sufficiently stable that it can be used in water, as in the high-yielding synthesis from L-ascorbic acid of its 2-sulphate, which is of interest as a more stable, bioavailable vitamin C equivalent (Lillard and Seib, 1978). Other sulphation reagents that can be used are chlorosulphonic acid or piperidine-*N*-sulphonic acid in a polar aprotic solvent.

Sulphation with pyridine–sulphur trioxide complex in pyridine is an equilibrium reaction, as demonstrated by the conversion of methyl β-D-galactopyranoside 2,3-disulphate into a mixture of mono-, di- and tri-sulphated species on exposure to this reagent system. The equilibrium is established quite rapidly (30 min at 60°C), even for hindered hydroxy groups (Harris and Turvey, 1969). Radiolabelled monosaccharide sulphates can be prepared using the pyridine–[$^{35}SO_3$] complex (Jatzkewitz and Nowoczek, 1967).

Sugars are sulphated in concentrated sulphuric acid alone (Turvey, 1965), although the reaction has found only occasional utility, e.g. in the preparation of L-ascorbic acid 6-sulphate accompanied by a small proportion of the 5-sulphate (Lillard and Seib, 1978). The combination of dicyclohexylcarbodiimide and sulphuric acid gave predominantly galactoside 6-sulphates (Takano *et al.*, 1992b).

A conceptually attractive strategy, that has received little attention, involves the formation, using phenylchlorosulphate and sodium hydride, of mixed sulphate diesters ($PhOSO_3$–sugar). These mixed diesters are non-ionic and sufficiently stable to survive further chemical reactions and be purified by conventional chromatography. They can be converted into carbohydrate sulphates on hydrogenolysis (over platinum) at the final stage (Penney and Perlin, 1981).

3.3.2 Sulphated free sugars and their glycosides

Sulphate monosaccharides and their glycosides have been prepared either by selective sulphation of the unprotected sugars or by a protection–sulphation–deprotection sequence in which only the hydroxy group to be sulphated is specifically exposed.

3.3.2.1 Selective sulphation

In reactions involving unprotected free sugars the anomeric hydroxy group is probably sulphated to some extent. The glycosyl sulphates, so produced, are spontaneously hydrolysed on exposure to water, however, so that such products are not isolated. Of the remaining hydroxy groups, the primary hydroxy group is the most reactive. 6-Sulphate esters of hexoses and hexopyranosides can be formed selectively, along with small amounts of other mono- and di-sulphates. The selectivity between the 3′- and 5′-hydroxy groups in 2′-deoxynucleosides is poor, and mixtures of mono- and di-esters result (Chang *et al.*, 1967). Chlorosulphonic acid in pyridine appears to be the most selective reagent for mono-sulphation.

In the absence of a primary hydroxy group a complex mixture usually results. This is the case for fucose, although under optimised conditions L-fucose 2-, 3- and 4-sulphates can be isolated by ion-exchange chromatography (Forrester *et al.*, 1976). α,α-Trehalose 2-sulphate, the sugar moiety of the principal sulphatide in *Mycobacterium tuberculosis*, was obtained as the major product, along with the 3-sulphate as the minor product, on sulphation (pyridine · SO_3 in pyridine) of trehalose bis-4,6-*O*-benzylidene acetal followed by deprotection (Liav and Goren, 1984). The amino group in free amino sugars can be preferentially *N*-sulphated.

The major difficulty associated with selective sulphation is the need to separate the major product from the accompanying regioisomers and under- and over-sulphated products. In the particular case of the model anionic surfactants, dodecyl α- and β-D-glucopyranoside 6-sulphate and tetradecyl β-maltopyranoside 6- and 6′-sulphate (prepared using pyridine · SO_3), it was possible to use reversed-phase HPLC (Böcker *et al.*, 1992).

3.3.2.2 Specific sulphation

For protection–sulphation–deprotection sequences to be successful, protecting groups for hydroxy, carboxylic acid and amino functions that are stable to the sulphation reagents and that can be removed under conditions that avoid cleavage of the sulphate ester must be used. Certain cyclic acetals (benzylidene, isopropylidene), esters (acetate, benzoate) and ether groups (trityl, benzyl) are favoured for *O*-protection; they can be removed under mild acid or alkaline conditions or by hydrogenolysis, respectively. *N*-Benzyloxycarbonylation or the use of azide groups is favoured for *N*-protection.

Trimethylamine-sulphur trioxide in dimethylformamide has been found to be particularly effective in the preparation of glycoside mono-sulphates (Rashid *et al.*, 1990). The purification of both protected and unprotected carbohydrate sulphates has been improved by the use of chromatography on silica gel with polar eluants to replace the traditional, more time-consuming, ion exchange with eluants containing volatile buffer salts (e.g. NH_4HCO_3) (Contreras *et al.*, 1991; Rashid *et al.*, 1990).

Numerous and increasingly complex examples of the use of this strategy for the synthesis of carbohydrate sulphates are catalogued in *Carbohydrate Chemistry*, a review published annually since 1966 by the Royal Society of Chemistry (see particularly the chapters on esters). Particularly noteworthy accomplishments are the synthesis of the highly sulphated heparin pentasaccharide that is the minimum ligand for antithrombin III binding (Figure 3.6) and related structures, and simplified analogues (e.g. Figure 3.12) as

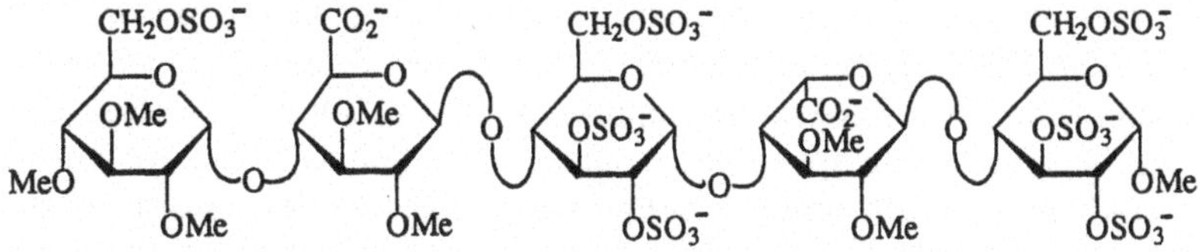

Figure 3.12 Simplified analogue of the minimum antithrombin III binding site of heparin.

potential synthetic antithrombotics (van Boeckel and Petitou, 1993), and the trisulphated tetrasaccharide fragment (Figure 3.13) of keratan sulphate (Kobayashi *et al.*, 1990).

3.3.2.3 Polysulphation of mono- and oligo-saccharides

The efficient preparation of derivatives with more than one sulphate group is potentially difficult because of the equilibrium nature of the sulphation reaction and the limited solubility of the sulphation reagent in some cases (e.g. pyridine·SO_3 in pyridine) (Harris and Turvey, 1969). Nevertheless, sulphated long-chain-alkyl glycosides of malto- and laminaro-oligosaccharides of 5–11 monosaccharide units (with between two and three sulphate groups per sugar residue) were obtained by use of either piperidine-*N*-sulphonic acid in dimethyl sulphoxide (DMSO) or pyridine–sulphur trioxide in pyridine. These materials were potent inhibitors of HIV (Katsuraya *et al.*, 1994). Similarly, crystalline sucrose octa-sulphate salts can be made using pyridine–sulphur trioxide complex in dimethyl formamide (DMF) (Ochi *et al.*, 1980), and deca- and nona-sulphates of rutin (a flavonoid disaccharide) have been reported (Nair *et al.*, 1987).

3.3.3 Sulphated polysaccharides

A wide variety of natural polysaccharides and quite a few chemically branched or modified polysaccharides have been subjected to partial or complete sulphation; for example, cellulose, xylan, dextran, amylose, curdlan

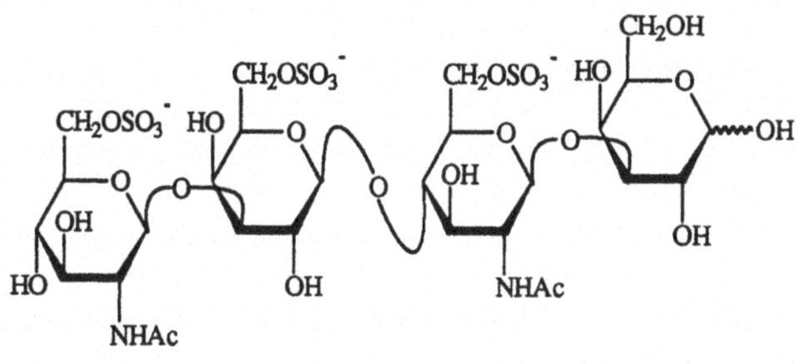

Figure 3.13 Synthetic keratan sulphate tetrasaccharide.

and even the gums frankincense and myrrh (see e.g. Doctor *et al.*, 1991; Osawa *et al.*, 1993; Yamamoto *et al.*, 1991; see also Sections 8.2 and 8.5).

Much of the effort devoted to preparing synthetic sulphated polysaccharides has been in an attempt to find a substance of defined structure with the anticoagulant or antithrombotic properties of the complex natural product heparin but without its adverse side effects (Casu, 1985). Additionally, many sulphated polysaccharides, both natural and synthetic, interfere with viral replication and infectivity in *in vitro* systems (see e.g. Osawa *et al.*, 1993). So far, synthetic products have not displaced heparin or its low molecular weight derivatives from clinical use and the challenge to translate *in vitro* antiviral activity into useful *in vivo* activity remains.

Chitosan 2-*N*,6-*O*-sulphate, partially *N*,*O*-sulphated derivatives of a 2-aminated polysaccharide prepared from amylose, and a C-6 oxidised chitosan *N*-sulphate were prepared because they have units that mimic the *N*-sulphated amino-sugar and sulphated uronic acid units of heparin (Horton and Usui, 1978). Dextran and xylan sulphates have useful antithrombotic properties, acting by a non-antithrombin-mediated mechanism; their greater effectiveness may be a result of the fact that they lack primary sulphate esters that tend to be involved in non-specific binding to a wide variety of proteins (Casu, 1994). Low molecular weight dextran sulphate has been given orally to patients as an anticoagulant and antilipemic agent in Japan for more than 25 years (Neville *et al.*, 1991).

Experimental difficulties are encountered in the sulphation of polysaccharides because of their poor solubility in most sulphation reagent systems and their susceptibility to chain cleavage during reaction. The reactivity of a polysaccharide can be enhanced, prior to exposure to a heterogeneous reaction system, by precipitation from aqueous solution with solvent, followed by distillative solvent drying (Guiseley, 1978). The use of the vigorous sulphation reagent chlorosulphonic acid increases the degree of sulphation but causes molecular weight reduction, particularly if some moisture is present. This is thought to be because of the acid generated in the reaction, even in the presence of pyridine as a base. Favoured reagents for polysaccharide sulphation are DMSO-SO_3, DMF-SO_3 or piperidine-*N*-sulphonic acid (Kennedy, 1974).

Sulphation of cellulose can be conducted in a homogeneous phase by first making a solution of cellulose trinitrite (by reaction with N_2O_4 in DMF), then substituting the labile nitrite esters for sulphate groups (by addition of DMF·SO_3 complex). In this way, high-viscosity cellulose sulphates with degrees of substitution in the range 0.3–2 have been made (Schweiger, 1978; Wagenknecht *et al.*, 1993).

As with monosaccharide glycosides, the primary hydroxy groups of polysaccharides are the most readily sulphated. Specific 6-*O*-sulphation of chitin, chitosan (in acidic media) and a number of glycosaminoglycans bearing no hydroxy group at C-2 of their amino-sugar residues, and selective 6-sulphation of other polymers such as cellulose, are possible. The order of reactivity of the hydroxy groups in cellulose (a 1,4-linked β-glucan) is $6 \gg 2 > 3$ for the heterogeneous reaction with sulphur trioxide and DMF (Yamamoto *et al.*, 1991). For curdlan (a 1,3-linked β-glucan), the 6-hydroxy group is still the most reactive; further sulphation occurs at both the 2- and 4-positions (Osawa *et al.*, 1993). Homogeneous phase sulphation of cellulose, by transesterification of its trinitrite with sulphur trioxide in DMF, favours 6-sulphation at 20°C, but 2-sulphation at −20°C (Wagenknecht *et al.*, 1993). Free amino groups, for example as in chitosan, can easily be *N*-sulphated.

3.3.4 Cyclic sulphates

Cyclic sulphates can be prepared directly from cyclic but not from acyclic carbohydrate diols by reaction with sulphuryl chloride in pyridine (Turvey, 1965) or phenyl chlorosulphate and sodium hydride (Abdel-Malik and Perlin, 1989). An alternative, that also works with acyclic diols, is to prepare, using thionyl chloride, the cyclic sulphite which is then oxidised to the cyclic sulphate (with MnO_4^- or RuO_4^-) (Lohray, 1992).

1,2-*cis*-Cyclic sulphates behave as epoxide equivalents, only they are more reactive. They can thus be used to form fluoro- and azido-sugars, as exemplified in Figure 3.14. Such reactions are often highly regiospecific, leading usually, but not always, to *trans*-diaxial products, from which the sulphate can be removed by mild acid-catalysed hydrolysis (Lohray, 1992; van der Klein *et al.*, 1992).

1,2-*Trans*-cyclic sulphates react with nucleophiles at sulphur, leading to intermediates that can either react further to form epoxides or hydrolyse on work-up to the corresponding 1,2-*trans*-diol (Berridge *et al.*, 1990).

3.3.5 Sulphate diesters

Mixed alkyl sugar sulphate diesters can be formed from sugar sulphate hemi-esters, as in the Hakamori permethylation reaction (Stevenson and Furneaux, 1991), but they readily hydrolyse on exposure to water. Mixed phenyl sugar sulphate diesters can be made by reaction of a free hydroxy group on a sugar with phenyl chlorosulphate; these are quite stable but can be converted into the corresponding carbohydrate sulphate hemi-esters on hydrogenolysis (Penney and Perlin, 1981). Symmetrical sulphate diesters

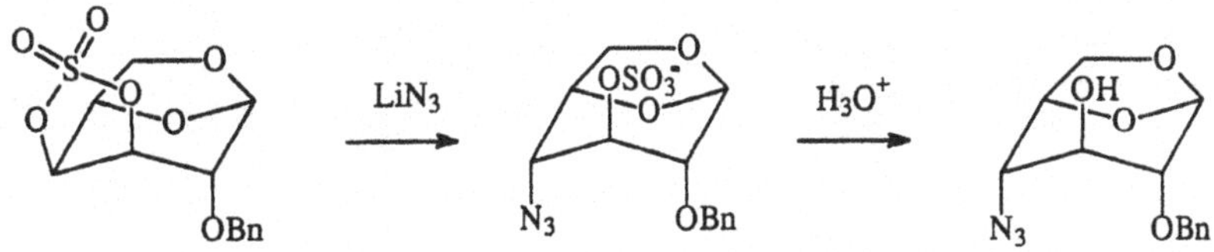

Figure 3.14 Reaction of cyclic sulphate.

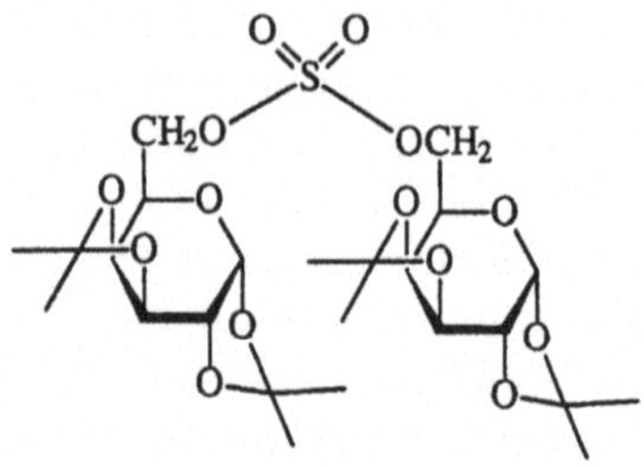

Figure 3.15 Sulphate diester.

with two identical carbohydrate moieties, for example the bis(galactos-6-yl) ester (Figure 3.15), have been obtained by condensation of the sugar alcohol with thionyl chloride and oxidation of the resulting sulphite with permanganate (Takiura and Honda, 1970).

3.4 Chemical stability

3.4.1 Alkali treatment and nucleophilic displacement reactions

Alkaline cleavage generally involves intramolecular participation of an oxyanion because direct cleavage of sulphate esters is very slow. Treatment with hot alkali causes the intramolecular displacement of sulphate from C-6 by O-3 in galacto-, gluco- and mannopyranosyl residues with corresponding ring closure to produce the 3,6-anhydride. The process also involves a conformational change from $^4C_1$ to $^1C_4$ (Figure 3.16). Such treatment is widely applied, particularly by industry, to improve the gelling properties of agars and some carrageenans, which are dependent on high levels of the 3,6-anhydride (Section 3.2.1). The kinetics of the cyclisation have been studied for carrageenans and found to be pseudo first order. The reaction is 20–60 times faster for conversion of μ/ν-carrageenans to κ-/ι-carrageenan than for λ-carrageenan to θ-carrageenan. This is thought to be because of steric hindrance and ionic interaction between the sulphate group present at C-2 of the adjoining 3-linked galactosyl unit in the λ-carrageenan structure, and the potentially nucleophilic hydroxyl group at C-3 of the 4-linked galactosyl unit (Ciancia *et al.*, 1993).

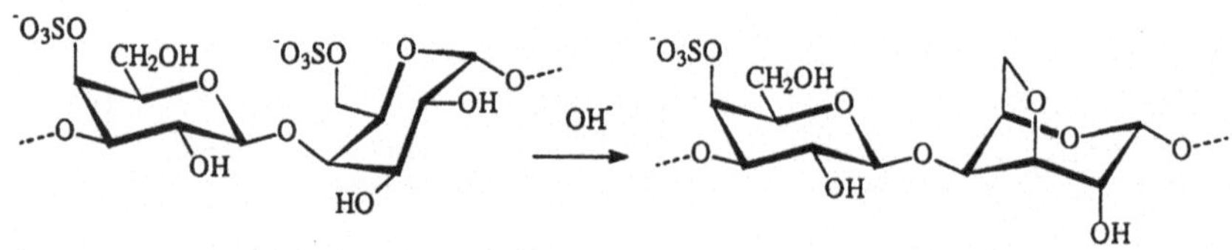

Figure 3.16 Intramolecular displacement of sulphate from C-6 of μ-carrageenan to give κ-carrageenan, showing conformational change of the pyranosyl ring from $^4C_1$ to $^1C_4$.

Less well studied is the intramolecular displacement of a sulphate by a vicinal *trans*-hydroxyl group which can lead to the formation of an epoxide intermediate. Thus, methyl-α-D-glucopyranose-4-sulphate is desulphated readily in aqueous alkali but the corresponding galactose isomer is stable (Roy and Turner, 1983). Depending on the point of nucleophilic attack, the epoxide may be opened to give the original sugar stereochemistry (but minus the sulphate ester) or the product epimeric at both centres. An example is the presence in heparin, after heating with dilute aqueous sodium carbonate, of α-L-galacturonic acid derived from α-L-iduronic acid 2-sulphate units (Rej and Perlin, 1990). Spectroscopic evidence of the intermediate 2,3-anhydro-L-iduronic acid was obtained after treatment of heparin with aqueous triethylamine (Fraidenraich y Waisman and Cirelli, 1992). Epoxide ring opening with methoxide ion gives a product with a methyl group in the position previously occupied by the sulphate ester. Evidence for such a process is the isolation of 2-*O*-methyl-D-xylose from xylose 2-sulphate units of a polysaccharide from *Ulva lactuca* following treatment with methoxide (Percival and Wold, 1963). The reactions outlined above may occur unintentionally during extraction or other chemical manipulations (King *et al.*, 1992).

Carbohydrate sulphates do not undergo the migration reactions commonly observed in the corresponding carboxylic acid ester derivatives.

3.4.2 *Acid hydrolysis*

Sulphate groups are easily removed by aqueous acid with S–O bond scission so that the configuration of the carbohydrate is retained. The general pattern of reactivity is as expected: glycosyl sulphate ≫ secondary equatorial sulphate > secondary axial sulphate > primary sulphate (Roy and Turner, 1983), although glycoside 2-sulphates (but not aldose 2-sulphates) are much more susceptible to acid catalysed ester hydrolysis than are the corresponding 3-, 4- or 6-sulphates (Forrester *et al.*, 1976). Glycopyranosidic bonds generally have a similar susceptibility to hydrolysis. The somewhat greater stability of 6-sulphate esters permits the isolation of L-galactose 6-sulphate in reasonable yield by controlled acid hydrolysis of porphyran, an agar-based polymer with a high proportion (*c*. 25%) of 4-linked α-L-galactopyranosyl 6-sulphate residues (Turvey and Rees, 1961). *N*-linked sulphates are much more labile than are *O*-linked sulphates.

3.4.3 *Methanolysis*

The selectivity for cleavage of sulphate esters over glycosidic cleavage is greater with methanolic hydrogen chloride at room temperature, which is a heterogeneous reaction in the case of polysaccharides. Susceptible glycoside bonds such as those of fucosides and 3,6-anhydrogalactosides found in agars and carrageenans are still cleaved, however.

3.4.4 *Solvolytic desulphation*

Desulphation of sulphated carbohydrates can be effected in hot DMSO by the presence of a proton source, for example water or methanol. Concomitant depolymerisation of polysaccharide samples can be reduced by addition of small amounts of pyridine (Usov *et al.*, 1971). For the reaction to work well, the sample must be in the pyridinium salt form. Such salts are not only DMSO-soluble but it is thought that the pyridinium ion is involved in the desulphation reaction since DMSO-soluble triethylammonium salt forms do not desulphate well (Nagasawa and Inoue, 1974; Nagasawa *et al.*, 1977). Certain conditions afford selective *N*-desulphation of heparin (Inoue and Nagasawa, 1976).

3.4.5 *Other methods of sulphate cleavage*

Surprisingly, silylating reagents have been found to cause regioselective desulphation at the 6-position of monosaccharides, red algal polysaccharides and GAGs (Matsuo *et al.*, 1993; Takano *et al.*, 1992a).

The mechanism of acetylative desulphation has been studied for peracetylated glucose-6-sulphate in refluxing ^{18}O-labelled acetic anhydride (Hyatt, 1993). Very little ^{18}O was found in the glucose obtained after deacetylation, suggesting that the reaction proceeds through the intermediate production of the free alcohol. Conditions have been found under which acid-catalysed acetylative desulphation of agarobiitols and carrabiitols is effected without concomitant glycosidic bond cleavage (Falshaw and Furneaux, 1995b). Hydrogen chloride in anhydrous acetone is useful for desulphation of sulphoglycerolipids without deacylation or degradation (Lingwood *et al.*, 1983).

3.4.6 *Effect of sulphate esters on other substituents*

Sulphate ester substituents can alter the rate of acid-catalysed hydrolysis of glycosidic bonds. Hexopyranosides are rendered more stable by addition of sulphate to the 6-position or an equatorial 3- or 4-position (Turvey, 1965). Hexopyranoside 2-sulphates, however, are unusually susceptible to glycosidic hydrolysis (Forrester *et al.*, 1976), although 3,6-anhydro-α-L-galactopyranosidic bonds appear to be stabilised by a 2-sulphate group (Stevenson and Furneaux, 1991). Acetal substituents, such as *O*-isopropylidene acetals, are rendered more labile by an adjacent sulphate substituent. Sulphation deactivates *O*-benzyl ethers towards hydrogenolysis (Turvey, 1965).

3.5 Methods of analysis

3.5.1 Separation techniques

The presence of sulphate groups on carbohydrates provides charge which can be exploited in the separation of carbohydrate sulphates from each other and also from other carbohydrates. The most widely used technique is anion-exchange chromatography, using either standard liquid chomatography media such as DEAE sepharose (Archibald *et al.*, 1981) or HPLC (LoGuidice *et al.*, 1994). The latter is particularly useful in the separation of glycosaminoglycan species (Volpi, 1994).

Charge is also necessary for electrophoretic mobility, although other factors such as gel medium, buffer, molecular weight, voltage and time affect precise species movement. Electrophoresis is often used to check the homogeneity of given fractions, for example, heteropolysaccharide sulphates from the brown seaweed *Undaria pinnatifida* (Mori *et al.*, 1982). Electrophoretic separation of various types of carrageenans has been achieved (Marrs *et al.*, 1984) but the technique has not been widely used. The electrophoretic separation of GAGs is complicated by their molecular weight inhomogeneity, which causes broad bands on agarose electrophoresis. All the different GAG species have been separated from each other by electrophoresis on cellulose acetate, however, which has sufficiently large pores so that size sieving does not occur (Bianchini *et al.*, 1980). Capillary-zone electrophoresis has been used to separate heparin oligosaccharides (Damm *et al.*, 1992) and also enzymically produced disaccharides from dermatan sulphate and chondroitin sulphate (Karamanos *et al.*, 1995).

3.5.2 Spectroscopic methods

3.5.2.1 NMR spectroscopy

^{13}C-NMR spectroscopy is a rapid non-destructive tool for the structural study of carbohydrate sulphates. The deshielding effect of a sulphate group generally leads to a downfield shift of 6–10 ppm in the signal of the carbon to which it is attached and an upfield shift of 1–2.5 ppm in the adjacent carbon signals (Archibald *et al.*, 1981). The magnitude of the shift is dependent on the steric crowding of the sulphate. Accurate prediction of ^{13}C chemical shifts is very difficult, as other factors, such as conformational distortions, may affect the molecular environment of individual carbon atoms. Well-resolved spectra are routinely obtained for molecules with a regular structure and of low viscosity. The spectra of polysaccharide sulphates may be enhanced by partial depolymerisation of the sample using ultrasound or hydrolytic techniques. ^{13}C-NMR techniques are not considered quantitative, although they can be used to determine relative degrees of sulphation at different hydroxyl groups where relaxation differences are likely to be minimal.

In ^{1}H-NMR, the presence of a sulphate group deshields both the geminal and vicinal proton resonances. A primary sulphate shifts the geminal protons upfield by 0.4–0.5 ppm whereas a secondary sulphate shifts its geminal proton by 0.6–0.7 ppm. Vicinal protons are shifted 0.1–0.4 ppm upfield, depending on the axial or equatorial positions of the sulphate and proton; an equatorial proton adjacent to an equatorial sulphate is shifted the most (Contreras *et al.*, 1988). This technique has been less widely used for the study of polymeric or complex carbohydrate sulphates owing to spectral complexity and poor resolution. Notable examples where ^{1}H-NMR spectroscopy has been of value are with echinodermal fucans (Mulloy *et al.*, 1994) and carrageenans (Knutsen and Grasdalen, 1992; Usov, 1984).

The NMR chemical shift data for a number of sulphated carbohydrates can be found in the general carbohydrate NMR databases accessible electronically via the World Wide Web (WWW). For general information see the Carbohydrate WWW page, address http://www.public.iastate.edu/~pedro/carbhyd/carbhyd.html. Two databases are Sugabase, WWW address http://boc.chem.uu.nl/sugabase/databases.html, and Carbbank, WWW address http://128.192.9.29/carbbank/default.htm.

3.5.2.2 *Infra-red spectroscopy*

Infra-red spectroscopy is useful for detecting the presence and type of carbohydrate sulphate ester groups. A strong absorption at *ca.* 1240 cm^{-1} is always observed, and in certain polysaccharides, notably carrageenans, peaks at 850 cm^{-1}, 830 cm^{-1} and 820 cm^{-1} are diagnostic for secondary axial, secondary equatorial and primary equatorial ester groups, respectively. There are exceptions to these rules with certain monosaccharides and particularly those containing 2-sulphates so that the use of this technique is generally confined to polysaccharide samples (Falshaw and Furneaux, 1994; Peat *et al.*, 1968; Turvey *et al.*, 1967).

3.5.2.3 *Mass spectrometry*

Structural information can be obtained from fast atom bombardment (FAB) mass spectra of both derivatised and underivatised sulphated oligosaccharides. Negative-ion FAB mass spectrometry (MS) of permethylated sulphated oligosaccharides from GAGs followed by positive-ion FABMS of the sample once the sulphate groups have been replaced by acetyl groups yields sequence data and defines the originally sulphated residues (Dell *et al.*, 1988). Molecular-weight-related and structurally significant fragmentations were observed for underivatised multisulphated disaccharides from mucus glycoproteins which differed by multiples of 102 amu owing to the presence or absence of sulphate ester(s) on certain sugar units (Mawhinney and Chance, 1994).

3.5.2.4 *Spectrofluorimetry and UV visible spectrometry*

The charge–charge binding with cationic dyes can be used to determine the quantity and type of sulphated polysaccharides in aqueous solution. A change in wavelength of maximum fluorescence emission can be detected in some cases, for example the fluorescence maximum of acridine orange shifts from 525 nm to 640 nm on binding (Cundall *et al.*, 1978). In a similar way, sulphatides can be specifically determined by formation of a coloured complex with azure A that is extractable into organic solution (Kean, 1968). GAGs also can be identified by interactions with cationic dyes (Homer *et al.*, 1993). Cationic dyes such as Alcian blue and Cupromeronic blue have been used to stain GAGs for histochemical studies by light and electron microscopy (Welsch *et al.*, 1992).

3.5.3 Chemical methods

3.5.3.1 Sulphate analysis

Most methods for determining sulphate content involve removal of sulphate from the carbohydrate by hydrolysis or combustion followed by detection of the sulphate by, for example, turbidimetry (Dodgson and Price, 1962), colourimetry (Silvestri *et al.*, 1982) or ion-exchange chromatography (Grotjan *et al.*, 1986). In all cases, it is important to remove extraneous sources of sulphate and all glassware must be nitric-acid-washed.

3.5.3.2 Methylation analysis

The nature of the carbohydrate moieties in a sample can be readily determined by a process of hydrolysis, reduction and acetylation to produce alditol acetates that are characterised by gas chromatography (GC)/MS. Sulphated polysaccharides are frequently analysed in this way. If the sulphated carbohydrate is chemically methylated prior to hydrolysis the products are partially methylated alditol acetates (PMAAs). The positions of the acetyl groups indicate glycosyl linkages or substitution with acid-labile groups such as sulphate esters or pyruvate ketals. Repetition of the process on a desulphated sample (Section 3.4.4) allows the positions of sulphate esters to be determined as those now bearing a methyl ether which were previously acetylated. Another approach is to methylate, desulphate then alkylate the sample using (trideuterio)methyl iodide or (trideuterio)ethyl iodide before hydrolysis, reduction and acetylation. The alkyl group marks the position of the original sulphate group. Numerous modifications have been made to this basic procedure (see e.g. Hanisch, 1994; Needs and Selvendran, 1993; 1994). The methylation reaction must be as complete as possible (often necessitating multiple treatments), otherwise false or ambiguous results are obtained.

Efficient methylation requires solubility of the carbohydrate in DMSO, which, in the case of anionic polysaccharides, is often achieved by conversion to an organic cation form. The methylation reaction involves the use of a base, commonly methyl sulphinyl carbanion (as the K^+, Na^+ or Li^+ salt), or sodium hydroxide. While the latter has gained recent popularity, the use of methyl sulphinyl carbanion has been recommended for some sulphated carbohydrates (Dell *et al.*, 1988). Undesirable reactions, such as intramolecular displacement of sulphate with consequent ring closure, may occur during these treatments (Section 3.4.1). The acid hydrolysis step is especially critical for acid-labile sugars such as 3,6-anhydrogalactose found in agars and carrageenans. A reductive hydrolysis procedure which combines acid hydrolysis with *in situ* reduction using the relatively acid-stable reductant 4-methylmorpholine-borane is now routinely used for analysis of sulphated red algal galactans (Stevenson and Furneaux, 1991). The position(s) of sulphate ester(s) on adjacent red algal galactose and 3,6-anhydrogalactose units can now be determined using a reductive partial-hydrolysis procedure (Falshaw and Furneaux, 1995b).

3.6 Conformational effects of sulphation

Crystallographic and modelling (Lamba *et al.*, 1994) studies show that sulphation increases the length of the C–O bond at the point of attachment and opens out the C–O–S angle slightly as compared with the C–O–H angle. A single sulphate group has little influence on the most favourable conformation of the sugar ring and does not hydrogen bond to carbohydrate hydroxyl groups. Few attempts at molecular modelling have been conducted because of the lack of forcefield parameters for charged groups such as sulphates but sets have recently been published (Huige and Altona, 1995). An NMR study of the Lewis A trisaccharide showed that mono-sulphation made only minor differences to the three-dimensional structure (Kogelberg and Rutherford, 1994).

Sulphation has a major effect on sugar conformation in multiply charged species, however, owing to electrostatic interactions. Thus, the pyranose ring of the 2,3,4-trisulphated β-D-glucopyranosyluronate residue in a highly sulphated disaccharide assumes a $^{3,O}B$ conformation, shown in Figure 3.17, to achieve maximal charge separation (Wessel, 1992). The repulsive interactions between sulphate residues and carboxyl groups in GAGs have a major effect on their conformation and hence on their biological activity (Ragazzi *et al.*, 1993). Heparin has been shown to present an approximately

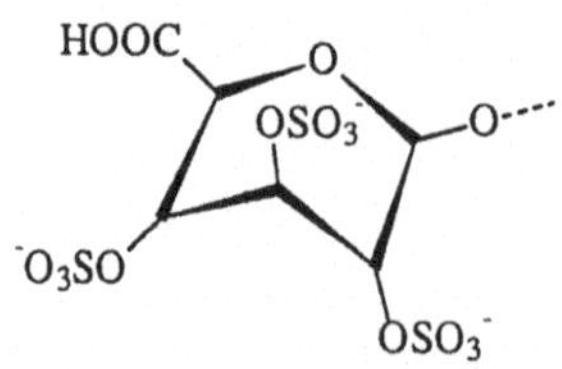

Figure 3.17 $^{3,O}B$ conformation of 2,3,4-trisulphated β-D-glucopyranosyluronate.

regular array of three sulphate groups separated by a carboxyl group down each side of the molecule (Mulloy *et al.*, 1993) which may account for many of its binding properties. As was described in Section 3.2.5.2, the increased flexibility of iduronic acid containing GAGs over glucuronic-acid-containing species allows them to better minimise these charge–charge interactions and adopt a wider variety of conformations on binding to positively charged species. The repulsive interactions between negatively charged groups on GAGs can be compensated by the formation of salt bridges by cations (Zsiška and Meyer, 1993).

References

Abdel-Fattah, A.F., Hussein, M.M.-D. and Salem, H.M. (1974) Studies of the purification and some properties of sargassan, a sulphated heteropolysaccharide from *Sargassum linifolium*. *Carbohydr. Res.*, **33**, 9–17.

Abdel-Malik, M.M. and Perlin, A.S. (1989) Reactions of phenyl chlorosulphate at OH-2 and OH-3 of aldohexopyranose derivatives. Competing substitution and displacement reactions. *Carbohydr. Res.*, **190**, 39–52.

Ahmad, V.U., Ghazala, S.U. and Ali, M.S. (1993) Saponins from *Zygophyllum propinquum*. *Phytochemistry*, **33**, 453–5.

Ahsan, A., Jeske, W., Mardiguian, J. and Fareed, J. (1994) Feasibility study of heparin mass calibrator as a GPC calibrator for heparins and low molecular weight heparins. *J. Pharm. Sci.*, **83**, 197–201.

Arad, S.M., Dubinsky, O. and Simon, B. (1993) A modified cell wall mutant of the red microalga *Rhodella reticulata* resistant to the herbicide 2,6-dichlorobenzonitrile. *J. Phycol.*, **29**, 309–13.

Archibald, P.J., Fenn, M.D. and Roy, A.B. (1981) ^{13}C NMR studies of D-glucose and D-galactose monosulphates. *Carbohydr. Res.*, **93**, 177–90.

Armisen, R. and Galatas, F. (1987) Production properties and uses of agar. In: McHugh, D.J. (Ed.) *Production and Utilization of Products from Commercial Seaweeds*, pp. 1–57, FAO Fish. Tech. Pap (288).

Baba, M., Snoeck, R., Pauwels, R. and De Clercq, E. (1988) Sulphated polysaccharides are potent and selective inhibitors of various enveloped viruses, including herpes simplex virus, cytomegalovirus, vesicular stomatitis virus, and human immunodeficiency virus. *Antimicrob. Agents Chemother.*, **32**, 1742–5.

Berridge, M.S., Franceschini, M.P., Rosenfeld, E. and Tewson, T.J. (1990) Cyclic sulphates: useful substrates for selective nucleophilic substitution. *J. Org. Chem.*, **55**, 1211–17.

Bianchini, P., Nader, H.B., Takahashi, H.K. *et al.* (1980) Fractionation and identification of heparin and other acidic mucopolysaccharides by a new discontinuous electrophoretic method. *J. Chromatogr.*, **196**, 455–62.

Bixler, H.J. (1995) Recent carrageenan developments. *Hydrobiologia*; XV International Seaweed Symposium, Valdivia, Chile, abstract 22.

Bociek, S.M., Darke, A.H., Welti, D. and Rees, D.A. (1980) The ^{13}C NMR spectra of hyalronate and chondroitin sulphates: further evidence of an alkali induced conformational change. *Eur. J. Biochem.*, **109**, 447–56.

Böcker, T., Lindhorst, T.K., Thiem, J. and Vill, V. (1992) Synthesis and properties of sulphated alkyl glycosides. *Carbohydr. Res.*, **230**, 245–56.

Bourgougnon, N., Lahaye, M., Chermann, J.-C. and Kornprobst, J.-M. (1993) Composition and antiviral activities of a sulphated polysaccharide from *Schizymenia dubyi* (Rhodophyta, Gigartinales). *Bioorg. Med. Chem. Lett.*, **3**, 1141–6.

Bourgougnon, N., Roussakis, C., Kornprobst, J.-M. and Lahaye, M. (1994) Effects in vitro of a sulphated polysaccharide from *Schizymenia dubyi* (Rhodophyta, Gigartinales) on a non-small-cell bronchopulmonary carcinoma line (NSCLC-N6). *Cancer Lett.*, **85**, 87–92.

Campbell, P., Hannesson, H.H., Sandbäck, D. *et al.* (1994) Biosynthesis of heparin/heparan sulphate. Purification of the D-glucuronyl C-5 epimerase from bovine liver. *J. Biol. Chem.*, **269**, 26953–8.

Cases, M.R., Stortz, C.A. and Cerezo, A.S. (1992) Methylated, sulphated xylogalactans from the red seaweed *Corallina officinalis*. *Phytochemistry*, **31**, 3897–900.

Casu, B. (1985) Structure and biological activity of heparin. *Adv. Carbohydr. Chem. Biochem.*, **43**, 51–134.

Casu, B. (1994) Heparin and heparin-like polysaccharides. In: Dumitriu, S. (Ed.) *Polymeric Biomaterials*, pp. 159–77. Marcel Dekker: New York.

Casu, B., Petitou, M., Provasoli, M. and Sinay, P. (1988) Conformational flexibility: a new concept for explaining binding and biological properties of iduronic acid-containing glycosaminoglycans. *Trends Biochem. Sci.*, **13**, 221–5.

Chang, P.K., Sciarini, L.J. and Cramer, J.W. (1967) Sulphate esters of 5-iododeoxyuridine and 5-iododeoxycytidine. *J. Med. Chem.*, **10**, 733–5.

Cheng, F., Heinegård, D., Malmström, A. *et al.* (1994) Patterns of uronosyl epimerization and 4-/6-*O*-sulphation in chondroitin/dermatan sulphate from decorin and biglycan of various bovine tissues. *Glycobiology*, **4**, 685–96.

Chong, A.S.F. and Parish, C.R. (1985) Nonimmune lymphocyte–macrophage interaction II. Evidence that the interaction involves sulphated polysaccharide recognition. *Cell. Immunol.*, **92**, 277–89.

Ciancia, M., Noseda, M.D., Matulewicz, M.C. and Cerezo, A.S. (1993) Alkali-modification of carrageenans: mechanism and kinetics in the kappa/iota-, mu/nu- and lambda-series. *Carbohydr. Polym.*, **20**, 95–8.

Cockin, G.H., Huckerby, T.N. and Nieduszynski, I.A. (1986) High-field NMR studies of keratan sulphate. *Biochem. J.* **236**, 921–4.

Conrad, H.E. (1993) Dissection of heparin – past and future. *Pure Appl. Chem.*, **65**, 787–91.

Contreras, R.R., Kamerling, J.P., Breg, J. and Vleigenthart, J.F.G. (1988) ^{1}H- and ^{13}C-NMR spectroscopy of synthetic monosulphated methyl α-D-galactopyranosides. *Carbohydr. Res.*, 411–18.

Contreras, R.R., Kamerling, J.P. and Vliegenthart, J.F.G. (1991) ^{1}H- and ^{13}C-NMR spectroscopy of synthetic monosulphated methyl α-D-mannopyranosides. *Recl. Trav. Chim. Pays-Bas.*, **110**, 85–8.

Craigie, J.S. (1990) Cell walls. In: Cole, K.M. and Sheath, R.G. (Eds) *Biology of the Red Seaweeds*, pp. 221–57. Cambridge: Cambridge University Press.

Cundall, R.B., Murray, D. and Phillips, G.O. (1978) Spectrofluorometric methods for estimating and studying the interactions of polysaccharides in biological systems. In: Schweiger, R.G. (Ed.) *Carbohydrate Sulphates*, pp. 67–94. Washington, DC: American Chemical Society.

Damm, J.B.L., Overklift, G.T., Vermeulen, B.W.M. *et al.* (1992) Separation of natural and synthetic heparin fragments by high-performance capillary electrophoresis. *J. Chromatogr.*, **608**, 297–309.

DeAngelis, P.L. and Glabe, C.G. (1987) Polysaccharide structural features that are critical for the binding of sulphated fucans to bindin, the adhesive protein from sea urchin sperm. *J. Biol. Chem.*, **262**, 13946–52.

Dell, A., Rogers, M.E., Thomas-Oates, J.E. *et al.* (1988) Fast-atom-bombardment mass-spectrometric strategies for sequencing sulphated oligosaccharides. *Carbohydr. Res.*, **179**, 7–19.

Dobashi, K., Nishino, T., Fujihara, M. and Nagumo, T. (1989) Isolation and preliminary characterization of fucose-containing sulphated polysaccharides with blood-anticoagulant activity from the brown seaweed *Hizikia fusiforme*. *Carbohydr. Res.*, **194**, 315–20.

Doctor, V.M., Lewis, D., Coleman, M. *et al.* (1991) Anticoagulant properties of semisynthetic polysaccharide sulphates. *Thrombosis Res.*, **64**, 413–25.

Dodgson, K.S. and Price, R.G. (1962) A note on the determination of the ester sulphate content of sulphated polysaccharides. *Biochem. J.*, **84**, 106–10.

Falshaw, R. and Furneaux, R.H. (1994) Carrageenan from the tetrasporic stage of *Gigartina decipiens* (Gigartinaceae, Rhodophyta). *Carbohydr. Res.*, **252**, 171–182.

Falshaw, R. and Furneaux, R.H. (1995a) Carrageenans from the tetrasporic stages of *Gigartina clavifera* and *Gigartina alveata* (Gigartinaceae, Rhodophyta). *Carbohydr. Res.*, **276**, 155–65.

Falshaw, R. and Furneaux, R.H. (1995b) The structural analysis of disaccharides from red algal galactans by methylation and reductive partial-hydrolysis. *Carbohydr. Res.*, **269**, 183–9.

Fernández, L.E., Valiente, O.G., Mainardi, V. *et al.* (1989) Isolation and characterization of an antitumour active agar-type polysaccharide of *Gracilaria dominguensis*. *Carbohydr. Res.*, **190**, 77–83.

Forrester, P.F., Lloyd, P.F. and Stuart, C.H. (1976) Synthesis of L-fucose 2-, 3-, and 4-sulphates. *Carbohydr. Res.*, **49**, 175–84.

Fraidenraich y Waisman, D. and Cirelli, A.F. (1992) Selective *O*-desulphation of heparin in triethylamine. *Carbohydr. Polym.*, **17**, 111–14.

Fransson, L.-A. (1985) Mammalian glycosaminoglycans. In: Aspinall, G.O. (Ed.) *The Polysaccharides*, vol. 3, pp. 337–415. Orlando, FL: Academic Press.

Freeze, H.H. and Wolgast, D. (1986) Structural analysis of *N*-linked oligosaccharides from glycoproteins secreted by *Dictyostelium discoideum*. *J. Biol. Chem.*, **261**, 127–34.

Furneaux, R.H. and Stevenson, T.T. (1990) The xylogalactan sulphate from *Chondria macrocarpa* (Ceramiales, Rhodophyta). *Hydrobiologia* **204/205**, 615–620.

Gallagher, J.T. (1989) The extended family of proteoglycans: social residents of the pericellular zone. *Curr. Opin. Cell Biol.*, **1**, 1201–18.

Gallagher, J.T. and Walker, A. (1985) Molecular distinctions between heparan sulphate and heparin. *Biochem. J.*, **230**, 665–74.

Gallagher, J.T., Lyon, M. and Steward, W.P. (1986) Structure and function of heparan sulphate proteoglycans. *Biochem. J.*, **236**, 313–25.

García-Villalón, D. and Gil-Fernández, C. (1991) Antiviral activity of sulphated polysaccharides against African swine fever virus. *Antiviral Res.*, **15**, 139–48.

Geresh, S., Dubinsky, O., Arad, S.M. *et al.* (1990) Structure of 3-*O*-α-D-glucopyranosyluronic acid-L-galactopyranose, an aldobiouronic acid isolated from the polysaccharides of various unicellular red algae. *Carbohydr. Res.*, **208**, 301–5.

Gigg, R. (1988) Studies on the synthesis of sulphur-containing glycolipids ('sulphoglycolipids'). In: Schweiger, R.G. (Ed.) *Carbohydrate Sulphates*, pp. 44–66. Washington, DC: American Chemical Society.

Grotjan Jnr, H.E., Padrnos-Hicks, P.A. and Keel, B.A. (1986) Ion chromatographic method for the analysis of sulphate in complex carbohydrates. *J. Chromatogr.*, **367**, 367–75.

Guezennec, J.G., Pignet, P., Raguenes, G. *et al.* (1994) Preliminary chemical characterisation of unusual eubacterial exopolysaccharides of deep sea origin. *Carbohydr. Polym.*, **24**, 287–94.

Guiseley, K.B. (1978) Some novel methods and results in the sulphation of polysaccharides. In: Schweiger, R.G. (Ed.) *Carbohydrate Sulphates*, pp. 148–62. Washington, DC: American Chemical Society.

Hanisch, F.-G. (1994) Methylation analysis of complex carbohydrates: overview and critical comments. *Biol. Mass Spectrom.*, **23**, 309–12.

Harris, M.J. and Turvey, J.R. (1969) Sulphates of monosaccharides and derivatives. Part VII. Synthesis of some disulphates and a new synthesis of D-galactose 4-sulphate. *Carbohydr. Res.*, **9**, 397–405.

Haug, A. (1976) The influence of borate and calcium on the gel formation of a sulphated polysaccharide from *Ulva lactuca*. *Acta Chem. Scand., Ser. B*, **30**, 562–6.

Hemmingson, J.A., Furneaux, R.H. and Murray-Brown, V.L. (1995) Biosynthesis of agar polysaccharides in *Gracilaria chilensis* Bird, McLachlan and Oliveira. *Carbohydr. Res.*, **287**, 101–115.

Hioki, M., Yamashita, T. and Harada, S. (1991) Polysaccharide sulphate salts with proteins as antiallergens. Jpn Kokai Tokkyo Koho JP 03,234,701.

Hoagland, K.D., Rosowski, J.R., Gretz, M.R. and Roemer, S.C. (1993) Diatom extracellular polymeric substances: function, fine structure, chemistry, and physiology. *J. Phycol.*, **29**, 537–66.

Homer, K.A., Denbow, L. and Beighton, D. (1993) Spectrophotometric method for the assay of glycosaminoglycans and glycosaminoglycan-depolymerising enzymes. *Anal. Biochem.*, **214**, 435–41.

Horton, D. and Usui, T. (1978) Sulphated glycosaminoglycans obtained by chemical modification of polysaccharides. In: Schweiger, R.G. (Ed.) *Carbohydrate Sulphates*, pp. 95–112. Washington, DC: American Chemical Society.

Huige, C.J.M. and Altona, C. (1995) Force field parameters for sulphates and sulphamates based on *ab initio* calculations: extensions of AMBER and CHARMM fields. *J. Comput. Chem.*, **16**, 56–79.

Hyatt, J.A. (1993) Acetylative desulphation of a glucose-6-sulphate: an oxygen isotope labelling study. *Carbohydr. Res.*, **239**, 291–6.

Ilyas, A.A., Quarles, R.H., MacIntosh, T.D. *et al.* (1984) IgM in human neuropathy related to paraproteinemia binds to a carbohydrate determinant in the myelin-associated glycoprotein and to a ganglioside. *Proc. Natl. Acad. Sci. USA*, **81**, 1225–9.

Imai, Y. and Irimura, T. (1994) Quantitative measurement of carbohydrate binding activity of mouse macrophage lectin. *J. Immunol. Methods*, **171**, 23–31.

Inoue, Y. and Nagasawa, K. (1976) Selective *N*-desulphation of heparin with dimethyl sulphoxide containing water or methanol. *Carbohydr. Res.*, **46**, 87–95.

Itoh, H., Noda, H., Amano, H. *et al.* (1993) Antitumor activity and immunological properties of marine algal polysaccharides, especially fucoidan, prepared from *Sargassum thunbergii* of Phaeophyaceae. *Anticancer Res.*, **13**, 2045–52.

Jandik, K.A., Gu, K. and Linhardt, R.J. (1994) Action pattern of polysaccharide lyases on glycosaminoglycans. *Glycobiology*, **4**, 289–96.

Jatzkewitz, H. and Nowoczek, G. (1967) Synthesis of ^{35}S-labelled D-galactose sulphate and ceramide D-galactose sulphate (cerebroside sulphate). *Chem. Ber.*, **100**, 1667–74.

Jones, R. (1991) Interaction of zona pellucida glycoproteins, sulphated carbohydrates and synthetic polymers with proacrosin, the putative egg-binding protein from mammalian spermatozoa. *Development*, **111**, 1155–63.

Kamel, M.S., Ohtani, K., Assaf, M.H. *et al.* (1992) Lignan glycosides from stems of *Salvadora persica*. *Phytochemistry*, **31**, 2469–71.

Karamanos, N.K., Axelsson, S., Vanky, P. *et al.* (1995) Determination of hyaluronan and galactosaminoglycan disaccharides by high-performance capillary electrophoresis at the attomole level. Applications to analyses of tissue and cell culture proteoglycans. *J. Chromatogr. A*, **696**, 295–305.

Katsuraya, K., Ikushima, N., Takahashi, N. *et al.* (1994) Synthesis of sulphated alkyl malto- and laminara-oligosaccharides with potent inhibitory effects on AIDS virus infection. *Carbohydr. Res.*, **260**, 51–61.

Kean, E.L. (1968) Rapid, sensitive spectrophotometric method for quantitative determination of sulphatides. *J. Lipid Res.*, **9**, 319–27.

Kennedy, J.F. (1974) Chemically reactive derivatives of polysaccharides. *Adv. Carbohydr. Chem. Biochem.*, **29**, 305–405 (see pp. 335–337).

King, K.R., Williams, J.M., Clamp, J.R. and Corfield, A.P. (1992) A study of possible sulphate loss during the chemical release of sulphated oligosaccharides from glycoproteins. *Carbohydr. Res.*, **235**, C9–C12.

Kjellen, L. and Lindahl, U. (1991) Proteoglycans: structures and interactions. *Ann. Rev. Biochem.*, **60**, 443–75.

Knutsen, S.H. and Grasdalen, H. (1992) The use of neocarrabiose oligosaccharides with different length and sulphate substitution as model compounds for ^{1}H-NMR spectroscopy. *Carbohydr. Res.*, **229**, 233–44.

Knutsen, S.H., Myslabodski, D.E., Larsen, B. and Usov, A.I. (1994) A modified system of nomenclature for red algal galactans. *Botanica Marina*, **37**, 163–9.

Kobata, A. (1993) Glycobiology: an expanding research area in carbohydrate chemistry. *Acc. Chem. Res.*, **26**, 319–24.

Kobayashi, M., Yamazaki, F., Ito, Y. and Ogawa, T. (1990) A regio- and stereo-controlled synthesis of β-D-Glc*p*NAc6SO_3-(1 → 3)-β-D-Gal*p*6SO_3-(1 → 4)-β-D-Glc*p*NAc6SO_3-(1 → 3)-D-Gal*p*, a linear acidic glycan fragment of keratan sulphate I. *Carbohydr. Res.*, **201**, 51–67.

Kogelberg, H. and Rutherford, T.J. (1994) Studies on the three dimensional behaviour of the selectin ligands Lewis A and sulphated Lewis A using NMR spectroscopy and molecular dynamics simulations. *Glycobiology*, **4**, 49–57.

Kornfeld, R. and Kornfeld, S. (1985) Assembly of asparagine-linked oligosaccharides. *Ann. Rev. Biochem.*, **54**, 631–64.

Krivan, H.C., Olson, L.D., Barile, M.F. *et al.* (1989) Adhesion of *Mycoplasma pneumonia* to sulphated glycolipids and inhibition by dextran sulphate. *J. Biol. Chem.*, **264**, 9283–8.

Lamba, D., Glover, S., Mackie, W. *et al.* (1994) Insights into stereochemical features of sulphated carbohydrates: X-ray crystallographic and modelling investigations. *Glycobiology*, **4**, 151–63.

Larsen, B. and Haug, A. (1963) Free-boundary electrophoresis of acidic polysaccharides from the marine alga *Ascophyllum nodosum*. *Acta Chem. Scand.*, **17**, 1646–52.

Lerouge, P., Roche, P., Faucher, C. *et al.* (1990) Symbiotic host-specificity of *Rhizobium meliloti* is determined by a sulphated and acylated glucosamine oligosaccharide signal. *Nature*, **344**, 781–4.

Liav, A. and Goren, M.B. (1984) Sulphatides of *Mycobacterium tuberculosis*. Synthesis of the core α,α-trehalose 2-sulphate. *Carbohydr. Res.*, **127**, 211–6.

Lillard, D.W. and Seib, P.A. (1978) Monosulphate esters of L-ascorbic acid. In: Schweiger, R.G. (Ed.) *Carbohydrate Sulphates*, pp. 1–18. Washington, DC: American Chemical Society.

Lindahl, U. and Kjellen, L. (1991) Heparin or heparan sulphate – what is the difference? *Thromb. Haemost.*, **66**, 44–8.

Lingwood, C., Sakac, D. and Vella, G.J. (1983) Desulphation of sulphoglycolipids by anchimeric assisted solvolysis. *Carbohydr. Res.*, **122**, 1–9.

Linhardt, R.J. (1991) Heparin: an important drug enters its seventh decade. *Chem. and Ind.*, 45–50.

Lo-Guidice, J.-M., Wieruszeski, J.-M., Lemoine, J. *et al.* (1994) Sialylation and sulphation of the carbohydrate chains in respiratory mucins from a patient with cystic fibrosis. *J. Biol. Chem.*, **269**, 18794–813.

Lohray, B.B. (1992) Cyclic sulphites and cyclic sulphates: epoxide like synthons. *Synthesis*, 1035–52.

Ludwig-Baxter, K.G. and Perlin, A.S. (1991) Dermatan sulphate of porcine mucosal tissue. NMR observations on its separation from heparin with the aid of heparinase, and its degradation by chondroitinase. *Carbohydr. Res.*, **217**, 227–36.

Luecke, H. and Quiocho, F.A. (1990) High specificity of a phosphate transport protein determined by hydrogen bonds. *Nature*, **347**, 402–6.

Lupescu, N., Arad, S.M., Geresh, S. *et al.* (1991) Structure of some sulphated sugars isolated after acid hydrolysis of the extracellular polysaccharide of *Porphyridium* sp., a unicellular alga. *Carbohydr. Res.*, **210**, 349–52.

Lyon, M. (1993) The isolation of membrane proteoglycans. *Methods Mol. Biol.*, **19**, 243–51.

Mabeau, S., Kloareg, B. and Joseleau, J.-P. (1990) Fractionation and analysis of fucans from brown algae. *Phytochemistry*, **29**, 2441–5.

McCandless, E.L., West, J.A. and Guiry, M.D. (1983) Carrageenan patterns in the Gigartinaceae. *Biochem. System. Ecol.*, **11**, 175–82.

McClure, M.O., Moore, J.P., Blanc, D.F. *et al.* (1992) Investigations into the mechanism by which sulphated polysaccharides inhibit HIV infection in vitro. *Aids Res. and Human Retroviruses*, **8**, 19–26.

Maeda, M., Uehara, T., Harada, N. *et al.* (1991) Heparinoid-active sulphated polysaccharides from *Monostroma nitidum* and their distribution in the chlorophyta. *Phytochemistry*, **30**, 3611–14.

Mahato, S.B., Ganguly, A.N. and Sahu, N.P. (1982) Steroid saponins. *Phytochemistry*, **21**, 959–78.

Mahato, S.B., Sarkar, S.K. and Poddar, G. (1988) Triterpenoid saponins. *Phytochemistry*, **27**, 3037–67.

Marcus, S.N., Marcus, A.J., Marcus, R. *et al.* (1992) The preulcerative phase of carrageenan-induced colonic ulceration in the guinea-pig. *Int. J. Exp. Path.*, **73**, 515–26.

Marrs, W.M., Kilcast, D. and Fry, J.C. (1984) Electrophoretic separation of carrageenans. In: Phillips, G.O., Wedlock, D.J. and Williams, P.A. (Eds.) *Gums and Stabilisers for the Food Industry 2: Applications of Hydrocolloids. Proc. 2nd Int. Conf.*, pp. 507–16. Oxford: Pergamon Press.

Matsuda, M., Hasui, M. and Okutani, K. (1993) Structural analysis of a sulphated polysaccharide from a marine *Pseudomonas*. *Nippon Suisan Gakkaishi*, **59**, 535–8.

Matsuhashi, T. (1990) Agar. In: Harris, P. (Ed.) *Food Gels*, pp. 1–51. London: Elsevier.

Matsuo, M., Takano, R., Kamei-Hayashi, K. and Hara, S. (1993) A novel regioselective desulphation of polysaccharide sulphates: specific 6-*O*-desulphation with *N*,*O*-bis(trimethylsilyl) acetamide. *Carbohydr. Res.*, **241**, 209–15.

Matulewicz, M.C., Haines, H.H. and Cerezo, A.S. (1994) Sulphated xylogalactans from *Nothogenia fastigiata*. *Phytochemistry*, **36**, 97–103.

Mawhinney, T.P. and Chance, D.L. (1994) Structural elucidation by fast atom bombardment mass spectrometry of multisulphated oligosaccharides isolated from human respiratory mucous glycoproteins. *J. Carbohydr. Chem.*, **13**, 825–40.

Medcalf, D.G. (1978) Sulphated fucose containing polysaccharides from brown algae: structural features and biochemical implications. In: Schweiger, R.G. (Ed.) *Carbohydrate Sulphates*, pp. 225–44. Washington, DC: American Chemical Society.

Millard, P. and Evans, L.V. (1982) The fate of sulphate within the capsular polysaccharide of the unicellular red alga *Rhodella maculata*. *J. Exper. Bot.*, **33**, 854–64.

Miller, I.J., Falshaw, R. and Furneaux, R.H. (1993) The chemical structure of the polysaccharide from *Dasyclonium incisum*. *Hydrobiologia*, **260/261**, 647–51.

Miller, I.J., Falshaw, R. and Furneaux, R.H. (1995) Structural analysis of the polysaccharide from *Pachymenia lusoria* (Cryptonemiaceae, Rhodophyta). *Carbohydr. Res.*, **268**, 219–32.

Miller, I.J., Falshaw, R. and Furneaux, R.H. (1996) The polysaccharide from the red seaweed *Champia novae-zelandiae*. *Hydrobiologia*, **326/327**, 505–9.

Mori, H., Kamei, H., Nishide, E. and Nisizawa, K. (1982) Sugar constituents of some sulphated polysaccharides of wakame (*Undaria pinnatifida*) and their biological activities. In: Hoppe, H.A. and Levring, T. (Eds.) *Marine Algae in Pharmaceutical Science*, pp. 109–21. Berlin: Walter de Gruyter.

Mori, H., Ohbayashi, F., Hirono, I. *et al.* (1984) Absence of genotoxicity of the carcinogenic sulphated polysaccharides carrageenan and dextran sulphate in mammalian DNA repair and bacterial mutagenicity assays. *Nutr. Cancer*, **6**, 92–7.

Mulloy, B., Forster, M.J., Jones, C. and Davies, D.B. (1993) NMR and molecular-modelling studies of the solution conformation of heparin. *Biochem. J.*, **293**, 849–58.

Mulloy, B., Ribero, A.-C., Alves, A.-P. *et al.* (1994) Sulphated fucans from echinoderms have a regular tetrasaccharide repeating unit defined by specific patterns of sulphation at the O-2 and O-4 positions. *J. Biol. Chem.*, **269**, 22113–23.

Murano, E., Gilli, R., Navarini, L. *et al.* (1995) Ion-driven gelation of highly sulphated agar. *J. Mar. Biotech.*, **3**, 143–5.

Nagasawa, K. and Inoue, Y. (1974) Solvolytic desulphation of 2-deoxy-2-sulphoamino-D-glucose and D-glucose 6-sulphate. *Carbohydr. Res.*, **36**, 265–71.

Nagasawa, K., Inoue, Y. and Kamata, T. (1977) Solvolytic desulphation of glycosaminoglycuronan sulphates with dimethysulphoxide containing water or methanol. *Carbohydr. Res.*, **58**, 47–55.

Nair, V., Joseph, J.P., Poletto, J.F. and Bernstein, S. (1987) A convenient procedure for the sulphation of flavonoid glycosides: preparation of rutin nono- and deca-sulphates. *J. Carbohydr. Chem.*, **6**, 639–44.

Nakano, T., Ito, Y. and Ogawa, T. (1993) Synthesis of sulphated glucuronyl glycosphingolipids: carbohydrate epitopes of neural cell-adhesion molecules. *Carbohydr. Res.*, **243**, 43–69.

Needs, P.W. and Selvendran, R.R. (1993) An improved methylation procedure for the analysis of complex polysaccharides including resistant starch and a critique of the factors which lead to undermethylation. *Phytochem. Anal.*, **4**, 210–16.

Needs, P.W. and Selvendran, R.R. (1994) A critical assessment of a one-tube procedure for the linkage analysis of polysaccharides as partially methylated alditol acetates. *Carbohydr. Res.*, **254**, 229–44.

Neier-Augenstein, W. and Schildknecht, H. (1990) Analytical and preparative high-performance liquid chromatographic systems for the separation of an anomeric mixture of 4-*O*-(D-glucopyranosyl)gallic acid. *J. Chromatogr.*, **518**, 254–57.

Neville, G.A., Rochon, P., Rej, R.N. and Perlin, A.S. (1991) Characterization and differentiation of some complex dextran sulphate preparations of medicinal interest. *J. Pharm. Sci.*, **80**, 239–44.

Nishino, T. and Nagumo, T. (1991) The sulphate-content dependence of the anticoagulant activity of a fucan sulphate from the brown seaweed *Ecklonia kurome*. *Carbohydr. Res.*, **214**, 193–7.

Nishino, T. and Nagumo, T. (1992) Anticoagulant and antithrombin activities of oversulphated fucans. *Carbohydr. Res.*, **229**, 355–62.

Nishino, T., Aizu, Y. and Nagumo, T. (1991a) The influence of sulphate content and molecular weight of a fucan sulphate from the brown seaweed *Ecklonia kurome* on its antithrombin activity. *Thrombosis Res.*, **64**, 723–31.

Nishino, T., Kiyohara, H., Yamada, H. and Nagumo, T. (1991b) An anticoagulant fucoidan from the brown seaweed *Ecklonia kurome*. *Phytochemistry*, **30**, 535–9.

Nishino, T., Nagumo, T., Kiyohara, H. and Yamada, H. (1991c) Structural characterization of a new anticoagulant fucan sulphate from the brown seaweed *Ecklonia kurome*. *Carbohydr. Res.*, **211**, 77–90.

Nishino, T., Takabe, Y. and Nagumo, T. (1994) Isolation and partial characterization of a novel β-D-galactan sulphate from the brown seaweed *Laminaria angustata* var. *longissima*. *Carbohydr. Polym.*, **23**, 165–73.

Ochi, Y., Watanabe, K., Okui, K. and Shindo, M. (1980) Crystalline salts of sucrose octasulphate. *Chem. Pharm. Bull.*, **28**, 638–41.

Osawa, Z., Morota, T., Hatanaka, K., Akaike, T. *et al.* (1993) Synthesis of sulphated derivatives of curdlan and their anti-HIV activity. *Carbohydr. Polym.*, **21**, 283–8.

Painter, T.J. (1983) Algal polysaccharides. In: Aspinall, G.O. (Ed.) *The Polysaccharides*, vol. 2, pp. 196–285. Orlando, FL: Academic Press.

Patankar, M.S., Oehninger, S., Barnett, T. *et al.* (1993) A revised structure for fucoidan may explain some of its biological activities. *J. Biol. Chem.*, **268**, 21770–6.

Paulsen, B.S., Vieira, A.A.H. and Klaveness, D. (1992) Structure of extracellular polysaccharides produced by a soil *Cryptomonas* sp. (Cryptophyceae). *J. Phycol.*, **28**, 61–3.

Pavão, M.S.G., Albano, R.M., Lawson, A.M. and Mourão, P.A.S. (1989) Structural heterogeneity among unique sulphated L-galactans from different species of ascidians (Tunicates). *J. Biol. Chem.*, **264**, 9972–9.

Peat, S., Bowker, D.M. and Turvey, J.R. (1968) Sulphates of monosaccharides and derivatives. Part VI. D-Glucose 2-sulphate, D-galactose 2- and 3-sulphate and D-galactose 2,3-disulphate. *Carbohydr. Res.*, **7**, 225–31.

Penney, C.L. and Perlin, A.S. (1981) A method for the sulphation of sugars, employing a stable, aryl sulphate intermediate. *Carbohydr. Res.*, **93**, 241–6.

Percival, E. and Wold, J.K. (1963) The acid polysaccharide from the green seaweed *Ulva lactuca*. Part II: The site of ester sulphate. *J. Chem. Soc.*, 5459–68.

Piotrowski, J., Murty, V.L.N., Czajkowski, A. *et al.* (1994) Association of salivary bacterial aggregating activity with sulphomucin. *Biochem. Mol. Biol. Int.*, **32**, 713–21.

Piotrowski, J., Slomiany, A., Murty, V.L.N. *et al.* (1991) Inhibition of *Helicobacter pylori* colonization by sulphated gastric mucin. *Biochem. Int.*, **24**, 749–56.

Poole, A.R. (1986) Proteoglycans in health and disease: structures and functions. *Biochem. J.*, **236**, 1–14.

Ragazzi, M., Ferro, D.R., Provasoli, A. *et al.* (1993) Conformation of the unsaturated uronic acid residues of glycosaminoglycan disaccharides. *J. Carbohydr. Chem.*, **12**, 523–5.

Rao, E.V. and Ramana, K.S. (1991) Structural studies of a polysaccharide isolated from the green seaweed *Chaetomorpha anteninna*. *Carbohydr. Res.*, **217**, 163–70.

Rao, E.V. Rao, N.V.S.A.V.P. and Ramana, K.S. (1991) Structural features of the sulphated polysaccharide from a green seaweed, *Spongomorphia indica*. *Phytochemistry*, **30**, 1183–6.

Rashid, A., Mackie, W., Colquhoun, I.J. and Lamba, D. (1990) Novel synthesis of monosulphated methyl α-D-galactopyranosides. *Can. J. Chem.*, **68**, 1122–7.

Rej, R.N. and Perlin, A.S. (1990) Base-catalysed conversion of the α-L-iduronic acid 2-sulphate unit of heparin into a unit of α-L-galacturonic acid, and related reactions. *Carbohydr. Res.*, **200**, 437–47.

Roberton, A.M., McKenzie, C.G., Sharfe, N. and Stubbs, L.B. (1993) A glycosulphatase that removes sulphate from mucus glycoprotein. *Biochem. J.*, **293**, 683–9.

Roux, L., Holojda, S., Sundblad, G., Freeze, H.H. and Varki, A. (1988) Sulphated *N*-linked oligosaccharides in mammalian cells. *J. Biol. Chem.*, **263**, 8879–89.

Roy, A.B. and Turner, J. (1983) On the desulphation of carbohydrate sulphates. *Carbohydr. Res.*, **124**, 338–43.

RSC (1966–present) *Carbohydrate Chemistry*, annual. London: Royal Society of Chemistry.

Sangadala, S., Bhat, U.R. and Mendicino, J. (1993) Structures of sulphated oligosaccharides in human trachea mucin glycoproteins. *Mol. Cell. Biochem.*, **126**, 37–47.

Schweiger, R.G. (1978) Sodium cellulose sulphate via cellulose nitrite. In: Schweiger, R.G. (Ed.) *Carbohydrate Sulphates*, pp. 163–72. Washington, DC: American Chemical Society.

Scott, J.E. (1993) The nomenclature of glycosaminoglycans and proteoglycans. *Glycoconjugate J.*, **10**, 419–21.

Selby, H.H. and Whistler, R.L. (1993) Agar. In: Whistler, R.L. and BeMiller, J.N. (Eds) *Industrial Gums: Polysaccharides and their Derivatives*, 3rd edn., pp. 87–103. San Diego, CA: Academic Press.

Shinagawa, S., Kasahara, F., Wada, Y. *et al.* (1984) Structures of bulgecins, bacterial metabolites with bulge-inducing activity. *Tetrahedron*, **40**, 3465–70.

Silvestri, L.S., Hurst, R.E., Simpson, L. and Settine, J.M. (1982) Analysis of sulphate in complex carbohydrates. *Anal. Biochem.*, **123**, 303–9.

Slim, G.C., Furneaux, R.H. and Yorke, S.C. (1994) A procedure for the analysis of glycosaminoglycan mixtures based on digestion by specific enzymes. *Carbohydr. Res.*, **255**, 285–93.

Soeda, S., Sakaguchi, S., Shimeno, H. and Nagamatsu, A. (1992) Fibrinolytic and anticoagulant activities of highly sulphated fucoidan. *Biochem. Pharmacol.*, **43**, 1853–8.

Stanley, N.F. (1990) Carrageenans. In: Harris, P. (Ed.) *Food Gels*, pp. 79–119, London: Elsevier.

Stevenson, T.T. and Furneaux, R.H. (1991) Chemical methods for the analysis of sulphated galactans from red algae. *Carbohydr. Res.*, **210**, 277–98.

Sugahara, K., Shigeno, K., Masuda, M. *et al.* (1994) Structural studies on the chondroitinase ABC-resistant sulphated tetrasaccharides isolated from various chondroitin sulphate isomers. *Carbohydr. Res.*, **255**, 145–63.

Sundblad, G., Holojda, S., Roux, L. *et al.* (1988a) Sulphated *N*-linked oligosaccharides in mammalian cells II. Identification of glycosaminoglycan-like chains attached to complex type glycans. *J. Biol. Chem.*, **263**, 8890–6.

Sundblad, G., Kaiji, K., Quaranta, V. *et al.* (1988b) Sulphated *N*-linked oligosaccharides in mammalian cells III. Characterisation of a pancreatic carcinoma cell surface glycoprotein with *N*- and *O*-sulphate esters on asparagine-linked glycans. *J. Biol. Chem.*, **263**, 8897–903.

Tailler, D., Jacquinet, J.-C. and Beau, J.-M. (1994) Total synthesis of NodRm IV (S): a sulphated lipotetrasaccharide symbiotic signal form *Rhizobium meliloti*. *J. Chem. Soc., Chem. Commun.*, 1827–8.

Takano, R., Matsuo, M., Kamei-Hayashi, K. *et al.* (1992a) A novel regioselective desulphation method specific to carbohydrate 6-sulphate using silylating reagents. *Biosci. Biotech. Biochem.*, **56**, 1577–80.

Takano, R., Ueda, T., Uejima, Y. *et al.* (1992b) Regioselectivity in sulphation of galactosides by sulphuric acid and dicyclohexylcarbodiimide. *Biosci. Biotech. Biochem.*, **56**, 1413–16.

Takiura, K. and Honda, S. (1970) Symmetric neutral sulphates of carbohydrates. *Chem. Pharm. Bull.*, **18**, 2125–7.

Telford, J.L., Covacci, A., Ghiara, P. *et al.* (1994) Unravelling the pathogenic role of *Helicobacter pylori* in peptic ulcer: potential new therapies and vaccines. *Tibtech.*, **12**, 420–6.

Therkelsen, G.H. (1993) Carrageenan. In: Whistler, R.L. and BeMiller, J.N. (Eds) *Industrial Gums: Polysaccharides and their Derivatives*, 3rd edn., pp. 145–80, San Diego, CA: Academic Press.

Thotakura, N.R. and Blithe, D.L. (1995) Glycoprotein hormones: glycobiology of gonadotrophins, thyrotrophin and free α subunit. *Glycobiology*, **5**, 3–10.

Turnbull, J.E. (1993) Oligosaccharide mapping and sequence analysis of glycosaminoglycans. *Methods Mol. Biol.*, **19**, 253–67.

Turvey, J.R. (1965) Sulphates of the simple sugars. *Adv. Carbohydr. Chem.*, **20**, 183–218.

Turvey, J.R. and Rees, D.A. (1961) Isolation of L-galactose-6-sulphate from a seaweed polysaccharide. *Nature*, **189**, 831–2.

Turvey, J.R., Bowker, D.M. and Harris, M.J. (1967) Infrared spectra of carbohydrate sulphates. *Chem. and Ind.*, 2081–2.

Tveter-Gallagher, E., Wight, T.N. and Mathieson, A.C. (1982) Effects of various types of carrageenans on human fibroblasts in vitro. *Mar. Algae Pharm. Sci.*, 51–64.

Tyrrell, D.J., Ishihara, M., Rao, N. *et al.* (1993) Structure and biological activity of a heparin-derived hexasaccharide with high affinity for basic fibroblast growth factor. *J. Biol. Chem.*, **268**, 4684–9.

Urch, U.A. and Patel, H. (1991) The interaction of boar sperm proacrosin with its natural substrate, the zona pellucida, and with polysulphated polysaccharides. *Development*, **111**, 1165–72.

Usov, A.I. (1984) NMR spectroscopy of red seaweed polysaccharides: agars, carrageenans, and xylans. *Botanica Marina*, **27**, 189–202.

Usov, A.I., Adamyants, K.S., Miroshnikova, L.I. *et al.* (1971) Solvolytic desulphation of sulphated carbohydrates. *Carbohydr. Res.*, **18**, 336–8.

Usov, A.I., Adamyants, K.S., Yarotsky, S.V. *et al.* (1973) The isolation of a sulphated mannan and a neutral xylan from the red seaweed *Nemalion vermiculare* sur. *Carbohydr. Res.*, **26**, 282–3.

van Boeckel, C.A.A. and Petitou, M. (1993) The unique antithrombin III binding domain of heparin: a lead to new synthetic antithrombotics. (Review). *Angew. Chem. Int. Ed. Engl.*, **32**, 1671–90.

Van Brunt, J. (1986) More to hyaluronic acid than meets the eye. *Biotechnology*, **4**, 780–2.

van der Klein, P.A.M., Filemon, W., Veeneman, G.H. *et al.* (1992) Highly regioselective ring-opening of five-membered cyclic sulphates with lithium azide: synthesis of azido sugars. *J. Carbohydr. Chem.*, **11**, 837–48.

Vischer, P. and Buddecke, E. (1991) Different action of heparin and fucoidan on arterial smooth muscle cell proliferation and thrombospondin and fibronectin metabolism. *Eur. J. Cell Biol.*, **56**, 407–14.

Volpi, N. (1994) Structural analysis and physico-chemical properties of charge-fractionated dermatan sulphate samples. *Carbohydr. Res.*, **260**, 159–67.

Vos, J.N., Westerduin, P. and van Boeckel, C.A.A. (1991) Synthesis of a 6-*O*-phosphorylated analogue of the antithrombin III binding sequence of heparin: replacement of one essential sulphate group by a phosphate group nullifies the biological activity. *Bioorg. Med. Chem. Lett.*, **1**, 143–6.

Vreeland, V., Zablackis, E. and Laetsch, W.M. (1992) Monoclonal antibodies as molecular markers for the intracellular and cell wall distribution of carrageenan epitopes in *Kappaphycus* (Rhodophyta) during tissue development. *J. Phycol.*, **28**, 328–42.

Wagenknecht, W., Nehls, I. and Philipp, B. (1993) Studies on the regioselectivity of cellulose sulphation in an N_2O_4-*N,N*-dimethylformamide-cellulose system. *Carbohydr. Res.*, **240**, 245–52.

Watanabe, Y., Sanada, S., Ida, Y. and Shoji, J. (1983) Comparative studies on the constituents of ophiopogonis tuber and its congeners. II. Studies on the constituents of the subterranean part of *Ophiopogon planiscapus* NAKAI. (1). *Chem. Pharm. Bull.*, **31**, 3486–95.

Welsch, U., Erlinger, R. and Storch, V. (1992) Glycosaminoglycans and fibrillar collagen in priapulida: a histo- and cytochemical study. *Histochemistry*, **98**, 389–97.

Wessel, H.P. (1992) Sulphated 1,6-anhydro-4-*O*-(β-D-glucopyranosyluronate)-β-D-glucopyranosyl derivatives: syntheses and conformations. *J. Carbohydr. Chem.*, **11**, 1039–52.

Witt, H.J. (1984) Sulphated seaweed polysaccharide conditioners for skin and hair. PCT Int. Appl. WO 84 04, 039.

Witvrouw, M., Este, J.A., Mateu, M.Q. *et al.* (1994) Activity of a sulphated polysaccharide extracted from the red seaweed *Aghardhiella tenera* against human-immunodeficiency-virus and other enveloped viruses. *Antiviral Chemistry and Chemotherapy*, **5**, 297–303.

Wlad, H., Maccarana, M., Eriksson, I. *et al.* (1994) Biosynthesis of heparin. Different molecular forms of *O*-sulphotransferases. *J. Biol. Chem.*, **269**, 24538–41.

Wong, C.-H., Gordon, J., Cooney, C.L. and Whitesides, G.M. (1981) Regeneration of NAD(P)H using glucose 6-sulphate and glucose-6-phosphate dehydrogenase. *J. Org. Chem.*, **46**, 4676–9.

Worawattanamateekul, W., Matsuda, M. and Okutani, K. (1993) Structural analysis of a rhamnose-containing sulphated polysaccharide from a marine *Pseudomonas*. *Nippon Suisan Gakkaishi*, **59**, 875–8.

Yamamoto, I., Takayama, K., Honma, K. *et al.* (1991) Structure and anti-viral activity of sulphates of cellulose and its branched derivatives. *Carbohydr. Polym.*, **14**, 53–63.

Yamashita, K., Ueda, I. and Kobata, A. (1983) Sulphated asparagine-linked sugar chains of hen egg albumin. *J. Biol. Chem.*, **258**, 14144–7.

Yoshida, T., Fennie, C., Lasky, L.A. and Lee, Y.C. (1994) A liquid-phase binding analysis for L-selectin. *Eur. J. Biochem.*, **222**, 703–9.

Yuen, C.-T., Bezouska, K., O'Brien, J. *et al.* (1994) Sulphated blood group Lewis A: a superior oligosaccharide ligand for human E-selectin. *J. Biol. Chem.*, **269**, 1595–8.

Zsiška, M. and Meyer, B. (1993) Influence of sulphate and carboxylate groups on the conformation of chondroitin sulphate related disaccharides. *Carbohydr. Res.*, **243**, 225–8.

4 Conjugation of monosaccharides – synthesis of glycosidic linkages in glycosides, oligosaccharides and polysaccharides

STEFAN OSCARSON

4.1 Introduction

Carbohydrates are heavily functionalised molecules. Each monosaccharide contains a number of hydroxyl groups to which another monosaccharide can be attached. This and the existence of other functional groups, substituents, two stereoisomers in the glycosidic linkage (α and β) and two ring forms (furanose and pyranose) render enormous possibilities of variation when coupling monosaccharides together to form oligo- and polysaccharides. Two D-glucopyranosyl moieties, for example, can be coupled to form 20 different products, which can be compared with only one possible product when forming a dipeptide from, for example, two L-alanyl residues. This means that a lot of biological information can be expressed in a rather short oligosaccharide sequence and the biological activity of carbohydrates are generally contained in structures smaller than a hexasaccharide. Therefore, carbohydrates in nature, compared with the other natural polymers, proteins and nucleic acids, consist of relatively short fragments, and when longer chains are formed these are built up by repeating units that are polymerised. From a synthetic point of view these limited lengths of biologically active carbohydrates are of course an advantage, but the complexity of carbohydrates makes oligosaccharide synthesis an area of craftsmanship, where a general automated approach, working well for most syntheses of oligopeptides and oligonucleotides, is still not available.

Three areas have to be mastered in oligosaccharide synthesis:

- synthesis of the starting monosaccharides, a problem with unusual, not commercially available, monosaccharide moieties, often found in bacterial and fungal carbohydrates;
- protecting group manipulation, to afford a regioselectively protected saccharide (an acceptor or aglycon) with a free hydroxyl group ready for glycosylation and also a suitably protected glycosyl donor;
- glycosidic linkage formation between the donor and the acceptor in a stereoselective manner.

When discussing oligosaccharide synthesis the main focus is often on the glycosylation reactions, whereas the protection and deprotection of intermediates are less discussed. When actually performing oligosaccharide synthesis, the problem is often to find a suitable protecting group strategy or, more precisely, to find a finely tuned 'teamwork' between the protecting group patterns and the glycosylation reactions, since the protecting groups not only protect the functionality not to be involved in the reactions but also render different properties and behaviour to the saccharides into which they are introduced. The methods used and problems encountered in glycoside synthesis will be discussed in this chapter. The main focus will be on chemical synthesis of *O*-glycosides, but the formation of *N*-, *S*- and *C*-glycosides as well as enzymatic synthesis of *O*-glycosides will also be discussed.

4.2 Chemical synthesis of *O*-glycosides

4.2.1 Glycosylation methods

In a synthesis of an *O*-glycoside, a glycosyl donor, so called since it donates a glycosyl moiety, is reacted with a free hydroxyl group in a glycosyl acceptor, generally in the presence of some promoter, to give a glycoside (Scheme 4.1).

In the formation of *O*-glycosides, methods that create an intermediate with a more or less pronounced cationic charge at the anomeric centre of the glycosyl donor through the promoter-induced departure of an anomeric leaving group have almost exclusively been used. This partly cationic intermediate, stabilised by resonance with the ring oxygen, is then attacked by the nucleophilic hydroxyl group of the acceptor, to form the glycosidic linkage (Scheme 4.1). Apart from the protecting group problem there are two primary questions to be asked when performing a glycoside synthesis: which leaving group and promoter are to be used and which anomeric configuration is to be synthesised, α or β, or, more important from a synthetic point of view, 1,2-*cis* or 1,2-*trans* configuration? In this chapter almost exclusively the formation of pyranosides will be discussed; however, furanoside formation follows the same principles,

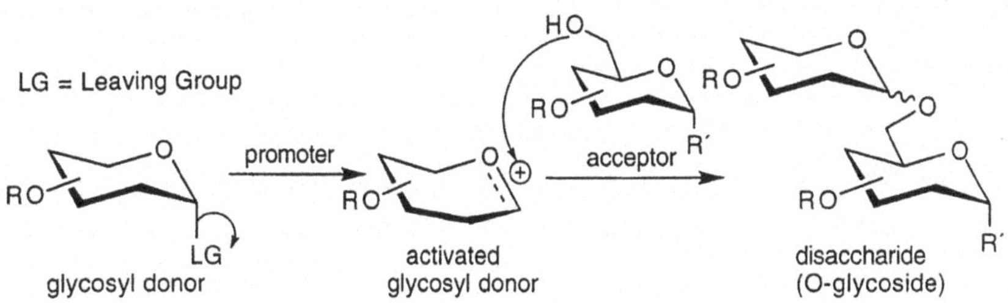

Scheme 4.1 General glycosylation reaction.

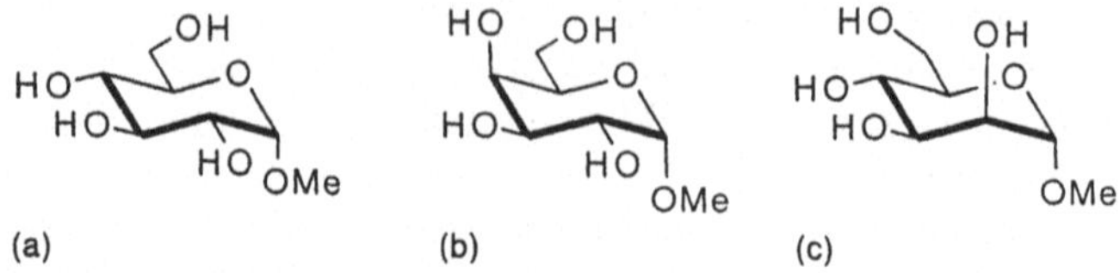

Figure 4.1 Methyl α-D-glucopyranoside; (b) methyl α-D-galactopyranoside; (c) methyl α-D-mannopyranoside.

and the same glycosylation methods can be and are being used. Furthermore, examples discussed will be with sugars having a D-*gluco*, D-*galacto* and D-*manno* configuration (Figure 4.1), since these are the most abundant in nature, especially in mammals, but once more the principles and methods are valid for other configurations as well.

As an answer to the first question above, one can choose between a vast number of methods, which have been developed during the years (for reviews and books, see Bochkov and Saikov, 1979; Garegg and Lindberg, 1988; Paulsen, 1982; Schmidt, 1986, 1991; Toshima and Tatsuta, 1993). Since the glycosidic linkage is part of an acetal, the most simple way to synthesise chemically a glycoside is to treat an unprotected saccharide (a hemiacetal) with an alcohol under acidic conditions (a Fischer glycosidation, Scheme 4.2). This works only with simpler alcohols, which can be used in excess to avoid glycosides formed from the saccharide's own hydroxyl groups. A further complication in this method is the simultaneous formation of both α- and β-glycosides as well as of both pyranosides and furanosides. Thus, for this glycosylation to be of value, either one of the products must dominate at equilibrium or a good method to isolate the wanted product from the product mixture, for example by crystallisation, has to be present. The main product is generally the α-pyranoside, since this is the thermodynamically most stable derivative of most hexoses because of the so-called anomeric effect (Juaristi and Cuevas, 1992). These limitations of the Fischer glycosidation method have made it necessary to develop other more general glycosidation methods in which protected donor and acceptor saccharides are used in more stoichiometric proportions.

Peracetylated, that is fully acetylated, saccharides are an easily accessible type of glycosyl donor, in which the anomeric acetyl group can be activated

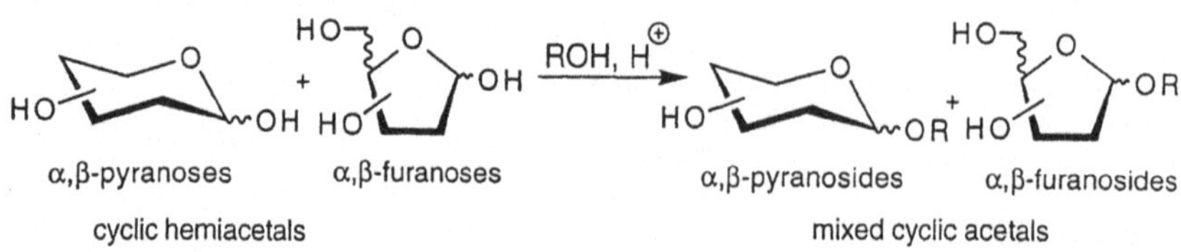

Scheme 4.2 Fischer glycoside synthesis.

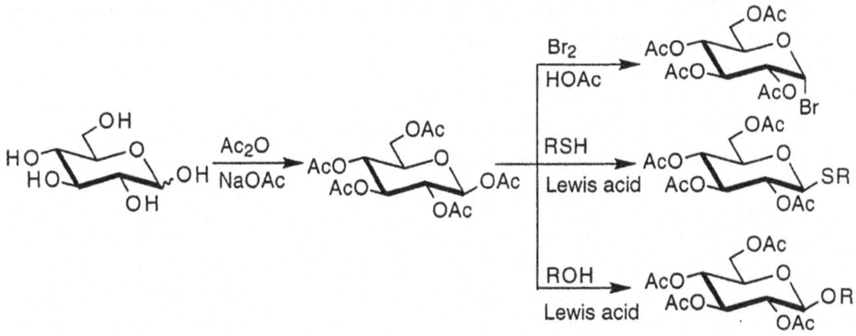

Scheme 4.3 Formation and some reactions of peracylated β-D-glucopyranose.

into a leaving group by Lewis acids. The anomeric acetates are good donors for simpler alcohol acceptors and give preferentially the 1,2-*trans*-glycosides (Section 4.2.2.1). If prolonged treatment with Lewis acid is performed the 1,2-*trans*-glycosides can anomerise to the 1,2-*cis*-glycosides if these are thermodynamically more stable. Anomeric acetates are also excellent precursors for many of the other types of donors discussed below (glycosyl halides, thioglycosides, pentenyl glycosides etc.; Scheme 4.3).

Although some oligosaccharide syntheses have been performed with anomeric acetates as donors, these are often too unreactive, and other types of methods generally have to be employed when the acceptor is another saccharide. The most frequently used methods are:

- glycosyl halides as donors and mercury and silver salts as promoters;
- trichloroacetimidates as donors and hard Lewis acids as promoters;
- thioglycosides as donors and thiophilic reagents (soft Lewis acids) as promoters (Baressi and Hindsgaul, 1995).

4.2.1.1 Glycosyl halides

These are the classical glycosyl donors, already in use in 1901 by Koenigs and Knorr, and still frequently in use. Glycosyl bromides or chlorides have been preferentially used. Acylated halides are conveniently made from the peracetylated saccharides by treatment with the corresponding hydrogen halide (HBr, HCl) in acetic acid (Scheme 4.3) and, because of their easy accessibility and high reactivity, they are still often the first choice of monosaccharide donors activated with various heavy metal salt promoters. Mercury bromide and cyanide have been extensively used, but recently trifluoromethanesulfonates (triflates) have been found to be the counterion of choice in glycosidation reactions and, if high reactivity is desired, silver triflate is now the promoter most often used.

The drawback of glycosyl halides as donors is that they are rather unstable, especially if the donor is alkylated and/or is a larger oligosaccharide. Therefore they have to be prepared in the last step before glycosylation. Elimination and hydrolysis are frequent side-reactions, which can:

- make glycosyl halides difficult to synthesise;
- lower the yields considerably in the glycosidations;
- mean that the donor must be added in large excess.

Glycosyl fluorides, which are more stable donors, are, then, an alternative used increasingly. In contrast to bromides and chlorides, some protecting group manipulations can be performed on fluorides, but most often they are prepared immediately prior to glycosidation, conveniently from, *inter alia*, thioglycosides (Scheme 4.5). Promoters used with fluorides are primarily metal salts, for example mixtures of tin and silver salts.

4.2.1.2 Trichloroacetimidates

These donors are made from the hemiacetal by treatment with trichloroacetonitrile and base [e.g. K_2CO_3, NaH or DBU (1,8-diazabicyclo[5,4,0]-undec-7-ene)], and the anomeric trichloroacetimidate is transformed into a leaving group by the addition of a Lewis acid, generally BF_3-etherate or trimethylsilyl triflate (Schmidt and Kinzy, 1994).

Although the trichloroacetimidates are reactive donors their formation under basic conditions is much less accompanied by side-reactions compared with glycosyl bromides and chlorides. Thus, donors even of large and alkylated oligosaccharide blocks can be prepared, and many syntheses of complex glycoconjugates have been performed using this approach.

Since the imidates are made from the hemiacetal some temporary protection of the anomeric function is necessary during hydroxyl protecting group manipulation and other glycosylations (Section 4.2.4.1). The hemiacetal function is then exposed by selective removal of the anomeric protecting group, whereafter the imidate is formed (Scheme 4.4) and used as a donor.

4.2.1.3 Thioglycosides

These donors are most conveniently prepared from peracetylated monosaccharides, which when treated with a mercaptan and a Lewis acid give

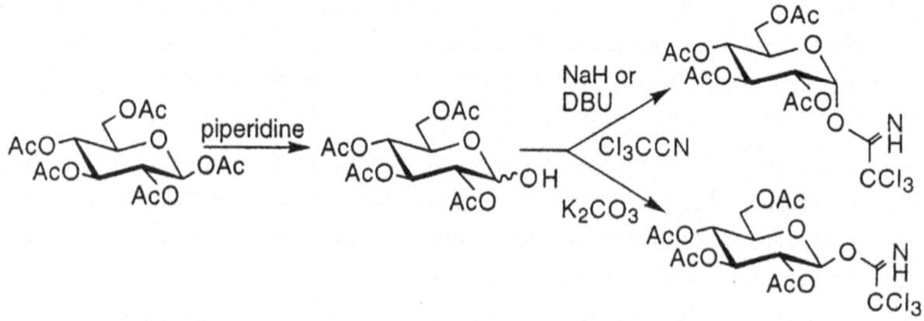

Scheme 4.4 Formation of trichloroacetimidate donors.

the corresponding 1,2-*trans* thioglycosides in high yields (Scheme 4.3). Activation of thioglycosides is performed by the addition of a thiophilic promoter, which activates the thioglycoside without affecting *O*-glycosidic linkages. These promoters are soft Lewis acids (C^+, S^+, I^+). The most common promoters used are methyl triflate, DMTST [dimethyl(methylthio)-sulfonium triflate], IDCP (iodonium dicollidine perchlorate) and NIS (*N*-iodosuccinimide), the latter often together with triflic acid or silver triflate (Norberg, 1995).

In contrast to the donors discussed above, the thioglycosides, if not activated, are stable during most protection group manipulations as well as under many glycosidation conditions using other types of donors, so-called orthogonal glycosylations, since the products are still donors. Thus, the thioglycoside can be introduced early, at the monosaccharide level, and then almost any type of glycosyl donor can be built up from this precursor using methods compatible with the thioglycoside function.

Another advantage with thioglycosides is that they are easily transformed into most of the other types of donors discussed. Treatment with chlorine or bromine gives the corresponding glycosyl chloride or bromide, which is, owing to the mild conditions, by far the best way to synthesise unstable glycosyl halides; treatment with, for example, DAST (diethylaminosulfur trifluoride) gives the glycosyl fluoride; treatment with a promoter in the presence of water gives the hemiacetal, which can be transformed to the imidate and, in the presence of acetic acid, the anomeric acetate is produced (Scheme 4.5).

4.2.1.4 Other types of donors

As mentioned, there are numerous other types of glycosyl donors (Figure 4.2), for example pentenyl glycosides (Fraser-Reid *et al.*, 1992), with properties similar to those of thioglycosides in that they are stable towards many other reaction conditions, both protecting group manipulations and glycosylations, and can be introduced early in the synthesis and then activated and function as donors at a suitable level in the synthesis. Another example is selenoglycosides (Mehta and Pinto, 1993), also fairly stable and which can be selectively activated in the presence of thio and pentenyl glycosides, adding flexibility to the synthetic schemes. Many other donors have been described (Toshima and Tatsuta, 1993) but have so far not found any wider applicability. Of interest in the future might be sulfoxide donors (Kahne *et al.*, 1989), reported to be more reactive than the corresponding thioglycoside from which they are derived by oxidation with MCPBA (*m*-chloroperbenzoic acid) and 1,2-anhydro donors (Danishefsky and Bilodeau, 1996), both of which also have been used as donors in solid-phase synthesis of oligosaccharides (Danishefsky *et al.*, 1993; Yan *et al.*, 1994). The 1,2-anhydro donors are obtained by a stereoselective epoxidation of glycals using 3,3-dimethyldioxirane. The glycals have also been frequently used

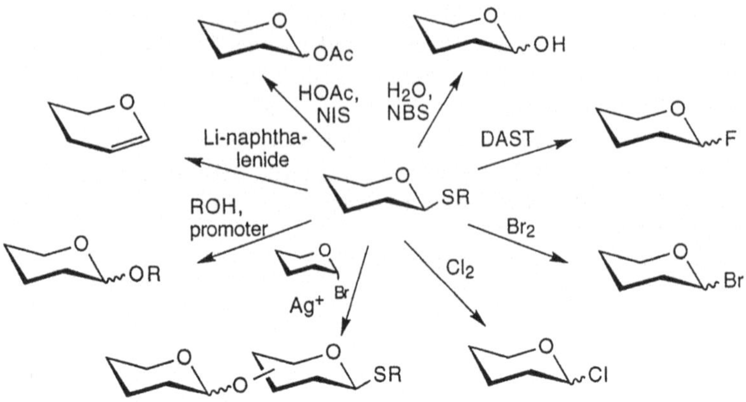

Scheme 4.5 Some reactions of thioglycosides.

directly as donors in the construction of 2-deoxysugars using NIS as promoter (Thiem and Klaffke, 1990). The method with glucals yields preferentially the diaxial 2-iodo glucoside, which is reduced to the 2-deoxy-glucoside. Other types of donors used are various phosphate or phosphite derivatives, resembling nature's own glycosyl donors (Section 4.4.1), and vinyl glycosides (Boons and Isles, 1996; Marra *et al.*, 1992). Most coupling reactions, independent of the method used, are accompanied by hydrolysis and 1,2-elimination of the donor (not the glycal!). This is especially prevalent with an unreactive acceptor and a reactive donor.

When choosing between different donors, not only the result in the actual glycosylation is important but also the preceding steps (i.e. the accessibility of

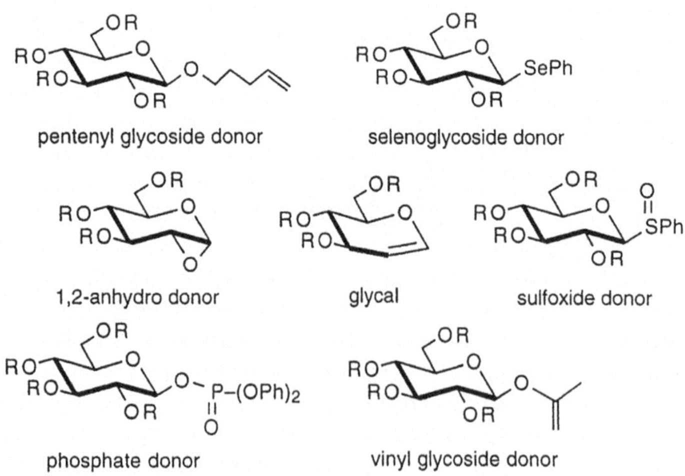

Figure 4.2 Various types of glycosyl donors (R = protecting groups).

the donor) and the following steps (i.e. the properties of the resulting glycoside) have to be taken into account. Another important factor is, as always in syntheses, familiarity with the method and thereby often the supply of useful starting materials and chemicals. When making larger oligosaccharides a combination of glycosylation methods usually has to be used, and often several donors and acceptors have to be tested to find optimum conditions. Of prime importance is, then, as mentioned, not only the glycosylation method but also the protecting group pattern of the donor and the acceptor.

4.2.2 Anomeric configuration

Of key importance is also the earlier mentioned question regarding the stereochemistry in the coupling: which anomeric configuration is to be synthesised?

*4.2.2.1 1,2-*Trans*-glycosides*

If a 1,2-*trans*-glycoside (i.e. a β-glycoside in the *gluco* and *galacto* series and the α-glycoside in the *manno* series; Figure 4.3) is the target molecule the stereoselectivity in the glycosidation reaction is usually accomplished by using a participating group as a protecting group at O-2.

With acyl protecting groups at O-2, the reaction is believed to proceed via a cyclic acyloxonium intermediate in which the anomeric centre can only be attacked from the opposite side by the acceptor to give the *trans*-glycoside (Scheme 4.6). This is normally also the case, and with 2-*O*-acyl protection *trans*-glycosides are produced exclusively or at least in large excess. Benzoates (R^1 = Ph) are often preferred over acetates (R^1 = Me) since they give fewer side-reactions. The same is true also for 1,2-anhydro donors, which can be attacked by the acceptor only from the opposite side of the 1,2-anhydro

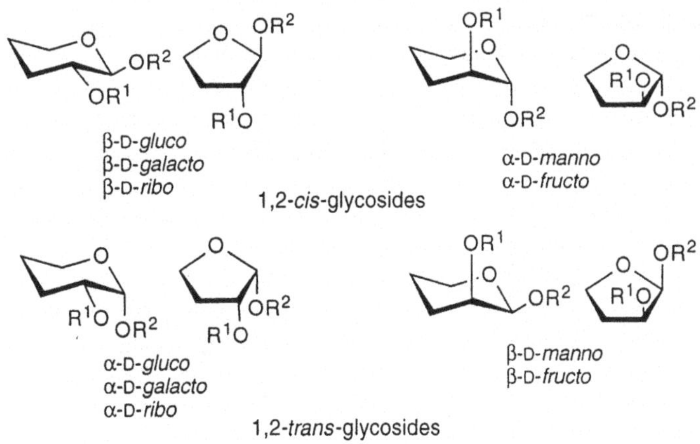

Figure 4.3

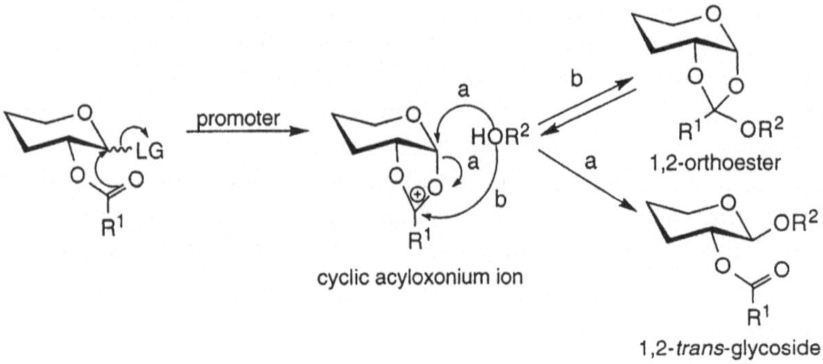

Scheme 4.6 Formation of 1,2-*trans*-linkages using neighbouring group participation.

bridge and thus give *trans*-glycosides. A possible side-reaction with participating 2-*O*-acyl groups is the formation of 1,2-orthoesters, which arise from the attack of the acceptor at the positively charged carbon in the cyclic intermediate instead of at the anomeric carbon (Scheme 4.6). If performed under basic conditions the orthoester will be the main product in the coupling reaction, but under acidic conditions the orthoester, if primarily formed, is rearranged to the glycoside. Orthoesters have also been used as glycosyl donors, with most success as the cyanoorthoester, to avoid competition from the orthoester alcohol in the formation of the glycoside, and then especially in polycondensation reactions (Scheme 4.7) (Backinovsky, 1994).

General glycosidation conditions are more or less strongly acidic, so an acid scavenger (a sterically hindered, non-nucleophilic amine) is sometimes added, especially if acid-labile protecting groups are present. To avoid orthoester formation, care has to be taken in these instances so as not to get too basic conditions. If orthoester formation is still a problem, the use of a pivaloyl ester as participating group at O-2 has been found to decrease the amount of orthoester product (Sato *et al.*, 1988).

Similar problems are encountered with 2-amino-sugars as donors. These are often present in nature as their 2-acetamido derivative, and to simplify protecting group manipulation it is an advantage to use this type of derivative as a donor in 1,2-*trans* glycosylations. However, 2-acetamido donors can after activation form, through the loss of the proton on the nitrogen,

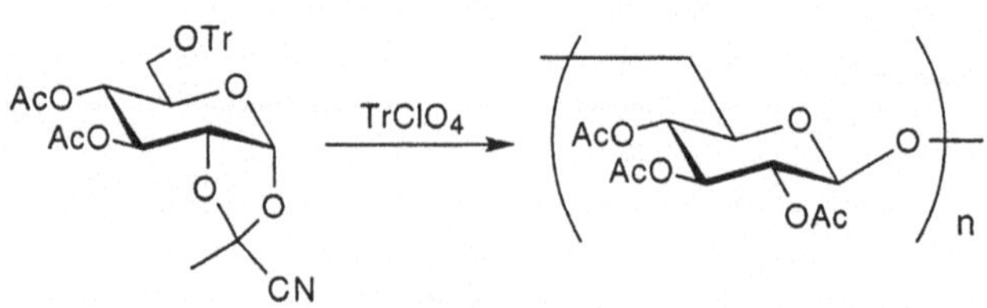

Scheme 4.7 Polycondensation of tritylated cyanoorthoesters.

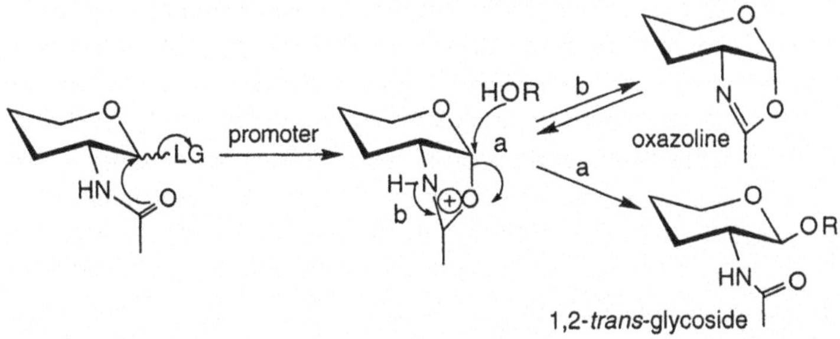

Scheme 4.8 Competing formation of an oxazoline or a 1,2-*trans*-glycoside from a 2-acetamido-2-deoxy donor.

oxazolines, which do not react further with unreactive acceptors even under acidic conditions (Scheme 4.8). Other protecting groups unable to form oxazolines, preferentially the phthalimido group and derivatives thereof, have therefore been developed as participating groups to create 1,2-*trans*-linkages with 2-amino-saccharides as donors (Figure 4.4) (Banoub *et al.*, 1992).

Recently, 2-*N*,*N*-di-acetates have been introduced as alternative donors. The advantage with these donors is their easy formation from and conversion back to the 2-acetamido derivative (by acetylation and mild base treatment, respectively) (Castro-Palomino and Schmidt, 1995).

Sometimes it is advantageous or even necessary to synthesise a 1,2-*trans* linkage without the use of a participating group, but this is considerably more difficult and almost never as stereoselective as with participating groups. There are few general rules (see discussion below) on how to perform these syntheses and an optimisation has to be made for each coupling. With homogenous promoters and non-participating groups a positively charged

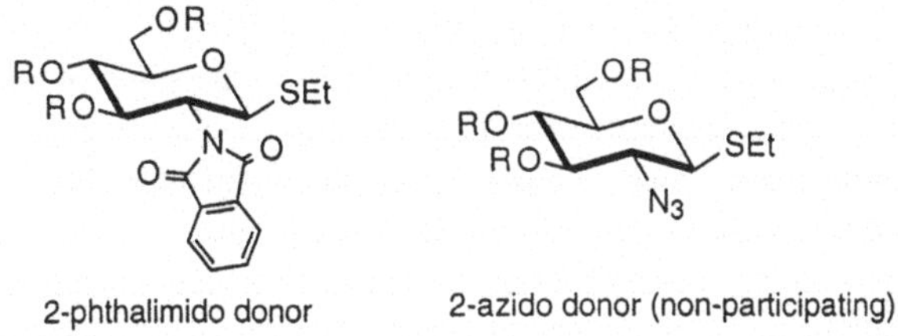

Figure 4.4 Various types of ethyl 2-amino-2-deoxy-1-thio-β-D-glucopyranoside donors (R = protecting group).

intermediate, more or less tightly bound to the counterion of the promoter, is usually first formed before the attack of the acceptor (Scheme 4.1). The stereochemistry of this intermediate is very difficult to predict and influence and therefore so also is the stereochemical outcome of the reaction. The thermodynamically more stable axial α-glycoside is normally formed in excess (Juaristi and Cuevas, 1992), especially if the acceptor is unreactive. This is especially marked with mannose donors, where a participating group is often not necessary to obtain 1,2-*trans*-α-D-manno linkages, and also, but less pronounced, with galactose donors (to give 1,2-*cis*-α-D-galacto derivatives). Both these saccharides have axial substituents (2 and 4, respectively) sterically covering the β side for an incoming aglycon.

*4.2.2.2 1,2-*Cis*-linkages*

When a 1,2-*cis*-linkage is to be synthesised no general solution like participating group for 1,2-*trans*-linkages is available, and a combination of empirically found rules are used together with trial and error to get an optimum yield of the desired product; one should not be surprised at exceptions in stereoselectivity from the discussed rules.

Addition of halide ions in couplings using halide donors has been found to enhance the α selectivity, and the halide-assisted glycosylation using gluco or galacto donors with non-participating groups and a tetraalkylammonium halide salt as sole promoter is a mild and highly stereoselective method for the formation of 1,2-*cis*-α linkages (Lemieux *et al.*, 1975). The addition of halide ions facilitates the anomerisation of the α-halide to the more reactive β-halide, which then can react, especially with reactive acceptors, in an S_N2 reaction to give the α-glycoside.

Since generally it is more difficult to make β linkages without the aid of participating groups, synthesis of 1,2-*cis*-β-D-mannosides, where participating groups cannot be used (Figure 4.3) is one of the most difficult glycosidic linkages to create. A main approach to solve this problem is to use a donor with α configuration and try to perform the reaction via an S_N2-type reaction, where the leaving group of the donor is directly displaced by the acceptor in a bimolecular reaction to give an inversion of configuration at the anomeric centre and, thus, the β-glycoside (Scheme 4.9). The α configuration is the stable form of most halide donors, and treatment of these with heterogeneous silver catalyst (e.g. silver zeolite or silver silicate) gives in a 'push–pull' S_N2 type of mechanism relatively high yields of the β-glycoside, especially with reactive acceptors (Paulsen and Lockhoff, 1981). This type of reasoning can also be used in the formation of α-glycosides from β donors.

Trichloroacetimidates can be synthesised either in the β form (kinetic product) or in the α form (thermodynamic product) (Scheme 4.4) and, by using BF_3-etherate as promoter at low temperatures, an excess of inversion product can be obtained, but the outcome is critically dependent on the reaction conditions (Schmidt and Kinzy, 1994). As discussed above,

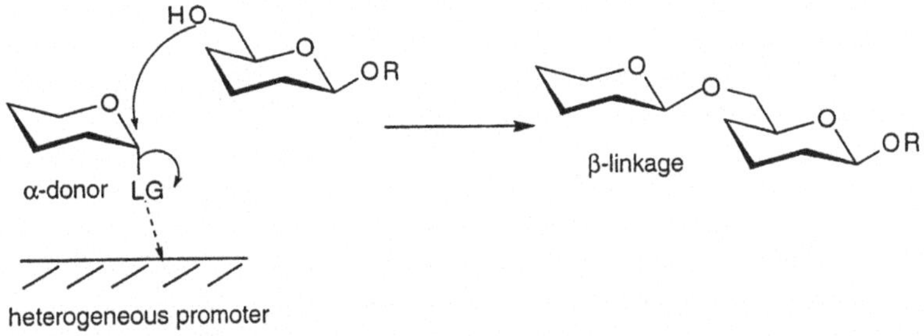

Scheme 4.9 Synthesis of glycosides with opposite configuration compared with the donor via an S_N2-type mechanism.

homogenous catalysts usually give an intermediate in which the stereochemistry of the donor is destroyed and the stereochemistry of which is difficult to influence. However, in special cases this might be possible, for example by using low temperature (−70°C) and conformationally restricted D-mannopyranose donors (4,6-*O*-benzylidene derivatives) it has been shown to be possible to form relatively stable α-triflates as intermediates. An S_N2 reaction with a later added acceptor gives β-D-mannopyranosides in high yield and with high stereoselectivity (Crich and Sun, 1996).

Solvent effects can also be used to influence the stereoselectivity. Acetonitrile has been found to increase the amount of β configuration, especially when the aglycon is a primary alcohol (Braccini *et al.*, 1993), probably through an α-nitrilium type of intermediate, and diethyl ether often increases the yield of the α-glycoside.

The protecting groups at the other positions apart from O-2 also affect the α/β ratio in a glycosylation, for example for a glucosyl bromide, when coupled to a 1,6-anhydro-manno derivative, it has been found that acyl groups at O-3 and O-6 raised the amount of α-linked glycoside compared with benzyl groups, whereas 4-*O*-acyl groups decreased the yield of α-linked product (van Boeckel, 1992). The actual use of these observations puts a lot of restraints on the protecting group patterns, but might be useful to take into consideration. It is important also to remember, as already pointed out, that there are often exceptions to these rules since every glycosylation is unique because of the complexity of carbohydrates.

A new approach towards 1,2-*cis* linkage formation is the idea of internal delivery of the aglycone. In this method the 2-substituent is once more used to control the stereoselectivity in the glycosidation, but now as an anchor for the aglycone, which then is delivered to the anomeric centre from the same side as the 2-substituent, to give a 1,2-*cis*-linkage. The aglycone is linked to the substituent at O-2 via an acetal linkage, either a carbon acetal or a silyl acetal. After activation of the anomeric site in the donor, usually a thioglycoside, the aglycone is transferred to the positively

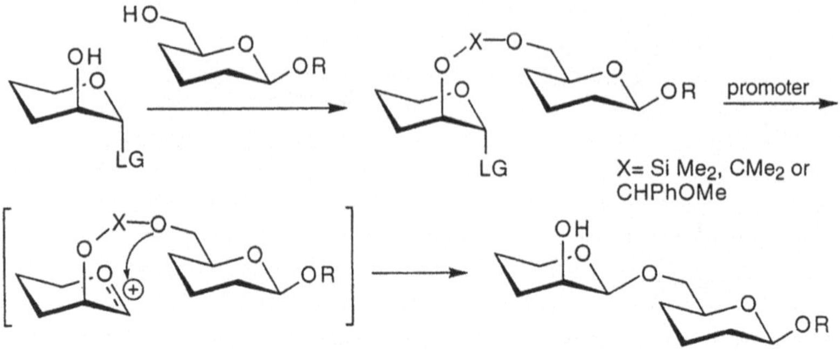

Scheme 4.10 Synthesis of 1,2-*cis*-glycosides using the internal aglycon delivery approach.

charged anomeric centre from the same side as the O-2 substituent to give good yields of 1,2-*cis*-glycosides (Scheme 4.10) (Baressi and Hindsgaul, 1994; Bols, 1993; Ito and Ogawa, 1994; Krog-Jensen and Oscarson, 1996; Stork and Kim, 1992). This method requires a selective protection of O-2, and the formation of the linked intermediate is not trivial, especially with complex carbohydrate donors and acceptors. Owing to these limitations it has not yet been used widely in oligosaccharide synthesis, and if the method is utilised it is early in the synthetic pathway (Dan *et al.*, 1995).

4.2.3 *Strategies*

Until now only the formation of a single glycoside linkage has been discussed, but when performing syntheses of oligosaccharides containing several glycosidic linkages, the order in which the linkages shall be formed must also be considered and decided. There are two major approaches: one may either perform a consecutive synthesis in which one monosaccharide at a time is introduced to give the desired oligosaccharide, or one may use a convergent (or block) synthesis in which larger blocks of oligosaccharides are first made and then connected to give the target oligosaccharide (Scheme 4.11).

The first approach has the advantage of an easy access to donors, since these are always monosaccharides. It also gives a number of intermediates, which can be deprotected and used in biological experiments to determine the minimum size of oligosaccharide needed for biological activity. The disadvantage is that a lot of reactions, both protecting group manipulations and glycosylations, have to be performed on large oligosaccharides, which often cause problems. On the other hand, in block syntheses, protecting group manipulation and glycosylations are mainly carried out on smaller fragments, so if (or perhaps one should say when) something goes wrong fewer synthetic steps are required to a new type of precursor. In convergent synthesis, the blocks can also be chosen so that difficult structural features

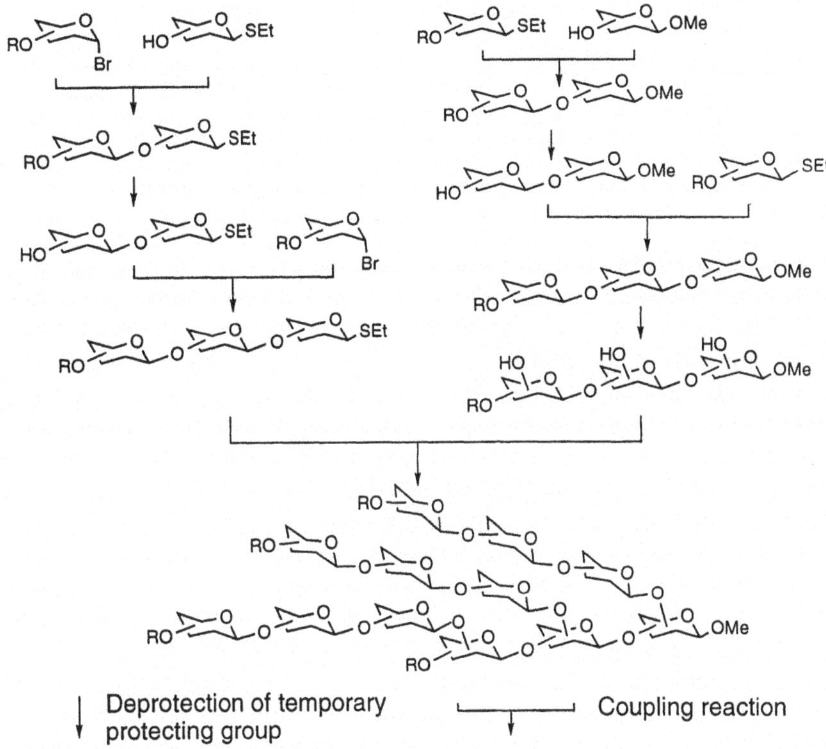

Scheme 4.11 Scheme of a block synthesis of a dodecasaccharide containing three identical trisaccharides.

(for example β-mannosides) can be synthesised at an early stage in the scheme. Earlier, the formation of stable oligosaccharide donors was a problem in block syntheses, but with the introduction of the new types of donors discussed (thioglycosides, pentenyl glycosides, trichloroacetimidates) the convergent type of synthesis is applicable to almost any type of structure and is now often the method of choice. The block type of synthesis is especially valuable when structures containing identical structural motifs are to be synthesised. A block corresponding to this structure is then first synthesised as a donor and then introduced at different positions into the target structure (Scheme 4.11). All larger oligosaccharide structures that have been made have been synthesised in this way, for example the present largest, a dodecapentasaccharide (Ogawa, 1994). The size of synthetic oligosaccharides not containing identical structural fragments has not yet exceeded the decasaccharide level.

When an oligosaccharide block donor is to be built up, a knowledge of which donors are compatible with which glycosylation methods is important.

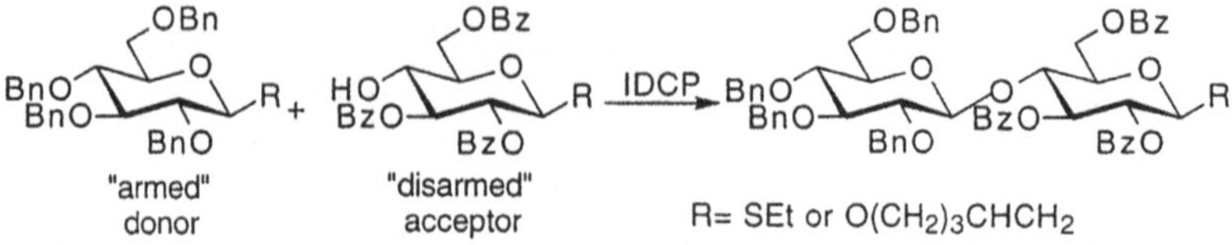

Scheme 4.12 Use of the armed–disarmed approach in glycoside synthesis.

If a thioglycoside block donor is to be used the block can be built up using most other glycosylation methods in orthogonal glycosylations, for example halide donors (silver salt promoters), trichloroacetimidates (hard Lewis acids) and seleno glycosides (silver salt promoters). Even thioglycoside donors can be used if these are activated ('armed') by alkyl protecting groups and the thioglycoside acceptor is deactivated ('disarmed') by acyl protecting groups. Then, if a mild promoter is used (for example IDCP) only the activated thioglycoside will function as a donor and no self-condensation of the deactivated acylated thioglycoside is observed (Scheme 4.12).

Another possibility is to use thioglycosides with different anomeric groups, since some of these can be activated, with a proper promoter, in the presence of the others. This can be performed using the same precursor thioglycoside, for example, a *p*-nitrophenyl thioglycoside or a phenyl thioglycoside, in which the anomeric group can be transformed into a more easily activated group, a *p*-acetamidophenyl thioglycoside or a phenyl sulfoxide donor, respectively, to allow coupling between the anomerically activated donor and the latent non-activated acceptor to form a *p*-nitrophenyl thioglycoside or a phenyl thioglycoside building block, respectively, which then can be used as donors in further glycosylations (Sliedregt *et al.*, 1994). This latent–active principle has also been used with glycals and 1,2-anhydro sugars (Danishefsky and Bilodeau, 1996) and with allyl/vinyl glycosides. In the latter a stable methallyl glycoside is isomerised at an appropriate stage to a dimethylvinyl glycoside, which can then act as a donor in the presence of methyl triflate in couplings with, for example, allyl glycosides as acceptors.

When the thioglycoside block is formed, it can, as mentioned, either be used as a donor directly, activated by thiophilic promoters, or first be converted to another type of donor (Scheme 4.5) and then activated with various promoters. Most of this is true also for pentenyl glycosides (Fraser-Reid *et al.*, 1992), but for trichloroacetimidate donors temporary protection of the anomeric centre during the build-up of the block is necessary. Some latent–active acceptor–donor pairs are illustrated in Figure 4.5.

Although this approach seems straightforward problems are often encountered in the formation of the different blocks apart from the earlier mentioned problems with low stereoselectivity. Unreactivity of acceptor hydroxyls, elimination and hydrolysis side-products, transacylation and transglycosidation side-products as well as a variety of other side-products

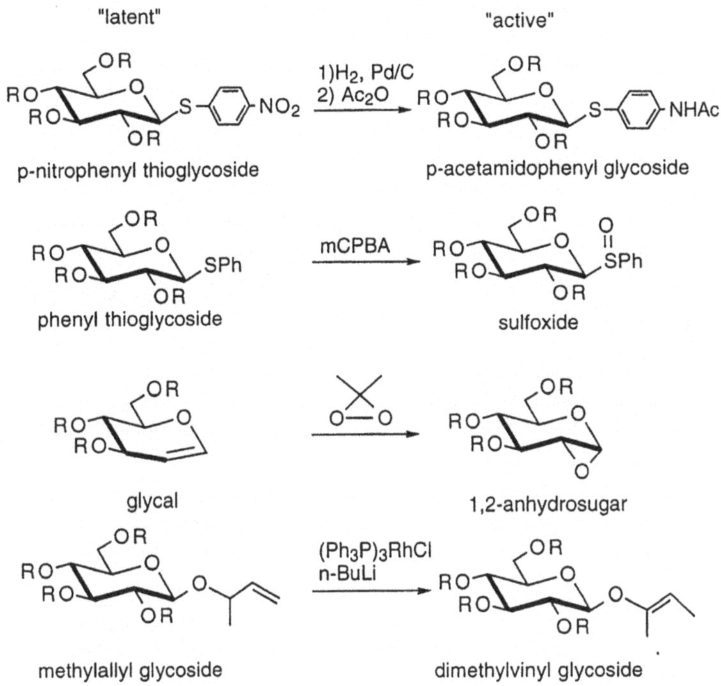

Figure 4.5 Some types of latent–active acceptor–donor pairs.

formed from unintentional reaction with various functionalities in the acceptor–donor pair often lower the yield in the coupling reactions (Scheme 4.13).

4.2.4 Protecting groups

The most abundant monosaccharides (pentoses and hexoses) are polyhydroxyaldehydes in the form of cyclic hemiacetals. Therefore hydroxyl protecting groups and hemiacetal protecting groups are of major interest in oligosaccharide synthesis. Other frequent functionalities in carbohydrates which need protection are amino groups (mainly in 2-amino-2-deoxy-hexoses) and carboxylic groups (e.g. in uronic acids in which the primary alcohol is oxidised) (Greene and Wuts, 1991; Kocienski, 1994).

In a normal glycosylation reaction, the glycosyl donor will have all hydroxyl groups protected and the hemiacetal hydroxyl group transformed into a leaving group, whereas the acceptor will have all the hydroxyl groups (including the hemiacetal) protected, except the one to be glycosylated (Scheme 4.1). The product can then either be deprotected to give an unprotected target oligosaccharide or be used as an intermediate in

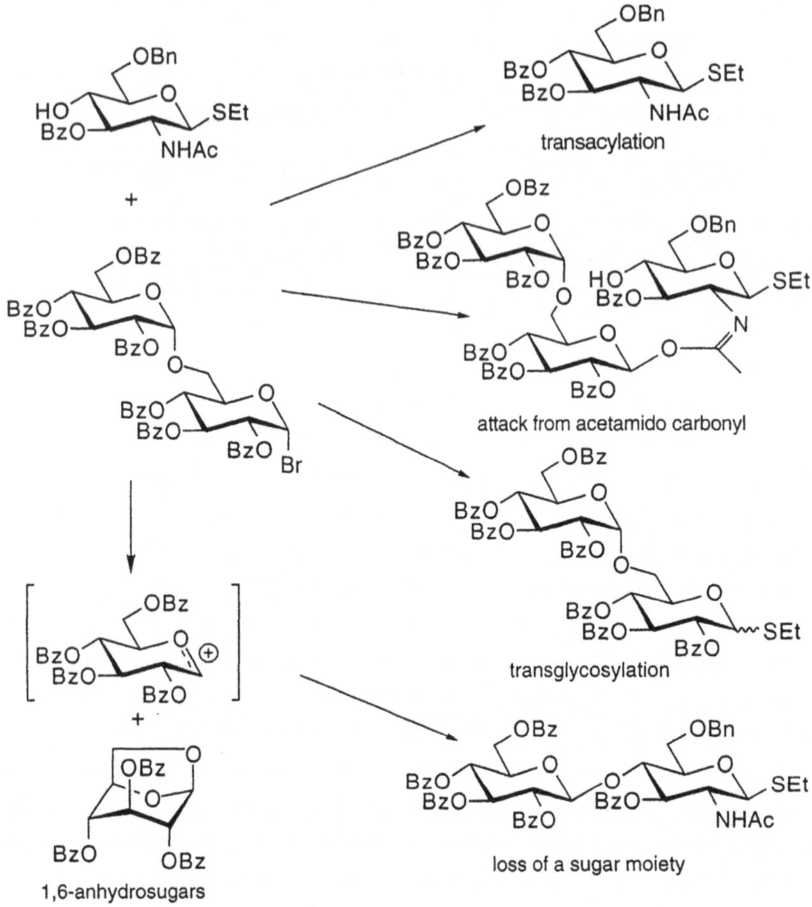

Scheme 4.13 Some possible side-reactions and products in a glycosylation reaction.

further glycosylations to give larger oligosaccharides. This can be performed in two ways. Either the product is used as an intermediate donor, which means that the hemiacetal protecting group must be transformed in a selective manner into a leaving group, or it is used as an intermediate acceptor, which means that one of the hydroxyl protecting groups must be removed selectively to afford a free hydroxyl group ready for glycosylation. Thus there is a need for various methods allowing regioselective introduction of protecting groups (Scheme 4.14) and also for both permanent and temporary blocking groups.

4.2.4.1 Hydroxyl protecting groups

By far the most used hydroxyl protecting groups are acetates, benzoates, benzyl ethers and benzylidene and isopropylidene acetals. Although many

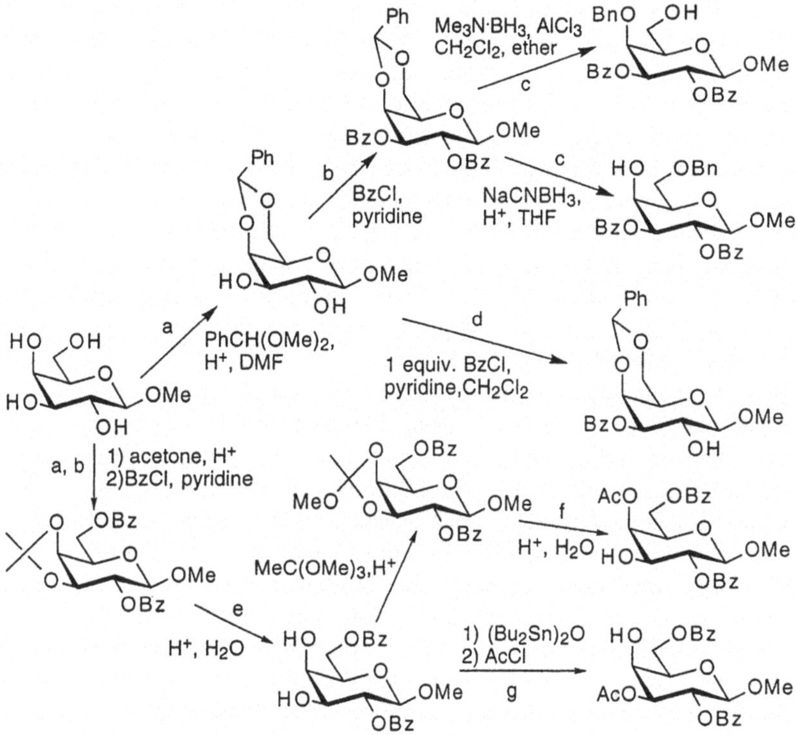

Scheme 4.14 Examples of regioselective protection of methyl β-D-galactopyranoside.

other new protecting groups have been developed, these are still the only ones that can be introduced and deprotected effectively on many positions at the same time. They also complement each other and can generally be put on and taken off in the presence of each other (so-called orthogonal protecting groups) and thus be used both as temporary and as persistent blocking groups. Furthermore, they also complement each other in the effect they have on the protected saccharide in glycosylation reactions; acyl groups are deactivating both on donors and acceptors, whereas ethers are activating.

Acetates are normally introduced using acetic anhydride or acetyl chloride with base (pyridine, sodium acetate) or acid catalysis, and benzoates using benzoyl chloride and pyridine (Scheme 4.14b). Peracylations are easily performed even on a large scale.

Regioselective acylation can be performed by using a deficit of acylating reagents or a less potent (e.g. acyl imidazoles) catalyst, to give directly suitable acceptors in which the least reactive hydroxyl group is unprotected (Scheme 4.14d) (Haines, 1976). Another approach is the use of tin activation prior to acylation. Dependent on whether dibutyl- or tributyltin oxide is used in the activation step, stannyl ethers or stannylidene acetals of the carbohydrate are produced, which enhance the nucleophilicity of different hydroxyl groups (Grindley, 1994). With dibutyltin oxide a general rule is that an equatorial hydroxyl group adjacent to an axial oxygen is preferentially acylated (Scheme 4.14g). Another useful method for the regioselective introduction of ester groups is the acidic opening of cyclic orthoesters. Five-membered cyclic orthoesters formed on *cis*-diols (eq, ax) give exclusively the axial ester (Scheme 4.14f) (Deslongchamps, 1983). With acylated derivatives with adjacent unprotected hydroxyl groups care has to be taken to avoid acyl migration. The migration is most prominent with acetates under slightly basic conditions but can also take place under acidic conditions and with benzoates. Deprotection of acyl groups is accomplished by treatment with base (MeO^-/MeOH, NH_3/MeOH, OH^-).

Benzyl protecting groups are best introduced under strongly basic conditions, for example NaH, using benzyl bromide as electrophile and dimethyl formamide (DMF) as solvent. Since acyl groups will not survive these conditions, milder basic conditions, for example Ag_2O, or other electrophiles, benzyl triflate or trichloroacetimidate, which can be introduced under neutral or slightly acidic conditions, have been used to allow benzylation of acyl protected or other base labile saccharides, but these latter methods are not as general and high yielding as the first mentioned.

Selective introduction of benzyl groups can be performed as for acyl groups by the use of molar equivalents of different reagents or by tin activation prior to benzylation (Grindley, 1994; Haines, 1976). The selectivity is comparable to those found for acylation. Other selective methods for the introduction of benzyl groups are prior formation of copper chelates, phase transfer catalysed benzylation and the reductive cleavage of benzylidene acetals (see below) (Stanek, 1990). When performed on diols, copper chelation gives mainly benzylation of the less acidic hydroxyl group (normally 4-*O*-benzylation of a 4,6-diol and 3-*O*-benzylation of a 2,3-diol) because of inactivation of the more acidic hydroxyl owing to chelation, whereas phase transfer catalysis gives the opposite selectivity. Benzyl groups are removed by catalytic hydrogenolysis using palladium on carbon as catalyst.

The introduction of benzylidene and isopropylidene acetals can be performed by the use of the corresponding aldehyde (benzaldehyde and acetone) and acid catalysis. More often the preformed dimethyl acetal of the carbonyl compound is used as reagent in a transacetalation reaction to give the more stable cyclic carbohydrate acetal (Scheme 4.14a). Other aldehyde-equivalent reagents used are *gem*-halogen derivatives, which are introduced under basic conditions, *gem*-haloethers (compare with glycosylation methods) and vinyl ethers. The advantages of the acetal protecting groups are that they protect two hydroxyl groups at a time, they are easy to introduce with a high degree of regioselectivity and the benzylidene and the isopropylidene acetals complement each other in selectivity (deBelder, 1977). Cyclic acetals are formed mainly between the 4- and 6-hydroxyl groups on hexopyranoses to form six-membered ring acetals (dioxanes) and between vicinal *cis*-hydroxyl groups to form five-membered ring acetals (dioxolanes). *Trans*-dioxolanes can be formed but this requires different conditions. Benzylidene acetals prefer to be in the dioxane form where the phenyl group can attain an equatorial position, whereas isopropylidene acetals with two methyl groups are more stable as dioxolane acetals. 4,6-*O*-benzylidene acetals are therefore easily formed from all hexopyranosides, whereas isopropylidenation of galactopyranosides gives preferentially the 3,4-*O*-acetal and of mannopyranosides the 2,3-*O*-acetal. If, however, kinetic control is used in the isopropylidenation reaction the 4,6-*O*-acetal will be formed, since the reaction with the primary hydroxyl is the fastest. Acetalation of reducing sugars with a free hemiacetal function sometimes complicates the picture but is also a way to many useful derivatives, for example furanose derivatives. The acetals are removed by acid hydrolysis (benzylidenes also by hydrogenolysis) and although the glycosidic linkage also is an acetal, the benzylidene and isopropylidene acetals can generally easily be removed without the hydrolysis of glycosidic linkages (Scheme 4.14e). Benzylidene acetals can also be transformed in a number of ways to give useful derivatives (Gelas, 1981). The most important of these are the reductive cleavage mentioned above to give regioselectively benzylated derivatives and oxidative cleavage using NBS in CCl_4 to give regioselectively benzoylated derivatives. If the latter reaction is performed under anhydrous conditions benzoylated bromo derivatives are obtained, which can be converted to deoxy sugars. In the reductive opening of 4,6-*O*-benzylidene derivatives, the use of $NaCNBH_3/HCl$ yields the 6-*O*-benzyl derivative, whereas $LiAlH_4/AlCl_3$ yields the 4-*O*-benzyl derivative. When using $(CH_3)_3N \cdot BH_3/AlCl_3$, tetrahydrofuran (THF) as solvent gives 6-*O*-selectivity whereas CH_2Cl_2/diethyl ether gives preferentially 4-*O*-selectivity (Scheme 4.14c). With five-membered ring acetals the selectivity is dependent not on the reagent or solvent but on the configuration of the acetal; *exo*-configuration gives the equatorial benzyl group, and *endo*-configuration the axial benzyl group. $NaCNBH_3$ and $(CH_3)_3N \cdot BH_3$ are compatible with ester groups, whereas $LiAlH_4$ is not.

These five protecting groups are the foundation upon which most protecting schemes are based, but there is often a need for additional orthogonal protecting groups with other properties. There is a long list of such groups, the most commonly used being various silyl groups, especially *tert*-butyldimethylsilyl and *tert*butyldiphenyl silyl groups, allyl ethers, trityl ethers, *p*-methoxybenzyl ethers, chloroacetates and levulinoyl esters (Greene and Wuts, 1991; Kocienski, 1994). All of these protecting groups are generally removable selectively in the presence of acetates, benzoates, benzyls and acetal groups and thus function well as temporary protecting groups at single hydroxyl groups. However, their introduction, stability and removal are not yet sufficiently optimised to allow their use as blocking groups at numerous positions. Their regioselective introduction can be performed using the same guidelines as outlined for benzyls, acetates and benzoates above. The silyl and trityl ethers, because of their bulkiness, can be used as selective protecting groups for primary hydroxyls.

Another approach to regioselective protection of carbohydrates is to use selective deprotection, chemical or enzymatic, of fully protected derivatives, this can sometimes be accomplished in a useful way using various protecting groups, but most often with acyl groups (Haines, 1981; Waldmann and Sebastian, 1994).

4.2.4.2 Anomeric (hemiacetal) protecting groups

Blocking groups for the hemiacetal function are mainly the same as those used for hydroxyl functions – benzyl, *p*-methoxybenzyl, allyl, acetyl, benzoyl groups and isopropylidene and benzylidene acetals – but also specific anomeric protecting groups such as thioglycosides, pentenyl glycosides and 2-(trimethylsilyl)ethyl glycosides (Jansson *et al.*, 1988) can be used. Anomeric acyl groups are introduced as for other acetates and benzoates, but they can be removed selectively in the presence of other acyl groups by the use of piperidine (or a similar weak base). The various glycosides are introduced either by a Fischer synthesis or by a glycosidation reaction using the peracetylated saccharide or the acetobromosugar as donor and the corresponding alcohol (or thiol) as acceptor (see Section 4.2.1).

The choice of anomeric protecting group is, *inter alia*, dependent on the fate of the anomeric centre. If it is to be used as an electrophile donor at some time during the synthesis and end up as part of an internal glycosidic linkage in the target product it can be an advantage to have it protected as a thio- or pentenyl glycoside, since these, as discussed, can be directly activated and used in glycosylation reactions, whereas the other protecting groups first have to be removed to give the hemiacetal, which then can be transformed into a donor. On the other hand, if the anomeric centre is to end up at the reducing end (i.e. as part of a hemiacetal) of the target oligosaccharide, it can be a disadvantage to use the thio- or pentenyl glycosides, since this restricts the use of other such compounds as donors in the synthesis.

The target oligosaccharide, however, is often not a reducing sugar but a glycoside. Frequently, the methyl glycoside of the oligosaccharide is chosen, because of its easy accessibility (monosaccharide methyl glycosides are often commercially available) and its usefulness in, for example, nuclear magnetic resonance (NMR) experiments and biological inhibition experiments. When the synthetic target oligosaccharides are to be used in other types of biological experiments, for example in affinity chromatography or as immunogens, they have to be coupled to different kinds of carriers to form various forms of glycoconjugates. This is usually accomplished with the use of a bifunctional linking arm. Examples of linking arms used are ethanolamine, *p*-nitro(or amino)phenol, 9-hydroxynonanoic acid, allyl alcohol and 2-[4-nitro(or amino)phenyl]ethanol. These are introduced into the saccharide as glycosides via their hydroxyl groups and their second functionality is then used for the coupling to the carrier; carboxyl groups via amide linkages, amino groups via amide or thiourea linkages, and double bonds via reaction with thioethanolamine and then amide or thiourea linkages (Aplin and Wriston, 1981; Garegg and Lindberg, 1988). In principle the linker can be introduced at any stage in the synthesis of the oligosaccharide but it is often advantageous to form the linker glycoside at the beginning, since it will then function as a permanent 'protecting group' for the reducing end throughout the synthesis.

4.2.4.3 Protection of other functional groups

In the protection of the amino function in 2-amino-2-deoxy-sugars, mainly two groups have been used apart from the naturally occurring acetamido function: as a participating group, the phthalimido group, and as a non-participating group, the azido group, an amino group equivalent (Figure 4.4) (Banoub *et al.*, 1992). The phthalimido group is formed through the reaction of the free amine with phthalic acid anhydride, whereas the azido group is introduced via azidonitration of a glycal derivative, azide displacement reaction of a triflate derivative or through an azide transfer reaction on the amino derivative (Vasella *et al.*, 1991). Treatment of the phthalimido function with strong base (e.g. hydrazine) regenerates the amino function whereas reduction of the azide also gives the amino derivative.

The most used protecting group for the carboxyl group is the methyl ester. If a base-labile protecting group cannot be used, the benzyl ester (deprotection: catalytic hydrogenolysis) or the *tert*-butyl ester (acid-labile) are alternatives.

4.2.5 Additional remarks

To avoid elaborate protecting group schemes regioselective glycosidation of polyols (so-called open glycosidation) might be considered. This approach is feasible if the reactivity difference between the hydroxyl groups is considerable and the stereoselectivity in the glycosylation reaction is good, otherwise the separation and characterisation problems of the different products as well as the loss in yield will outweigh the benefits. As an alternative for a fast formation of useful intermediates from unprotected monosaccharides the method has obvious advantages (Scheme 4.15) (Garegg *et al.*, 1995; Karhta *et al.*, 1995).

In conclusion, with the use of a few basic protecting groups together with a few basic glycosylation methods, sometimes perhaps 'spiced' with some alternative temporary protecting groups and glycosylation method, the synthesis of oligosaccharides of common monosaccharides, although still a

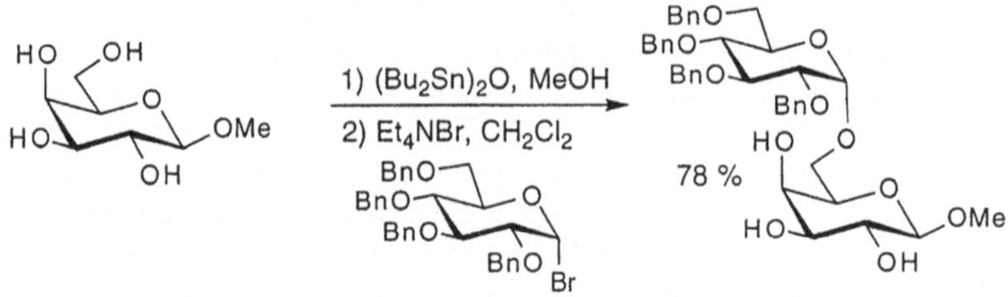

Scheme 4.15 Open glycosidation using stannylene activation of methyl β-D-galactopyranoside.

multistep synthesis, has become fairly straightforward, although problems with low stereoselection in some couplings and low accessibility of some acceptor hydroxyl groups, especially in the syntheses of branched oligosaccharides, can arise. However, though not discussed in this chapter, the introduction of new functionalities (amino groups, deoxy functions, carboxyl groups, etc.) and substituents (sulphates, phosphates, acetates, etc.) into the target molecules, features often present in biologically active carbohydrates and glycoconjugates, complicates the picture considerably and can make the syntheses of even a disaccharide a real challenge (and sometimes also a nuisance!).

4.3 Chemical synthesis of *S*-, *N*- and *C*-glycosides

4.3.1 S-glycosides

S-glycosyl derivatives are not so common in nature, but occur, *inter alia*, as glucosinolates in mustard-oil glucosides (Horton and Hutson, 1963). From a synthetic point of view they are perhaps most interesting as excellent glycosyl donors (see previous section), but thiooligosaccharides, in which the interglycosidic oxygen atoms have been replaced by sulfur, are also of interest as *O*-glycosyl analogues with useful properties, with, *inter alia*, increased resistance to glycosidases.

Thioglycosides of simple thiols are, as already discussed, most conveniently prepared from the peracetylated saccharide using a hard Lewis acid as promoter (Scheme 4.3). The acetobromosugar can also be used as a precursor in a reaction with a thiolate anion as nucleophile. Another approach is first to create the anomeric C–S linkage, through the reaction of a halide sugar with various thionucleophiles, for example thioacetic acid or thiourea. These derivatives are then transformed into the anomeric thioanion, which are alkylated to give the thioglycoside (Norberg, 1995).

This approach has also been used in the synthesis of thiooligosaccharides. The anomeric thioanion is allowed to displace, with inversion, a sugar triflate to yield the desired thiooligosaccharide (Scheme 4.16). In this way a number

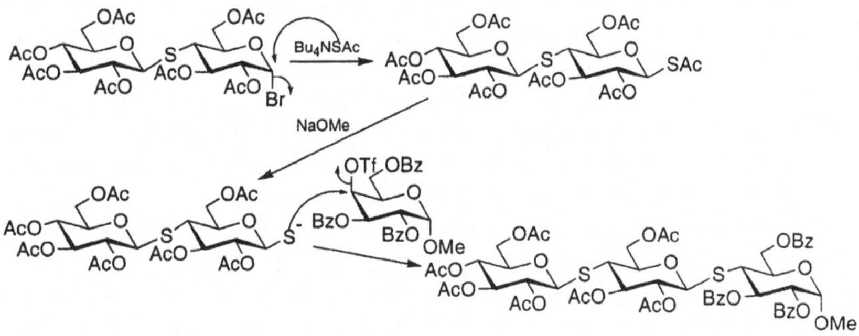

Scheme 4.16 Synthesis of a dithiocellotrioside.

of thiooligosaccharides have been built up and used as *O*-glycoside analogues in biological experiments (Defaye and Gelas, 1991; Orgeret *et al.*, 1992).

4.3.2 N-*glycosides*

Compounds with a nitrogen atom linked to the anomeric centre are found in nature, *inter alia* as parts of nucleic acids and *N*-linked glycoproteins, and the syntheses of such structures have been primarily directed towards the construction of these linkages. Once more, methods have been used involving an electrophilic anomeric centre of the glycosyl donor, which is attacked by a nucleophile, this time a nitrogen.

In the construction of nucleosides and nucleoside analogues, glycosyl halides or anomeric acylates have been used as donors in combination with various purine and pyrimidine base derivatives (Walker, 1979). With halide sugars, chloromercuri-, alkoxy- or silylated derivatives of the bases have been used as acceptors, the latter two methods promoted preferentially by mercuric bromide. With peracylated sugars, silylated bases in the presence of a Lewis acid are the favoured acceptors and promoter, although free bases also can be used in a fusion procedure *in vacuo*. The former method, the Vorbrüggen procedure (Niedballa and Vorbrüggen, 1974), seems to be the general approach and has been used extensively in the synthesis of many nucleoside and nucleoside analogues (Scheme 4.17). Thioglycosides promoted with NIS/triflic acid or NBS have also been employed as donors in this procedure (Knapp and Shieh, 1991; Sugimura *et al.*, 1993). With ribose, constituent of RNA, acylated derivatives give mainly the β configuration owing to participation of the 2-*O*-acyl group, whereas 2-deoxy-ribose (in DNA) gives an α/β mixture. With 2-deoxy-derivatives also *O*-glycosides, for example methyl glycosides, can function as donors in the Vorbrüggen method, owing to the fact that the glycosidic linkage in these derivatives is quite acid labile. Other nitrogen nucleophiles, for example azide derivatives (see below), urea and isocyanates, have also been utilised to form the

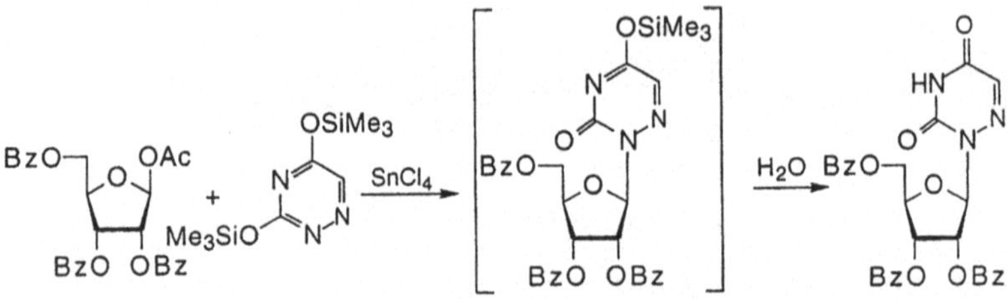

Scheme 4.17 Synthesis of a 6-aza-uridine analogue using the Vorbrüggen procedure.

anomeric C–N linkages from the earlier mentioned type of donors. The heterocyclic base is then constructed starting from the anomeric nitrogen precursor.

The linkage connecting the carbohydrate parts to the protein in *N*-linked glycoproteins is a β-anomeric linkage between *N*-acetylglucosamine and asparagine (see Chapter 5). This linkage is generally synthesised by esterification of the glycosyl amine with a protected aspartic acid derivative (Scheme 4.18).

Glycosylamines can be synthesised directly from the unprotected sugar by treatment with ammonia (Isbell and Frush, 1980), which gives an anomeric mixture, or by treatment with aqueous ammonium bicarbonate, which gives exclusively the β-anomer (Kallin *et al.*, 1989; Likhosherstov *et al.*, 1986). Glycosylamines are sensitive to hydrolysis and anomerise easily, especially under acidic conditions, but are stable after *N*-acylation. Another route to glycosylamines is through the reduction of glycosyl azides. These are synthesised either from glycosyl halides by displacement with sodium azide or from anomeric acylates by treatment with trimethylsilyl azide under Lewis acid catalysis (Györgydeák *et al.*, 1993). 1,2-*Trans*-configuration is obtained with participating groups and preferentially α-azides with non-participating groups at O-2. Care has to be taken in the reduction step, since the glycosylamine readily anomerises and thus the configuration introduced into the

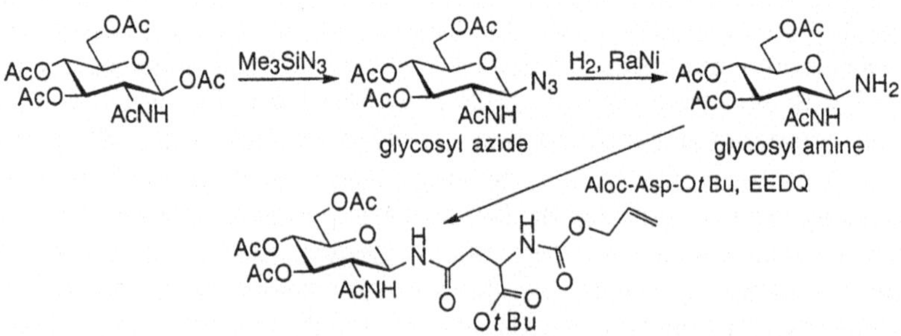

Scheme 4.18 Synthesis of an *N*-linked glycoamino acid via the glycosyl azide and the glycosyl amine.

glycosyl azide can easily be destroyed. Glycosyl azides can also be formed from protected hemiacetal derivatives by the Mitsonobu reaction (HN_3/diethyl azodicarboxylate/triphenyl phosphine) (Mitsonobu, 1981; Schörkhuber and Zbiral, 1990).

4.3.3 C-glycosidic compounds

C-glycosides are abundant in nature. The sugar moiety is often a deoxysugar and the aglycon part an aryl compound, but a variety of other structures are also found, *inter alia C*-oligosaccharides, in which the interglycosidic oxygen is replaced by a methylene group. The latter type of compounds are also of interest as unnatural, stable analogues to various *O*-glycosides, which are valuable in many biological applications. *C*-glycosides are furthermore attractive as chiral starting materials. The interest in syntheses of *C*-glycosides has increased dramatically during recent years. Since here it is not a question of formation of an acetal but of a carbon–carbon linkage, the synthesis of *C*-glycosides has become an interest not only for carbohydrate chemists but also for many other types of organic chemists. A wide variety of known methods for carbon–carbon bond formation have been used (Herscovici and Antonakis, 1992; Postema, 1992). Here only methods which start from carbohydrate-like precursors will be discussed, and not those that build up the carbon skeleton first and then through some cyclisation procedure form the pyranose ring. As examples, methods utilised in the synthesis of *C*-disaccharides have been chosen. Unlike *O*-, *N*- and *S*-glycosylation methods, which almost invariably use an electrophilic anomeric centre, *C*-glycosylation methods involve positive, negative and radical anomeric centres, which are reacted with carbon nucleophiles, electrophiles or radical receptors, respectively.

4.3.3.1 Electrophilic anomeric centre

When using an electrophilic anomeric centre many of the glycosyl donors used in *O*-glycosylation can and have been used, *inter alia* halogen sugars, thioglycosides, trichloroacetimidates, glycals, 1,2-epoxides and anomeric acetates and benzoates, as well as hemiacetals, lactones and *O*-glycosides. After activation these donors have been reacted with a variety of carbon nucleophiles: silyl derivatives, Grignard reagents, cuprates, aluminium complexes, enolates, activated aromates, Wittig reagents, etc., to give *C*-glycosides. The stereochemical outcome of the reactions varies and is dependent both on steric and on stereoelectronic effects. Axially substituted *C*-glycosides at C-1 are often more accessible. The participating effect of neighbouring groups is not as pronounced as with *O*-glycosides. A stereoselective route to the β-glycoside is to start from the sugar lactone, since the reduction of the resulting hemiacetal gives exclusively the equatorial *C*-glycoside (Scheme 4.19) (Rouzaud and Sinay, 1983).

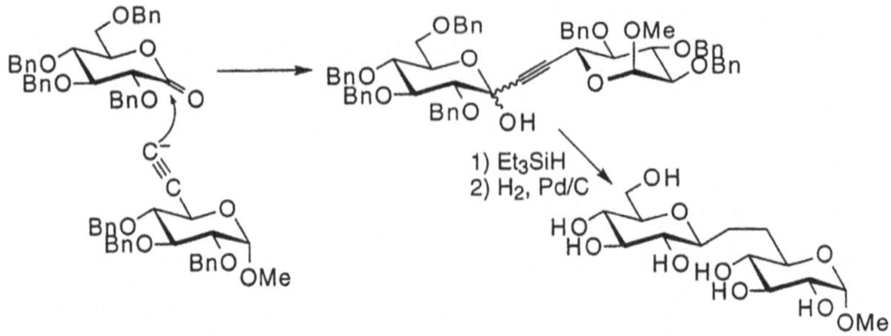

Scheme 4.19 Synthesis of a β-(1 → 6)-*C*-disaccharide using an electrophilic anomeric centre.

4.3.3.2 Nucleophilic anomeric centre

When a nucleophilic anomeric centre is to be created, some auxiliary group that stabilises the anion is generally needed; examples used are sulfone, sulfoxide, sulfide, carboxyl and nitro groups. These groups attached at C-1 or at C-2 in glycals activate the anomeric hydrogen to strong bases and the anomeric anion can be formed. Glycosyl lithium (and samarium) compounds without auxiliary groups have also been made by reduction from chloro sugars or glycosyl phenylsulfones and by direct metal exchange from glycosyl stannyl compounds. A limitation to the use of anomeric anions is that they readily eliminate C-2 substituents, so only 2-deoxy sugars or glycals can be used as starting materials. Examples of carbon electrophiles used in couplings with anomeric anions are alkyl halides and aldehydes (Scheme 4.20) (Schmidt and Preuss, 1989).

4.3.3.3 Radical anomeric centre

Anomeric radicals are formed by the addition of a radical initiator [e.g. azobisisobutyronitrile (AIBN)] to a glycosyl halide or a thio- or selenoglycoside. The

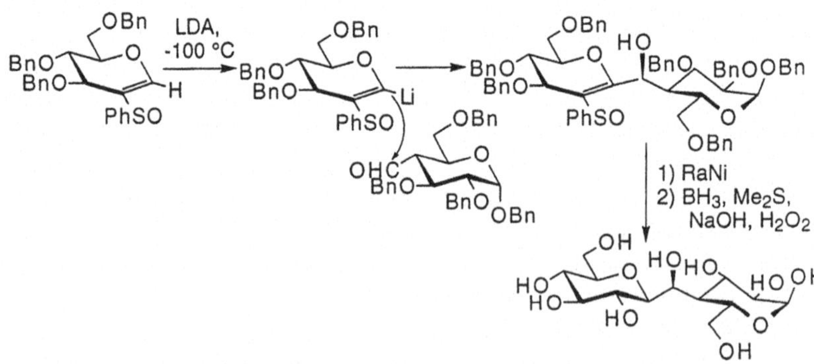

Scheme 4.20 Synthesis of a β-(1 → 4)-*C*-disaccharide using a nucleophilic anomeric centre.

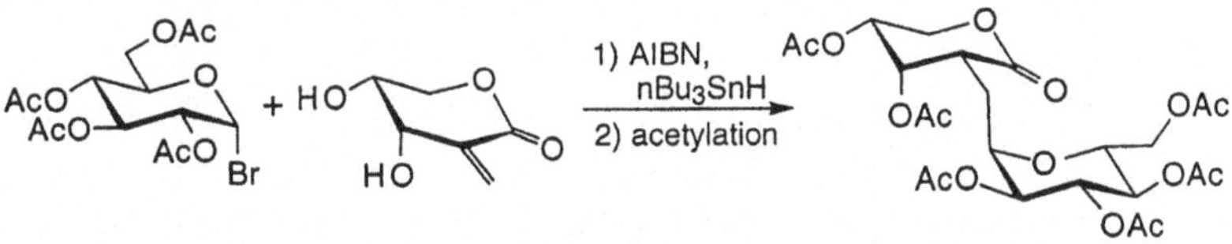

Scheme 4.21 Synthesis of an α-(1 → 2)-*C*-disaccharide using a radical anomeric centre.

axial anomeric radical is preferentially formed. Without the addition of a radical receptor, acyl groups at C-2 migrate to give C-2 radicals, and in the presence of *n*-Bu_3SnH, 2-deoxy sugars. Reactions with other C radicals or double bonds give, *inter alia*, *C*-glycosides (Scheme 4.21) (Giese and Witzel, 1986).

The concept of internal aglycon delivery used for the stereoselective synthesis of 1,2-*cis*-*O*-glycosides has also been used in the formation of *C*-glycosides. A silyl acetal bridge is formed between a hydroxyl group in a selenoglycoside donor and a hydroxyl group in the acceptor. In the formation of *O*-glycosides it is this latter oxygen that reacts with the activated anomeric centre of the donor, but in the formation of *C*-glycosides it is an adjacent multiple bond that reacts in a stereospecific way with the anomeric radical obtained through the activation of the selenoglycoside (Scheme 4.22) (Vauzeilles *et al.*, 1993). The stereoselectivity obtained is dependent on the site of the silyl tethering and also on the protecting groups used in the donor.

4.3.3.4 Anomeric carbenes

Recently the synthesis of another type of anomerically activated glycosyl derivative has been described, namely, glycosyl carbenes. They are made from glycosyl diazides or glycosylidine diazirines. The carbenes can be

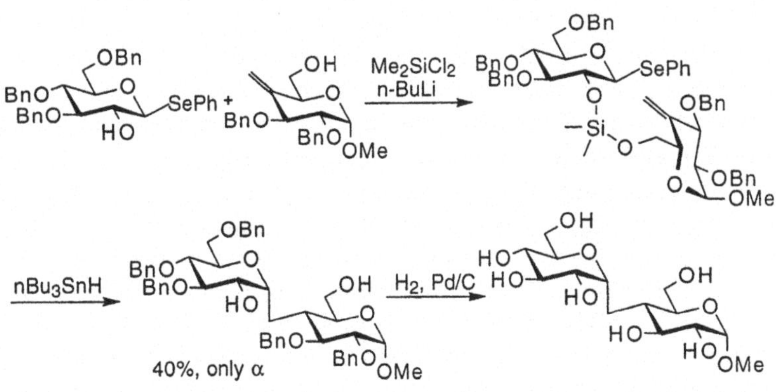

Scheme 4.22 Synthesis of an α-(1 → 2)-*C*-disaccharide using silicon tethering.

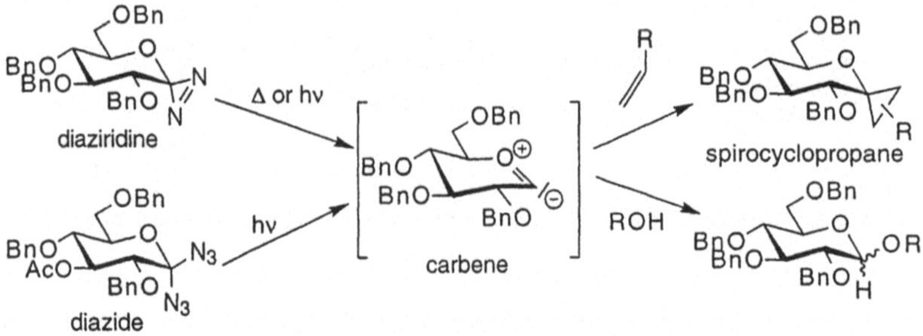

Scheme 4.23 Syntheses and reactions of glycosyl carbenes.

used to form *C*-glycosidic compounds (spirocyclopropanes) and *O*-glycosides (Scheme 4.23) (Praly *et al.*, 1990; Vasella, 1993; Vasella and Waldraff, 1991).

4.4 Enzymatic synthesis of *O*-glycosides

Enzymatic syntheses of oligosaccharides are performed using two different approaches. Either glycosyl transferases, which are the enzymes used by nature in the formation of *O*- and *N*-glycosides, or glycosyl hydrolases, which in nature hydrolyse *O*-glycosides but can be made to catalyse the reverse reaction, are used (Bednarski and Simon, 1991; Toone *et al.*, 1989; Wong and Whitesides, 1994; Wong *et al.*, 1995).

4.4.1 Glycosyl transferases

Because of the laborious protecting group manipulations and the problem with stereoselectivity involved in the chemical synthesis of oligosaccharides, it would be a great advantage to have a catalyst that could catalyse the formation of an *O*-glycosidic bond in a regiospecific and stereospecific manner starting from unprotected saccharide moieties. Glycosyl transferases are good candidates for such catalysts. Since carbohydrates are secondary gene products there are no templates for their biosynthesis and therefore the enzymes involved have to be very specific to avoid mistakes. With the transferase in hand and using the natural acceptor–donor pair, high yields of the coupling products are produced. However, this specificity, which definitely is an asset in nature, raises problems in *in vitro* synthesis, once more because of the complexity of carbohydrates. As pointed out, saccharides can be linked to each other in innumerable ways, and if the concept of 'one enzyme, one linkage' was absolutely true this would mean the need of a huge arsenal of different transferases to be able to perform enzymatic synthesis of various oligosaccharides, and unfortunately at present there are only a few available. The synthesis of unnatural analogues would also be strictly limited. However, although few systematic investigations of the specificity of glycosyl transferases have been carried out, it has been shown that the specificity is not always absolute and that they can accept different acceptor–donor pairs, without the loss of regiospecificity and stereospecificity in the bond formation, but often with a loss of efficiency. This enables the synthesis of more than one structure from a certain transferase, for example, a fucosyl transferase has been described which allows almost any manipulations at *C*-6 of the fucose donor moiety giving the possibility of synthesising a variety of structures using the same enzyme. This approach has been used to alter the structure and thereby the properties of various cell-surface glycoproteins (Srivastava *et al.*, 1992). Still, the need of numerous transferases is a problem, especially considering the synthesis of oligosaccharides of bacterial and fungal origin, owing to the enormous variety of structures found there (Lindberg, 1990). Oligosaccharides of mammalian origin are better targets since these are built up mainly from only eight different monosaccharides (see below) and often contain common structural features.

There are very few commercially available transferases for the carbohydrate chemist and therefore there is a need to isolate others from different sources. This is definitely not trivial; the transferases are present in low concentrations in nature and the instability of many of these enzymes is a further complication. Owing to these problems not many glycosyl transferases have been purified and used in synthesis. The main ones are galactosyl-, fucosyl-, *N*-acetylglucosylaminyl and sialyl transferases, involved mainly in the synthesis of structures of human interest (e.g. blood group substances and *N*-linked glycoprotein structures). The stability of the transferases can be improved by thorough purification and/or by immobilisation of the enzyme on some solid support, for example silica gel. The immobilisation also makes recycling of the enzymes possible. The problem with a limited amount of transferase can be solved by genetic engineering, that is, by identification of the glycosyl transferase gene, followed by cloning and overexpression of this gene in a suitable system, for example yeast or bacteria, but this, as well as the purification of the transferases, requires special skills and facilities normally not present in an organic chemistry laboratory.

Most glycosyl transferases use the Leloir pathway, that is, they employ sugar nucleoside phosphates as donors, generally the diphosphates (UDP-Gal, UDP-Glc, UDP-GlcNAc, UDP-GalNAc, UDP-GlcA, GDP-Man, GDP-Fuc but CMP-NeuAc). These can be synthesised from the sugar 1-phosphate, the donor used by transferases of the non-Leloir pathway, either by enzymatic or chemical methods using either nucleoside triphosphates or activated nucleoside monophosphates, respectively, as reagent (Scheme 4.24). The sugar 1-phosphate can be chemically synthesised by phosphorylation of OH-1 in a suitably protected hemiacetal derivative or by treatment of activated glycosyl donors with phosphates, followed by deprotection. Enzymatic methods using different

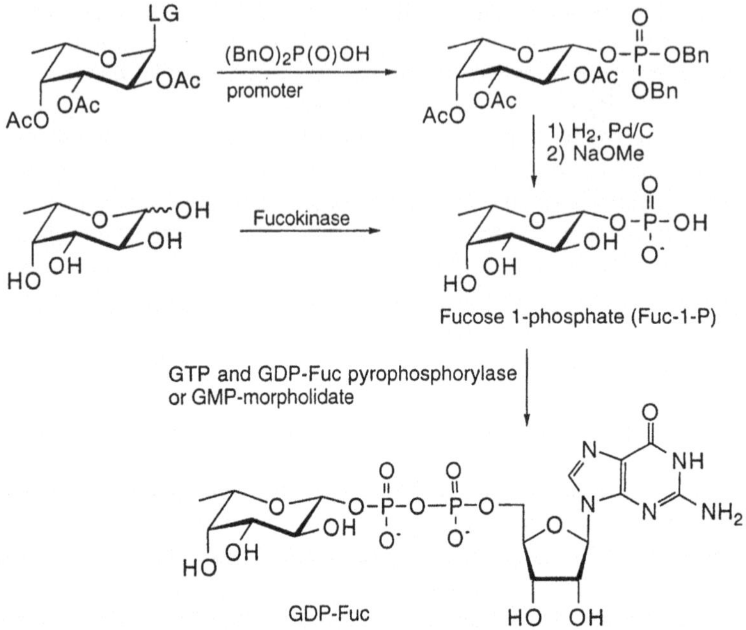

Scheme 4.24 Chemical and enzymatic synthesis of a sugar nucleoside diphosphate.

phosphorylases are also possible if the enzymes are available (Scheme 4.24). Some of the sugar nucleoside phosphates and 1-phosphates are also commercially available, but are generally quite expensive.

Since there are not so many glycosyl transferases available a combined chemical–enzymatic approach, which combines the efficiency of the transferases with a diversity of chemical methods, can be an efficient solution to the synthesis of many complex oligosaccharides. A striking example is the synthesis of a sialyl Le^{x} containing hexasaccharide involving the consecutive use of four transferases on a chemically synthesised spacer disaccharide (Scheme 4.25) (Oehrlein *et al.*, 1993). These types of structures, known to

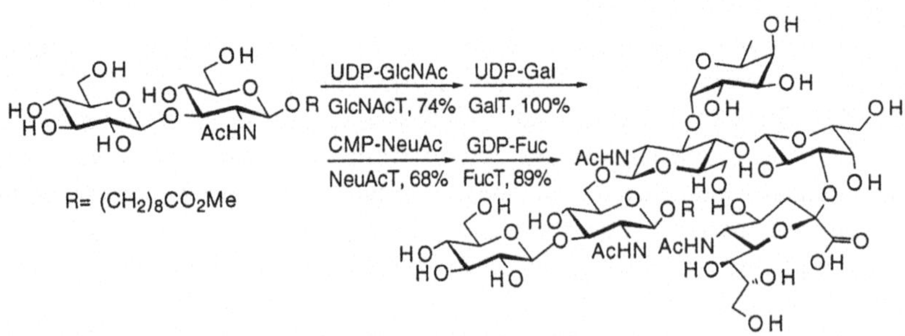

Scheme 4.25 Combined chemical–enzymatic synthesis of a sialyl-Le^{x}-containing hexasaccharide.

play a role in inflammation processes and therefore of major biological interest, have also been synthesised using a multi-enzyme approach. The need for sugar nucleoside phosphates as precursors was overcome by enzymatic *in situ* preparation of these from simple sugars or sugar phosphates, and all enzymes were cloned to allow large-scale production (Ichikawa *et al.*, 1992).

4.4.2 Glycosyl hydrolases

The problems encountered with glycosyl transferases, few easily available enzymes and the need for sugar nucleoside phosphates as donors have made the use of glycosyl hydrolases (glycosidases) for enzymatic synthesis of *O*-glycosides an interesting alternative (Nilsson, 1995). Some of these glycosidases are present in relatively high concentrations in accessible sources and are therefore commercially available at a moderate price. By using a high substrate concentration, the hydrolases can be made to catalyse the reverse reaction and construct a glycosidic linkage instead of hydrolysing it. This can be performed under either thermodynamically or kinetically controlled conditions. In the former method reducing sugars (hemiacetals) are used as donors and in the latter method aryl glycosides, glycosyl fluorides or di- or oligosaccharides are used as donors in a transglycosylation reaction. If in the kinetic approach the reaction is monitored and stopped when the donor is consumed to avoid reverse reaction and hydrolysis, the oligosaccharide product can be produced in much higher concentrations than in the normal equilibrium distribution. Thus, this method gives better yields than does the thermodynamic method and is often the method of choice, but still the yields are moderate, seldom exceeding 30%.

The stereoselectivity of the glycosidases is high. Most glycosidases are of the retaining type, that is, they preserve the stereochemistry of the anomeric centre of the glycosyl donor used, but inverting glycosidases are also known and have been used in synthesis.

The regioselectivity of glycosidases is lower and less predictable than that of the transferases. Generally, the primary hydroxyl of the acceptor reacts preferentially, but the selectivity differs between enzymes from different sources and can also be dependent on the type (anomeric configuration and structure) of acceptor glycoside used and often a mixture of products is obtained.

Owing to the moderate yields and the sometimes low regioselectivity in glycosidase-catalysed reactions this is not a method suitable for the synthesis of large complex oligosaccharides but is a way to synthesise valuable smaller oligosaccharides and glycoconjugates from cheap and easily available saccharide precursor. Homopolysaccharides can also be obtained with this approach. Using β-cellobiosyl fluoride in a cellulase-catalysed reaction, crystalline cellulose was obtained with a polymerisation number larger than 22 and in the exceptional yield of 77% (Kobayashi *et al.*, 1991).

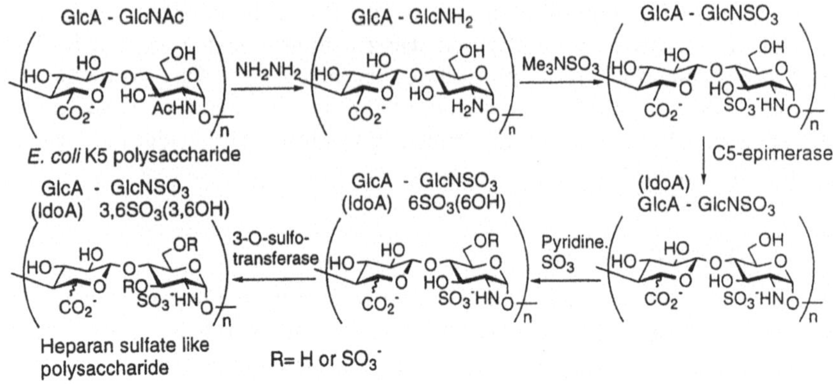

Scheme 4.26 Combined chemical–enzymatic synthesis of a heparan sulfate-like polysaccharide from the *E. coli* K5 polysaccharide.

4.4.3 Other enzymes

Manipulation of protecting groups can be performed using enzymatic protecting group techniques. The major ones in carbohydrate chemistry are regioselective esterification or de-esterification using various esterases and lipases (Waldmann and Sebastian, 1994).

Another approach to the synthesis of biologically interesting carbohydrates is to start from a saccharide, which is easy to isolate from some natural sources and which has a structure similar to the one that is to be synthesised. Modification of this saccharide is then performed to give the desired product. An excellent example is the transformation of a capsular polysaccharide from an *E. coli* bacterium to give biologically active heparin-like species through chemical and enzymatic modification involving, *inter alia*, epimerisation of glucuronic acid residues into iduronic acid residues with a C-5-epimerase and selective sulfation both chemically and with a sulfotransferase (Scheme 4.26) (Casu *et al.*, 1994). In the future, when more and more enzymes will become available, this combined enzymatic–chemical approach will probably be more and more frequently used as an elegant solution to the complicated problem of complex oligosaccharide synthesis, and thereby the need will arise for synthetic chemists to master both enzymatic and chemical methods.

References

Aplin, J.D. and Wriston, J.C., Jr (1981) Preparation, properties, and applications of carbohydrate conjugates of proteins and lipids. *CRC Crit. Rev. Biochem.*, **10**(4), 259–306.

Backinovsky, L.V. (1994) Sugar cyanoethylidene derivatives: useful tools for the chemical synthesis of oligosaccharides and regular polysaccharides, in *Synthetic oligosaccharides* (ed. P. Kovác), *ACS Symposium Series*, **560**, 36–50.

Banoub, J., Boullanger, P. and Lafont, D. (1992) Synthesis of oligosaccharides of 2-amino-2-deoxy sugars. *Chem. Rev.*, **92**, 1167–95.

Baressi, F. and Hindsgaul, O. (1994) The synthesis of β-mannopyranosides by intramolecular aglycon delivery: scope and limitations of the existing methodology. *Can. J. Chem.*, **72**, 1447–65.

Baressi, F. and Hindsgaul, O. (1995) Chemically synthesized oligosaccharides, 1994. A searchable table of glycosidic linkages. *J. Carbohydr. Chem.*, **14**(8), 1043–87.

Bednarski, M.D. and Simon, E.S. (eds) (1991) *Enzymes in carbohydrate synthesis. ACS Symposium Series*, **466**.

Bochkov, A.-F. and Saikov, G.E. (1979) *Chemistry of the O-glycosidic bond, formation and cleavage*, Pergamon Press, Oxford.

Bols, M. (1993) Efficient stereocontrolled glycosidation of secondary sugar hydroxyls by silicon tethered intramolecular glycosidation. *Tetrahedron*, **49**(44), 10049–60.

Boons, G.-J. and Isles, S. (1996) Vinyl glycosides in oligosaccharide synthesis. 2. The use of allyl and vinyl glycosides in oligosaccharide synthesis. *J. Org. Chem.*, **61**, 4262–71.

Braccini, I., Derouet, C., Esnault, J. *et al.* (1993) Conformational analysis of nitrilium intermediates in glycosylation reactions. *Carbohydr. Res.*, **246**, 23–41.

Castro-Palomino, J.C. and Schmidt, R.R. (1995) *N,N*-diacetyl-glucosamine and -galactose-amine derivatives as glycosyl donors. *Tetrahedron Lett.*, **36**, 6871–4.

Casu, B., Grazioli, G., Hannesson, H.H. *et al.* (1994) Biologically active, heparan sulfate-like species by combined chemical and enzymic modification of the *Escherichia coli* polysaccharide K5. *Carbohydr. Lett.*, **1**, 107–14.

Crich, D. and Sun, S.X. (1996) Formation of β-mannopyranosides of primary alcohols using the sulfoxide method. *J. Org. Chem.*, **61**, 4506–7.

Dan, A., Ito, Y. and Ogawa, T. (1995) A convergent and stereocontrolled synthetic route to the core pentasaccharide structure of asparagine-linked glycoproteins. *J. Org. Chem.*, **60**, 4680–1.

Danishefsky, S.J. and Bilodeau, M.D. (1996) Glycals in organic synthesis: the evolution of comprehensive strategies for the assembly of oligosaccharides and glycoconjugates of biological consequence. *Angew. Chem. Int. Ed. Engl.*, **35**, 1380–419.

Danishefsky, S.J., McClure, K.F., Randolph, J.T. and Ruggeri, R.B. (1993) A strategy for the solid-phase synthesis of oligosaccharides. *Science*, **260**, 1307–9.

deBelder, A.N. (1977) Cyclic acetals of the aldoses and aldosides. *Adv. Carbohydr. Chem. Biochem.*, **39**, 179–209.

Defaye, J. and Gelas, J. (1991) Thio-oligosaccharides: their synthesis and reactions with enzymes. *Stud. Nat. Prod. Chem.*, **8**, 315–57.

Deslongchamps, P. (1983) *Stereoelectronic Effects in Organic Chemistry*, Pergamon Press.

Fraser-Reid, B., Udodong, U.E., Wu, Z. *et al.* (1992) *n*-Pentenyl glycosides in organic chemistry: a contemporary example of serendipity. *Synlett*, 927–42.

Garegg P.J. and Lindberg A.A. (1988) The synthesis of oligosaccharides for biological and medical applications, in *Carbohydrate Chemistry* (ed. J.F. Kennedy), Clarendon Press, Oxford, pp. 500–53.

Garegg, P.J., Maloisel, J.-L. and Oscarson, S. (1995) Stannylene activation in glycoside synthesis: regioselective glycosidations at the primary position of galactopyranosides unprotected in the 2-, 3-, 4-, and 6-positions. *Synthesis*, 409–14.

Gelas, J. (1981) Reactivity of cyclic acetals. *Adv. Carbohydr. Chem. Biochem.*, **39**, 71–156.

Giese, B. and Witzel, T. (1986) Synthesis of '*C*-disaccharides' by radical C–C bond formation. *Angew. Chem. Int. Ed. Engl.*, **25**(5), 450–1.

Greene, T.W. and Wuts, P.G.M. (1991) *Protective Groups in Organic Synthesis*, 2nd edn, John Wiley, New York.

Grindley, T.B. (1994) Applications of stannyl ethers and stannylene acetals to oligosaccharide synthesis, in *Synthetic oligosaccharides* (ed. P. Kovác), *ACS Symposium Series*, **560**, 51–76.

Györgydeák, Z., Szilágyi, L. and Paulsen, H. (1993) Synthesis, structure and reactions of glycosyl azides. *J. Carbohydr. Chem.*, **12**(2), 139–63.

Haines, A.H. (1976) Relative reactivities of hydroxyl groups in carbohydrates. *Adv. Carbohydr. Chem. Biochem.*, **33**, 11–109.

Haines, A.H. (1981) The selective removal of protecting groups in carbohydrate chemistry. *Adv. Carbohydr. Chem. Biochem.*, **39**, 13–70.

Herscovici, J. and Antonakis, K. (1992) Recent developments in *C*-glycoside synthesis. *Stud. Nat. Prod. Chem.*, **10**, 337–403.

Horton, D. and Hutson, D.H. (1963) Developments in the chemistry of thio sugars. *Adv. Carbohydr. Chem.*, **18**, 123–99.

Ichikawa, Y., Lin, Y.-C., Dumas, D.P. *et al.* (1992) Chemical–enzymatic synthesis and conformational analysis of sialyl Lewis X and derivatives. *J. Am. Chem. Soc.*, **114**, 9283–98.

Isbell, H.S. and Frush, H.L. (1980) Primary glycopyranosylamines. *Methods Carbohydr. Chem.*, **8**, 255–9.

Ito, Y. and Ogawa, T. (1994) A novel approach to the stereoselective synthesis of β-mannosides. *Angew. Chem. Int. Ed. Engl.*, **33**(17), 1765–7.

Jansson, K., Ahlfors, S., Frejd, T. *et al.* (1988) 2-(Trimethylsilyl)ethyl glycosides. Synthesis, anomeric deblocking, and transformation into 1,2-trans 1-*O*-acyl sugars. *J. Org. Chem.*, **53**, 5629–47.

Juaristi, E. and Cuevas, G. (1992) Recent studies of the anomeric effect. *Tetrahedron*, **48**(24), 5019–87.

Kahne, D., Walker, S., Cheng, Y. and Van Engen, D. (1989) Glycosylation of unreactive substrates. *J. Am. Chem. Soc.*, **111**, 6881–2.

Kallin, E., Lönn, H., Norberg, T. and Elofsson, M. (1989) Derivatization procedures for reducing oligosaccharides, part 3: preparation of oligosaccharide glycosylamines, and their conversion into oligosaccharide–acrylamide copolymers. *J. Carbohydr. Chem.*, **8**(4), 597–611.

Karhta, K.P.R., Kiso, M., Hasegawa, A. and Jennings, J.J. (1995) Simple and efficient strategy for making β-(1 → 6)-linked galactooligosaccharides using 'naked' galactopyranosides as acceptors. *J. Chem. Soc. Perkin Trans.*, **1**, 3023–6.

Knapp, S. and Shieh, W.-C. (1991) Nucleoside synthesis from thioglycosides. *Tetrahedron Lett.*, **32**(30), 3627–30.

Kobayashi, S., Koshiwa, K., Kawasaki, T. and Shoda, S.-I. (1991) Novel method for polysaccharide synthesis using an enzyme: the first in vitro synthesis of cellulose via a nonbiosynthetic path utilizing cellulase as catalyst. *J. Am. Chem. Soc.*, **113**(8), 3079–84.

Kocienski, P.J. (1994) *Protecting Groups*, Georg Thieme, Stuttgart.

Koenigs, W. and Knorr, E. (1901) Über einige Derivate des Traubenzuckers und der Galactose. *Chem. Ber.*, **34**, 957–84.

Krog-Jensen, C. and Oscarson, S. (1996) Stereospecific synthesis of β-D-fructofuranosides using the internal aglycon delivery approach. *J. Org Chem.*, **61**, 4512–13.

Lemieux, R.U., Hendricks, K.B., Stick, R.V. and James, K. (1975) Halide ion catalyzed glycosidation reactions. Syntheses of α-linked disaccharides. *J. Am. Chem. Soc.*, **97**(14), 4056–62.

Likhosherstov, L.M., Novikova, O.S., Derevitskaja, V.A. and Kotchetkov, N.K. (1986) A new simple synthesis of amino sugar β-D-glycosyl amines. *Carbohydr. Res.*, **146**, C1–C5.

Lindberg, B. (1990) Components of bacterial polysaccharides. *Adv. Carbohydr. Chem. Biochem.*, **50**, 279–318.

Marra, A., Esnault, J., Veyrieres, A. and Sinay, P. (1992) Isopropenyl glycosides and congeners as novel classes of glycosyl donors: theme and variations. *J. Am Chem. Soc.*, **114**, 6354–60.

Mehta, S. and Pinto, B.M. (1993) Novel glycosylation methodology. The use of phenyl selenoglycosides as glycosyl donors and acceptors in oligosaccharide synthesis. *J. Org. Chem.*, **58**, 3269–76.

Mitsunobu, O. (1981) The use of diethyl azodicarboxylate and triphenylphosphine in synthesis and transformation of natural products. *Synthesis*, 1–28.

Niedballa, U. and Vorbrüggen, H. (1974) A general synthesis of *N*-glycosides. I–V. *J. Org. Chem.*, **39**(25), 3654–74.

Nilsson, K.G.I. (1995) Synthesis with glycosidases, in *Modern Methods in Carbohydrate Synthesis* (eds S.H. Khan and R.A. O'Neill), Harwood Academic Publishers, London, pp. 518–47.

Norberg, T. (1995) Glycosylation properties and reactivity of thioglycosides, sulfoxides, and other *S*-glycosides: current scope and future prospects, in *Modern Methods in Carbohydrate Synthesis* (eds S.H. Khan and R.A. O'Neill), Harwood Academic Publishers, London, pp. 82–106.

Oehrlein, R., Hindsgaul, O. and Palcic, M.M. (1993) Use of the 'core-2'-*N*-acetylglucosaminyltransferase in the chemical–enzymatic synthesis of a sialyl-Le^x-containing hexasaccharide found on *O*-linked glycoproteins. *Carbohydr. Res.*, **244**, 149–59.

Ogawa, T. (1994) Experiments directed towards glycoconjugate synthesis. *Chem. Soc. Rev.*, 397–407.

Orgeret, C., Seillier, E., Gautier, C. *et al.* (1992) 4-Thiocellooligosaccharides. Their synthesis and use as ligands for the separation of cellobiohydrolases of *Trichoderma reesei* by affinity chromatography. *Carbohydr. Res.*, **224**, 29–40.

Paulsen, H. (1982) Advances in selective chemical syntheses of complex oligosaccharides. *Angew. Chem. Int. Ed. Engl.*, **21**(3), 155–224.

Paulsen, H. and Lockhoff, O. (1981) Neue effektive β-glycosidsynthese für mannose-glycoside synthesen von mannose-haltigen oligosacchariden. *Chem. Ber.*, **114**, 3102–14.

Postema, M.H.D. (1992) Recent developments in the synthesis of *C*-glycosides. *Tetrahedron*, **48**(40), 8545–99.

Praly, J.-P., El Kharraf, Z. and Descotes, G. (1990) Syntheses of anomeric glucopyranosylidene diazides. *J. Chem. Soc. Chem. Commun.*, 431–2.

Rouzaud, D. and Sinay, P. (1983) The first synthesis of a '*C*-disaccharide'. *J. Chem. Soc. Chem. Comm.*, 1353–4.

Sato, S., Nunomura, S., Nakano, T. *et al.* (1988) An efficient approach to stereoselective glycosylation of ceramide derivatives: use of pivaloyl group as a stereocontrolling auxiliary. *Tetrahedron Lett.*, **29**(33), 4097–100.

Schmidt, R.R. (1986) New methods for the synthesis of glycosides and oligosaccharides – are there alternatives to the Koenigs–Knorr method? *Angew. Chem. Int. Ed. Engl.*, **25**, 212–35.

Schmidt, R.R. (1991) Synthesis of glycosides, in *Comprehensive Organic Chemistry*, volume 6, Pergamon Press, Oxford, pp. 33–64.

Schmidt, R.R. and Kinzy, W. (1994) Anomeric-oxygen activation for glycoside synthesis: the trichloroacetimidate method. *Adv. Carbohydr. Chem. Biochem.*, **50**, 21–117.

Schmidt, R.R. and Preuss, R. (1989) Synthesis of carbon bridged *C*-disaccharides. *Tetrahedron Lett.*, **30**(26), 3409–12.

Schörkhuber, W. and Zbiral, E. (1990) Glykosylazide als ausgangsbasis zur gewinnung von nucleosidanalogen. *Liebigs Ann. Chem.*, 1455–69.

Sliedregt, L.A.J.M., van der Marel, G.A. and van Boom, J.H. (1994) Potential usefulness of 4-nitrophenylthio-β-D-glycosides in the chemoselective synthesis of oligosaccharides. *Proc. Indian Acad. Sci (Chem. Sci.)*, **106**, 1213–24.

Srivastava, G., Kaur, K.J., Hindsgaul, O. and Palcic, M.M. (1992) Enzymic transfer of a preassembled trisaccharide antigen to cell surfaces using a fucosyltransferase. *J. Biol. Chem.*, **267**(31), 22356–61.

Stanek J., Jr (1990) Preparation of selectively alkylated saccharides as synthetic intermediates, in *Topics in Current Chemistry*, volume 154 (ed. J. Thiem), Springer, Berlin, pp. 209–56.

Stork, G. and Kim, G. (1992) Stereocontrolled synthesis of disaccharides via the temporary silicon connection. *J. Am. Chem. Soc.*, **114**, 1087–8.

Sugimura, H., Muramoto, I., Tadashi, K. and Osumi, K. (1993) Stereoselective synthesis of 1,2-*cis-N*-glycosides by the *N*-bromosuccinimide promoted reaction of thioglycosides with silylated pyrimidine bases. *Chem. Lett.*, 169–72.

Thiem, J. and Klaffke, W. (1990) Syntheses of deoxy oligosaccharides, in *Topics in Current Chemistry*, volume 154 (ed. J. Thiem), Springer, Berlin, pp. 285–332.

Toone, E.J., Simon, E.S., Bednarski, M.D. and Whiteside, G.M. (1989) Enzyme-catalyzed synthesis of carbohydrates. *Tetrahedron*, **45**(17), 5365–422.

Toshima, K. and Tatsuta, K. (1993) Recent progress in *O*-glycosylation methods and its application to natural products synthesis. *Chem. Rev.*, **93**(4), 1503–31.

van Boeckel, C.A.A. (1992) Protective group strategies in the synthesis of functionalized carbohydrates, in *Modern Synthetic Methods* (ed. R. Scheffold), Verlag Helvetica Chimica Acta, Basel, pp. 439–81.

Vasella, A. (1993) A new approach to the synthesis of glycosides. *Pure & Appl. Chem.*, **65**(4), 731–52.

Vasella, A. and Waldraff, C.A.A. (1991) Glycosylidene carbenes. Part 3. Synthesis of spirocyclopropanes. *Helv. Chim. Acta*, **74**(3), 585–93.

Vasella, A., Witzig, C., Chiara, J.-L. and Martin-Lomas, M. (1991) Convenient synthesis of 2-azido-2-deoxy-aldoses by diazo transfer. *Helv. Chim. Acta*, **74**(8), 2043–53.

Vauzeilles, B., Cravo, D., Mallet, J.-M. and Sinay, P. (1993) An expeditious synthesis of a *C*-disaccharide using a temporary ketal connection. *Synlett*, 522–4.

Waldmann, H. and Sebastian, D. (1994) Enzymatic protecting group techniques. *Chem. Rev.*, **94**(4), 911–37.

Walker, R.T. (1979) Nucleosides, in *Comprehensive Organic Chemistry*, volume 5 (eds. D. Barton and W.D. Ollis), Pergamon Press, Oxford, pp. 55–104.

Wong, C.H. and Whitesides, G.M. (eds) (1994) *Enzymes in Synthetic Organic Chemistry*, Pergamon Press, Oxford.

Wong, C.-H., Halcomb, R.L., Ichikawa, Y. and Kajimoto, T. (1995) Enzymes in organic synthesis: application to the problems of carbohydrate recognition (parts 1 and 2). *Angew. Chem. Int. Ed. Engl.*, **34**, 412–32 and 521–46.

Yan, L., Taylor, C.M., Goodnow R., Jr and Kahne, D. (1994) Glycosylation on the Merrifield resin using anomeric sulfoxides. *J. Am. Chem. Soc.*, **116**(15), 6953–54.

5 Chemistry of glycopeptides

HORST KUNZ, BIRGIT LÖHR and JÖRG HABERMANN

5.1 Introduction

For a long time, peptides and proteins on the one hand and carbohydrates on the other have been considered separate classes of natural products. The rather strict distinction between these major fields of natural product chemistry is still apparent not only in the organisation of chemistry text books but also in the different approaches to immunological and cell biological recognition phenomena.

In this context it is interesting to remember that the chemical methodology of both carbohydrate and peptide chemistry was founded by Emil Fischer (Fischer, 1893; 1906). However, the analytical tools available at the end of the nineteenth and the beginning of the twentieth century did not reveal that most natural proteins, in particular those of multicellular organisms, are glycoproteins and that many typical polysaccharides contain covalently linked polypeptide portions. As a consequence, two almost separate schools emerged from Emil Fischer's research group, one dedicated to peptide chemistry and the other engaged in carbohydrate chemistry.

With the help of new analytical methods, for example the different forms of chromatography, electrophoresis, mass spectrometry and nuclear magnetic resonance (NMR) spectroscopy, as well as new biological assays, the ubiquity of protein glycosylation was recognised during the past three decades (Lis and Sharon, 1993). Most membrane proteins and secretory proteins of eukaryotic cells contain covalently bound glycan side-chains. The saccharide portions of natural glycoproteins can exhibit very different functions. Of course, they influence the physiochemical properties of these macromolecules, for example their solubility and their preferred conformation (Hounsell, 1994). In a number of cases, carbohydrates of glycoproteins are involved in biological recognition phenomena, for example in immunological differentiation (Kobata, 1993) or infectious processes (Lis and Sharon, 1993; Olofsson, 1991). Since glycoproteins obviously exhibit crucial roles in pathological processes neither completely understood nor sufficiently curable, for example tumor development and virus infections, they are receiving increasing interest in the investigation of regulatory processes. Therefore, model glycopeptides having both a defined saccharide and peptide structure are attractive as ligands of membrane receptors, lectins, toxins, and substrates or inhibitors of enzymes.

5.2 Saccharide-peptide linkages of natural glycoproteins

5.2.1 The N*-glycosyl asparagine linkage*

The β-*N*-glycosidic linkage between the amide group of asparagine and *N*-acetylglucosamine **1** (Scheme 5.1) is the most common saccharide–peptide linkage in natural glycoproteins (Neuberger *et al.*, 1972). This type of connection is found, for example, in hen egg albumin, in human transferrin, in many membrane glycoproteins and also in virus glycoproteins, for example in gp120 of HIV-1. The sites of *N*-glycosylation of these *N*-glycoproteins typically contain the consensus sequon Asn–X–Ser/Thr in which X may be any amino acid except proline.

The hydroxyl function of serine or threonine obviously plays a crucial role (Bause and Legler, 1981) during the transfer of the oligosaccharide side-chain from the glycosyl dolichol–diphosphate precursor to the growing peptide chain in the biosynthesis of *N*-glycoproteins (Parodi *et al.*, 1972). However, not every occurring sequon structure may actually be *N*-glycosylated in glycoproteins.

In the majority of *N*-glycoproteins, the saccharide portion contains a pentasaccharide consisting of two mannose residues α-glycosidically linked to a third mannose which is β-glycosidically linked to a chitobiose asparagine conjugate (Kornfeld and Kornfeld, 1980).

Only a few years ago, the *N*-acetylglucosamine–asparagine bond was the only known *N*-glycosidic linkage in glycoproteins. However, in recent years, other types of *N*-glycoproteins which contain *N*-glucosyl-(**2**) or *N*-acetylgalactosaminyl asparagine **3** have been found in bacteria (Paul *et al.*, 1986). Apparently, these types of glycoproteins also contain asparagine in a sequon structure Asn–X–Ser/Thr.

5.2.2 O*-Glycosidic linkages of glycoproteins*

In comparison with the *N*-glycosidic bonds, *O*-glycosidic linkages of glycoproteins show a distinctly greater variety. Most important glycoproteins of mammals, for example mucin glycoproteins, are characterised by an α-*O*-glycosidic bond between *N*-acetyl-galactosamine and serine or threonine

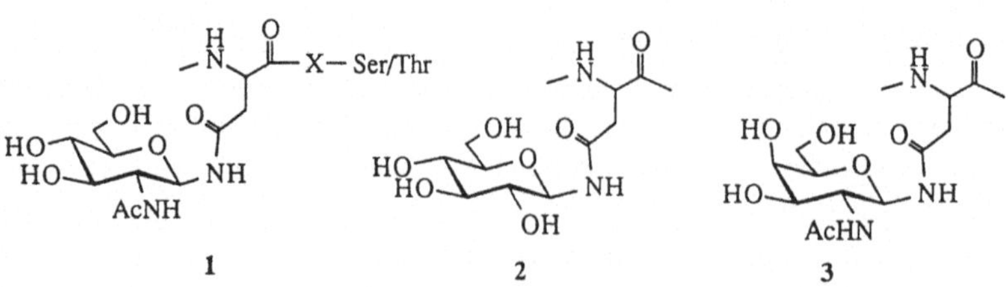

Scheme 5.1.

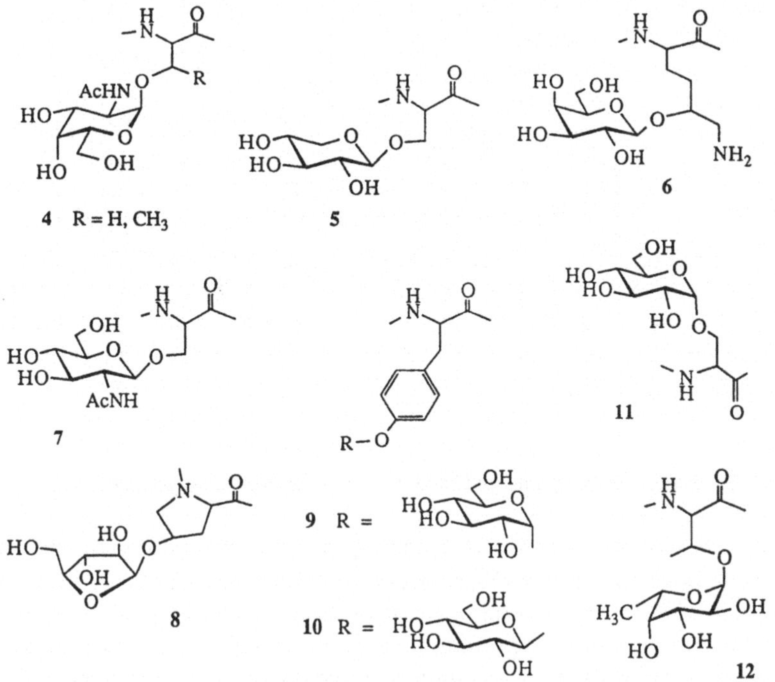

Scheme 5.2 *O*-glycosidic bonds in glycoproteins.

4 (Scheme 5.2; Kornfeld and Kornfeld, 1976). A great number of membrane glycoproteins, for example human glycophorin (Tomita *et al.*, 1978), also contain this type of *O*-glycosidic linkage.

A β-xylosyl serine linkage **5** is found in proteoglycans, the typical carbohydrate-rich connective tissue glycoproteins (Lindahl and Rodin, 1964). The typical glycan peptide linkage in collagen consists of a β-galactosyl hydroxylysine bond **6** (Butler and Cunningham, 1966).

Only a few years ago, the β-glycosidic bond between *N*-acetylglucosamine and serine/threonine **7** (Hart *et al.*, 1989) was recognised as a common structural feature of glycoproteins in nuclear membranes and cytoplasmic compartments. The linkage between β-L-arabinofuranose and hydroxyproline (**8**) is typical for extensin-type *O*-glycoproteins, which are required for the formation of cell walls in plants and algae (Lamport and Clark, 1969).

Among the recently discovered *O*-linked glycoproteins, the *O*-glucosyl tyrosine containing structures **9** and **10** should be mentioned. The α-glucosyl tyrosine linkage **9** occurs in the glycogenin which is the starting conjugate for the glycogen biosynthesis (Lomako *et al.*, 1990; Smythe and Cohen, 1991). The corresponding β-glucosyl tyrosine structure **10** constitutes the linkage element of the crystalline surface glycoprotein of *Clostridium thermohydrosulfuricum* (Messner *et al.*, 1992).

The *O*-α-glucosyl serine structure **11**, mainly as a part of a Xylα1–3Xylα1–3Glc saccharide side-chain, has been found in epithermal growth factor (EGF) modules of human and bovine blood clotting factor IX (Hase *et al.*, 1990). The linkage was observed exclusively in a consensus peptide sequence Cys–Xaa–Ser–Xaa–Pro–Cys. The α-*O*-fucosyl serine/threonine linkage **12** was also found in EGF modules of factors VII, IX and XII (Harris *et al.*, 1992). The consensus sequence in which this linkage occurs was determined as Cys–Xaa–Xaa–Gly–Gly–Ser/Thr–Cys. For other recently discovered and rare glycosidic bonds, relevant biochemical reviews and literature should be consulted (Harris and Spellman, 1993; Hayes and Hart, 1994; Lis and Sharon, 1993).

5.3 Problems of isolation and structure elucidation of glycoproteins

In many cases, the isolation of glycoproteins from biological material is only possible in small amounts. Structural elucidations must be carried out using enzymes for the degradation of the peptide chain in combination with site-specific and stereoselective glycosidases (Scheme 5.3). The terminal monosaccharides are identified by carbohydrate-specific lectins (Sharon and Lis, 1989). Structural elucidations of smaller glycopeptides and oligosaccharides are nowadays achieved using modern methods of mass spectrometry, for example electrospray or laser-desorption methods (Egge *et al.*, 1991). The regio- and stereochemistry of the intersaccharide bonds can be clarified in many cases by the use of either stereo- and regioselective glycosidases or lectins. A combination of both methods is also possible in the sequential use of glycosidases and purification via affinity chromatography with immobilised lectins (Takahashi and Muramatsu, 1992).

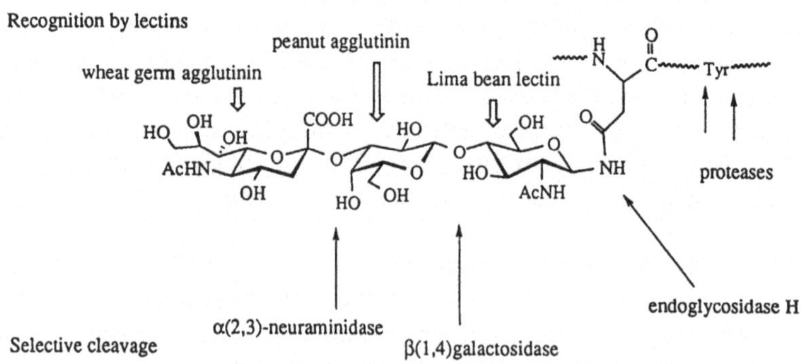

Scheme 5.3 Enzymatic degradation and structure analysis using (immobolised) enzymes.

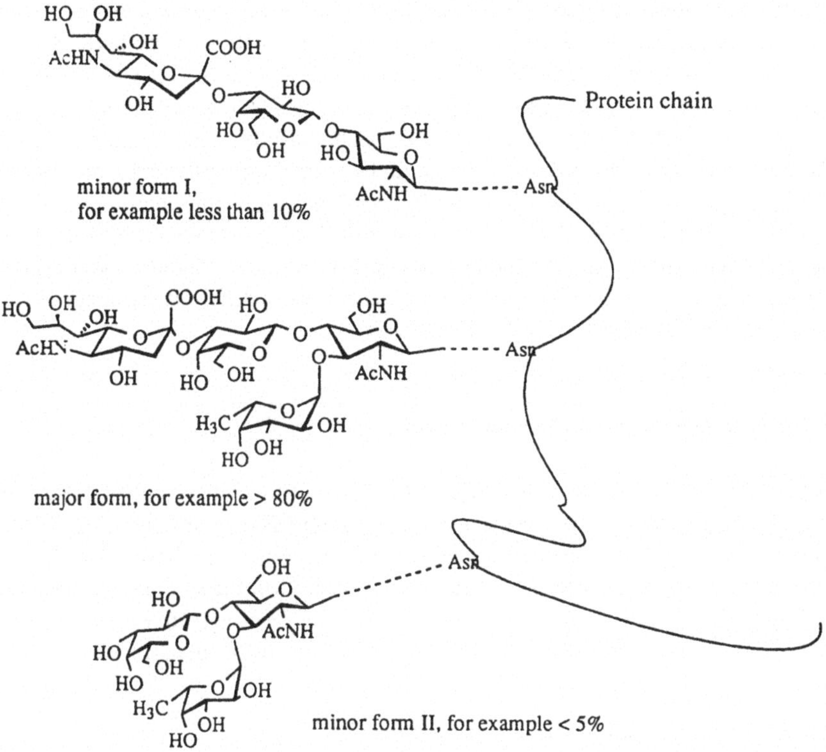

Scheme 5.4 Illustration of glycoprotein microheterogeneity; one peptide chain (or different chains of an identical protein) can carry different glycan side-chains.

If these tools of bioanalytical recognition are not available for a certain analysis, model compounds are required for a reliable structural assignment. NMR spectroscopy allows the determination of the connectivity as well as the stereochemistry of oligosaccharides and glycopeptides and is, therefore, considered a very valuable tool in the structural analysis of glycoproteins (Vliegenthart *et al.*, 1983).

A major problem of the isolation and purification of natural glycoproteins consists in the biological microheterogeneity of these complex glycoconjugates (Scheme 5.4). Since the saccharide side chains are biosynthetically constructed by enzymes, failure sequences and incomplete chains are often found. As a consequence, a protein may carry carbohydrate portions which deviate from the structure of the main saccharide component.

In all cases in which the carbohydrate is involved in the selective biological function of a certain glycoprotein, either via its influence on the peptide structure or more directly as a part of the recognised ligand, the biological microheterogeneity of glycoproteins renders the elucidation of the regulatory mechanisms more difficult. These problems may lead to wrong assignments of immunological effects and contradictory observations in the field of biological recognition in general. A clarification of structure–activity relationships often requires saccharide-, peptide- and glycopeptide model ligands of exactly specified structure regarding connectivity, regiochemistry, and stereochemistry of all linkages and sequences in the peptide portion. These model compounds can be provided by synthesis applying chemical as well as enzymatic methods. These model compounds also provide an opportunity to investigate the influence which glycosylation exhibits on peptide secondary structure and, vice versa, the effect which peptides of different sequences may exert on the oligosaccharide conformation.

5.4 Synthesis of glycosyl amino acids

A prerequisite of a successful synthesis of glycopeptides is the construction of the glycosidic linkage to the required amino acid (Garg and Jeanloz, 1985; Garg *et al.*, 1994). As outlined in Section 5.2, there are principally two different types of these connections, the *N*-glycosidic bond and the *O*-glycosidic linkage. The synthesis of these structures will be demonstrated on examples based on the most frequently found and most important glycoprotein types.

5.4.1 The glucosamine asparagine linkage

The *N*-glycosidic bond between the amide group of asparagine and *N*-acetylglucosamine **1** (Scheme 5.1) is the most frequently found in nature. This may not only be because of the natural abundance of this type of bond, but also because of the stability of the *N*-glycosyl asparagine linkage towards acids and bases, rendering it stable to common work-up procedures in protein biochemistry. The stability of the *N*-glycosidic bond also facilitates the chemical synthesis of *N*-glycopeptides. The construction of the *N*-glycosidic linkage is generally not carried out via a *N*-glycosylation of asparagine but by the condensation of a 1-amino-glucosamine derivative with a suitably protected aspartic acid derivative (Garg and Jeanloz, 1974). The method was originally introduced by Marks and Neuberger (1961) for the synthesis of N^4-acetylglucosaminyl asparagine affording the basic linkage element and was elaborated later on by Jeanloz and co-workers (Garg and Jeanloz, 1972) for the construction of *N*-glycosyl asparagine conjugates with larger oligosaccharides.

The glycosyl azide **13** (Scheme 5.5) is obtained from the corresponding glycosyl chloride and sodium azide in formamide solution (Yamamoto *et al.*, 1965). However, already the formation of the analogous chitobiose azide could only be achieved using (the hazardous) silver azide (Spinola and Jeanloz, 1970).

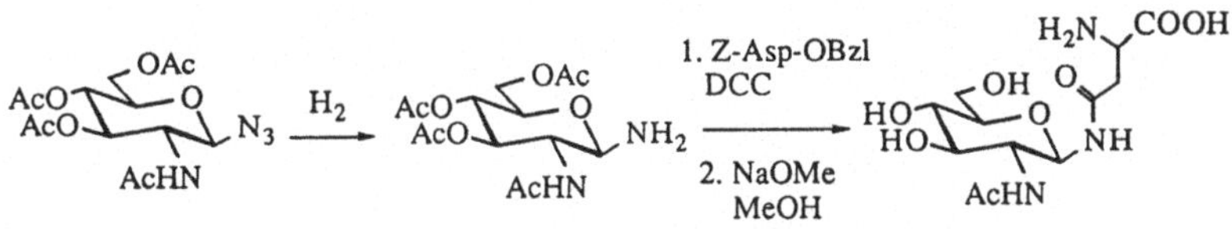

13 **14** **15**

Scheme 5.5 Ac = acetyl; Bzl = benzyl; Me = methyl; DCC = *N*,*N′*-dicyclohexyl carbodiimide.

A more general method for the synthesis of oligosaccharide azides with *N*-acetylglucosamine at the reducing end consists in the phase-transfer-catalysed reaction of glycosyl chlorides with sodium azide as shown for the chitobiose derivatives **16–17** (Scheme 5.6; Kunz and Waldmann, 1985; Kunz *et al.*, 1989). Condensation of the chitobiose amine **18** with *tert*-butyloxycarbonyl (Boc) aspartic acid α-allyl ester gave the *N*-glycosidic asparagine derivative **19**, from which the protecting groups in the amino acid part can be selectively removed either by acidolytic cleavage of the Boc group or, alternatively, by palladium(0)-catalysed removal of the allylic protecting group under practically neutral conditions (Kunz and Waldmann, 1984).

Glycosyl azides are also accessible from GlcNAc-derived oxazolines **20** through their reaction with trimethylsilyl (TMS) azide in the presence of $SnCl_4$ (Scheme 5.7; Nakabayashi *et al.*, 1988). This method was successfully applied to the conversion of a peracetylated core-region heptasaccharide isolated from biological material into the corresponding asparagine conjugate.

The success of the oxazoline reaction is apparently dependent upon the saccharide structure. An alternative formation of asparagine N^4-glycosides was found by Khorlin *et al.* (1980) who employed the reaction of glycosyl

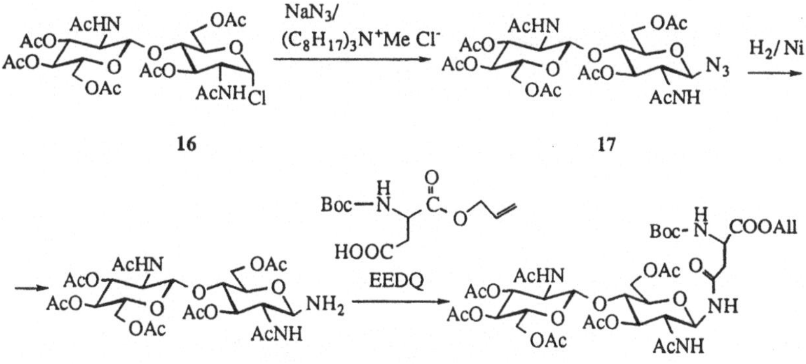

16 **17** **18** **19**

Scheme 5.6 All = allyl; Boc = *tert*-butyloxycarbonyl; EEDQ = ethyl 2-ethoxy-1,2-dihydroquinoline-1-carboxylate.

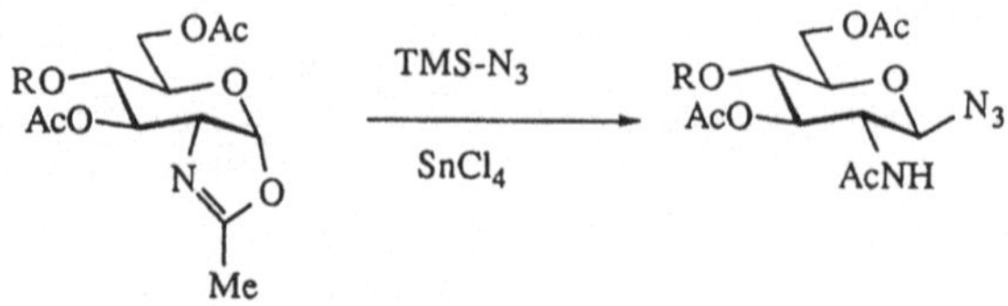

20 R = hexasaccharide

Scheme 5.7 TMS = trimethylsilyl.

isothiocyanates with aspartic acid derivatives having a free β-carboxy function. While the β-mannosyl chitobiose oxazoline did not result in the formation of the trisaccharide azide according to the above-mentioned reaction with TMS azide, the isothiocyanate **21**, obtained from the oxazoline, reacted with allyloxycarbonyl aspartic acid *tert*-butyl ester to yield the desired conjugate **22** (Scheme 5.8; Günther and Kunz, 1990). Similar to the selective deblocking of **19**, the functional groups of **22** can be selectively deprotected either by acidolytic cleavage of the *tert*-butyl ester using trifluoroacetic acid (TFA) or by palladium(0)-catalysed removal of the allyloxycarbonyl (Aloc) group (Kunz and Unverzagt, 1984; Unverzagt and Kunz, 1994).

The reaction of the unprotected carbohydrate **23** with saturated ammonium bicarbonate solution affords the glycosylamines of mono- and oligosaccharides (Scheme 5.9; Likhosherstov *et al.*, 1986). However, this reaction must be controlled by high-performance liquid chromatography (HPLC) monitoring of a subsequent condensation reaction of the formed glycosylamine (Cohen-Anisfeld and Lansbury, 1993). Nevertheless, this direct access to glycosylamines is promising for the synthesis of *N*-glyco-conjugates

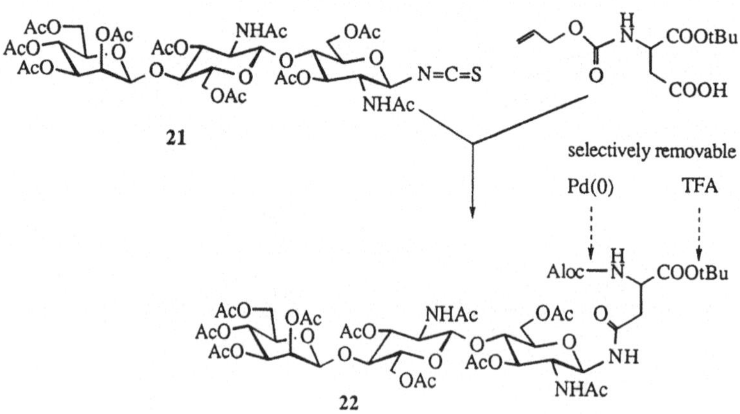

Scheme 5.8 Aloc = allyloxycarbonyl; TFA = trifluoroacetic acid; Pd(0) = $[Ph_3P]_4Pd$ and scavenger.

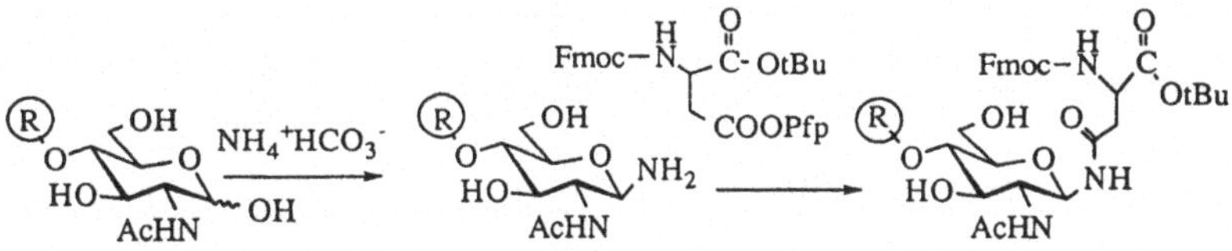

23 24 25

Scheme 5.9 Ⓡ= unprotected saccharide; Pfp = pentafluorophenyl.

from oligosaccharides which are isolated and purified from biological sources.

Condensation of the obtained glycosylamines **24** with suitably protected aspartic acid derivatives gave *N*-glycosyl asparagine building blocks which can be selectively deprotected and have been successfully used in the solid-phase syntheses of glycopeptides (Meldal and Bock, 1990; Otvos *et al.*, 1990). On the basis of this chemistry, Cohen-Anisfeld and Lansbury (1990; 1993) have developed a convergent synthesis of glycopeptides in which the aspartic acid-containing peptide, for example **26**, is presynthesised, then subjected to a side-chain deprotection of the aspartic acid units to yield **27**, which was subsequently condensed with glycosylamines, for example **28**, to furnish the *N*-glycopeptide, in this case **29** (Scheme 5.10). Since more complex saccharides are available from biological sources only in small amounts or, alternatively, must be synthesised in demanding multistep procedures, the introduction of the saccharide according to this convergent concept appears favourable. However, the amide condensation with a preformed aspartic acid peptide (incorrectly designated glycosylation) bears problems which mainly arise from the activation of the aspartic β-carboxylic function. This gives rise to a competing formation of succinimide intermediates and can lead to isoaspartic acid-containing *N*-glycopeptides (transpeptidation). Logically, the yields of *N*-glycopeptide drop significantly for peptides which contain amino acids less sterically demanding than leucine in the C-terminal position of aspartic acid (Cohen-Anisfeld and Lansbury, 1993).

The direct introduction of the anomeric amino group is limited to *O*-unprotected carbohydrates which cannot be subjected to further modifications afterwards, for example introduction of protecting groups or extensions of the saccharide structure.

In contrast to glycosylamines, glycosyl azides have been shown to be valuable synthons allowing a versatile manipulation of the protecting group pattern and oligosaccharide synthesis (Bröder and Kunz, 1993; Kunz and Unverzagt, 1988; Unverzagt and Kunz, 1992; 1994). As an illustration, the synthesis of a protected Lewis[a] antigen azide (März and Kunz, 1992) is shown in Scheme 5.11. Starting from glucosaminyl azide **13**, removal of the *O*-acetyl groups and introduction of the 4,6-(4-methoxybenzylidene)acetal protecting groups (Johansson and Samuelsson,

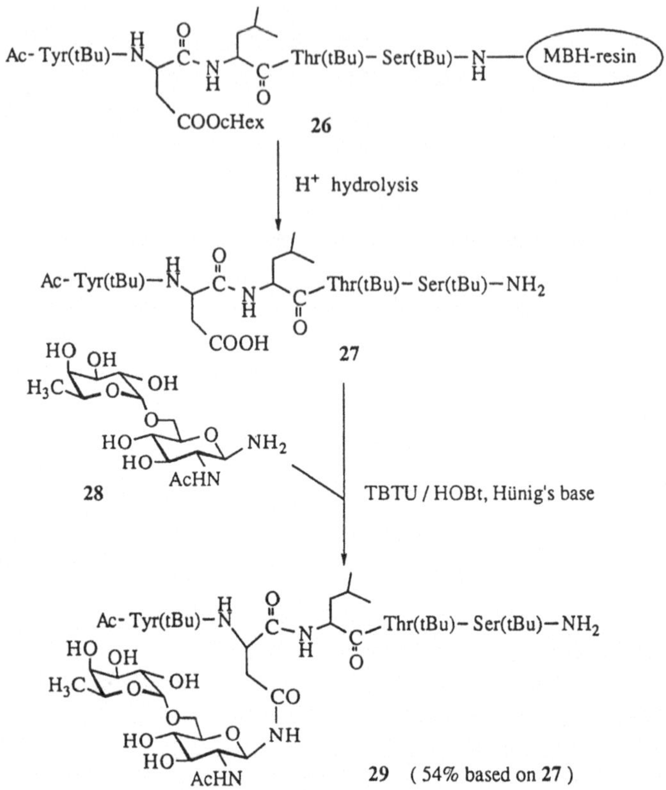

Scheme 5.10 cHex = cyclohexyl; TBTU =2-(1H-benzotriazol-1-yl)-1,1,3,3-tetramethyl-uronium tetrafluoroborate; HOBt = 1-hydroxybenzotriazol; Hünig's base = $EtiPr_2N$; MBH resin = 4-methoxy-4′-alkoxy-benzhydryl (amine) resin.

1984) gave the monofunctional glycosyl acceptor **30** without affecting the anomeric azide. Galactosylation under Helferich conditions furnished the type 1 lactosamine azide **31**, which was subjected to regioselective reductive opening of the 4,6-acetal (Garegg *et al.*, 1982) to yield the disaccharide acceptor **32**. Introduction of the α-fucosyl unit was achieved by *in situ* anomerisation (Lemieux and Haymi, 1965) of the *O*-(4-methoxybenzyl)-(Mpm)-protected fucopyranosyl bromide **33** (Kunz and Unverzagt, 1988; März and Kunz, 1992; von dem Bruch and Kunz, 1994) and resulted in the formation of the Lewis[a] trisaccharide derivative **34**. Because the α-fucoside linkages are quite sensitive to treatment with acids, for example trifluoroacetic acid (TFA), formic acid (Kunz and Unverzagt, 1988), the ether-type protecting groups in **34** were exchanged for *O*-acetyl groups (**35**) which exert a marked stabilising effect on the glycosidic bonds. Throughout the synthesis up to this point, the anomeric azido function remained intact. Its final reduction gave *O*-acetyl-protected Lewis[a] antigen amine **36** which

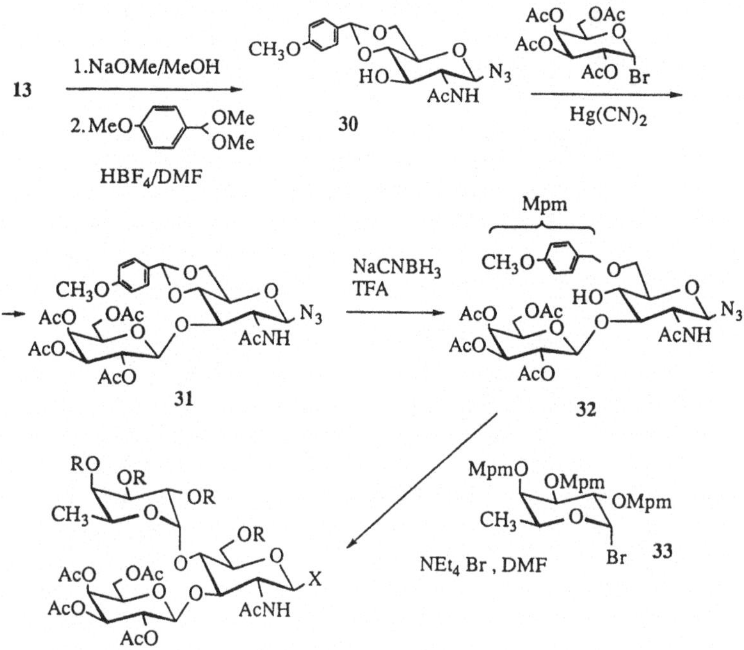

1. $Ce(NH_4)_2(NO_3)_6$
2. Ac_2O, pyridine

34 R = Mpm, X = N_3 95%
35 R = Ac, X = N_3 93% —H_2/Ni→ **36** R = Ac, X = NH_2 quant.

Scheme 5.11 Mpm = 4-methoxybenzyl; TFA = fluoroacetic acid; DMF = dimethylformamide.

was shown to be very suitable for the synthesis of Lewis[a] antigen glycopeptides (Kunz and März, 1992).

An interesting method to generate glycosyl azides at a late stage in the synthesis, that is after the completion of the oligosaccharide, was developed by Danishefsky (Danishefsky and Roberge, 1995) on the basis of the electrophilic activation of glycals (Lemieux and Levine, 1964). Activation of the protected galactal **37** using dimethyldioxirane to form the intermediate epoxide (Danishefsky and Holcomb, 1989) and subsequent reaction with galactal **38** resulted in a disaccharide glycal which, after *O*-acetylation, was subjected to an analogous coupling with glucal **39** to furnish the trisaccharide glycal **40** (Scheme 5.12). After *O*-acetylation of the 2′-OH group, iodo-initiated *N*-glycosidic formation of **40** with anthracene sulfonamide gave the 2-iodo-α-*N*-glycoside **41** of the trisaccharide. Reaction of **41** with tetrabutylammonium azide resulted in an intermediate aziridine (which has gluco configuration) and its subsequent nucleophilic opening to yield the trisaccharide azide **42**. This elegant double S_N2-type reaction cascade not only introduces the anomeric azide but also inverts the configuration at C-2 owing to the intramolecular presentation of the nucleophile. Intermolecular

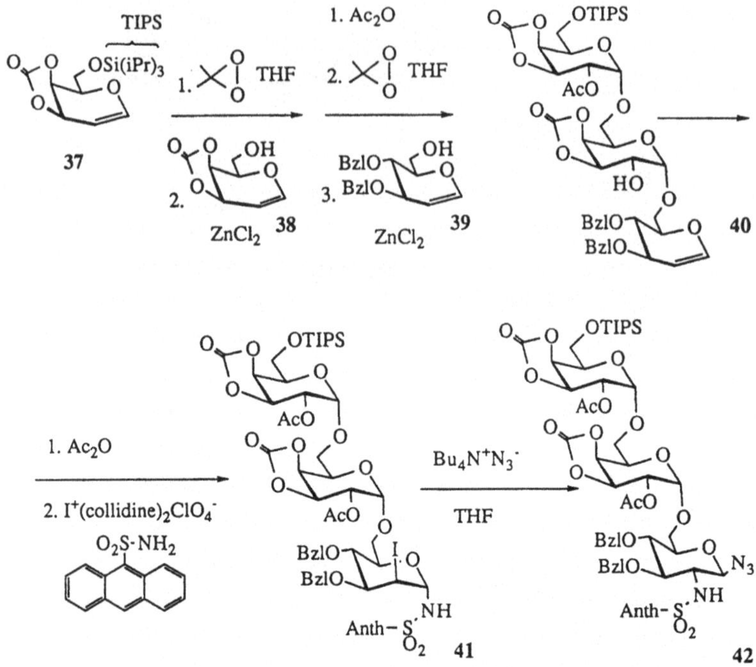

Scheme 5.12 Anth = 9-anthryl; Bzl = benzyl; THF = tetrahydrofuran.

S_N2-type substitutions of 2-iodo-glycosides are known to be very difficult or often impossible. After *N*-acetylation of the sulfonamide **42** the reductive treatment with aluminium amalgam resulted in a simultaneous removal of the sulfonamide group and the formation of the trisaccharide amine, which was successfully used in a glycopeptide synthesis (Danishefsky and Roberge, 1995). The same methodology also proved efficient in an oligosaccharide and glycopeptide synthesis on solid phase (Roberge *et al.*, 1995).

5.4.2 O-Glycosyl serine and threonine linkages

5.4.2.1 General problems in the chemistry of O-*glycosyl serine and threonine derivatives*

The *O*-glycosyl serine or threonine containing glycoproteins are not only formed via biosynthetic pathways quite different from those of *N*-glycoproteins (Beyer *et al.*, 1981; Schwartz and Datema, 1982) but are also more sensitive to basic and acidic influences. Actually, the base-sensitivity of the *O*-glycosyl serine/threonine bond is generally exploited for the determination of this type of glycan–protein linkage in glycoproteins (Neuberger *et al.*, 1972).

In alkaline solution, these *O*-glycoproteins **43** are subject to an easy β-elimination of the glycan side-chain to form the protein **44** containing a

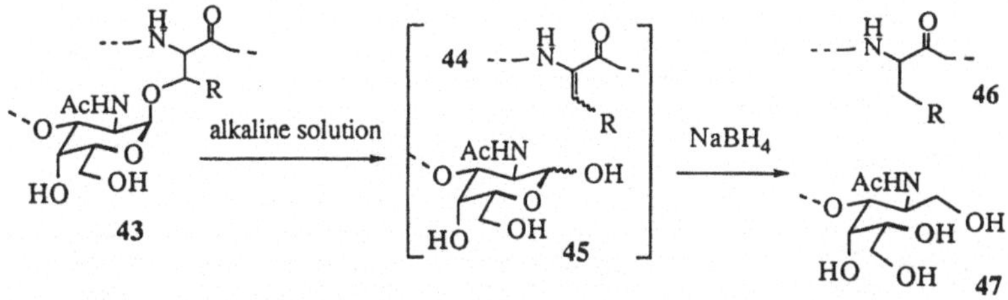

Scheme 5.13 β-Elimination of the glycan from *O*-glycosyl serine and threonine glycoproteins under alkaline conditions, and subsequent reduction to yield alditols.

dehydroamino acid and the free saccharide **45** (Scheme 5.13). Reduction of the latter yields the oligosaccharide **47** with an alditol at the terminal position. *O*-glycosidic bonds (e.g. to serine and threonine) are aldols and therefore are generally sensitive to acids. This sensitivity to acids, however, depends upon the nature of the saccharide as well as on the peptide structure. Acylamino groups vicinal to the glycosidic bond and ammonium groups (in hydroxylysine glycosides) generally stabilise the carbohydrate–peptide linkage towards proton attack. Glycopeptides lacking these stabilising functions may not only be cleaved by treatment with acids but can also sustain an acid-catalysed anomerisation of glycosidic bonds (Scheme 5.14). Attention must be paid to these common properties of *O*-glycosidic bonds throughout the chemical synthesis of glycopeptides and in glycosylation reactions as well as in deprotection and chain-extension reactions (Kunz, 1987; see also Danishefsky and Roberge, 1995).

The stereochemistry of glycosylation reactions (Paulsen, 1982; Schmidt, 1986) constitutes another important aspect of both oligosaccharide and glycopeptide syntheses. This will be outlined below with examples of different classes of *O*-glycosyl amino acid and peptide derivatives.

5.4.2.2 *The* O-*glycosyl serine bond of proteoglycans*

The first chemical synthesis of β-*O*-xylosyl serine was reported by Lindberg and Silvander (1965), who reacted *N*-toluenesulfonyl serine methyl ester **48**

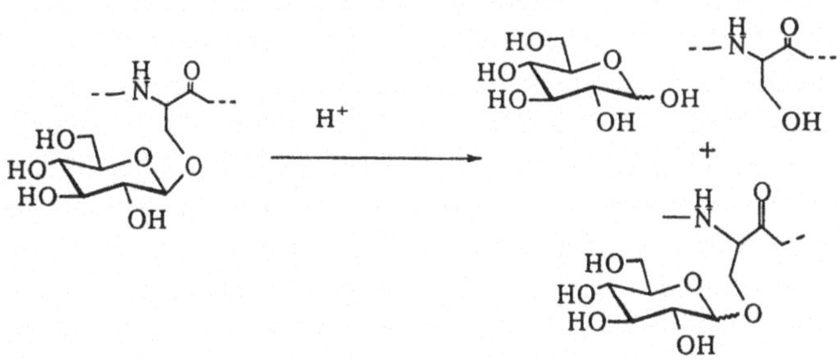

Scheme 5.14 Acid-catalysed cleavage and/or anomerisation of *O*-glycopeptides.

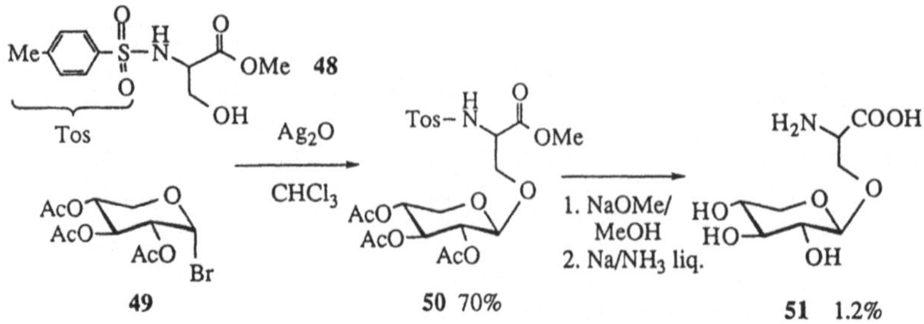

Scheme 5.15 Tos = *p*-toluenesulfonyl.

with 2,3,4-tri-*O*-acetyl-α-D-xylopyranosyl bromide **49** by promotion with silver oxide (Scheme 5.15). The neighbouring group participation of the 2-*O*-acyl protecting group ensures a stereoselective formation of the β-xyloside **50** (Paulsen, 1982; Schmidt, 1986).

As expected, problems occurred during the deprotecting procedures. Treatment of **50** with sodium hydroxide in methanol to saponify the serine methyl ester and the carbohydrate ester groups obviously resulted in the β elimination of the xylose as outlined above (Section 5.4.2.1). After subsequent reductive removal of the toluenesulfonyl group, the free xylosyl serine was obtained in only poor yield.

The sensitivity of *O*-glycosidic bonds towards base-catalysed β elimination is the major source of accompanying degradations during the deblocking reactions. Of course, this sensitivity depends upon the carbonyl activity of the ester carbonyl which acidifies the αCH group. The amido group (in serine glycopeptides) and even more so the free carboxylate have much lower acidifying effects, and, therefore, *O*-glycopeptides are considerably more stable to bases than are *O*-glycosyl serine/threonine esters. Nevertheless, glycopeptide chemistry had to begin with amino acid building blocks, and logically, a mildly cleavable protecting group became essential (Kunz, 1987; 1993).

O-benzylic groups cleavable by hydrogenation under neutral conditions are useful in glycopeptide chemistry. Garegg *et al.* (1979) carried out the glycosylation of benzyloxycarbonyl (Z) serine benzyl ester with xylosyl bromide **52** in the presence of the reactive, soluble silver triflate as the promoter (Scheme 5.16; Hanessian and Banoub, 1977). This afforded the β-xylosyl serine derivative **53**. Selective cleavage of the 4′-*O*-allyl ether of **53** allowed a saccharide chain extension of **54** with an *O*-benzoyl-protected galactosyl β(1-3)galactosyl bromide to furnish the trisaccharide serine derivative **55**. From this compound, the Z and the benzyl ester group were removed by hydrogenation. Owing to the lack of carbonyl activity of the carboxylate, the ester protecting groups of the saccharide portion were now cleavable by Zemplén transesterification to give **56** in good yield.

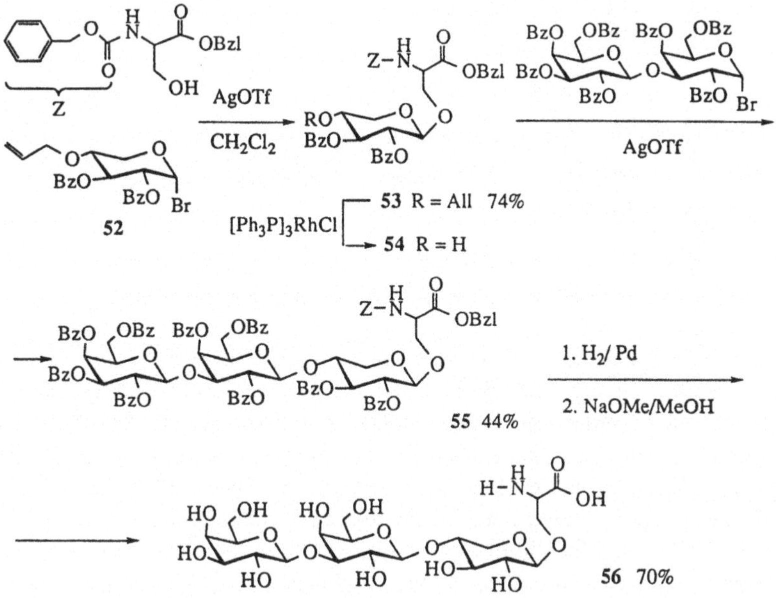

Scheme 5.16 Z = benzyloxycarbonyl; All = allyl; Bz = benzoyl.

While the selective liberation of a saccharide hydroxyl group was achieved in the synthesis of **54**, the selective deprotection of the amino acid functionalities remained unsolved.

Xylosylation of 9-fluorenylmethoxycarbonyl (Fmoc) serine benzyl ester with *O*-benzoyl-protected xylosyl bromide gave the β-xylosyl serine ester **57** (Scheme 5.17; Schultheiss-Reimann and Kunz, 1983). Selective and

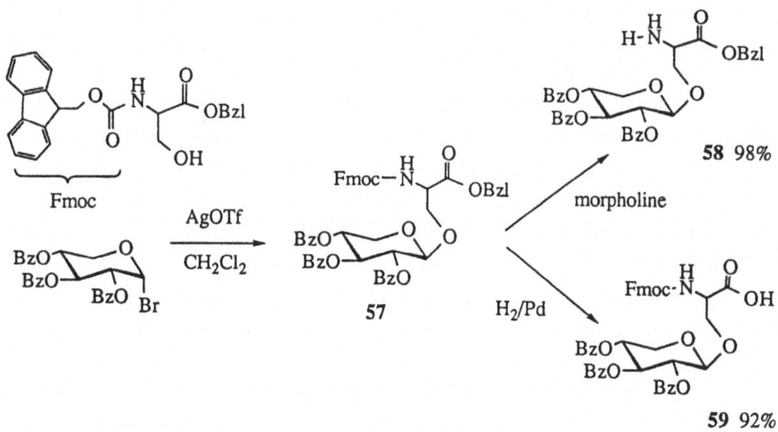

Scheme 5.17 Fmoc = 9-fluorenylmethoxycarbonyl; Bz = benzoyl; OTf = trifluoromethanesulfonate.

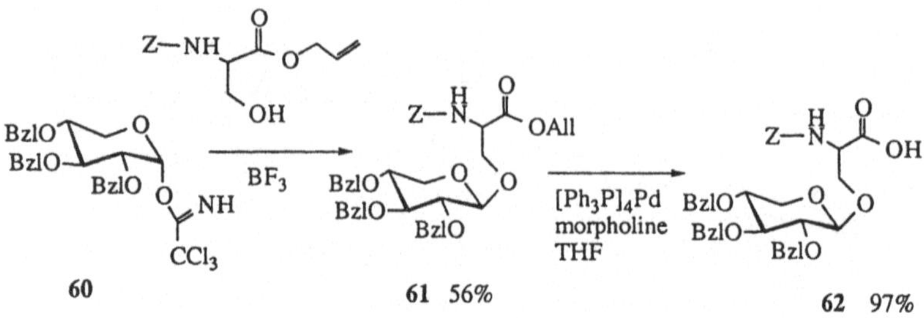

Scheme 5.18 All = allyl; Bzl =benzyl; THF = tetrahydrofuran.

alternative deprotection of **57** was accomplished by either removal of the Fmoc group using morpholine to give the amino component **58** (Schultheiss-Reimann and Kunz, 1983) or hydrogenolytic cleavage of the benzyl ester to form the carboxylic component **59** (Kunz, 1987). Both deprotection reactions resulted in almost quantitative yields. The selectively deprotected glycosyl amino acid derivatives **58** and **59** constitute basic building blocks for peptide chain elongation in either the C- or the N-terminal direction. In particular, *N*-protected *O*-glycosyl amino acid derivatives of the type **59** are useful in Fmoc-based syntheses of glycopeptides in solution and on solid phase.

The removal of the *O*-benzoyl groups can be achieved with hydrazine in methanol, via a hydrazine-catalysed methanolysis (Schultheiss-Reimann and Kunz, 1983). Less-sensitive *O*-glycosyl amino acid carboxylates and *O*-glycopeptides can be subjected to dilute sodium methoxide in methanol, resulting in *O*-deacetylation in the carbohydrate portion (Paulsen *et al.*, 1988).

Trichloroacetimidates of mono- and oligosaccharides have been demonstrated to be very efficient glycosyl donors (Schmidt, 1986; Schmidt and Michel, 1980). Trichloroacetimidate **60** of *O*-benzyl protected xylopyranose reacted with benzyloxycarbonyl serine allyl ester in the presence of borontrifluoride to give the β-xylosyl serine conjugate **61** in good yield and stereoselectivity (Scheme 5.18; Kunz and Waldmann, 1984). Since the xylose portion of **61** carries ether-type protection, its glycosidic bond is sensitive both to bases and to acids. Nevertheless, its selective carboxy deprotection can be achieved by palladium(0)-catalysed allyl transfer to a nucleophile, for example morpholine, to yield **62**.

This palladium(0)-catalysed cleavage of allyl esters or allyloxycarbonyl (Aloc) groups (Kunz and Unverzagt, 1984) proceeds under practically neutral conditions and is a very useful tool in the synthesis of peptides and glycopeptides (Kunz, 1993). Depending upon the type of nucleophile, which must irreversibly trap the allylic moiety from the intermediate π-allyl palladium complex, the allyl deprotection conditions can be adjusted around the neutral point (Ciommer and Kunz, 1991; Kunz and März,

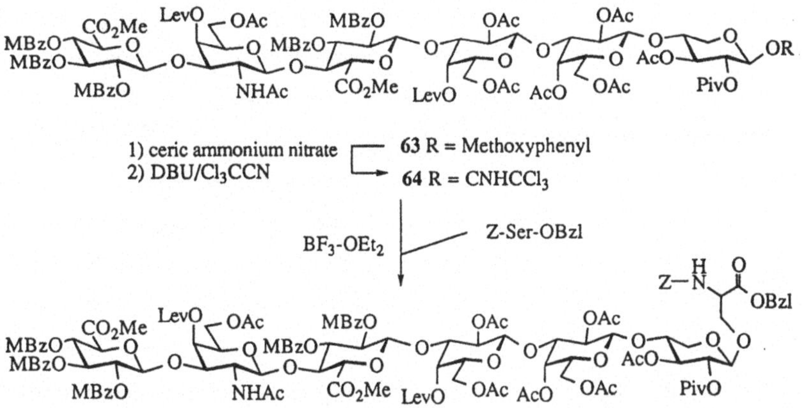

65

Scheme 5.19 MBz = 4-methylbenzoyl; Piv = pivaloyl; Lev = levulinoyl; DBU = diazabicyclo-undecene.

1988) and are compatible with the demands of syntheses of complex sensitive glycopeptides (Günther and Kunz, 1990; Kunz and März, 1992; Kunz and Unverzagt, 1988).

The efficiency of trichloroacetimidates in *O*-glycosylations of hydroxyamino acid derivatives was demonstrated in the synthesis of the hexasaccharide core structure of proteoglycans (Goto and Ogawa, 1992). After oxidative cleavage of the 4-methoxyphenyl hexasaccharide **63**, the free anomeric hydroxyl group was transformed into the trichloroacetimidate **64**, which was reacted with Z-serine benzyl ester to furnish the saccharide–serine conjugate **65** (Scheme 5.19). The protecting group pattern within the saccharide portion allowed the selective removal of the levulinoyl groups in the 4-position of the galactosyl units and a subsequent regioselective sulfation prior to the complete removal of all remaining protecting groups in the conjugate.

β-D-Galactopyranosyl- and β-D-glucopyranosyl serine derivatives were generally obtained by procedures analogous to the outlined syntheses of the xylosyl serine components. This holds true for glycoconjugates containing the simultaneously removable Z and benzyl ester protecting groups (Kum and Roseman, 1966) as well as for the selectively deprotectable Fmoc (*O*-glucosyl) serine derivatives (Schultheiss-Reimann and Kunz, 1983) and the corresponding oligosaccharide conjugates with glucose as the linkage sugar (Fukase *et al.*, 1992).

*5.4.2.3 The α-*N*-acetylgalactosamine serine/threonine linkage*

The α-*O*-galactosamine-serine/threonine linkage is most frequently found in mammalian *O*-glycoproteins. This type of glycosidic bond is typical for

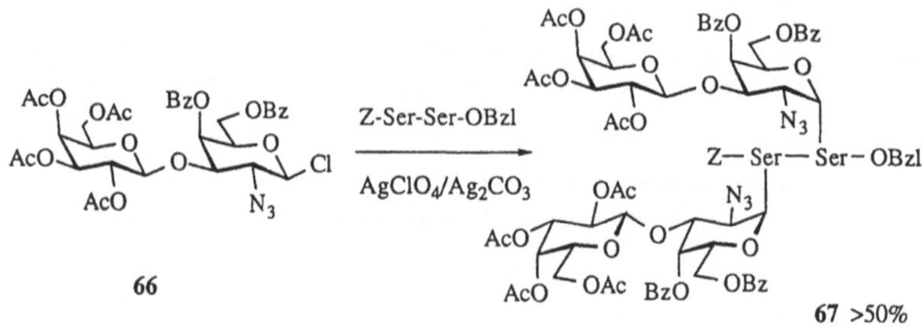

Scheme 5.20.

mucins, highly *O*-glycosylated proteins expressed on the membranes of various epithelial cell types (Hilkins, 1988). During tumorgenesis, mucin-mediated cell adhesion and anti-adhesion processes appear to be out of control. The glycosylation profile of the cell membrane and the balance of these adhesion phenomena are obviously closely related to each other. Glycoproteins with plain αGalNAc–Ser/Thr (T_N-antigen) and βGal(1-3)GalNAcα–Ser/Thr (T-antigen) structure have been found to be tumor-associated (Springer, 1984) and in some cases even tumor-specific (Karsten *et al.*, 1993). The α-*O*-GalNAc–Ser/Thr linkage is also present in the extracellular portion of transmembrane glycoproteins, for example in glycophorin of erythrocytes (Tomita *et al.*, 1978) and in gp120 of HIV-1 (Hansen *et al.*, 1991). The stereoselective formation of the α-glycosidic linkage between *N*-acetylgalactosamine and hydroxy amino acids at first presented a problem, because the neighbouring-group-active *N*-acetyl group of GalNAc strongly favours the formation of β-galactoside bonds. The problem was solved by Paulsen and Hölck (1982), who applied 2-azido-galactosyl donors, readily accessible via azidonitration (Lemieux and Ratcliffe, 1979), in order to stereoselectively form α-glycosides of GalNAc. The method was adopted to the glycosylation of Z-serine benzyl ester (Ferrari and Pavia, 1983) and was also successfully used for the simultaneous glycosylation of a protected serine–serine dipeptide with two Gal(1-3)$GalN_3$ donors **66** to give the precursor of the T-antigen structure **67** (Scheme 5.20; Paulsen and Paal, 1984).

The synthesis of T_N-antigen and T-antigen conjugates, which are selectively deprotectable in either amino acid functionality, was achieved with T_N-antigen or T-antigen conjugates containing the Z-/*tert*-butyl ester protecting group combination **68** (Verez-Bencomo and Sinay, 1983), the Fmoc-/benzyl ester combination **69** (Kunz and Birnbach, 1986), the Fmoc/active ester combination **70** and **71** (Ferrari and Pavia, 1983; Peters *et al.*, 1992), the Fmoc-/*tert*-butyl ester combination **72** (Paulsen and Adermann, 1989), all illustrated in Scheme 5.21, or the Fmoc-/allyl ester combination (Ciommer and Kunz, 1991), for which the selective deprotection of the T-antigen serine conjugate **73** is demonstrated in Scheme 5.22.

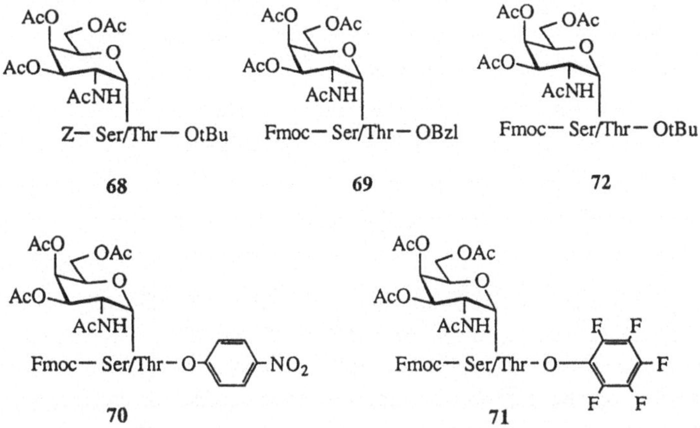

Scheme 5.21 Fmoc = 9-fluorenylmethoxycarbonyl.

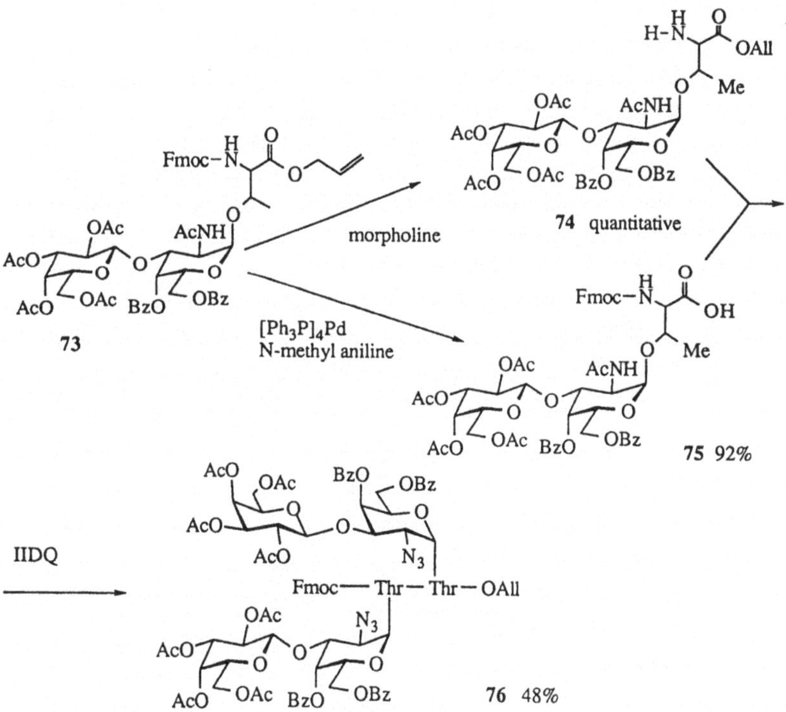

Scheme 5.22 IIDQ = isobutyl 2-isobutyloxy-1,2-dihydroquinoline-1-carboxylate; Fmoc = 9-fluorenylmethoxycarbonyl.

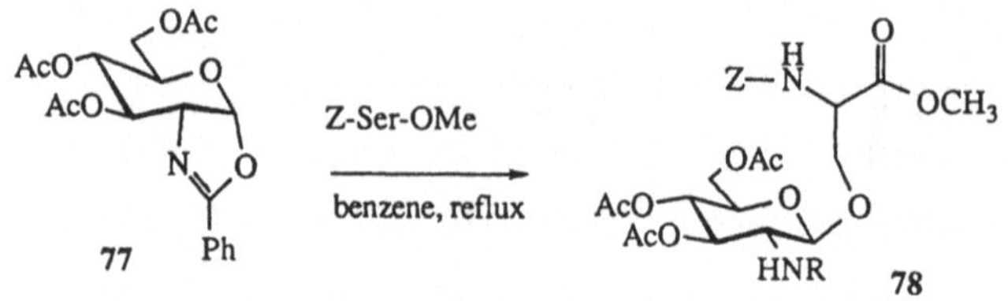

Scheme 5.23.

Selective removal of the Fmoc group was carried out using morpholine (Schultheiss-Reimann and Kunz, 1983) to give the deblocked amino component **74** (Ciommer and Kunz, 1991) almost quantitatively. Alternative deprotection of the carboxylic function to give **75** was accomplished by Pd(0)-catalysed allyl transfer to *N*-methylaniline (Kunz and Waldmann, 1984), which irreversibly traps the allyl moiety but is not basic enough to affect the Fmoc group. Condensation of the two monofunctional conjugates **74** and **75** gave the clustered glycopeptide **76**, containing a protecting group combination identical to the starting conjugate **73** and therefore being useful for peptide chain extension in either direction.

5.4.2.4 The β-glucosamine serine/threonine linkage

Glycoproteins containing the β-D-glucosaminyl serine/threonine structure have been recognised only recently (Hart *et al.*, 1989). They are ubiquitous in subcellular compartments and occur, particularly abundantly, in nuclear membranes. Nevertheless, their biological function still remains unclear. In contrast to the late discovery of this linkage structure, β-glucosaminyl serine/threonine conjugates were among the first to be constructed by chemical synthesis (Micheel and Köchling, 1958).

Besides glucosamine-derived oxazolines (Garg and Jeanloz, 1985; Micheel and Köchling, 1958), glycosyl halides (Vercellotti and Luetzow, 1966) have been used as glycosyl donors in the syntheses of β-glucosaminyl serine/threonine derivatives (Schemes 5.23 and 5.24). Both glycosylation methods

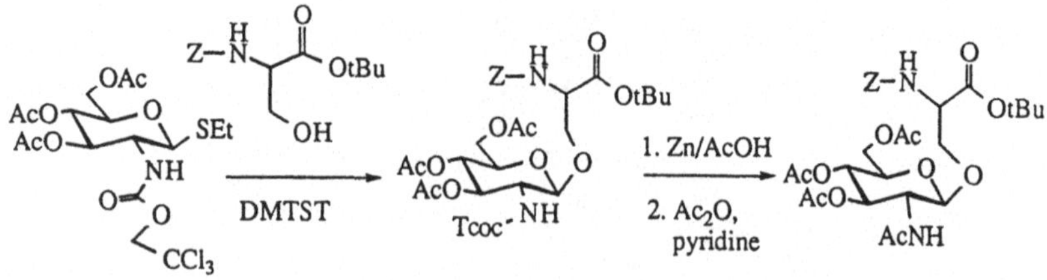

79 **80** 85% **81** 85-90%

Scheme 5.24 DMTST = dimethyl methylthiosulfonium trifluoromethanesulfonate; Tcoc = trichloroethoxycarbonyl.

often require long reaction times under acidic conditions and give low yields of the desired product. Progress in this field was achieved using thioglycosides as the glycosyl donors together with the *N*-trichloroethoxycarbonyl (Tcoc or Troc) protecting group in the glucosamine unit (Schultz and Kunz, 1993). The Tcoc group of **79** exhibits the desired neighbouring-group effect during the glycosylation reaction but minimises the formation of an oxazoline. Moreover, the Tcoc group can be selectively removed from the β-glucoside **80** by reductive elimination and thus exchanged for the *N*-acetyl group. In this context it is noteworthy that the activation of an *N*-allyloxycarbonyl (Aloc) protected glucosaminyl bromide (Boullanger *et al.*, 1987) under Helferich conditions [$Hg(CN)_2$] gave the *O*-glucosamine serine conjugate in moderate yields. Subsequent Pd(0)-catalysed removal of the Aloc group (Kunz and Unverzagt, 1984) and *N*-acetylation alternatively led to **81** (Schultz and Kunz, 1993).

5.5 Synthesis of oligosaccharides and glycopeptides

In this section, the construction of biologically interesting oligosaccharide side-chains of glycoproteins will be briefly illustrated with the use of several selected examples.

5.5.1 The Thomsen–Friedenreich antigen disaccharide

The disaccharide Galβ(1-3)GalNAc is found as a structural element in gangliosides and, α-glycosidically linked to serine or threonine, as a core region of mucin glycoproteins. Glycoproteins of this type have been described as tumor-associated antigens (Samuel *et al.*, 1990; Springer, 1984). The synthesis of the Thomsen–Friedenreich (T) antigen disaccharide was efficiently achieved starting from *O*-acetylated galactal **82** (Paulsen and Hölck, 1982) which was subjected to the azidonitration reaction using sodium azide and ceric ammonium nitrate (Lemieux and Ratcliffe, 1979). After removal of the *O*-acetyl groups from **83**, selective 4,6-di-*O*-benzoylation resulted in a monofunctional acceptor (**84**), which was galactosylated with penta-*O*-acetyl-galactopyranose via activation with trimethylsilyl triflate (Ogawa *et al.*, 1981) to give the protected T antigen precursor **85** (Scheme 5.25).

Formation of the T antigen disaccharide donor **86** was achieved by treatment of **85** with acetic anhydride/H_2SO_4 and subsequent conversion of the formed anomeric acetate into the glycosyl bromide using $TiBr_4$ (Paulsen and Paal, 1984). After glycosylation to give the T antigen precursors of type **67**, the azido group can be reacted with thioacetic acid (Rosen *et al.*, 1988).

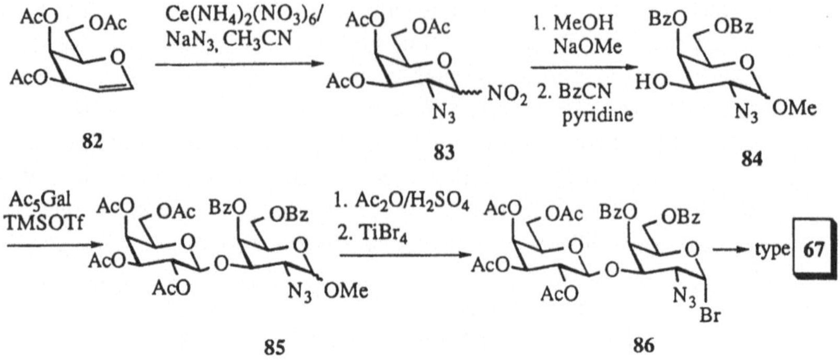

Scheme 5.25 TMSOTf = trimethylsilyl trifluoromethanesulfonate.

5.5.2 *Sialyl T_N disaccharide*

Sialyl T_N antigen glycoproteins have also been described as tumor-associated antigens (Kurosaka *et al.*, 1988). Two strategies have been applied to synthesise sialyl T_N serine/threonine conjugates: either the 2-azido-precursor of the NeuNAcα(2-6)GalNAc disaccharide was preformed and then coupled to the hydroxyamino acid, or a $GalN_3$-serine/threonine conjugate was synthesised first and subsequently sialylated. The former method takes advantage of the less demanding acceptor **87** in the sialylation reaction, which was carried out using the 2-chloro-2-deoxy-sialic acid methyl ester derivative **88** (Scheme 5.26; Iijima and Ogawa, 1988).

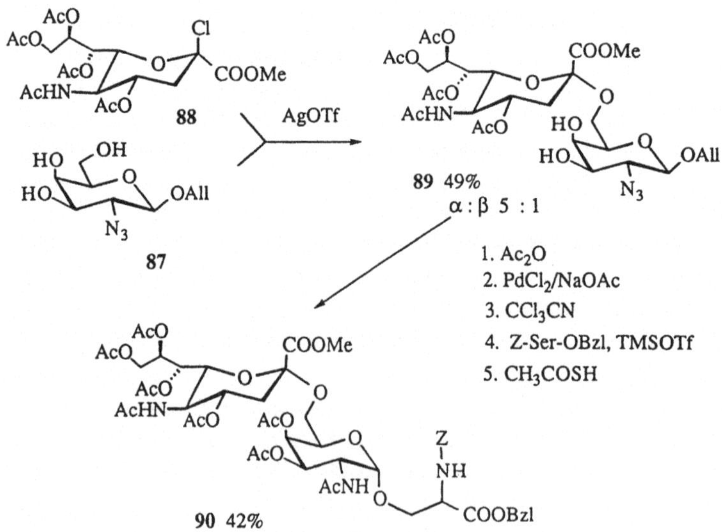

Scheme 5.26.

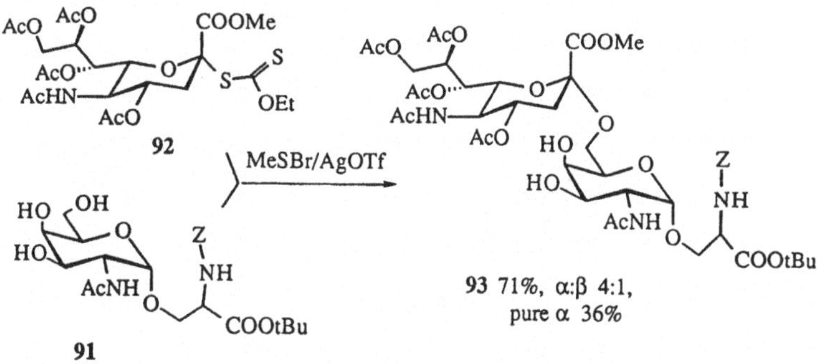

Scheme 5.27 tBu = *tert*-butyl.

The glycosylation proceeded with moderate yield, complete regioselectivity at the 6-OH group and a high stereoselectivity in favour of the desired α-sialoside. After chromatographic separation and deallylation using $PdCl_2$/NaOAc, the anomeric trichloroacetimidate (Grundler and Schmidt, 1984) was formed and activated by TMS-triflate for coupling with Z-serine benzyl ester. The stereoselectivity of this glycosylation was low (α:β = 3:2). The separation of the diastereomers was possible after reduction of the azido group and subsequent *N*-acetylation to afford the target conjugate **90** in a yield of 42%. From these results, it was concluded that the formation of the sialyl T_N disaccharide prior to glycosylation of the amino acid derivative involves two problematic glycosylation steps.

The authors, therefore, developed an alternative concept and first synthesised the GalNAc-serine/threonine (T_N antigen) conjugate containing free 4- and 6-hydroxy groups in the monosaccharide portion (Nakahara *et al.*, 1990). In the subsequent sialylation reaction, they introduced the 3-phenylthio group in the neuraminic acid donor as a temporary stereodifferentiating tool which favours the stereoselective formation of the desired α-sialoside.

However, a more straightforward stereoselective and regioselective sialylation is possible with the T_N antigen conjugate **91** (Scheme 5.27) which contains three free hydroxy groups (Liebe and Kunz, 1994). The 6-hydroxy group of the T_N antigen structure **91** was glycosylated using the neuraminic acid xanthogenate **92** (Marra and Sinay, 1989) under promotion by methylsulfenyl triflate (Dasgupta and Garegg, 1988).

The reaction proceeded with high yield (71%), showed excellent regioselectivity and a stereoselectivity of α:β of 4:1. After chromatographic separation, the sialyl T_N serine conjugate **93**, which can be selectively deprotected at either amino acid function, was isolated in a yield of 36% (Liebe and Kunz, 1994). Deprotection of the sialyl T_N conjugate **93** was carried out by hydrogenolysis of the Z group, *N*-acetylation, cleavage of the *tert*-butyl ester

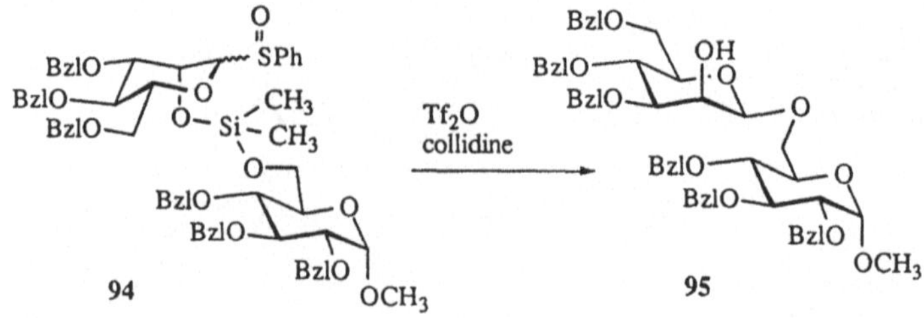

Scheme 5.28.

using formic acid and final treatment with 1 N NaOH in methanol in order to remove the *O*-acetyl groups and the neuraminic acid methyl ester.

5.5.3 β-Mannosyl chitobiose asparagine

Among the glycosidic bonds found in natural saccharides and glycoconjugates, the β-mannoside bond is one of the most difficult bonds to synthesise by chemical methods. Both the anomeric effect and neighbouring-group participations strongly favour the formation of α-mannosides (Toshima and Tatsuta, 1993). Successful syntheses of this type of saccharide include the epimerisation of β-glucosides via an oxidation–reduction reaction sequence (Augé *et al.*, 1980) and the use of silver silicate to promote mannosylation procedures under heterogeneous conditions (Paulsen and Lockhoff, 1981). In both procedures, significant amounts of undesired diastereomers are obtained, in particular, if more demanding glycosyl acceptors are to be mannosylated.

More stereoselective, multistep β-mannoside syntheses have been developed by Barresi and Hindsgaul (1991) and Stork and Kim (1992), who constructed preformed mannose–acceptor conjugates containing an acetal- or silane-type tether between the 2-oxygen of the mannosyl donor and the acceptor, as, for example, in **94** (Scheme 5.28). Upon activation of the donor, the acceptor was introduced from the β-direction, favoured by the preformed tether, to give the β-mannoside **95**. A similar concept has been reported (Ito and Ogawa, 1994) and involved a mixed 4-methoxybenzylidene acetal tether.

Alternatively, the β-mannoside linkage can be stereoselectively constructed by a sterically controlled conversion of the corresponding β-glucosides via an S_N2-type inversion of configuration at the 2 position. Commonly, S_N2 reactions at the 2-position of glucosides and hexapyranosides in general are considered difficult or even impossible because of the Coulomb repulsion between the approaching nucleophile and the lone pairs of the ring oxygen. However, by exploiting the neighbouring-group effect of a carbamoyl group in the 3-position of the glucose in the disaccharide **96**, the S_N2-reaction was efficiently accomplished, after conversion of the 2-OH group into the trifluoromethylsulfonate, to yield the desired β-mannosyl glucosamine

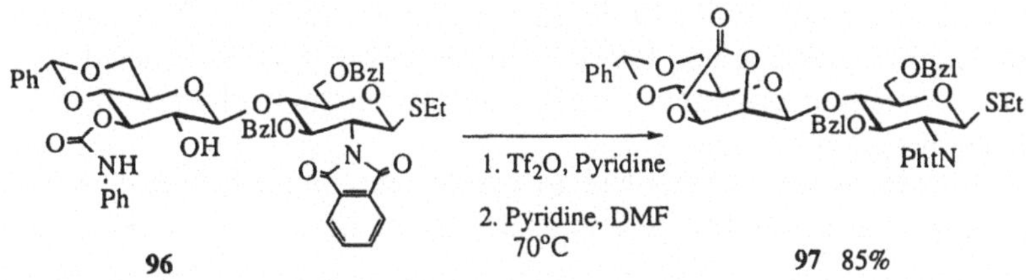

Scheme 5.29 Bzl = benzyl; Et = ethyl; Pht = phthaloyl; DMF = dimethylformamide.

structure **97** (Scheme 5.29; Kunz and Günther, 1988). Extension of the product to the β-mannosyl chitobiose **21** (Scheme 5.8) was achieved by activation of the thioglycoside. Finally, the complete core region β-mannosyl chitobiosyl asparagine **22** was synthesised (Günther and Kunz, 1990; 1992).

An excellent synthesis of a complete biantennary *N*-glycoprotein linkage conjugate has been reported (Unverzagt, 1994). After hydrolysis of the carbonate **98**, the first glucosamine mannose disaccharide was coupled α-glycosidically to the O-3 of the β-mannoside unit of **99** (Scheme 5.30). Hydrolytic removal of the benzylidene acetal made the regioselective introduction of a second glucosamine mannose unit at the O-6 possible, yielding the biantennary heptasaccharide **100**.

5.5.4 *Lewis^X antigen glycopeptides*

The Lewis^X trisaccharide has been found in both glycoproteins and glycolipids of mammalian cells and was described as a status-dependent antigen (SSEA-1, stage-specific embryonic antigen) which is later masked by further

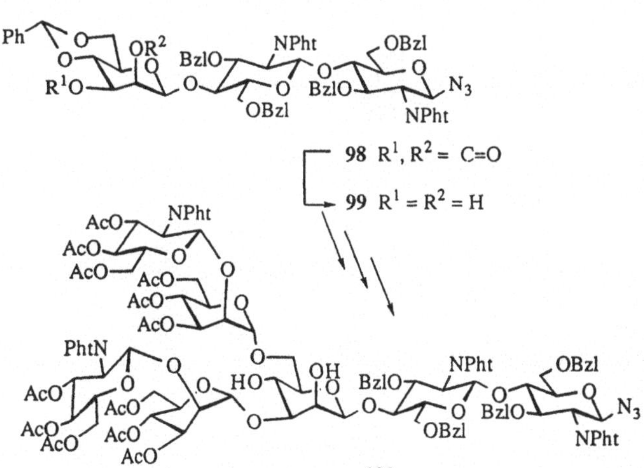

100

Scheme 5.30 Bzl = benzyl; Ph = phenyl; Pht = phthaloyl.

glycosylation (Gooi *et al.*, 1981). In the synthesis of Lewis[X] glycopeptides (von dem Bruch and Kunz, 1994), the trisaccharide was constructed according to a strategy analogous to the one employed in the synthesis of the Lewis[a] antigen **36** (Scheme 5.11).

The azido group was applied to protect the anomeric position in the glucosamine acceptor **101** (Scheme 5.31). After stereoselective introduction

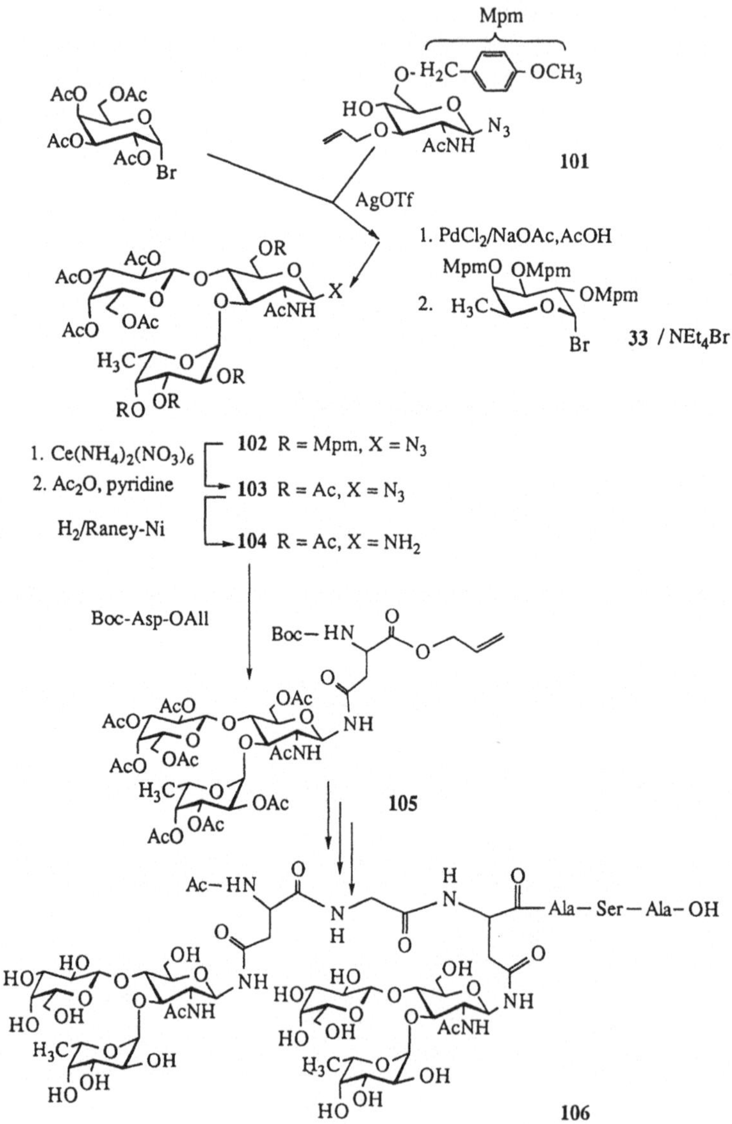

Scheme 5.31 Boc = *tert*-butyloxycarbonyl; Mpm = 4-methoxybenzyl.

of the β-galactose at O-4, the 3-*O*-allyl ether was cleaved using palladium chloride. Stereoselective α-fucosylation to give **102** was achieved with the Mpm ether-protected fucosyl bromide **33**. Subsequent exchange of the Mpm groups for the *O*-acetyl groups rendering the fucoside bond of **103** more acid-stable was followed by hydrogenation of the azide to yield the protected Lewis^X trisaccharide amine **104**. Its condensation with Boc aspartic acid α-allyl ester furnished the Lewis^X asparagine conjugate **105**. Owing to the increased acid-stability of the saccharide portion, the Boc group can be removed from **105** selectively even by application of HCl/diethyl ether. On the other hand, the allyl ester can be cleaved selectively either via the palladium(0)-catalysed allyl transfer (Kunz and Waldmann, 1984) or the rhodium(I)-catalysed isomerisation reaction (Waldmann and Kunz, 1983). Chain extensions and complete deprotection reactions furnished a glycohexapeptide **106** which carries two Lewis^X antigen side-chains. This synthetic clustered Lewis^X antigen glycopeptide was coupled to carrier proteins such as bovine serum albumin and keyhole limpet hemocyanin to furnish Lewis^X antigen neoglycoproteins (von dem Bruch and Kunz, 1994).

5.5.5 Sialyl Lewis^X glycopeptides: chemical and enzymatic synthesis

The sialyl Lewis^X tetrasaccharide has been recognised as a very important ligand for selectins which are decisive receptors in cell adhesion processes (Philips *et al.*, 1990). Because of this, glycoconjugates containing sialyl Lewis^X structures are of particular interest for the investigation of cell adhesion phenomena (Springer, 1994). Although a high standard has been reached in the chemical synthesis of sialyl saccharides (Danishefsky *et al.*, 1992; Hasegawa *et al.*, 1993; Kameyama *et al.*, 1991; Nicolaou *et al.*, 1991), enzymatic processes appear particularly attractive for the construction of these compounds. Enzymatic syntheses avoid extensive protection and deprotection reactions and are therefore favourable for glycosylations with the complex and expensive neuraminic acid.

Valuable progress in enzymatic glycosyltransfer reactions was achieved when the glycosyltransferases were applied in combination with an alkaline phosphatase which hydrolyses the nucleoside diphosphate liberated during the glycosyltransfer. The product inhibition of the glycosyltransferase by nucleoside diphosphate is thus eliminated (Unverzagt *et al.*, 1990).

The efficiency of this type of glycosyltransfer reaction was demonstrated on the *N*-glycopeptide **107** (Scheme 5.32). Sequential enzymatic galactosylation and sialylation in 'one-pot' yielded the trisaccharide glycopeptide **108** in an overall yield of 83%. In the absence of phosphatase, the inhibition of the transferases by the nucleoside diphosphate limits the yield to about 30%.

An excellent application of enzymatic glycosyltransfer reactions was recently published by Wong *et al.* (Schuster *et al.*, 1994), who carried out

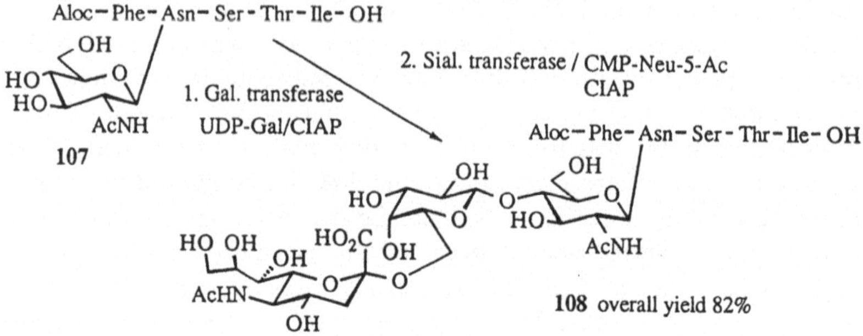

Scheme 5.32 CIAP = calf intestine alkaline phosphatase; CMP = cytidine-5′-monophosphate; UDP = uridine-5′-monophosphate.

sequential glycosyltransfer reactions with a solid phase-bound glycosamine-asparagine substrate **109** (Scheme 5.33).

The galactosyltransferase-catalysed reaction with UDP-galactose was followed by α-2,3-sialyltransferase-catalysed sialylation. Both glycosylations were conducted in the presence of alkaline phosphatase. The release of the formed trisaccharide peptide from the controlled-pore glass support was accomplished by chymotrypsin-catalysed hydrolysis of the phenylalanine carboxylic amide bond. The soluble glycopeptide was subsequently subjected to enzymatic regio- and stereoselective fucosyltransfer to give the sialyl Lewis^X glycopeptide **110** (Schuster *et al.*, 1994).

In a chemical synthesis of sialyl Lewis^X glycopeptides, the azido group again served as the anomeric protecting group and the precursor of the anomeric amino function in the glucosamine derivative **111** (Sprengard *et al.*, 1995). α-Fucosylation to give **113** was carried out with the benzylether-protected

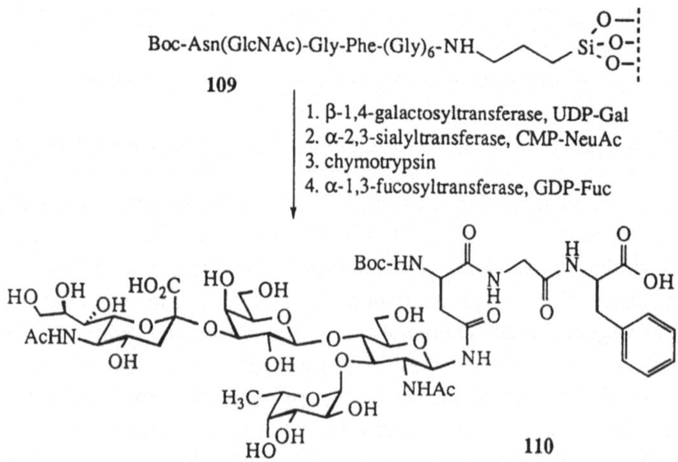

Scheme 5.33 Solid support CPG (controlled-pore glass). Boc = *tert*-butyloxycarbonyl.

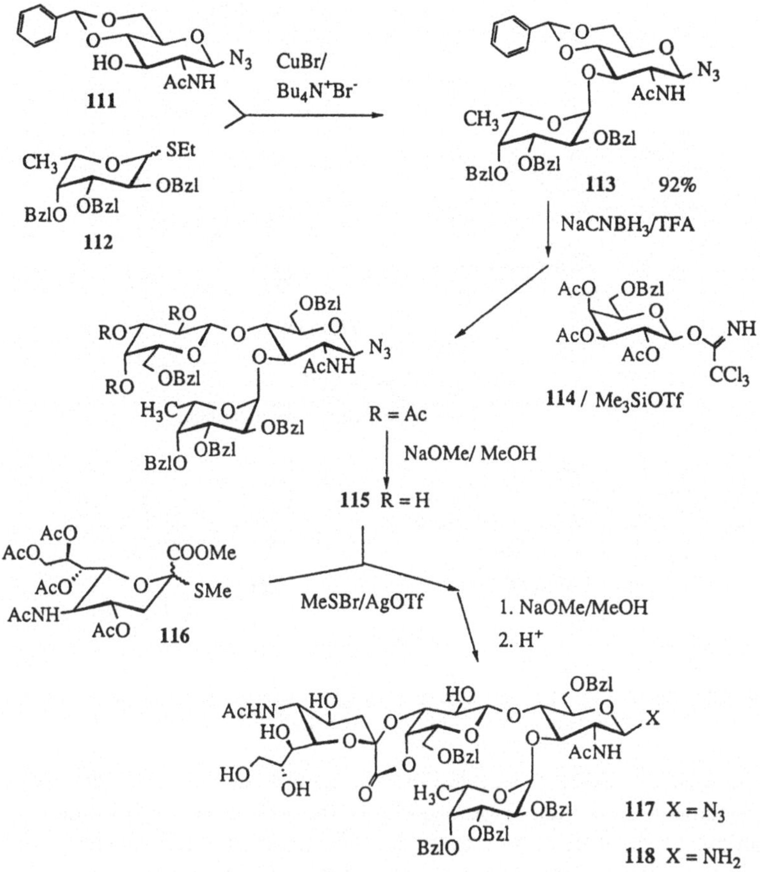

Scheme 5.34 TFA = trifluoroacetic acid.

thiofucoside **112** (Scheme 5.34). After the reductive opening of the benzylidene acetal (Garegg *et al.*, 1982), β-galactosylation with the 6-*O*-benzyl-protected galactose was possible only with the trichloroacetimidate **114** (Schmidt and Michel, 1980). Removal of the *O*-acetyl groups from the obtained Lewis^X structure led to the acceptor **115**. The sialylation of **115** was performed with the methylthioglycoside of neuraminic acid methyl ester **116** (Hasegawa *et al.*, 1993) under promotion by methylsulfenyl triflate (Dasgupta and Garegg, 1988) to furnish the sialyl Lewis^X tetrasaccharide with high α-stereoselectivity and regioselectivity. The crude product was subjected to *O*-deacetylation and saponification of the methyl ester and then converted into the 4′-lactone **117** by treatment with acidic ion-exchange resin. This lactone served as an internal protecting group of the neuraminic acid carboxylic group during the construction of glycopeptides. To this end, the anomeric azido group was hydrogenated to give the sialyl Lewis^X amine **118**, which was then condensed with a

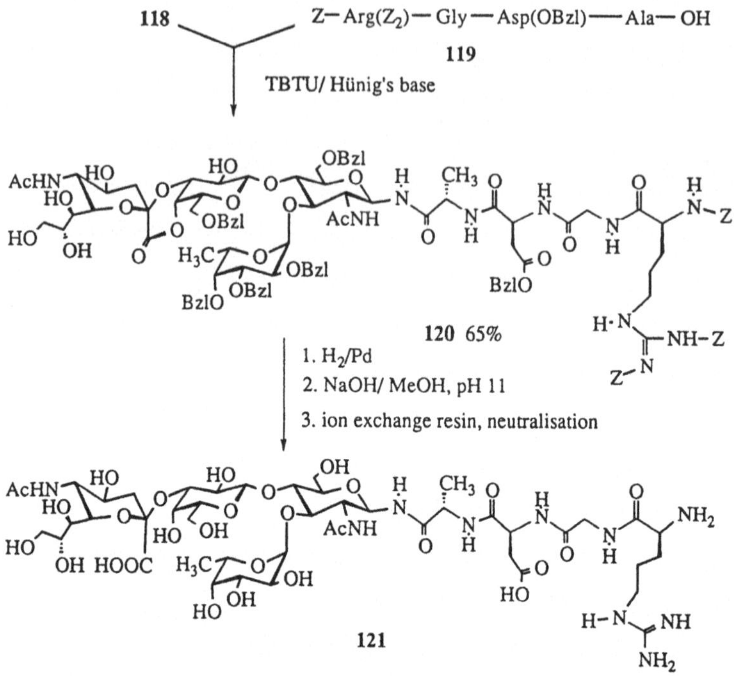

Scheme 5.35 TBTU = [2-(1H-benzoltriazol-1-yl)-1,1,3,3-tetramethyl-uronium tetrafluoroborate]; Hünigs base = $EtiPr_2N$; Ac = acyl; Bzl = benzyl.

presynthesised RGD-tetrapeptide **119** to furnish the sialyl Lewis^X glycopeptide **120** (Scheme 5.35). Under these conditions, the lactone remained intact.

Complete deprotection was carried out by hydrogenolytic cleavage of the benzyl ether, benzyl ester and benzyloxycarbonyl protecting groups prior to the lactone opening under mildly alkaline conditions and gave the free sialyl Lewis^X RGD-peptide conjugate **121**. Compound **121** combines two ligand structures of cell adhesion molecules: the sialyl Lewis^X as the ligand of selectins and the RGD motif for the fibronectin receptor gpIIb/IIIa. It proved to be a high-affinity ligand to P-selectins (Sprengard *et al.*, 1995). It should also be mentioned that the strategy pursued in the synthesis of the sialyl Lewis^X saccharide was also successful in the synthesis of sulphated Lewis^X derivatives (Stahl *et al.*, 1995).

5.6 Solid-phase synthesis of glycopeptides

Solid-phase synthesis is the method of choice when a great variety of different peptides is to be synthesised on a small scale (Barany *et al.*, 1987) and it is of particular value for combinatorial chemistry (Jung and Beck-Sickinger, 1992). In the solid-phase synthesis of glycopeptides the presence of glycosidic bonds demands special attention throughout the synthesis, since these bonds may be sensitive to acids and in some cases to bases (Kunz, 1987; 1993). Provided that water was strictly excluded, solid-phase synthesis of glycopeptides was successfully accomplished according to the classical Merrifield technique, even including the final detachment from the solid support by cleavage of the benzyl ester anchor with hydrogen fluoride (Lavielle *et al.*, 1981). In the meantime, solid-phase methods based on highly acid-sensitive anchoring groups and the Fmoc group as the temporary amino protection have been developed (Mergler *et al.*, 1988) and efficiently employed in the solid-phase syntheses of glycopeptides (Lüning *et al.*, 1989; Meldal and Jensen, 1990; Paulsen *et al.*, 1988). As an example, the multiple solid-phase

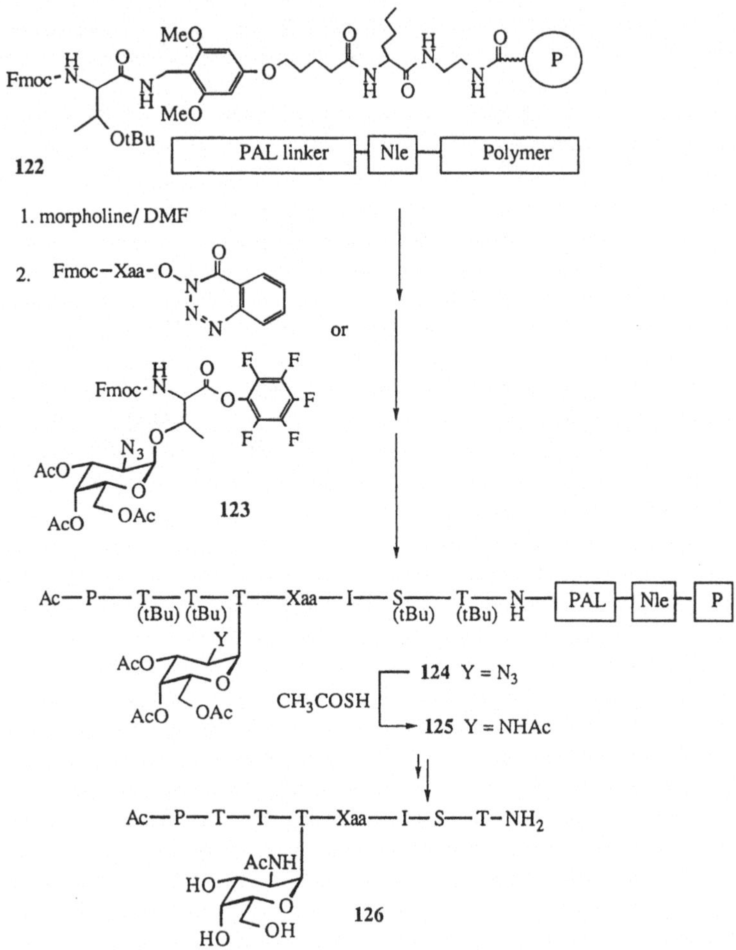

Scheme 5.36 One letter code of amino acids; Ac = acyl; DMF = dimethylformamide; Fmoc = 9-fluorenylmethoxycarbonyl; Nle = norleucine; PAL = peptide amide linker; tBu = *tert*-butyl.

synthesis of a great variety of mucin-type *O*-glycopeptides is shown in the Scheme 5.36 (Paulsen *et al.*, 1994).

Starting from the polymer **122**, equipped with norleucine as the standard amino acid, the peptide amide linker (PAL) and loaded with the Fmoc protected C-terminal amino acid, chain extension after Fmoc removal using morpholine was carried out with Fmoc amino acid or glycosylated amino acid active esters such as **123** to give the resin-linked glycopeptide **124**.

The T_N antigen saccharide structure was also formed in the solid-phase by conversion of the 2-azido-galactose **124** into the 2-acetamido galactose

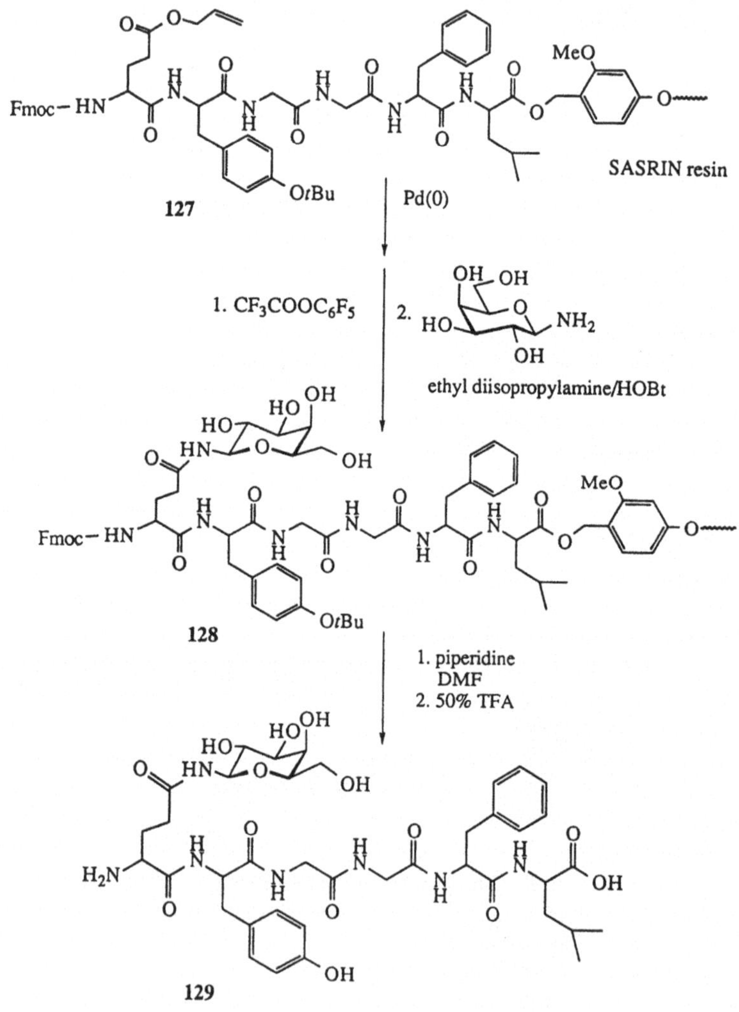

Scheme 5.37 DMF = dimethylformamide; Fmoc = 9-fluorenylmethoxycarbonyl; *t*Bu = *tert*-butyl; TFA = trifluoroacetic acid.

structure **125**. After detachment of the glycopeptides **126**, with various amino acids in the position Xaa, the purified compounds were obtained in a yield of 12%–74% relative to the loaded polymer **122**.

A convergent synthesis of glycopeptides (see Section 5.4.1; Cohen-Anisfeld and Lansbury, 1990), on the solid-phase was conducted by a preceding solid-phase synthesis of the [Leu]enkephaline derivative **127** (Scheme 5.37) on the acid-labile SASRIN-resin (Mergler *et al.*, 1988) combining the Fmoc amino protection with the allyl ester as the side-chain protection of the glutamic acid (Vetter *et al.*, 1995). Selective removal of the side-chain allyl ester via Pd(0)-catalysed cleavage (Kunz *et al.*, 1985) was followed by the condensation with unprotected galactosylamine to give the resin-linked glycopeptide **128**. Removal of the protecting groups and detachment from the resin via acidolysis yielded the free glycopeptide **129** (Vetter *et al.*, 1995).

An inverse strategy of convergent glycopeptide synthesis in the solid-phase involved the solid-phase synthesis of a polymer-linked trisaccharide **130** according to the glycal methodology and the subsequent condensation with an aspartic acid peptide **131** to give the polymer-linked glycopeptide **132** (Scheme 5.38; Roberge *et al.*, 1995). The protecting group combination of **132** allowed the alternative extension of the peptide chain in either the N- or the C-terminal direction. Final release of the synthesised glycopeptide was performed with HF-pyridine in dichloromethane to cleave the silyl ether.

Solid-phase syntheses of glycopeptides using glycosylated amino acid building blocks which may even contain acid-labile structures, for example

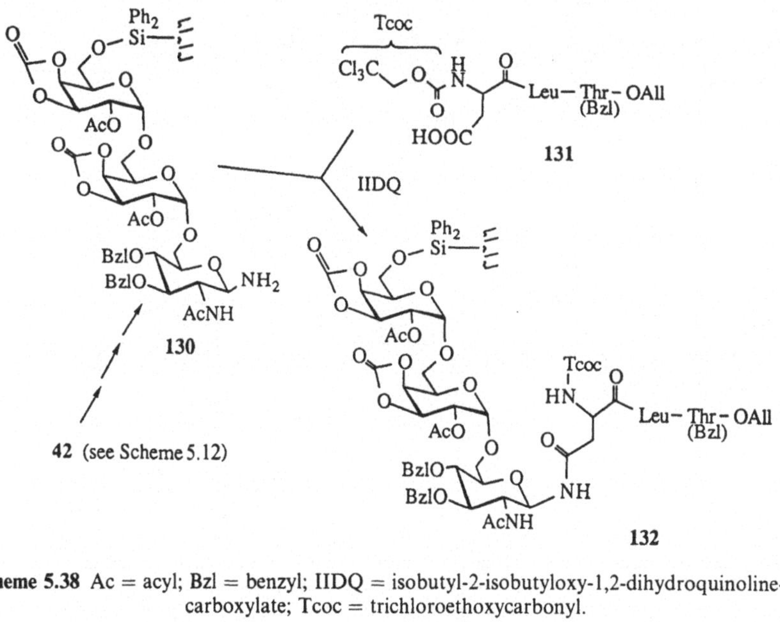

Scheme 5.38 Ac = acyl; Bzl = benzyl; IIDQ = isobutyl-2-isobutyloxy-1,2-dihydroquinoline-1-carboxylate; Tcoc = trichloroethoxycarbonyl.

fucoside bonds (Kunz and von dem Bruch, 1994), were achieved with allylic anchoring groups (HYCRAM and analogous groups) (Kosch *et al.*, 1994; Kunz and Dombo, 1988). Owing to the stability of the allylic anchoring ester to both acids and bases, the solid-phase synthesis on allyl anchor resins can be carried out according to the Fmoc as well as the Boc strategy. This is illustrated for the solid-phase synthesis of a mucin-type glycononapeptide on a solid phase equipped with a combination of the hydroxycrotyl anchor and an oligoethyleneglycol spacer (HYCRON-resin) (Seitz and Kunz, 1995). After removal of the Fmoc group from the resin-linked glycin **133**, the first chain-extension was carried out with Boc-proline to give the resin-bound Boc-dipeptide **134** (Scheme 5.39).

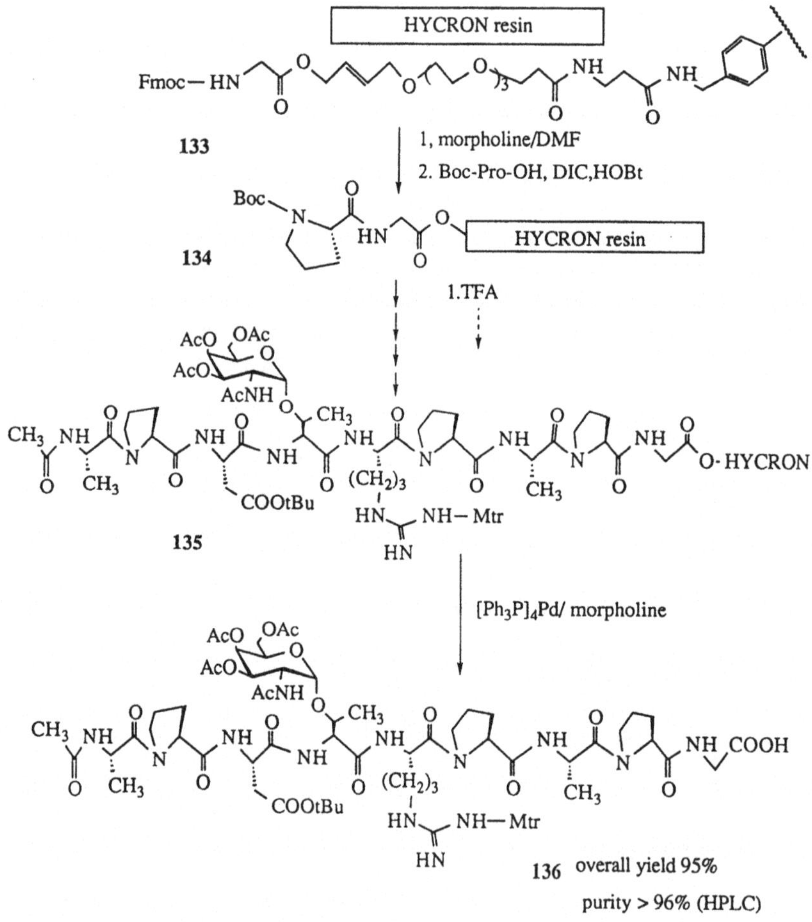

Scheme 5.39 Ac = acyl; Boc = *tert*-butyloxycarbonyl; DMF = dimethylformamide; Fmoc = 9-fluorenylmethoxycarbonyl; HPLC = high-performance liquid chromatography; Mtr = 4-methoxy-2,3,6-trimethyl-phenylsulfonyl; tBu = *tert*-butyl; TFA = trifluoroacetic acid.

Acidolytic cleavage of the Boc group proceeded without cleavage of the allyl anchor and without undesired formation of diketopiperazine. Therefore, the glycopeptide synthesis using Fmoc protected compounds in the following steps was achieved with high efficiency and yielded **135**. Final release by palladium(0)-catalysed allyl transfer to morpholine gave the glycononapeptide **136** in an overall yield of 95% relative to **133**. According to its analytical HPLC, this product had a purity of 96% (Seitz and Kunz, 1995).

Glycononapeptide **136** was completely deprotected by acidolytic removal of the *tert*-butyl ester and methoxy-trimethylphenylsulfonyl (Mtr) groups using trifluoroacetic acid and subsequent transesterification of the *O*-acetyl groups with dilute sodium methoxide in methanol. The complete MUC-1 tandem repeat unit was synthesised as a glyco-eicosapeptide using the same methodology. These results illustrate the efficiency of glycopeptide syntheses using allylic anchoring groups.

5.7 Conclusions

The synthetic methodology developed during the past 15 years provides access to synthetic glycopeptides of complex structures containing numerous functional groups in both the carbohydrate and the peptide portion. Recently developed mild chemoselective reactions are compatible with sensitive structures, for example of fucose-containing glycopeptides or *O*-glycosyl serine and threonine derivatives. In addition to chemical methods, enzymatic transformations are promising tools in glycopeptide synthesis, not only for glycosyltransfer processes (Section 5.5.5), but also for mild and selective deprotection reactions (Braun *et al.*, 1993). Chemical and enzymatic syntheses yield glycopeptides of exactly specified structure in preparative amounts, making even gram-scale quantities accessible. These defined model compounds can be used to study their biological effects and to investigate the influence of glycosylation on the conformation of the peptide backbone (Otvos *et al.*, 1995). Coupling the synthetic glycopeptides to carrier proteins affords neoglycoproteins which are interesting as selective ligands or antigens in biological and immunological investigations (Kunz and Birnbach, 1986; Kunz *et al.*, 1994; von dem Bruch and Kunz, 1994).

References

Augé, C., Warren, C.D., Jeanloz, R.W. *et al.* (1980) The synthesis of *O*-β-D-mannopyranosyl-(1 → 2)-*O*-(2-acetamido-2-deoxy-β-D-glucopyranosyl)-(1 → 4)-2-acetamido-2-deoxy-D-glucopyranose. *Carbohydr. Res.*, **82**, 71–95.

Barany, G., Kneib-Cordonier, N. and Mullen, D.G. (1987) Solid-phase peptide synthesis: a silver anniversary report. *Int. J. Peptide Protein Res.*, **30**, 705–39.

Barresi, F. and Hindsgaul, O., (1991) Synthesis of β-mannopyranosides by intramolecular aglycon delivery. *J. Am. Chem. Soc.*, **113**, 9376–7.

Bause, E. and Legler, G. (1981) The role of the hydroxy amino acid in the triplet sequence Asn–Xaa–Thr(Ser) for the *N*-glycosylation step during glycoprotein biosynthesis. *Biochem. J.*, **195**, 639–44.

Beyer, T.A., Sadler, J.E., Rearick, J.I. *et al.* (1981) Glycosyltransferases and their use in assessing oligosaccharide structure and structure–function relationships. *Adv. Enzymol.*, **52**, 23–175.

Boullanger, P., Banoub, J. and Descotes, G. (1987) *N*-allyoxycarbonyl derivatives of D-glucosamine as promotors of 1,2-*trans*-glycosylation in Koenigs–Knorr reactions and in Lewis acid catalysed condensations. *Can. J. Chem.*, **65**, 1343–8.

Braun, P., Waldmann, H. and Kunz, H. (1993) Chemoenzymatic synthesis of *O*-glycopeptides carrying the tumor associated TN-antigen structure. *Bioorg. Med. Chem.*, **1**, 197–207.

Bröder, W. and Kunz, H. (1993) A new method of anomeric protection and activation based on the conversion of glycosyl azides into glycosyl fluorides. *Carbohydr. Res.*, **249**, 221–41.

Butler, W.T. and Cunningham, L.W. (1966) Evidence for the linkage of a disaccharide to hydroxylysine in tropocollagen. *J. Biol. Chem.*, **241**, 3882–8.

Ciommer, M. and Kunz, H. (1991) Synthesis of glycopeptides with partial structure of human glycophorin using the fluorenylmethoxycarbonyl/allyl ester protecting group combination. *Synlett*, 593–5.

Cohen-Anisfeld, S.T. and Lansbury, P.T. (1990) A convergent approach to the chemical synthesis of asparagine-linked glycopeptides. *J. Org. Chem.*, **55**, 5560–2.

Cohen-Anisfeld, S.T. and Lansbury, P.T. (1993) A practical, convergent method for glycopeptide synthesis. *J. Am. Chem. Soc.*, **115**, 10531–7.

Danishefsky, S.J. and Holcomb, R.L. (1989) On the direct epoxodation of glycals: application of a reiterative strategy for the synthesis of β-linked oligosaccharides. *J. Am. Chem. Soc.*, **111**, 6661–6.

Danishefsky, S.J. and Roberge, J.Y. (1995) Advances in the development of convergent schemes for the synthesis of biologically important glycoconjugates. *Pure Appl. Chem.*, **67**, 1647–62.

Danishefsky, S.J., Gervay, J., Peterson, J.M. *et al.* (1992) Remarkable regioselectivity in the chemical glycosylation of glycal acceptors: a concise solution to the synthesis of sialyl-Lewis^X glycal. *J. Am. Chem. Soc.*, **114**, 8329–31.

Dasgupta, F. and Garegg, P.J. (1988) Alkyl sulfenyl triflate as activator in the thioglycoside-mediated formation of 3-glycosidic linkages during oligosaccharide synthesis. *Carbohydr. Res.*, **177**, C13–C17.

Egge, H., Peter-Katalinic, J., Karas, M. and Stahl, B. (1991) The use of fast atom bombardment and laser desorption mass spectrometry in the analysis of complex carbohydrates. *Pure Appl. Chem.*, **63**, 491–8.

Ferrari, B. and Pavia, A.A. (1983) Blood group antigens: synthesis of T_N glycopeptide related to human glycophorin A^M. *Int. J. Pept. Protein Res.*, **22**, 549–59.

Fischer, E. (1893) Über die Glucoside der Alkohole. *Ber. Dtsch. Chem. Ges.*, **26**, 2400–12.

Fischer, E. (1906) Untersuchungen über Aminosäuren, Polypeptide und Proteine. *Ber. Dtsch. Chem. Ges.*, **39**, 530–610.

Fukase, K., Hase, S., Ikenaka, T. and Kusumoto, S. (1992) Synthesis of new serine-linked oligosaccharides in blood-clotting factors VII and IX and protein 2. *Bull. Chem. Soc. Jpn.*, **65**, 436–45.

Garegg, P.J., Hultberg, H. and Wallin, S. (1982) A novel ring-opening of carbohydrate benzylidene acetals. *Carbohydr. Res.*, **108**, 97–101.

Garegg, P.J., Lindberg, B. and Norberg, T. (1979) Synthesis of *O*-β-D-galactopyranosyl-(1-3)-*O*-β-D-galactopyranosyl-(1-4)*O*-β-D-xylopyranosyl-(1-3)-L-serine. *Acta Chem. Scand. B*, **33**, 449–52.

Garg, H.G. and Jeanloz, R.W. (1972) The synthesis of 2-acetamido-3,4,6-tri-*O*-acetyl-*N*-[*N*-(benzyloxycarbonyl)-L-aspart-1- and -4-oyl]-2-deoxy-β-d-glucopyranosylamine. *Carbohydr. Res.*, **23**, 437–9.

Garg, H.G. and Jeanloz, R.W. (1974) The synthesis of protected glycopeptides containing the amino acid sequences 34–37 and 34–38 of bovine ribonuclease B. *Carbohydr. Res.*, **32**, 37–46.

Garg, H.G. and Jeanloz, R.W. (1985) Synthetic *N*- and *O*-glycosyl derivatives of L-asparagine, L-serine and L-threonine. *Adv. Carbohydr. Chem. Biochem.*, **43**, 135–201.

Garg, H.G., von dem Bruch, K. and Kunz, H. (1994) Recent developments in the synthesis of glycopeptides containing glycosyl L-asparagine, L-serine and L-threonine. *Adv. Carbohydr. Chem. Biochem.*, **50**, 277–310.

Gooi, H.C., Feizi, T., Kapadia, A. *et al.* (1981) Stage specific embryonic antigen involves α(1-3)fucosylated type 2 blood group chains. *Nature*, **292**, 156–8.

Goto, F. and Ogawa, T. (1992) Synthesis of a sulfated glycopeptide corresponding to the carbohydrate–protein linkage region of proteoglycans. *Tetrahedron Lett.*, **33**, 5099–102.

Grundler, G. and Schmidt, R.R. (1984) Anwendung des Trichloracetimidat-Verfahrens auf 2-Azidoglucose- und 2-Azidogalactose-Derivate. *Liebigs Ann. Chem.*, 1826–47.

Günther, W. and Kunz, H. (1990) Synthesis of a β-mannosyl chitobiosyl asparagine conjugate – a central element of the core region of *N*-glycoproteins. *Angew. Chem. Int. Ed. Engl.*, **29**, 1068–9.

Günther, W. and Kunz, H. (1992) Synthesis of β-D-mannosides from β-D-glucosides via an intramolecular S_N2 reaction at C-2. *Carbohydr. Res.*, **228**, 217–41.

Hanessian, S. and Banoub, J. (1977) Chemistry of the glycosidic linkage. An efficient synthesis of 1,2-*trans*-disaccharides. *Carbohydr. Res.*, **53**, C13–C16.

Hansen, J.-E.S., Nielsen, C., Arendrup, M. *et al.* (1991) Broadly neutralizing antibodies targeted to mucin-type carbohydrate epitopes of human immunodeficiency virus. *J. Virol.*, **65**, 6461–7.

Harris, R.J. and Spellman, M.W. (1993) *O*-linked fucose and other post-translational modifications unique to EGF modules. *Glycobiology*, **3**, 219–24.

Harris, R.J., Ling, V.T. and Spellman, M.W. (1992) *O*-linked fucose is present in the first epidermal growth factor domain of human factor XII but not protein C. *J. Biol. Chem.*, **267**, 5102–7.

Hart, G.W., Haltiwanger, R.S., Holt, G.D. and Kelly, W.G. (1989) Nucleoplasmic and cytoplasmic glycoproteins. *Ann. Rev. Biochem.*, **58**, 785–838.

Hase, S., Nishimura, H., Kawabata, S. *et al.* (1990) The structure of $(xylose)_2$glucose-*O*-serine 53 found in the first epidermal growth factor-like domain of bovine blood clotting factor IX. *J. Biol. Chem.*, **265**, 1858–61.

Hasegawa, A., Fushimi, K., Ishida, H. and Kiso, M. (1993) Synthetic studies on sialoglycoconjugates. 52. Synthesis of sialyl $Lewis^X$ analogs containing azidoalkyl groups at the reducing end. *J. Carbohydr. Chem.*, **12**, 1203–16.

Hayes, B.K. and Hart, G.W. (1994) Novel forms of protein glycosylation. *Curr. Opin. Struct. Biol.*, **4**, 692–6.

Hilkins, J. (1988) Biochemistry and functions of mucins in malignant disease. In *Cancer Reviews*, vol. 11–12, J. Hilgers and S. Zotter (eds), Munksgaard, Copenhagen, pp. 25–54.

Hounsell, E.F. (1994) Physicochemical analyses of oligosaccharide determinants of glycoproteins. *Adv. Carbohydr. Chem. Biochem.*, **50**, 311–50.

Iijima, H. and Ogawa, T. (1988) Synthetic studies on cell-surface glycans. Part 52. Total synthesis of 3-*O*-[2-acetamido-6-*O*-(*N*-acetyl-α-D-neuraminyl)-2-deoxy-α-D-galactosyl]-L-serine and a stereoisomer. *Carbohydr. Res.*, **172**, 183–95.

Ito, Y. and Ogawa, T. (1994) A new approach for the stereoselective synthesis of β-mannosides. *Angew. Chem. Int. Ed. Engl.*, **33**, 1765–7.

Johansson, R. and Samuelsson, B. (1984) Regioselective reductive ring-opening of 4-methoxybenzylidene acetals of hexopyranosides. Access to a novel protecting-group strategy. Part 1. *J. Chem. Soc. Perkin Trans I*, 2371–4.

Jung, G. and Beck-Sickinger, A.G. (1992) Methods for multiple peptide syntheses and their application. *Angew. Chem. Int. Ed. Engl.*, **31**, 367–83.

Kameyama, A., Ishida, H., Kiso, M. and Hasegawa, A. (1991) Total synthesis of sialyl Lewis X. *Carbohydr. Res.*, **209**, C1–C4.

Karsten, U., Papsdorf, G., Pauly, A. *et al.* (1993) Subtypes of non-transformed human mammary epithelial cells cultured in vitro: histo-blood group antigen H type 2 defines basal cell-derived cells. *Differentiation*, **54**, 55–66.

Khorlin, A.Y., Zurabyan, S.E. and Macharadze, R.G. (1980) Synthesis of glycosylamides and 4-*N*-glycosyl-L-asparagine derivatives. *Carbohydr. Res.*, **85**, 201–8.

Kobata, A. (1993) Glycobiology: an expanding research area in carbohydrate chemistry. *Acc. Chem. Res.*, **26**, 319–24.

Kornfeld, R. and Kornfeld, S. (1976) Comparative aspects of glycoprotein structure. *Ann. Rev. Biochem.*, **45**, 217–37.

Kornfeld, R. and Kornfeld, S. (1980) Structure of glycoproteins and their oligosaccharide units. In *The Biochemistry of Glycoproteins and Proteoglycans*, W.J. Lennarz (ed.), Plenum Press, New York, pp. 1–34.

Kosch, W., März, J. and Kunz, H. (1994) Synthesis of glycopeptide derivatives of peptide T on a solid phase using an allylic anchor. *Reactive Polymers*, **22**, 181–94.

Kum, K. and Roseman, S. (1966) The synthesis of *O*-serine glycosides. *Biochemistry*, **5**, 3061–5.

Kunz, H. (1987) New synthetic methods. 67. Synthesis of glycopeptides: partial structures of biological recognition systems. *Angew. Chem. Int. Ed. Engl.*, **26**, 294–308.

Kunz, H. (1993) Glycopeptides of biological interest: a challenge for chemical synthesis. *Pure Appl. Chem.*, **65**, 1223–32.

Kunz, H. and Birnbach, S. (1986) Synthesis of tumor-associated T_N and T antigen type *O*-glycopeptides and their conjugation to bovine serum albumin. *Angew. Chem. Int. Ed. Engl.*, **25**, 360–2.

Kunz, H. and Dombo, B. (1988) Solid-phase synthesis of peptides and glycopeptides on polymeric support with allylic anchor groups. *Angew. Chem. Int. Ed. Engl.*, **27**, 711–13.

Kunz, H. and Günther, W. (1988) β-Mannoside synthesis via inversion of configuration on β-glucosides by intramolecular nucleophilic substitution. *Angew. Chem. Int. Ed. Engl.*, **27**, 1188–9.

Kunz, H. and März, J. (1988) The *p*-nitrocinnamoyl (Noc) moiety – an acid-stable amino protecting group removable under neutral conditions for peptide and glycopeptide synthesis. *Angew. Chem. Int. Ed. Engl.*, **27**, 1375–7.

Kunz, H. and März, J. (1992) Synthesis of glycopeptides with Lewis^a antigen side chain and HIV peptide T sequence using the trichloroethoxycarbonyl/allyl ester protecting group combination. *Synlett*, 591–3.

Kunz, H. and Unverzagt, C. (1984) Allyloxycarbonyl (Aloc) moiety – conversion of an unsuitable into a valuable useful amino protecting group for peptide synthesis. *Angew. Chem. Int. Ed. Engl.*, **23**, 436–7.

Kunz, H. and Unverzagt, C. (1988) Protecting group dependent stability of intersaccharide bonds – synthesis of a fucosyl-chitobiose glycopeptide. *Angew. Chem. Int. Ed. Engl.*, **27**, 1697–9.

Kunz, H. and von dem Bruch, K. (1994) Neoglycoproteins from synthetic glycopeptides. In *Methods in Enzymology*, vol. 247, Y.C. Lee and R.T. Lee (eds), Academic Press, San Diego, CA, pp. 3–30.

Kunz, H. and Waldmann, H. (1984) The allyl group as mildly and selectively removable carboxy-protecting group for the sythesis of labile *O*-glycopeptides. *Angew. Chem. Int. Ed. Engl.*, **23**, 71–2.

Kunz, H. and Waldmann, H. (1985) Construction of disaccharide *N*-glycopeptides – synthesis of the linkage region of the transmembrane neuraminidase of an influenza virus. *Angew. Chem. Int. Ed. Engl.*, **24**, 883–5.

Kunz, H., von dem Bruch, K., Unverzagt, C. and Kosch, W. (1994) Synthesis of glycopeptides: a challenge for chemical synthesis. In *Peptides: Chemistry, structure and biology*, R.S. Hodges and J.A. Smith (eds), ESCOM, Leiden, pp. 321–3.

Kunz, H., Waldmann, H. and März, J. (1989) Synthese von *N*-Glycopeptid-Partialstrukturen der Verknüpfungsregion sowohl der Transmembran-Neuramidase eines Influenza-Virus als auch des Faktors B des menschlichen Komplementsystems. *Liebigs Ann. Chem.*, 45–9.

Kunz, H., Waldmann, H. and Unverzagt, C. (1985) The allyl ester as a temporary protecting group for the β-carboxy function of aspartic acid. *Int. J. Pept. Prot. Res.*, **26**, 493–7.

Kurosaka, A., Kitagawa, H., Fukui, S. *et al.* (1988) A monoclonal antibody that recognizes a cluster of a disaccharide, NeuAcα2-6GalNAc, in mucin-type glycoproteins. *J. Biol. Chem.*, **263**, 8724–6.

Lamport, D.T.A. and Clark, L. (1969) The isolation and partial characterization of hydroxyproline-rich glycopeptides obtained by enzymatic degradation of primary cell walls. *Biochemistry*, **8**, 1155–63.

Lavielle, S., Ling, N.C., Saltman, R. and Guillemin, R.C. (1981) Synthesis of a glycotripeptide and a glycosomatostatin containing the 3-*O*-(2-acetamido-2-deoxy-β-D-glucopyranosyl)-L-serine residue. *Carbohydr. Res.*, **89**, 229–36.

Lemieux, R.U. and Haymi, J.I. (1965) The mechanism of the anomerization of the tetra-*O*-acetyl-D-glucopyranosyl chlorides. *Can. J. Chem.*, **43**, 2162–73.

Lemieux, R.U. and Levine, S. (1964) Synthesis of alkyl 2-deoxy-α-D-glycopyranosides and their 2-deuterio derivatives. *Can. J. Chem.*, **42**, 1473–80.
Lemieux, R.U. and Ratcliffe, R.M. (1979) The azidonitration of tri-*O*-acetylgalactal. *Can. J. Chem.*, **57**, 1244–51.
Liebe, B. and Kunz, H. (1994) Synthesis of sialyl-Tn antigen. Regioselective sialylation of a galactosamine threonine conjugate unblocked in the carbohydrate portion. *Tetrahedron Lett.*, **35**, 8777–8.
Likhosherstov, L.M., Novikova, O.S., Derevitskaja, V.A. and Kochetkov, N.K. (1986) A new simple synthesis of amino-sugar β-D-glycosylamines. *Carbohydr. Res.*, **146**, C1–C5.
Lindahl, N. and Rodin, L. (1964) The linkage of heparin to protein. *Biochim. Biophys. Res. Commun.*, **17**, 254–6.
Lindberg, B. and Silvander, B.-G. (1965) Synthesis of *O*-β-D-xylopyranosyl-L-serine. *Acta Chem. Scand.*, **19**, 530–1.
Lis, H. and Sharon, N. (1993) Protein glycosylation – structural and functional aspects. *Eur. J. Biochem.*, **218**, 1–27.
Lomako, J., Lomako, W.M. and Whelan, W.J. (1990) The biogenesis of glycogen-nature of the carbohydrate in the protein primer. *Biochem. Int.*, **21**, 251–60.
Lüning, B., Norberg, T. and Tejbrant, J. (1989) Synthesis of mono- and disaccharide amino acid derivatives for use in solid phase peptide synthesis. *Glycoconjugate J.*, **6**, 5–19.
Marks, G.S. and Neuberger, A. (1961) Synthetic studies relating to the carbohydrate–protein linkage in egg albumin. *J. Chem. Soc.*, 4872–9.
Marra, A. and Sinay, P. (1989) A stereoselective synthesis of 2-thioglycosides of *N*-acetyl-α-neuraminic acid. *Carbohydr. Res.*, **187**, 35–42.
März, J. and Kunz, H. (1992) Synthesis of selectively deprotectable asparagine glycoconjugates with a Lewis^a antigen side chain. *Synlett*, 589–90.
Meldal, M. and Bock, K. (1990) Pentafluorophenyl esters for temporary carboxyl group protection in solid-phase synthesis of *N*-linked glycopeptides. *Tetrahedron Lett.*, **31**, 6987–90.
Meldal, M. and Jensen, K.J. (1990) Pentafluorophenyl esters for the temporary protection of the α-carboxy group in solid-phase glycopeptide synthesis. *J. Chem. Soc. Chem. Commun.*, 483.
Mergler, M., Tanner, R., Gosteli, J. and Grogg, P. (1988) Peptide synthesis by a combination of solid-phase and solution methods. I. A new very acid-labile anchor group for the solid-phase synthesis of fully protected fragments. *Tetrahedron Lett.*, **29**, 4009–12.
Messner, P., Christian, R., Kolbe, J. *et al.* (1992) Analysis of a novel linkage unit of *O*-linked carbohydrates from the crystalline surface layer glycoprotein of clostridium thermohydrosulfurin. *J. Bacteriol.*, **174**, 2236–40.
Micheel, F. and Köchling, H. (1958) Über die Reaktionen des Glucosamins. *Chem. Ber.*, **91**, 673–6.
Nakabayashi, S., Warren, C.D. and Jeanloz, R.W. (1988) Amino sugars 135. The preparation of a partially protected heptasaccharide-asparagine intermediate for glycopeptide synthesis. *Carbohydr. Res.*, **174**, 219–89.
Nakahara, Y., Iijima, H., Sibayama, S. and Ogawa, T. (1990) A highly stereoselective synthesis of di- and trimeric sialosyl-Tn epitope: a partial structure of glycophorin A1. *Tetrahedron Lett.*, **31**, 6897–900.
Neuberger, A., Gottschalk, A., Marshall, R.D. and Spiro, R.G. (1972) Carbohydrate–peptide linkages in glycoproteins and methods for their elucidation. In *Glycoproteins, Part A*, A. Gottschalk (ed.), Elsevier, Amsterdam, pp. 450–90.
Nicolaou, K.C., Hummel, C.W., Bockovich, N.J. and Wong, C.-H. (1991) Stereocontrolled synthesis of sialyl Le^X, the oligosaccharide binding ligand to ELAM-1 (Sialyl = *N*-acetylneuramin). *J. Chem. Soc. Chem. Commun.*, 870–2.
Ogawa, T., Beppu, K. and Nakabayashi, S. (1981) Trimethylsilyl trifluoromethane sulfonate as an effective catalyst for glycoside synthesis. *Carbohydr. Res.*, **93**, C6–C9.
Olofsson, S. (1991) Old and new issues regarding the function of *O*-linked oligosaccharides. *Trends Biochem. Sci.*, **16**, 11–12.
Otvos Jr, L., Krivulka, G.R., Urge, L. *et al.* (1995) Comparison of the effect of amino acid substitutions and β-*N*- vs. α-*O*-glycosylation on the T-cell stimulatory activity and conformation of an epitope on the rabies virus glycoprotein. *Biochim Biophys. Acta*, **1267**, 55.
Otvos Jr, L., Urge, L., Hollosi, M. *et al.* (1990) Automated solid-phase synthesis of glycopeptides. Incorporation of unprotected mono- and disaccharides of *N*-glycoprotein antennae into T cell epitopic peptides. *Tetrahedron Lett.*, **31**, 5889–92.

Parodi, A.J., Behrens, N.H., Leloir, L.F. and Carminatti, H. (1972) The role of polyprenol bound saccharides as intermediates in glycoprotein synthesis in liver. *Proc. Nat. Acad. Sci. USA*, **69**, 3268–72.

Paul, G., Lottspeich, F. and Wieland, F. (1986) Asparaginyl-*N*-acetylgalactosamine. Linkage unit of halobacterial glycosaminoglycan. *J. Biol. Chem.*, **261**, 1020–4.

Paulsen, H. (1982) Progress in the selective chemical synthesis of complex oligosaccharides. *Angew. Chem. Int. Ed. Engl.*, **21**, 155–224.

Paulsen, H. and Adermann, K. (1989) Synthese von variierten *O*-Glycopeptidsequenzen von Interleukin-2. *Liebigs Ann. Chem.*, 771–80.

Paulsen, H. and Hölck, J.-P. (1982) Building blocks of oligosaccharides. Part XV. Synthesis of *O*-β-D-galactopyranosyl-(1-3)-*O*-(α-D-2-acetamido-2-deoxy-α-D-galactopyranosyl)-(1-3)-L-serine and L-threonine glycopeptides. *Carbohydr. Res.*, **109**, 89–107.

Paulsen, H. and Lockhoff, O. (1981) Neue effektive β-Glycosidsynthese für Mannose-Glycoside. Synthesen von Mannose-haltigen Oligosacchariden. *Chem. Ber.*, **114**, 3102–14.

Paulsen, H. and Paal, M. (1984) Blocksynthese von *O*-Glycopeptiden und anderen T-Antigen Strukturen. *Carbohydr. Res.*, **135**, 71–84.

Paulsen, H., Bielfeldt, T., Peters, S. *et al.* (1994) Anwendung des Azid-Glycopeptidsynthese-Verfahrens in der multiplen Festphasensynthese zur Gewinnung von *O*-Glycopeptiden des Mucin-Typs. *Liebigs Ann. Chem.*, 381–7.

Paulsen, H., Merz, G. and Weichert, U. (1988) Solid-phase synthesis of *O*-glycopeptide sequences. *Angew. Chem. Int. Ed. Engl.*, **27**, 1365–7.

Paulsen, H., Rauwald, W. and Weichert, U. (1988) Glycosidation of oligosaccharide thioglycosides to *O*-glycoprotein segments. *Liebigs. Ann. Chem.*, 75–88.

Peters, S., Bielfeldt, T., Meldal, M. *et al.* (1992) Multiple-column solid-phase glycopeptide synthesis. *J. Chem. Soc. Perkin Trans.*, **1**, 1163–71.

Philips, M.L., Nudelman, E., Gaeta, F.C.A. *et al.* (1990) ELAM 1 mediates cell adhesion by recognition of a carbohydrate ligand. Sialyl Lewis X. *Science*, **250**, 1130–2.

Roberge, J.Y., Beebe, X. and Danishefsky, S.J. (1995) A strategy for a convergent synthesis of *N*-linked glycopeptides on a solid support. *Science*, **269**, 202–4.

Rosen, T., Lico, I.M. and Chu, D.T.W. (1988) A convenient and highly chemoselective method for the reductive acetylation of azides. *J. Org. Chem.*, **53**, 1580–2.

Samuel, J., Nonjaim, A.A., McLean, G.D. *et al.* (1990) Analysis of human tumor associated Thomsen–Friedenreich antigen. *Cancer Res.*, **50**, 4801–8.

Schmidt, R.R. (1986) New methods of glycoside and oligosaccharide synthesis – are there alternatives to the Koenigs–Knorr method. *Angew. Chem. Int. Ed. Engl.*, **25**, 212–35.

Schmidt, R.R. and Michel, J. (1980) Simple synthesis of α- and β-*O*-glycosyl imidates; preparation of glycosides and disaccharides. *Angew. Chem. Int. Ed. Engl.*, **19**, 731–2.

Schultheiss-Reimann, P. and Kunz, H. (1983) *O*-glycopeptide synthesis by 9-fluorenylmethoxycarbonyl(Fmoc)-protected blocks. *Angew. Chem. Int. Ed. Engl.*, **22**, 62.

Schultz, M. and Kunz, H. (1993) Synthetic *O*-glycopeptides as model substrates for glycosyltransferases. *Tetrahedron Asymm.*, **4**, 1205–20.

Schuster, M., Wang, P., Paulson, J.C. and Wong, C.-H. (1994) Solid-phase chemical–enzymatic synthesis of glycopeptides and oligosaccharides. *J. Am. Chem. Soc.*, **116**, 1135–6.

Schwartz, R.T. and Datema, R. (1982) The lipid pathway of protein glycosylation and its inhibitors: the biological significance of protein bound carbohydrates. *Adv. Carbohydr. Chem. Biochem.*, **40**, 287–379.

Seitz, O. and Kunz, H. (1995) A novel allylic anchor for solid-phase synthesis of protected and unprotected *O*-glycosylated mucin-type glycopeptides. *Angew. Chem. Int. Ed. Engl.*, **34**, 803–5.

Sharon, N. and Lis, H. (1989) *Lectins*, Chapman and Hall, London.

Smythe, C. and Cohen, P.P. (1991) The discovery of glycogenin and the priming mechanism for glycogen biogenesis. *Eur. J. Biochem.*, **200**, 625–31.

Spinola, M. and Jeanloz, R.W. (1970) The synthesis of a di-*N*-acetylchitobiose asparagine derivative, 2-acetamido-4-*O*-(2-acetamido-2-deoxy-β-D-glucopyranose)-1-*N*-(4-L-aspartyl)-2-deoxy-β-D-glucopyranosylamine. *J. Biol. Chem.*, **245**, 4158–68.

Sprengard, U., Kretzschmar, G., Bartnik, E. *et al.* (1995) Synthesis of a RGD–Sialyl–Lewis^{X}–glycoconjugate: a new highly active ligand for P-selectin. *Angew. Chem. Int. Ed. Engl.*, **34**, 990–3.

Springer, G.F. (1984) T and T_N, general carcinoma autoantigens. *Science*, **224**, 1198–206.
Springer, T.A. (1994) Traffic signals for lymphotcyte recirculation and leukocyte emigration: the multistep paradigm. *Cell*, **76**, 301–14.
Stahl, W., Sprengard, U., Kretzschmar, G. *et al.* (1995) Synthesis of sulfated Le^X-trisaccharides. *J. Prakt. Chem.*, **337**, 441–5.
Stork, G. and Kim, G. (1992) Stereocontrolled synthesis of disaccharides via the temporary silicon connection. *J. Am. Chem. Soc.*, **114**, 1087–8.
Takahashi, N. and Muramatsu, T. (1992) *CRC Handbook of Endoglycosidases and Glycoamidases*, CRC Press, Boca Raton, FL.
Tomita, M., Furthmayr, H. and Marchesi, V.T. (1978) Primary structure of human erytrocyte glycophorin. A. Isolation and characterisation of peptides and complete amino acid sequence. *Biochemistry*, **17**, 4756–70.
Toshima, K. and Tatsuta, K. (1993) Recent progress in *O*-glycosylation methods and its application to natural products synthesis. *Chem. Rev.*, **93**, 1503–31.
Unverzagt, C. (1994) Synthesis of a branched heptasaccharide by regioselective glycosylation. *Angew. Chem. Int. Ed. Engl.*, **33**, 1102.
Unverzagt, C. and Kunz, H. (1992) Stereoselective synthesis of glycosides and anomeric azides of glucosamine. *J. Prakt. Chem.*, **334**, 570–8.
Unverzagt, C. and Kunz, H. (1994) Synthesis of glycopeptides and neoglycoproteins containing the fucosylated linkage region of *N*-glycoproteins. *Bioorg. Med. Chem.*, **2**, 1189–201.
Unverzagt, C., Kunz, H. and Paulson, J.C. (1990) High-efficiency synthesis of sialyloligosaccharides and sialoglycopeptides. *J. Am. Chem. Soc.*, **112**, 9308–9.
Vercellotti, J.R. and Luetzow, A.E. (1966) β-Elimination of glycoside monosaccharide from 3-*O*-(2-amino-2-deoxy-D-glucopyranosyl)serine – evidence for an intermediate in glycoprotein hydrolysis. *J. Org. Chem.*, **31**, 825–30.
Verez-Bencomo, V. and Sinay, P. (1983) Synthesis of glycopeptides having clusters of *O*-glycosylic disaccharide chains [β-D-Gal-(1-3)-α-D-GalNAc] located at vicinal amino acid residues of the peptide chain. *Carbohydr. Res.*, **116**, C9–C12.
Vetter, D., Tumelty, D., Singh, S.K. and Gallop, M.A. (1995) A versatile solid-phase synthesis of *N*-linked glycopeptides. *Angew. Chem. Int. Ed. Engl.*, **34**, 60–2.
Vliegenthart, J.F.G., Dorland, L. and van Halbeck, H. (1983) High-resolution, ^{1}H-nuclear magnetic resonance spectroscopy as a tool in the structural analysis of carbohydrates related to glycoproteins. *Adv. Carbohydr. Chem. Biochem.*, **41**, 209–374.
von dem Bruch, K. and Kunz, H. (1994) Synthesis of *N*-glycopeptide clusters with $Lewis^X$ antigen side-chains and their coupling to carrier proteins. *Angew. Chem. Int. Ed. Engl.*, **33**, 101–3.
Waldmann, H. and Kunz, H. (1983) Allylester als selektiv abspaltbare Carboxylschutzgruppen in der Peptid- und *N*-Glycopeptidsynthese. *Liebigs Ann. Chem.*, 1712–25.
Yamamoto, A., Miyashita, C. and Tsukamoto, H. (1965) Preparation of *N*-[L-α(and β)-aspartyl]-2-acetamido-2-deoxy-β-D-glucosylamine and *N*-(L-γ-glutamyl)2-acetamido-2-deoxy-*N*-β-D-glucosylamine. *Chem. Pharm. Bull.*, **13**, 1041–6.

6 Oligosaccharide geometry and dynamics

JOHN BRADY

6.1 Introduction

As with proteins and nucleic acids, the structure and dynamics of oligosaccharides can be quite important to their biological function. Many oligosaccharides, such as the carbohydrate portions of glycoproteins, are involved in molecular recognition, in which their particular structures and conformations are recognized by other proteins which specifically bind to them. Furthermore, since many of the disaccharides are the repeat units of polysaccharides, their dynamical behavior and conformational preferences largely shape the properties of these larger polymers as well. For these reasons, it would be most desirable to develop a general understanding of oligosaccharide conformations and dynamics. While a great deal is known about the primary structures of oligosaccharides (Doubet *et al.*, 1989), it has unfortunately proven to be quite difficult to study their conformations and dynamics experimentally, and this class of molecules is generally much more poorly known than other types of biopolymers.

Describing the conformational structure of oligosaccharides and polysaccharides is in some ways analogous to the problem of specifying the conformations of polypeptides. In the case of polypeptides, Ramachandran observed that the description of the polymer backbone conformation could be reduced to the specification of two torsion angles, ϕ and ψ which determine the relative orientations of successive planar peptide groups (Ramachandran *et al.*, 1963). Using this description of the backbone polypeptide conformation, most (ϕ, ψ) conformations can be ruled out as unlikely to occur because they result in sterically-disallowed atomic overlaps. Since the ring structures adopted by the sugar monomers in a polysaccharide are relatively rigid, the description of the conformation of a carbohydrate polymer often can also be reduced to the specification of two torsional angles ϕ and ψ which determine the orientation of adjacent sugar rings across the two single bonds of their connecting glycosidic linkage (Figure 6.1). Describing the total conformation of a carbohydrate polymer chain then reduces to listing these angles for each successive linkage. The regular propagation of disaccharide repeat units with the same conformation about the glycosidic linkage produces helices, and many polysaccharides are believed or are known to adopt regular helical conformations. Nonrepeating combinations of glycosidic torsion angles will produce irregular

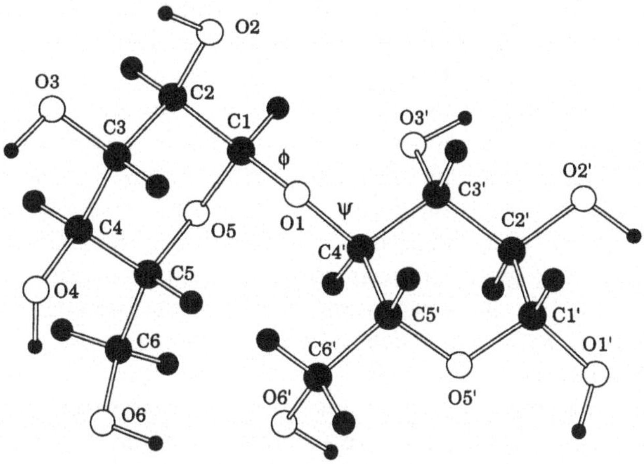

Figure 6.1 The typical disaccharide maltose, indicating the torsional angles ϕ and ψ which specify the conformation of the dimer about the glycosidic oxygen. Both monomer rings are in the $^4C_1$ conformation. It should be noted that these torsion angles can be specified in more than one way; crystallographers often use the heavy atom sequences O5–C1–O1–C4′ and C1–O1–C4′–C5′ for ϕ and ψ, while NMR spectroscopists often employ the definitions H1–C1–O1–C4′ and C1–O1–C4′–H4′. In the maps displayed in this chapter, the latter definition has been used. By IUPAC-IUB convention [*Eur. J. Biochem.* (1983) **131**, 5–7] ϕ is zero when the H1–C1 and O1–C4′ bonds eclipse one another and ψ is zero when the C1–O1 and C4′–H4′ bonds eclipse, the angles have a positive sign when a bond viewed from the front must be rotated clockwise to achieve the eclipsed conformation along the central bond. Angles ϕ and ψ have values between −180° and +180°. A positive value requires a clockwise rotation of the remote reference atom from the eclipsed conformation. Reprinted with permission (Ha *et al.*, 1988b).

coils. The structurally more complex branched oligosaccharides may have quite irregular overall shapes, yet their conformations can still be described in this fashion by listing their linkage torsion angles.

In spite of this structural analogy between these two types of polymers, there are a number of significant ways in which oligosaccharides are different from polypeptides. Carbohydrate polymers are sometimes thought of as being less interesting than proteins, since they contain less sequence variation than the typical polypeptide. However, oligosaccharides have a greater range of possible variations in their linkage types than do polypeptides. For the case of pyranoid hexoses, the C1 anomeric center can be linked to a second sugar through a glycosidic linkage to either the C1, C2, C3, C4 or C6 positions, resulting in (1 → 1), (1 → 2), (1 → 3), (1 → 4) and (1 → 6) linkages. This diversity of possible linkages allows numerous oligosaccharide types even for a homopolymer (that is, one in which all of the monomer units are the same sugar). Furthermore, as there are also two possible stereochemistries at the anomeric center, α and β, two distinct geometries are possible for each type of glycosidic linkage even between the same two sugars. Also, unlike the linear polypeptides, sugars can make more than two linkages, leading to the possibility of branched chains. In general, however, naturally occurring polysaccharides do not randomly combine all of these possible structural elements, in the way that a protein might be made up of any combination of amino acids. So, for example, there are a number of possible homopolymers of glucose, but only a few occur naturally in any significant amounts. The linear $\alpha(1 \rightarrow 4)$-linked homopolymer of glucose is amylose, while the linear $\beta(1 \rightarrow 4)$-linked homopolymer produces cellulose. The $\alpha(1 \rightarrow 6)$ linked dimer is isomaltose, and the occurrence of such $\alpha(1 \rightarrow 6)$ linkages on a linear amylosic chain gives rise to the type of branching found in amylopectin and glycogen; the bacterial dextrans are glucose polysaccharides in which the backbone linkages are predominantly of the $\alpha(1 \rightarrow 6)$ type.

In addition to this diversity of linkage types, some variability does arise due to sequence. Copolymers are common in naturally occurring polysaccharides, such as the alternating galactose and 3,6-anhydrogalactose monomers of the carrageenans, with their alternating $(1 \rightarrow 3)$ and $(1 \rightarrow 4)$ linkages, and the random block copolymers of the alginates. There are several examples of important mixed disaccharides, such as lactose and sucrose. Sucrose, of course, is an interesting case of mixing an aldose and a ketose, as well as a pyranoid and a furanoid ring. The most structurally complex carbohydrates are the oligosaccharide components of glycoconjugates, which often are highly branched and may contain a variety of linkage types and monomers. Often these oligosaccharides also contain sugar derivatives such as *N*-acetyl glucosamine (NAG or GlcNAc). One of the more typical structural motifs for these oligosaccharides involves a central core consisting of the sequence Man$\beta(1 \rightarrow 4)$NAG$\beta(1 \rightarrow 4)$NAG-Asn, with various other sugars then branching off from these three residues. Also of particular interest are the high-mannose oligomers of fungi and bacteria, since high-mannose glycoproteins do not occur in mammalian cells and thus serve as an immunological marker of foreign pathogens. The Man$\alpha(1 \rightarrow 3)$Man linkage found in the core of these oligomers has been the focus of extended debate concerning disaccharide flexibility (Carver, 1991; Homans, 1990; Stevens, 1994). The conformations and dynamics of glycoconjugate oligosaccharides are of clear importance because of their role in molecular recognition in such processes as cell adhesion, host invasion by pathogens and immunological response and self-recognition.

As with other types of polymers, some broad general statements about the conformations and dynamical fluctuations of oligosaccharides can be easily made. In general, disaccharides should have a preference for staggered conformations about the two linkage bonds, although this preference could be overridden, for example, by the geometric constraints imposed by a hydrogen bond between the two rings. It is also generally believed that a so-called 'exo-anomeric effect' (Lemieux *et al.*, 1969) favors *gauche* conformations about the glycosidic linkage angle ϕ (Pérez and Marchessault,

1978), essentially owing to a donation of oxygen atom lone pair electronic density to an adjacent antibonding C–O σ^* orbital, with the extent of this effect being conformation dependent (Tvaroska and Bleha, 1989). *Ab initio* calculations and experimental observation support this model. However, there are still questions about the magnitude of this effect and its importance in determining glycosidic conformation. An interesting report by Duda and Stevens (1993) finds that the solution conformation of the β,β-trehalose analog in which the glycosidic oxygen is replaced by a CH_2 group, which should have no exo-anomeric effect, is apparently the same as for β,β-trehalose itself, which should experience such stabilization, implying that the effects of this exo-anomeric term are not necessarily dominant in determining conformation.

The gross topological characteristics of the different linkage types also impose certain general features on dimer conformations. Disaccharide linkages where both groups involved in the linkage are equatorial (β in the case of the anomeric group) produce flat, extended, ribbon-like helices similar in some qualitative features to the cellulose molecule ($\phi = 28°$, $\psi = -30°$; Aabloo and French, 1994), even when the sugars are not glucose or when the linked group of the non-reducing ring is on the C2 or C3 carbon. Similarly, disaccharides in which the anomeric configuration of the non-reducing sugar is axial (α) and the reducing sugar's linkage O–C bond is equatorial tend qualitatively to resemble maltose ($\phi = 5°$, $\psi = 13°$; Gress and Jeffrey, 1977) and the amylose chain, even if the sugars are not glucose. When both of the linked groups are axial, the rings are tilted almost perpendicular to one another. Such an arrangement exists in α,α-trehalose ($\phi = -60°$, $\psi = -43°$; Brown *et al.*, 1972), or when there is an α linkage to an axial group, such as the C4 position of galactose or the C2 group of mannose. The special case of the (1 → 6)-linkage, as in isomaltose, results in significant molecular flexibility. This type of linkage introduces a third single bond between the two monomer rings (Figure 6.2), producing more

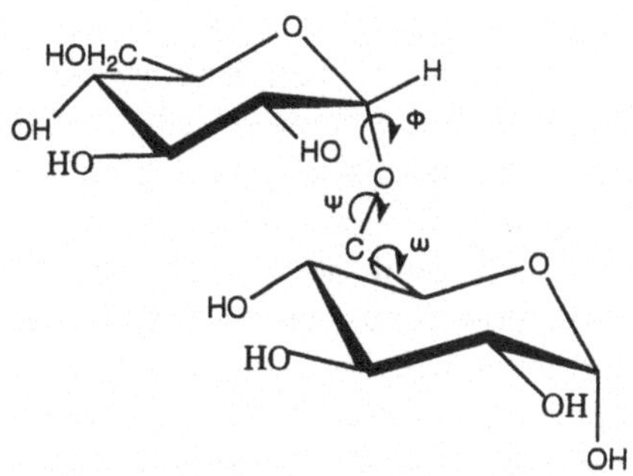

Figure 6.2 The isomaltose molecule, indicating the three bonds between the two sugar rings about which rotation can occur.

conformational possibilities and further separating the rings, so that fewer conformational restrictions occur as the result of steric clashes between them. Indeed, isomaltose is so flexible in solution that it cannot be easily crystallized as can other disaccharides.

Beyond these broad general features, however, the structural behavior of the oligosaccharides is actually rather complex. The basic problem of disaccharide and oligosaccharide conformational chemistry is then to be able to characterize the conformations and dynamics of these molecules in various types of biological systems and to be able to relate these properties to an understanding of the underlying energetics of the molecules. Although experimental characterizations have been problematic, considerable effort has been devoted to their study using nuclear magnetic resonance (NMR) spectroscopy, diffraction, and various forms of spectroscopy. In addition, computer calculations have also been used extensively to model carbohydrates. Often, theoretical and experimental approaches have been combined, as in the analysis of NMR, fiber diffraction, and optical rotation measurements. The success achieved by all of these approaches to date has been limited, but it would appear that in each area the groundwork has been laid for substantial progress in the near future.

6.2 Experimental characterization of oligosaccharide conformations

6.2.1 X-ray and electron diffraction

Recent years have seen an unprecedented explosion in our knowledge of protein structural biochemistry, owing largely to the extraordinary wealth of information provided by high-resolution X-ray diffraction experiments on single crystals. Similar progress has been made in our knowledge of nucleic acid conformations. Unfortunately, there has not been a comparable advance in the understanding of oligosaccharide conformations, since the experimental characterization of their structures has proven to be more difficult than for other types of biopolymers. Perhaps because of their inherent flexibility or their heterogeneous distribution of chain lengths, usefully diffracting single crystals of polysaccharides cannot be produced. Crystal structures reported to date of unbranched oligosaccharides have been limited, with very few exceptions, to chains of four or fewer monomer units. For example, such structures have recently been published for the important case of β-D-cellotetrose, a model for cellulose (Gessler *et al.*, 1995; Raymond *et al.*, 1995). However, high-resolution diffraction structures have been determined for nearly all of the important monosaccharides (Jeffrey and Sundaralingam, 1985) and many have been deposited with the Cambridge Data Base. The availability of these structures allows the monomer units of oligosaccharides to be characterized with some degree of confidence.

Single crystals can also be prepared for many disaccharides, and X-ray diffraction structures are available for most of these molecules as well. Structures have been reported for cellobiose (Chu and Jeffrey, 1968), β-maltose (Gress and Jeffrey, 1977; Quigley *et al.*, 1970), α-maltose (Takusagawa and Jacobson, 1978), methyl β-maltopyranoside (Chu and Jeffrey, 1967) and lactose (Fries *et al.*, 1971). In addition, crystal structures are available for several interesting non-reducing disaccharides, which, although not repeat units of larger oligosaccharides and polymers, are important in themselves. The most significant examples of this class of disaccharides are sucrose (Brown and Levy, 1973) and trehalose (Brown *et al.*, 1972; Taga *et al.*, 1972). Structures are known for several trisaccharides, including the sucrose derivatives 1-kestose (Jeffrey and Park, 1972), 6-kestose (Ferreti *et al.*, 1984), melezitose I and II (Avenel *et al.*, 1976; Becquart *et al.*, 1982) and raffinose (Jeffrey and Huang, 1990); the linear trimer mannotriose (Mackie *et al.*, 1986); and the important model for the $\alpha(1 \rightarrow 6)$ branch in amylopectin, panose (Imberty and Pérez, 1988). At least two tetrasaccharides are known, nystose (Jeffrey and Huang, 1993) and stachyose (Gilardi and Flippen-Anderson, 1986). The best characterized of the larger oligomers are the cyclodextrins, including the cyclic hexasaccharide α-cyclodextrin (Harata, 1977), which is increasingly important in separation technology, and the heptasaccharide β-cyclodextrin, for which there are two neutron diffraction structures, one at 293 K (Betzel *et al.*, 1984) and one at 120 K (Zabel *et al.*, 1986). The lower temperature crystal represents a different solid phase, with different conformations for the hydroxyl groups and the exocyclic primary alcohol. There have been a few structures determined for linear synthetic oligosaccharide derivatives, including a hexasaccharide analog of amylose (Hinrichs and Saenger, 1990). The atomic-level detail available from these structures is extremely valuable in understanding polysaccharide conformations. Unfortunately, there is no guarantee that the crystal conformation of a particular disaccharide is the same as that adopted in solution or when it is part of a polymer; both conformations could be affected by interactions such as intermolecular contacts or intrapolymer interactions along the chain which might be unique to that particular environment.

The conformations of the complex branched oligosaccharides of glycoproteins have also proven difficult to characterize by crystallography, even in cases where the structures of the proteins to which they are linked are known. In some cases it is necessary to cleave off the carbohydrate oligomer before the protein will crystallize into a diffracting crystal lattice. Even in cases where this procedure has not proven necessary, there is frequently insufficient electronic density for the oligosaccharide to determine its conformation, perhaps as the result of dynamical flexibility or sequence heterogeneity. Recently, this situation has begun to change, and several crystal structures have been reported of glycoproteins in which the structure of the oligosaccharide is determined to high resolution. More structures are known for *N*-linked cases, where the oligosaccharide is bound to an Asn residue, than for *O*-linked molecules, where the point of attachment is a Thr or Ser residue. Interesting examples of the former class include the EcorL lectin from *Erythrina corallodendron* (Shaanan *et al.*, 1991) and human leukocyte elastase (Bode *et al.*, 1986). Imberty and Pérez have summarized the results for 26 *N*-linked glycoproteins available in the Brookhaven Data Bank in a recent review (Imberty and Pérez, 1995). Such structures provide the best source of detailed information about oligosaccharide conformations currently available.

X-ray fiber diffraction has been used extensively to study polysaccharide conformations (French and Gardner, 1980). Polysaccharides generally have extended helical structures, rather than compact globular conformations like proteins, and diffraction from aligned fibers of these polymers can yield information about their helical parameters. Unfortunately, these experiments produce limited molecular-level detail, and the results must be supplemented with modeling analysis in order to be interpreted. A number of helical conformations have been proposed for polysaccharides based upon fiber diffraction data, but in those cases where other supporting data are unavailable, these structures have often been controversial. While single crystals of polysaccharides large enough for X-ray diffraction cannot be produced, small crystals of some of these polymers can be made and used in electron diffraction experiments. Progress in the application of these methods to carbohydrates has recently been reviewed (Pérez, 1991).

All of these diffraction studies suffer from the possibility that the structure in a crystal or fiber environment might be different from that adopted in aqueous solution, or that the molecule may have several conformations rather than just one. For example, in the EcorL glycoprotein mentioned above, the molecule crystallizes as a dimer, with the oligosaccharides trapped in the interface between the two proteins (Shaanan *et al.*, 1991). This situation could constrain the oligomer conformation (which might explain why the oligosaccharide is even visible at all in this structure) and thus may not be completely representative of the possible conformations for this oligosaccharide in solution. Several important disaccharides, including maltose, cellobiose and sucrose, have been found to have an intramolecular hydrogen bond between the two sugar rings of the molecule in their crystal structures. Because these hydrogen bonds appear to stabilize particular conformations and could not be maintained in various alternate conformations, each case has occasioned much discussion and study to determine whether or not these hydrogen bonds survive in aqueous solution or are exchanged for hydrogen bonds to solvent water molecules. The sucrose example has probably been the subject of the most extensive debate (Adams and Lerner, 1992; Bock and Lemieux, 1982; Engelsen *et al.*, 1995; McCain and Markley, 1986a, 1986b; Mathlouthi, 1981; Mathlouthi and Luu, 1980; Mathlouthi *et al.*, 1980). Early work (Bock and Lemieux, 1982) suggested that sucrose was a very rigid molecule with the same conformation in solution as in the crystal; more recent work (Adams and Lerner, 1992; Engelsen *et al.*, 1995) has concluded that there are no persistent intraring hydrogen bonds in solution. Because of the possibility of solvent-induced conformational shifts, it would be most desirable to find experimental techniques capable of probing carbohydrate structures in solution as a supplement to the information available from crystallography.

6.2.2 *Nuclear magnetic resonance of oligosaccharides*

NMR spectroscopy has been an invaluable tool in the study of molecular structure and dynamics and has produced a wealth of information about the conformations of monosaccharides. The measurement of proton–proton coupling constants has provided a means to probe the ring puckering of sugars and the conformations of their exocyclic primary alcohol groups. Intramolecular nuclear Overhauser effect (NOE) measurements have been used to study the conformations of monosaccharide rings and their flexibility and the rates of rotational transitions for hydroxymethyl groups (Hajduk *et al.*, 1993). NMR has also been used for sugars in dimethylsulfoxide (DMSO) to determine hydroxyl orientations (Christofides and Davies, 1985; 1987), but the relevance of these measurements to aqueous solutions is not clear. Most of our information about monosaccharide geometries in solution comes from such NMR experiments. In recent years, through-bond correlations determined by multinuclear NMR experiments have been used to determine efficiently the covalent chemical structure of complex oligosaccharides and polysaccharides (Abeygunawardana *et al.*, 1991), further demonstrating the great power of these methods.

NMR techniques have also often been applied to the problem of determining oligosaccharide conformations. For example, the ω angle of $(1 \rightarrow 6)$ linkages in disaccharides can be determined from coupling constant measurements, when the linkage protons can be stereospecifically assigned, in the same manner as is done for free hydroxymethyl groups. Low temperature ^1H NMR studies of supercooled solutions of methyl-β-lactoside have been used to study the crystallographic intraring hydrogen bond between OH3 and O5′, from which it was suggested that under these conditions this hydrogen bond persists in solution (Poppe and van Halbeek, 1994). As discussed above, many such studies of sucrose have been conducted. The number of NMR studies of disaccharides and oligosaccharides are far too numerous to discuss completely here. However, as is the case with crystallography, the application of NMR techniques to carbohydrates has lagged behind the progress achieved for proteins and nucleic acids because of the greater experimental difficulties for saccharides. For example, sugar hydroxyl hydrogen atoms generally exchange too rapidly for study by NMR, and the aliphatic protons often have similar chemical shifts and overlapping peaks which are difficult to assign. Homonuclear coupling constants cannot be used to measure glycosidic torsion angles because there are no vicinal proton coupling constants across the glycosidic linkage. Karplus-type equations employing the long-range heteronuclear J-coupling constants to estimate glycosidic conformation have been developed (Tvaroska *et al.*, 1988) but are not always sufficiently accurate for this purpose and suffer from degeneracy, and the $^3J_{C,H}$ values are often small and difficult to measure. The principal source of information about the relative conformations of linked sugar rings is through-space NOE interactions. However, there are usually too few NOE measurements between the sugars of a disaccharide to determine unambiguously glycosidic conformational angles (Wooten *et al.*, 1990). The nearly complete conformation of the high-mannose *N*-linked oligosaccharide of the human CD2 glycoprotein has recently been reported from NMR data (Wyss *et al.*, 1995) but such structural determinations are still uncommon. Nevertheless, NMR measurements offer the chance to probe carbohydrate conformations in solution, free of crystal constraints, and for this reason have been widely applied to oligosaccharide problems. These studies hold considerable promise for future progress in structural analysis.

At present, the limited amount of data available from NOE experiments generally must be supplemented with additional information in order to specify a disaccharide's conformation. Often the necessary additional information is provided by molecular mechanics modeling of the disaccharide, and studies combining NMR experiments with theoretical calculations for carbohydrates have become quite common (Adams and Lerner, 1992; Carver *et al.*, 1990; Edge *et al.*, 1990; Engelsen *et al.*, 1995; Hajduk *et al.*, 1993; Homans, 1990; Stevens, 1994). In such a combined approach, a conformation is judged to be most likely the correct one if it is both consistent with the known NOE constraints and corresponds to a minimum energy structure using molecular mechanics energy calculations. Theoretical analysis of a variety of disaccharides has led to the conclusion that many of these molecules have more than one possible low-energy conformation (Carver, 1991; Cumming and Carver, 1987), and some may be quite flexible. Thus, it is likely that in solution these disaccharides establish a dynamic equilibrium between their various possible low-energy states, and that polymers made up of these repeat units are also unlikely to have a single conformation in solution but may actually exist as an ensemble of conformers. In such cases, the analysis of NMR data will be further complicated by the conformational averaging between the accessible conformers. If so, the dynamics of the transitions between these states is of considerable interest in understanding the overall conformational behavior of these polymers. However, in spite of the limitations, NMR techniques remain the best experimental probe of oligosaccharide dynamics (Xu and Bush, 1996).

6.2.3 Other experimental methods

An interesting alternative source of information about disaccharide conformations in aqueous solution has been developed by Stevens and co-workers

(Stevens and Sathyanarayana, 1987). This chiroptical technique is based upon estimates of the variation in the optical activity of a particular disaccharide as a function of its glycosidic conformation. In this semi-empirical model, the optical activity is calculated from the transition moments for various structural and conformational features of the molecule, with a solvent correction included. Using this model, Ramachandran-like maps of optical rotation as a function of the glycosidic conformation can be prepared and compared with the observed optical rotation. It is not always possible to identify the glycosidic conformation from such a map alone, although sometimes particular conformations can be ruled out or judged to contribute little to the conformational equilibrium owing to rotation values inconsistent with that which is observed. A number of disaccharides have been characterized using this method, including the important cases of maltose and cellobiose (Stevens and Sathyanarayana, 1988) and sucrose (Stevens and Duda, 1991). Since this algorithm allows the conformationally averaged optical rotation of a molecule to be calculated from theoretical models, it provides a useful measure of the accuracy of potential energy functions developed for disaccharides (see below). This method is proving particularly valuable in resolving debates arising when results from different studies or methods are at variance (Stevens, 1994).

6.3 Disaccharide energetics

Because the various experimental techniques for studying disaccharides in solution are often insufficient to determine accurately their conformations alone, they are increasingly coupled with computer modeling. While these simulations suffer from inadequacies of their own (see below) they are often correct in certain broad overall features and help to select those structures, from the range of structures found to be energetically possible, which are also consistent with the known experimental data. These energy calculations fall into two broad categories: quantum mechanical studies, which calculate molecular energies directly from the Schrödinger equation, and molecular mechanics calculations using semi-empirical energy equations. Only the latter approach is truly practical for disaccharides and oligosaccharides, although it is somewhat less satisfying conceptually.

6.3.1 Quantum mechanical studies

The logical jumping-off point for discussion of polymer conformations is an analysis of how the molecular energy changes as a function of the linkage geometry. Unfortunately, this conformational energy cannot be measured directly in experiments, and it has also been difficult to calculate such energies from *ab initio* quantum mechanics, although *ab initio* calculations have been used to prepare conformational energy maps for peptides (Head-Gordon *et al.*, 1991; Jensen and Gordon, 1991). The principal problem for the study of carbohydrates is one of computational resources. Disaccharides generally contain 20 or more heavy atoms, which is more than can be routinely handled by presently available computers. Moreover, their tendency to hydrogen bond requires that electron correlation be effectively dealt with in the calculations. Even more disturbingly, recent evidence suggests that very high basis set levels (triple-ζ quality or better) and electron correlation treatment beyond second-order many-body perturbation theory may be required for calculational convergence (Barrows *et al.*, 1995), although this question is still a subject of debate (Tvaroska and Carver, 1995). Adequate convergence does appear to have been achieved in *ab initio* studies of the prototypical anomeric model compounds dimethoxymethane and dimethoxyethane at the 6–31G*/MP2 level, with the effects of larger basis set and higher order electron correlation canceling one another (Wiberg and Murcko, 1989).

While converged *ab initio* calculations of a pyranoid pentose such as xylose will probably appear in the near future, the difficulties of using such an approach to study disaccharides would be enormous. The problems mentioned above are limitations which would be encountered in calculating the energy of a single disaccharide conformation. Unfortunately, sugar energies change significantly when even a single hydroxyl group is rotated (Cramer and Truhlar, 1993). Even for the pentose monosaccharide xylose in the $^4C_1$ ring conformation, there are 81 possible combinations of staggered conformations for the four hydroxyl groups (without considering the additional states which arise from the mixture of anomeric forms). In order to explore the glycosidic angle conformations of a disaccharide such as maltose, 59 049 hydroxyl and primary alcohol conformational substates would have to be calculated for each (ϕ, ψ) conformation investigated; clearly an impossible task when even one such calculation is presently not feasible. Furthermore, these calculations would necessarily be for an isolated molecule, because of the impossibility of including dozens, if not hundreds, of solvent water molecules. However, it is probable that in many cases there might be significant coupling between solvent and a sugar solute (see below) which would render a vacuum calculation suspect.

High-level *ab initio* calculations have been applied to various compounds which serve as models for disaccharides (Wiberg and Murcko, 1989). Tvaroska and Carver (1995) have reported extensive studies of the exo-anomeric effect in 2-methoxytetrahydropyran and its dependence upon basis set. A double-ζ basis set has been used to study a somewhat different model of the glycosidic linkage, 2-cyclohexooxytetrahydropyran (Odelius *et al.*, 1995). Semi-empirical molecular orbital calculations have been more widely applied to disaccharides. For example, an AM1 calculation of maltose has recently been reported by Brewster *et al.* (1993). Although this study did not attempt to calculate the glycosidic conformational energy surface for the disaccharide, it did determine that the optimum structure found was similar in conformation to the crystallographic structure. The conformational energies of a number of disaccharides have been examined by Tvaroska and co-workers using the PCILO method (Tvaroska, 1982, 1984; Tvaroska and Kozar, 1980). An advantage of such semi-empirical calculations is that they can easily include a reaction field treatment of solvation (Cramer and Truhlar, 1993; Tvaroska, 1982, 1984; Tvaroska and Kozar, 1980).

6.3.2 Empirical energy calculations

While sufficiently accurate quantum mechanical calculations of disaccharides are not presently feasible, it is possible to estimate their conformational energy using empirical force fields which have been parameterized to reproduce experimental properties for certain test molecules. These force fields allow the specification of the molecular energy as a function of the atomic coordinates. Such calculations are called molecular mechanics (MM) calculations (Brooks *et al.*, 1988; Burkert and Allinger, 1982). With this information it is then possible to map out the allowed low-energy conformations of a disaccharide. A basic simplifying assumption used in this approach is that there are only a limited number of degrees of freedom which are important to the dimer's conformation as a result of geometric constraints in the monomer structure. Ramachandran and co-workers first applied these ideas to the analysis of polypeptide structures. In polypeptides, the geometry of the peptide linkage is nearly constant along a polymer chain, with the carbonyl and amide units having a roughly planar *trans* conformation as the result of large rotational barriers arising from the partial double bond character of the peptide bond. Bond lengths and angles are approximately constant in these peptide units, regardless of the sequence (apart from proline), which is determined by the chemical identity of the sidechain of the C^{α} atom. The problem of specifying the backbone conformation of a protein then reduces to specifying the two torsional angles (ϕ, ψ) which determine the orientation relative to each other of the two successive peptide planes around a C^{α} atom. It is then a simple matter to search for those peptide conformations which result in unacceptable steric clashes between the atoms of the peptide linkage, and to map out the regions of allowed structures in a two-dimensional conformation space using the linkage torsion angles ϕ and ψ as the independent coordinates (Ramachandran *et al.*, 1963). This approach is equivalent to employing a hard-sphere energy function where the energy is either zero if there are no atomic overlaps or infinite if any two atoms approach closer than the sum of their van der Waals radii.

Although there are more atoms in the 'backbone' of an oligosaccharide than there is in a peptide, a similar simplification can be applied to carbohydrates as a result of the relative rigidity of pyranoid rings. For most of the sugars, the pyranose ring forms primarily adopt the $^4C_1$ conformation.

Although such rings are not completely rigid, and undergo a variety of small deformations and fluctuations (Brady, 1986, 1989; Joshi and Rao, 1979), transitions to twist-boats or to the alternate $^{1}C_{4}$ chair are generally rare (Hajduk *et al.*, 1993). For this reason, the disaccharide conformational problem reduces to the specification of the two torsion angles, again usually called ϕ and ψ, which describe the relative orientations of the two rings about the glycosidic oxygen atom (see Figure 6.1). It is then possible to map out the molecular energy as a function of these two angles. In the earliest energy mapping studies, a simple hard-sphere repulsive interaction was used; shortly after first using this method to analyze polypeptide structures, Ramachandran, Rao and co-workers applied the same approach to the study of polysaccharides (Rao *et al.*, 1967), and such conformational energy maps have come to be called Ramachandran maps.

A simple early potential energy function which has been widely used to study carbohydrates is the so-called HSEA (hard-sphere, exo-anomeric) force field (Lemieux *et al.*, 1980). In spite of its name, this function is not actually a hard-sphere model, but rather employs a purely repulsive Kitaygorodski function to model steric repulsions. These terms were augmented by a periodic function of the linkage torsion angles alone, which favored *gauche* conformers and which was intended to represent the exo-anomeric contribution to the conformational energy. The individual sugar rings were kept rigidly fixed in their known crystallographic conformations. Success in rationalizing crystal structures and NMR data seems to have been obtained for some disaccharides, such as blood group oligosaccharides, using only this energy function and ignoring other types of contributions (Lemieux *et al.*, 1980; Thogerson *et al.*, 1982).

Later, more complete force fields were developed to model carbohydrates (Brant, 1972; Dauchez *et al.*, 1995; Glennon *et al.*, 1994; Ha *et al.*, 1988a; Homans, 1990; Niketic and Rasmussen, 1977; Woods *et al.*, 1995), and several general molecular mechanics force fields not specifically developed for carbohydrates, such as GROMOS (Koehler *et al.*, 1987a), TRIPOS (Dauchez *et al.*, 1992), DISCOVER (Hardy and Sarko, 1993a), and MM3 (Aabloo *et al.*, 1994) have also been used successfully to model oligosaccharides. Not only are these various newer functions more realistic than the earlier simplified models such as HSEA, but, since they are compatible with similar functions developed for proteins, it is now possible to study mixed protein–carbohydrate systems, such as glycoconjugates, lectins or enzymes with carbohydrate substrates. In addition, most also permit the explicit inclusion of solvent water molecules, which, as will be seen below, can be quite important in some cases.

Molecular mechanics energy functions for carbohydrates differ in their details and often even in their functional form, but most treat the molecular potential energy as a sum of valence terms and non-bonded interactions (Brooks *et al.*, 1988; Burkert and Allinger, 1982). A typical potential energy function of the internal coordinates $\boldsymbol{q}$ might include harmonic terms to represent bond stretching and bending, a periodic term to mimic the energy associated with hindered rotations about bonds, and electrostatic and van der Waals interactions between non-bonded atoms,

$$V(\boldsymbol{q}) = \sum k_b(b - b_0)^2 + \sum k_\theta(\theta - \theta_0)^2 + \sum k_\phi[1 + \cos(n\phi - \delta)]$$
$$+ \sum \frac{A_{ij}}{r_{ij}^{12}} - \frac{B_{ij}}{r_{ij}^{6}} + \frac{q_i q_j}{r_{ij}} \tag{6.1}$$

where k_b and k_θ are bond-stretching and angle-bending force constants, ϕ is a torsion angle with a force constant k_ϕ, periodicity n and phase δ, and q_i and q_j are the atomic partial charges for atoms separated by a distance r_{ij}. The various constants which appear in such equations are generally determined either experimentally by fitting calculated properties of small molecules to measured values, or are obtained from high-level *ab initio* calculations on such molecules. For example, equilibrium bond lengths and angles are generally obtained by fitting to crystal structures determined by diffraction, and stretching and bending force constants are obtained by fitting vibrational spectra obtained from infra-red (IR) or Raman spectroscopy. Some potential energy functions include additional terms, such as a specific term to represent hydrogen bonding, or a specific exo-anomeric term. More sophisticated energy functions explicitly include anharmonicity and the coupling of internal energy terms (Dauchez *et al.*, 1995), as in Urey–Bradley or valence force fields.

In the first applications of modern-type energy functions to disaccharides, the calculation procedure simply involved rigidly rotating the disaccharide crystal structure to each (ϕ, ψ) point on a regular grid and evaluating the resulting energy (Brant, 1972). Such rigid rotation maps have been calculated for many disaccharides and have the considerable advantage of being quick and cheap to prepare in terms of computer time. Figure 6.3 presents such a map for maltose calculated with the CHARMM force field (Brooks *et al.*, 1983; Ha *et al.*, 1988a, 1988b). This map is qualitatively similar to maps made with a variety of other force fields, particularly in having a central low-energy region and another low-energy polar region around (0°, 180°), with significant barriers between the two. The reason that various force fields give similar results is that the broad outlines of this map are determined by non-bonded repulsions as the rings clash for certain combinations of the glycosidic torsion angles away from the allowed regions. The similarity of the various maps is encouraging since it is evidence that in this essential feature all of the models are basically correct. The finer-scale details of the maps differ somewhat, which is a result both of the differences in the force fields and of the way in which the maps are prepared.

A basic assumption made in calculating Ramachandran energy maps is that the conformational energy is primarily a function of only these two

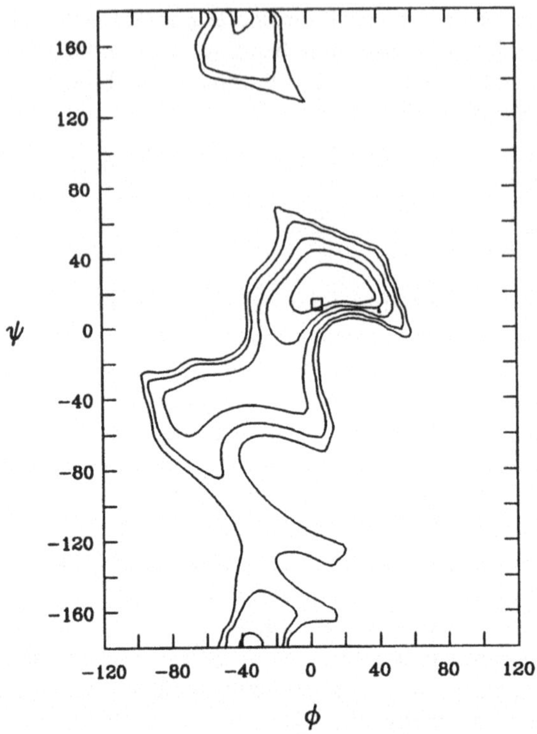

Figure 6.3 A 'rigid' Ramachandran-like conformational energy map for maltose, computed without allowing relaxation of the ring geometries, which are held fixed in their crystallographic structures. Contours are indicated at 2, 5, 10, 25 and 50 kcal/mol above the global minimum. The square symbol indicates the crystal conformation for this disaccharide determined by diffraction. Note: 1 cal = 4.1868 J. Reprinted with permission (Ha *et al.*, 1988b).

internal variables ϕ and ψ, and that all of the other degrees of freedom are unimportant. The maps may thus be thought of as a two-dimensional section through the hyperdimensional internal coordinate space. In reality, however, many of these other internal coordinates can significantly affect the molecular energy. For example, in a crystal structure, each hydroxyl group and primary alcohol has some particular orientation, generally one which produces the most favorable interactions with crystal neighbors. However, rigidly rotating such a crystal structure about ϕ and ψ to some arbitrary conformation might produce an overlap of hydroxyl groups on different rings which could be relaxed by rotating one or both of these hydroxyl groups such as to move the clashing hydrogen atoms away from one another. There is thus more than one possible conformational energy for a given (ϕ, ψ) point. In fact, owing to the complex way in which the molecular energy depends on the interaction of these hydroxyl dipoles with the field of the remainder of the molecule, and the many possible intramolecular hydrogen bond-like interactions which can be formed, there are a great many such distinct energy levels for given values of ϕ and ψ, even without considering the possibility of alternate ring forms (twist-boats, etc.). For a typical disaccharide composed of two hexose monomers, there will be at least 3^{10}, or 59 049, possible energies for each (ϕ, ψ) point, even if only one ring conformation is considered for each ring, and only one anomeric configuration at the reducing end. A relaxed or adiabatic energy map would plot the lowest of these possible energy values for each grid point on the Ramachandran map, which is a daunting computational task.

One might attempt to construct an adiabatic conformational map of the lowest value of these energies for each point in Ramachandran space by rigidly rotating the crystal structure to the desired grid point, constraining ϕ and ψ to remain at these values, and then using energy minimization algorithms to reduce the energy of the molecule as a function of the remaining coordinates to its lowest value. Unfortunately, this approach, while producing a relaxed map (French, 1988), will not in general identify the lowest energy value at that point because of the so-called 'multiple minimum problem' which also frustrates attempts to predict protein tertiary structures by energy minimization (Scheraga, 1983). This situation arises as the result of the tendency of minimization algorithms to become trapped in local minima. For example, going from one structure to another of lower energy might require rotating a hydroxyl group by 120° to an alternate rotameric conformation. However, during this process the energy will initially increase as the intrinsic torsional barrier is surmounted. Unfortunately, conventional minimization algorithms will not follow paths in which the energy increases.

In order to be certain that all lowest-energy structures have been located in an adiabatic energy mapping, it would be necessary to perform minimizations for all 59 049 possible conformations of the hydroxyl and primary alcohol torsional angles of both rings at each mapping grid point. Such a calculation is prohibitively expensive for most purposes at present, although the rapid decrease in the cost of high-speed workstation computers may make massive conformational searches of this type nearly routine in only a few years. In the interim, a number of relaxed energy maps have been prepared by restricting the search of possible conformers to those most likely actually to have low energies. In some cases, the savings involved in such an approach can be considerable. For example, in the absence of any type of environment (a vacuum calculation), it has been repeatedly found that the equatorial hydroxyl groups of glucose tend to align themselves around the periphery of the sugar rings such that they all point in the same direction, either clockwise or counterclockwise relative to the numbering of the ring atoms, in order to make very strained hydrogen bond interactions to the neighboring hydroxyl groups (Ha *et al.*, 1988b). Such artificial conformers would not be expected to arise in aqueous solution, where the solvent

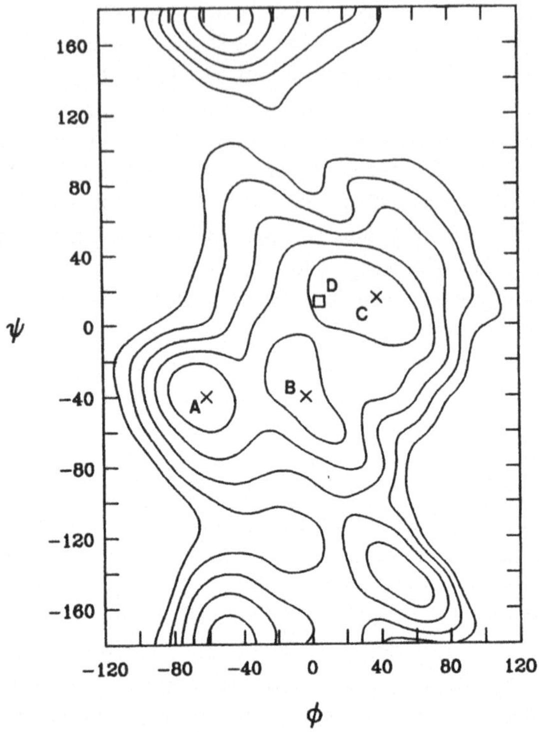

Figure 6.4 A 'relaxed' or 'adiabatic' conformational energy map for maltose calculated using the same potential energy function as in Figure 6.3. Contours are indicated at 2, 4, 6, 8 and 10 kcal/mol above the global minimum at 'C' which is very slightly lower in energy than the minimum labeled 'A', which is the global minimum on relaxed energy surfaces calculated using other energy functions (e.g. Tran *et al.*, 1989). The crystal conformation is in the global minimum energy well and is labeled 'D'. (Note: 1 cal = 4.1868 J.) Reprinted with permission (Ha *et al.*, 1988b).

water molecules would offer alternate hydrogen bond partners, but this type of arrangement has been observed experimentally for sugars in crystals and DMSO solution using NMR (Christofides and Davies, 1985, 1987). Restricting the minimization search to geometries of this type greatly reduces the number of conformations which must be minimized for each (ϕ, ψ) point, since there are only two such orientations, plus the three hydroxymethyl conformers, for each ring, or 36 structures. Figure 6.4 presents an adiabatic energy map for maltose calculated using this approach (Ha *et al.*, 1988b). As can be seen, it is qualitatively similar to the rigid rotation maps in the high-energy regions (Figure 6.3), since, as already noted, the overall shape of energetically allowed regions should be primarily determined by the absence of steric overlaps. The barriers between the central and polar regions have been lowered from more than 50 kcal/mol (209 kJ/mol) to around 6 kcal/mol (25 kJ/mol), since hydroxyl groups could rotate, and bonds and angles could distort, in order to relieve any strains which might develop during a transition between these two regions. This barrier lowering would make it physically possible for transitions to occur at thermal energies, although they would still be very unlikely and infrequent. On the relaxed surface, the central low-energy region is divided into three smaller wells, labeled A, B and C, separated by low-energy barriers. The lowest energy geometry is the 'C' well, and it is gratifying to note that the crystal structure of maltose is located close to this structure and in the same well. While this fine-scale detail varies somewhat with potential functions (Brant, 1972; Tran *et al.*, 1989), it is also encouraging that various force fields and calculational procedures produce broad general agreement. Unfortunately, the conformational difference between the 'A' and 'C' geometries is still large enough to be of practical importance, and the various conformational energy mappings which have been prepared for different disaccharides often differ at this level (see below).

In molecular mechanics calculations, the results, of course, can be no better than the models used to derive them. For this reason, it is desirable to have the most realistic functional form and parameterization possible. Significant effort must go into the development of these force fields. A number of problems are known to affect the energy functions used in previous studies. Simplistic forms such as the HSEA force field ignore too many of the important features of carbohydrate energetics such as hydrogen bonding and dipole interactions, while probably overemphasizing exo-anomeric effects. Potential energy functions such as GROMOS (Koehler *et al.*, 1987a) and CHARMM (Ha *et al.*, 1988a) have been demonstrated to do a poor job of representing hydroxyl group rotations (Kouwijzer *et al.*, 1993). Molecular dynamics simulations of glucose crystals using both force fields found that the hydroxyl groups rotated frequently, apparently as a result of rotational barriers which are too low. In addition, these functions often produce the wrong distribution of exocyclic primary alcohol conformations. This problem has been particularly vexing for Ha *et al.*'s surface (1988a), which incorrectly predicts a 50%–60% population for the TG conformer in glucose, while NMR experiments generally detect essentially none of this conformer in solution. The reason for these problems lies in electrostatic repulsions between the O5 and O6 oxygen atoms in *gauche* conformations which are not otherwise compensated for in the torsional term; these functions also give the wrong torsional energy profile for ethylene glycol, which serves as a model for this type of system. Possibly because it has smaller oxygen charges, the GROMOS force field generally does a better job for this term. Neither of these general force fields contains an explicit treatment of the exo-anomeric energy contribution to total conformational energy, but Homans (1990) has developed a variation of the AMBER and CHARMM functions which includes an added term meant to account for the exo-anomeric energy. New force fields for carbohydrates are in development and should be available soon (Hwang *et al.*, 1997; Liang and Brady, unpublished).

Often the conformational energy maps calculated for disaccharides have more than one local minimum on the two-dimensional energy surface, as is the case for the maltose map illustrated in Figure 6.4. If these minima are close in energy, more than one of them might contribute significantly to the equilibrium properties of the molecule, provided that the barriers between the wells are not so large as to prohibit transitions. The actual molecule would then establish a dynamic equilibrium between these conformations, with the populations determined by their integrated Boltzmann-weighted probabilities and the details of the dynamics determined by the contours of the energy surface. In such a case, measured properties, such as NOEs, would be characteristic of a 'virtual', average structure rather than any single conformation, and this possibility has occasioned considerable discussion (Carver, 1991). If the force fields were sufficiently accurate to be quantitatively reliable, such averaging could be accomplished directly from the surface, and estimates of the likelihood of such averaging established from the barrier heights.

6.4.4 Conformational fluctuations and dynamics

An important feature of the newer carbohydrate energy functions is that they represent the molecular energy as an analytic function of the atomic coordinates, from which derivatives can be calculated. This property allows the numerical integration of the Newtonian equations of motion for each atom in molecular dynamics (MD) simulations, since the force acting on each atom is the negative gradient of the potential energy. Molecular dynamics calculations have become routine for protein and nucleic acid systems (Brooks *et al.*, 1988), and over the past decade they have become increasingly common for carbohydrates (Brady, 1986, 1990). Dynamics simulations have the advantage of allowing the study of the time evolution of systems, including fluctuations and diffusion, and the calculation of rates. High-temperature dynamics simulations are also used to explore conformation space, particularly in structural refinement, since in such trajectories energy barriers between low-energy states can be more easily surmounted. It is becoming quite common for MD simulations to be performed in conjunction with NMR studies to help rationalize the experimental results (Adams and Lerner, 1992; Engelsen *et al.*, 1995; Hajduk *et al.*, 1993).

Figure 6.5 displays two typical trajectories in glycosidic angle space followed by maltose molecules in two vacuum MD simulations at 300 K. These trajectories were computed using Ha *et al.*'s (1988a) version of the CHARMM force field (Brooks *et al.*, 1983) and are superimposed upon the relaxed vacuum energy map for maltose calculated with the same energy function. As can be seen, the relaxed energy surface is a good description of the energy experienced by the molecule, since the trajectories closely conform to the topology of the energy surface and generally remain within

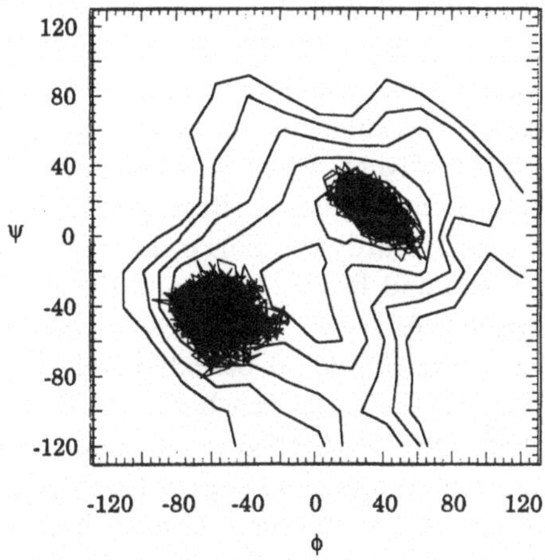

Figure 6.5 Two typical trajectories calculated for maltose in vacuum, projected onto the relaxed energy map for this molecule. Contours are indicated at 2, 4, 6, 8 and 10 kcal/mol above the global minimum. Note: 1 cal = 4.1868 J.

the 2 kcal/mol (8 kJ/mol) contours above their respective minima. These simulations were integrated for relatively short periods, only 110 ps, and each remained confined to the well in which it began. Longer simulations would be more likely to experience the dynamical fluctuations necessary to allow transitions between the wells, and simulations in the nanosecond range or much longer would be needed to ensure proper conformational averaging between the wells. Thus, it is not correct to argue on the basis of one or a few short simulations in which no transitions occur that the subject disaccharide is rigid, as has sometimes been done in the past. Thermodynamic convergence requires that if there is more than one minimum on an energy surface, each must contribute to observed properties proportional to its Boltzmann weight, unless the barriers between wells are so high as to prevent any transitions on the time scale of the experiment. For this reason, the principal use of MD simulations of disaccharides is to model more complex systems than isolated molecules, such as solutions or complexes with proteins, since the vacuum disaccharide can be well described by its relaxed energy surface. However, an example will be discussed below of the use of MD simulations to evaluate the entropic contribution to the vacuum free energy of a disaccharide.

6.5 Hydration of oligosaccharides

It is quite possible that the presence of aqueous solvent could affect the conformational energy of disaccharides. As has already been observed, the numerous hydroxyl groups of carbohydrates will strongly hydrogen bond with solvent water molecules, and the geometric requirements of these solute–solvent hydrogen bonds could favor one conformation over another. A distinct advantage of molecular dynamics simulations is that they allow the explicit inclusion of solvent water molecules. A number of solution simulations have now been conducted (Brady, 1989; Brady and Schmidt, 1993; Engelsen *et al.*, 1995; Hardy and Sarko, 1993; Homans, 1990; Koehler *et al.*, 1988a and b; Ueda and Brady, 1996) and the results of these studies have been quite revealing. As would be expected, sugar hydroxyl groups in such simulations have been found to be extensively hydrogen bonded to solvent. Also as expected, this hydrogen bonding allows the disruption of the intramolecular hydrogen bond networks around the periphery of sugar rings which occur in vacuum modeling studies (see above). The presence of hydrogen bond partners in the solvent promotes hydroxyl rotations both by lowering the energetic cost of rotations relative to the vacuum case and by providing random thermalizing 'kicks' to initiate rotations. In MD simulations of sufficient length most or all possible combinations of hydroxyl states arise with frequencies determined by their Boltzmann probabilities (Schmidt, 1995).

Carbohydrates contain both polar hydroxyl groups and non-polar components such as the aliphatic carbon and hydrogen atoms, with both types of functionalities in close proximity to one another. Close analysis of solvent structuring in MD simulations has revealed the different types of structuring imposed upon adjacent water molecules by each type of group (Brady, 1989; Brady and Schmidt, 1993; Liu and Brady, 1996; Schmidt *et al.*, 1996). Hydroxyl groups have been found to make on average between two and three hydrogen bonds with solvent, as would be expected since each hydroxyl group can serve once as a donor in a hydrogen bond and twice as an acceptor. Because of the proximity of adjacent hydroxyl groups, many water molecules were found simultaneously to hydrogen bond to two hydroxyl groups. The collective structure imposed by a sugar molecule upon the nearby solvent was found to be highly anisotropic and to vary significantly with configuration in some cases (Liu and Brady, 1996; Schmidt *et al.*, 1996).

Water molecules doubly hydrogen bonded to adjacent hydroxyl groups of sugar solutes (Brady and Schmidt, 1993) can be shared between hydroxyl groups on different rings of disaccharides for certain conformations and thus affect the conformational equilibrium. A particularly dramatic case is that of the β-carrageenan analog neocarrabiose. A crystal structure for this molecule is known from diffraction experiments (Lamba *et al.*, 1990) and a structure has also been proposed from fiber-diffraction studies of ι-carrageenan (Arnott *et al.*, 1974). However, when a relaxed Ramachandran energy map was prepared for neocarrabiose (Ueda and Brady, 1996) neither of these experimental geometries was found to correspond to a minimum on

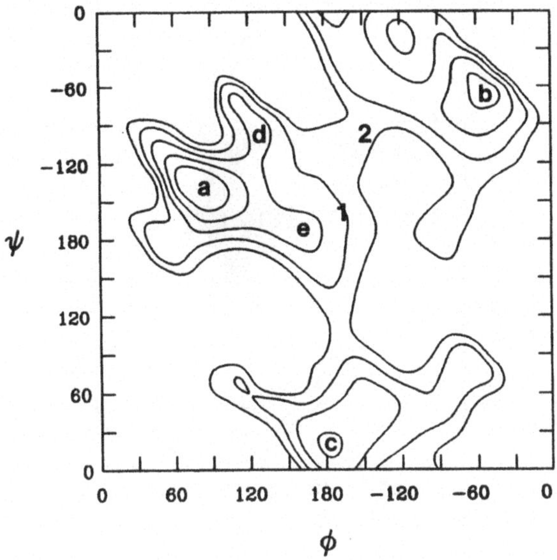

Figure 6.6 A relaxed conformational energy map for neocarrabiose. Contours are in 2 kcal/mol (*c.* 8 kJ/mol) intervals above the global minimum at 'a'. A fiber diffraction structure for the disaccharide repeat unit of the related ι-carrageenen is indicated by the '1', and the crystal structure for this dimer is indicated by a '2'. Reprinted with permission.

the calculated surface (Figure 6.6). In the lowest energy structure from this vacuum surface a hydrogen bond between the O2 hydroxyl group of the non-reducing ring and the O2′ hydroxyl group of the reducing ring stabilizes the conformation. MD simulations for isolated molecules *in vacuo* closely followed the low-energy regions of this map and did not adopt the experimental geometries; during these simulations the stabilizing intraring hydrogen bond remained intact. However, when an MD simulation of neocarrabiose in aqueous solution was calculated a dramatic solvent stabilization occurred (Ueda and Brady, 1996). The solution trajectory divided its time between the fiber and crystal conformations (Figure 6.7), and appeared to be indefinitely stable in these experimental conformations. Upon detailed examination, these conformations were found to be stabilized by a water molecule which replaced the intraring hydrogen bond by hydrogen bonding to both hydroxyl groups simultaneously, 'bridging' between the two rings. Figure 6.8 shows a typical configuration for the apparent minimum free energy structure, in the vicinity of the fiber diffraction structure, illustrating the bridging water molecule.

Another important example is sucrose. As already discussed, it has long been debated whether the molecule is relatively rigid in solution, with the same conformation as in the crystal structure, or is sufficiently flexible to undergo fluctuations large enough to break the two intraring hydrogen

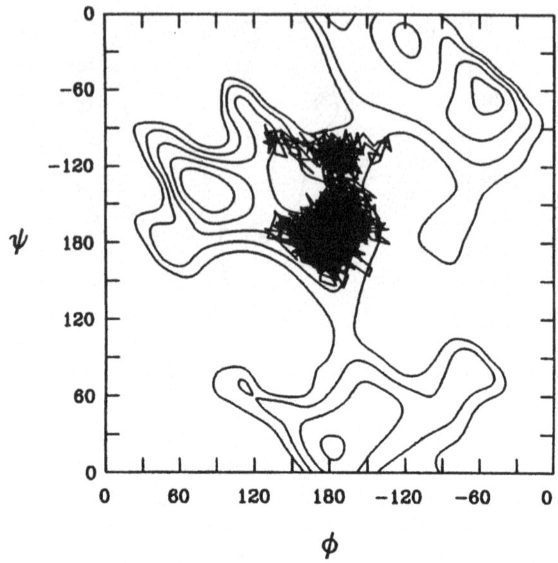

Figure 6.7 A calculated trajectory for neocarrabiose in aqueous solution projected onto the vacuum relaxed energy map. Reprinted with permission (Ueda and Brady, 1996).

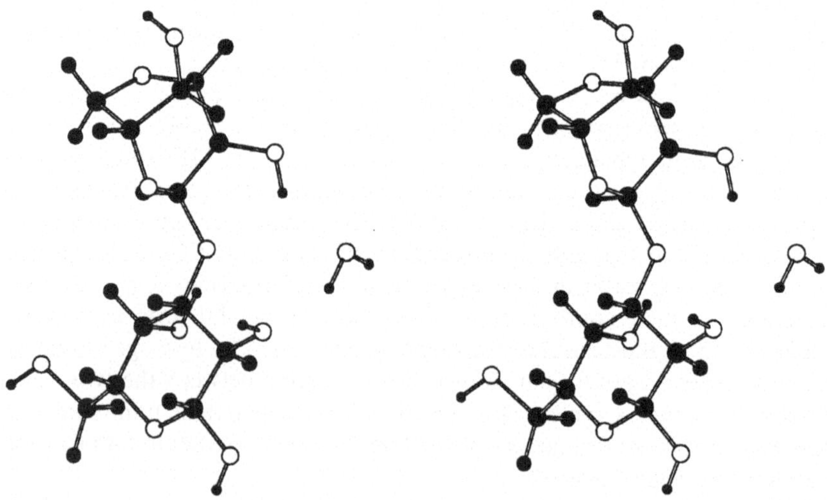

Figure 6.8 A stereo pair illustrating a typical conformation for neocarrabiose in solution, approximately in the fiber structure, illustrating a bridging water molecule hydrogen bonded to the O2 hydroxyl groups of both sugar rings, stabilizing the conformation. Reprinted with permission (Ueda and Brady, 1996).

bonds. Recently, Engelsen *et al.* (1995) used nanosecond time scale MD simulations of sucrose in aqueous solution to interpret NMR data for this molecule. From these simulations they calculated glycosidic heteronuclear coupling constants and NOESY volumes in good agreement with the experimental values, in contrast to the results from a rigid model. Because the internal motions of the disaccharide occurred on the same time scale as overall tumbling, a new motional model for spectral densities was developed which may be generally applicable to small carbohydrates. The calculated rotational tumbling time, translational diffusion coefficient and radius of gyration also agreed well with experiment. Interestingly, in the MD simulations, the disaccharide exhibited considerable conformational flexibility, as suggested from its relaxed conformational energy map (Tran and Brady, 1990). In particular, the two intraring hydrogen bonds of the crystal did not persist in solution. While they were present intermittently for a small fraction of the simulation time, most of the time they were exchanged for hydrogen bonds to solvent molecules. As in the case of neocarrabiose, a conformation was found in which a water molecule bridged between the two rings by hydrogen bonding to a hydroxyl group from each; this conformation was found to be present about 25% of the simulation time. It is probable that this type of study, which combines simulations with experimental data, will become increasingly common in the future, particularly as force fields are refined to be more realistic and as increases in computer speeds allow simulations to be extended to substantially longer times.

6.6 Conformational free energies

Relaxed or adiabatic Ramachandran maps prepared for isolated disaccharide molecules in principle should provide good models for the conformational behavior of these disaccharides. However, in addition to any errors resulting from inaccuracies in the potential energy functions used, such vacuum maps may incorporate problems arising from several other sources. These problems result from the fact that it is the conformational free energy, and not simply the internal mechanical energy which determines the preferred conformational structure. This free energy surface will also contain entropic contributions to the free energy, which are not included in the conformational energy mapping even if one approximated the enthalpy by the mechanical energy. These entropic contributions could affect the topology of disaccharide energy surfaces since the entropy will in general be a function of the conformation. In addition, highly polar molecules such as carbohydrates cannot be put into the vapor state because of the numerous and strong interactions which they make with their environment. In particular, as has already been discussed, interactions with aqueous solvent molecules could result in some conformations being favored over others in order to maximize solvent interactions. For these reasons, the total free energy of a system, including the solvent, would be a more appropriate quantity to contour as a function of glycosidic conformation. Such free energy surfaces are sometimes referred to as potentials of mean force (PMFs).

The calculation of the potential of mean force for a system is quite difficult because this implies a direct calculation of the free energy of a system, which is inherently problematic from molecular mechanics simulations since this involves estimating the system partition function. Statistically meaningful estimates of this fundamental quantity from MD and MC simulations are troublesome because these types of simulations by their nature avoid high-energy regions which must be included in the integral. A number of special techniques have been developed to get around this problem, but all are hard to implement and generally costly in terms of computer time (Brooks *et al.*, 1988; McCammon and Harvey, 1987). Attempts have been made to prepare conformational free energy maps for biological molecules using these methods. Hermans (1989) has reported free energy Ramachandran surfaces for dipeptides, and van Eijck and co-workers have reported the calculation of the one-dimensional potential of mean force for the rotation of the primary alcohol in glucose using a modified, 'adaptive' version of a technique called umbrella sampling (van Eijck *et al.*, 1993).

In principle, the Ramachandran PMF map can be calculated from the relative probabilities of occurrence of each (ϕ, ψ) point during a molecular mechanics simulation, $P(\phi, \psi)$, since this probability is determined by the free energy $W(\phi, \psi)$.

$$P(\phi, \psi) = C \exp[-\beta W(\phi, \psi)] \tag{6.2}$$

and

$$W(\phi, \psi) = -kT \ln P(\phi, \psi) + C' \tag{6.3}$$

where

$$P(\phi, \psi) = \frac{\int \delta(\phi, \psi) \exp[-\beta V(q)]\, dq}{\int \exp[-\beta V(q)]\, dq} \tag{6.4}$$

In practice this approach does not work well since the system will rarely sample the high-energy regions necessary for the evaluation of the normalizing integral in the denominator of equation (6.4). However, umbrella sampling methods have been developed to circumvent this difficulty by augmenting the system potential energy with an additional 'umbrella' potential energy term which has the effect of increasing the probability of high-energy conformations. The effects of this additional energy term can then be subsequently removed through the use of a computational trick. If the probability density resulting from the molecular mechanics simulation using the biased potential energy function (that is, including the umbrella potential term V_U), as calculated directly from the simulation, is $P^*(\phi, \psi)$, then

$$P(\phi,\psi) = \frac{\int \delta(\phi,\psi)\exp[-\beta(V+V_U)]\{1/\exp[-\beta V_U]\}\,dq}{\int \exp[-\beta(V+V_U)]\{1/\exp[-\beta V_U]\}\,dq} \cdot \frac{\int \exp[-\beta(V+V_U)]\,dq}{\int \exp[-\beta(V+V_U)]\,dq} \qquad (6.5)$$

$$= \langle P^*(\phi,\psi)\exp[\beta V_U(\phi,\psi)]\rangle_U / \langle \exp[\beta V_U]\rangle_U \qquad (6.6)$$

where $\langle\rangle_U$ indicates averaging over the ensemble of trajectory-generated configurations using the umbrella potential V_U. Once $P(\phi, \psi)$ has been determined, equation (6.2) can be used to find the potential of mean force.

Figure 6.9 displays a calculated vacuum conformational free energy map for the maltose disaccharide computed using umbrella sampling techniques (Schmidt *et al.*, 1995) for the same potential energy function used to compute the adiabatic surface presented in Figure 6.4. In this study, the negative of the adiabatic energy map in vacuum was used as the umbrella potential function, since this choice on average would have the effect of canceling the potential energy experienced at each point. In the adiabatic map, the global

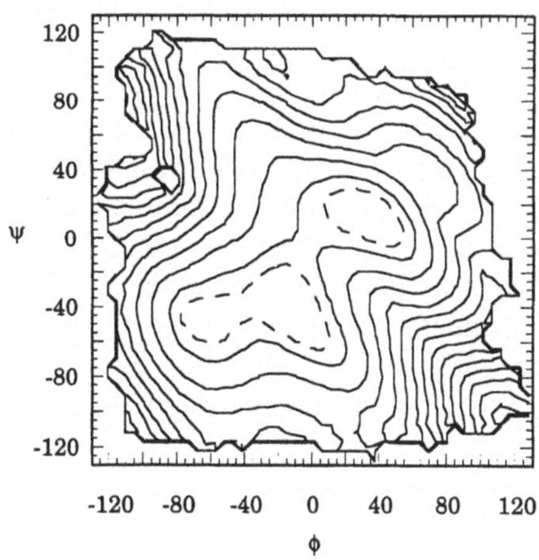

Figure 6.9 A calculated potential of mean force, or free energy map, for the disaccharide maltose in vacuum calculated from umbrella-sampling-weighted molecular dynamics trajectories using the same potential energy function employed for Figures 6.3 and 6.4. Reprinted with permission (Schmidt *et al.*, 1995).

minimum energy conformation is located in the well labeled 'C' (Figure 6.4), close to the structure of maltose in the crystal. However, as noted above, this map differs somewhat from the experimental observations from NMR and optical rotation studies which indicate that the molecule in aqueous solution undergoes a conformational shift from the crystal structure in the 'C'-region to a conformation in the lower-left region of the map, near the 'A' well. As can be seen from Figure 6.9, the vacuum PMF map is quite similar to the adiabatic energy map except in the 'A'–'B' region. These two wells have coalesced, with a significant lowering of the barrier between these two local minima as the result of a larger number of states of the same energy in this region. The global minimum on this PMF is also in this 'A'–'B' region. This change in surface topology is the result of internal entropic effects, since there are no solvent contributions included in this calculation. It is interesting that these entropic effects alone are sufficient to bring the calculation into agreement with experiment. Although no other PMF maps are yet available for disaccharides, the importance of entropic effects in the present case would indicate that such free energy maps should be prepared for other disaccharides as well. Experience with this vacuum case indicates that PMFs including solvent will be much more difficult to calculate reliably, but attempts to compute such a surface from MD simulations are underway for several different disaccharides (Naidoo and Brady, unpublished results).

6.7 Conclusions

At present the conformations of disaccharides and oligosaccharides in solution remain difficult to characterize. The best source of information about these molecules is still crystallography for those cases in which diffracting single crystals can be produced. Where available, the detail provided by this type of experiment exceeds all other methods. Unfortunately, for flexible molecules the conformation in solution may not be the same as the crystal conformation, and for these cases other methods must be used. Quite possibly some of the most important problems of oligosaccharides in molecular recognition will fall into this class. Fortunately, the steady progress being made in combining various types of modeling studies with NMR and chiroptical experiments should provide the necessary tools for studying these glycoconjugates.

References

Aabloo, A. and French, A.D. (1994) *Macromol. Theory Simul.* **3**, 185–91
Aabloo, A., French, A.D., Mikelsaar, R.-H. and Pertsin, A.J. (1994) *Cellulose* **1**, 161–8.
Abeygunawardana, C., Bush, C.A. and Cisar, J.O. (1991) *J. Biochem.* **30**, 6528–40.
Adams, B. and Lerner, L. (1992) *J. Am. Chem. Soc.* **114**, 4827–9.

Arnott, S., Scott, W.E., Rees, D.A. and McNab, C.G.A. (1974) *J. Mol. Biol.* **45**, 85–99.
Avenel D., Neuman, H. and Gillier-Pandraud, H. (1976) *Acta Crys.* **B32**, 2598–605.
Barrows, S.E., Dulles, F.J., Cramer, C.J. *et al.* (1995) *Carbohydr. Res.* **276**, 219–51.
Becquart, J., Neuman, A. and Gillier-Pandraud, H. (1982) *Carbohydr. Res.* **11**, 9–21.
Betzel, C., Saenger, W., Hingerty, B.E. and Brown, G.M. (1984) *J. Am. Chem. Soc.* **106**, 7545–56.
Bock, K. and Lemieux, R.U. (1982) *Carbohydr. Res.* **100**, 63–74.
Bode, W., Wei, A.Z., Huber, R. *et al.* (1986) *EMBO J.* **5**, 2453–8.
Brady, J.W. (1986) *J. Am. Chem. Soc.* **108**, 8153–60.
Brady, J.W. (1989) *J. Am. Chem. Soc.* **111**, 5155–65.
Brady, J.W. (1990) *Adv. Biophys. Chem.* **1**, 155–202.
Brady, J.W. and Schmidt, R.K. (1993) *J. Phys. Chem.* **97**, 958–66.
Brant, D.A. (1972) *Annu. Rev. Biophys. Bioeng.* **1**, 369–408.
Brewster, M.E., Huang, M.-J., Pop, E. *et al.* (1993) *Carbohydr. Res.* **242**, 53–67.
Brooks, B.R., Bruccoleri, R.E., Olafson, B.D. *et al.* (1983) *J. Comput. Chem.* **4**, 187–217.
Brooks, C.L., Karplus, M. and Pettitt, B.M. (1988) *Proteins: A Theoretical Perspective of Dynamics, Structure and Thermodynamics.* Adv. Chem. Phys. Vol. LXXI. Wiley-Interscience, New York.
Brown, G.M. and Levy, H.A. (1973) *Acta Crys.* **B29**, 790–7.
Brown, G.M., Rohrer, D.C., Berking, B. *et al.* (1972) *Acta Crys.* **B28**, 3145–58.
Burkert, U. and Allinger, N.L. (1982) *Molecular Mechanics.* ACS Monograph 177. American Chemical Society, Washington, DC.
Carver, J.P. (1991) *Curr. Opinion Struct. Biol.* **1**, 716–20.
Carver, J.P., Mandel, D., Michnick, S.W. *et al.* (1990) in *Computer Modeling of Carbohydrates.* ACS Symposium Series 430, A.D. French and J.W. Brady (eds). American Chemical Society, Washington, DC.
Christofides, J.C. and Davies, D.B. (1985) *J. Chem. Soc., Chem. Commun.* 1533–4.
Christofides, J.C. and Davies, D.B. (1987) *J. Chem. Soc. Perkin Trans.* **2**, 97–102.
Chu, S.S.C. and Jeffrey, G.A. (1967) *Acta Crys.* **23**, 1038–49.
Chu, S.S.C. and Jeffrey, G.A. (1968) *Acta Crys.* **B24**, 830–8.
Cramer, C.J. and Truhlar, D.G. (1993) *J. Am. Chem. Soc.* **115**, 5745–53.
Cumming, D.A. and Carver, J.P. (1987) *Biochemistry* **26**, 6664–76.
Dauchez, M., Derreumaux, P., Lagant, P. and Vergoten, G. (1995) *J. Comp. Chem.* **16**, 188–99.
Dauchez, M., Mazurier, J., Montreuil, J. *et al.* (1992) *Biochimie* **74**, 63–74.
Doubet, S., Bock, K., Smith, D. *et al.* (1989) *TIBS* **14**, 475–7.
Duda, C.A. and Stevens, E.S. (1993) *J. Am. Chem. Soc.* **115**, 8487–8.
Edge, C.J., Singh, U.C., Bazzo, R. *et al.* (1990) *Biochemistry* **29**, 1971–4.
Engelsen, S.B., Hervé du Penhoat, C. and Pérez, S. (1995) *J. Phys. Chem.* **99**, 13334–51.
Ferreti, V., Bertolasi, C., Gilli, G. and Accorsi, C.A. (1984) *Acta Crys.* **C40**, 531–5.
French, A.D. (1988) *Biopolymers* **27**, 1519–25.
French, A.D. and Gardner, W.A. (eds) (1980) *Fiber Diffraction Methods.* ACS Symposium Series 141. American Chemical Society, Washington, DC.
Fries, D.C., Rao, S.T. and Sundaralingam, M. (1971) *Acta Crys.* **B27**, 994–1005.
Gessler, K., Krauss, N., Steiner, T. *et al.* (1995) *J. Am. Chem. Soc.* **117**, 11397–406.
Gilardi, R. and Flippen-Anderson, J.L. (1986) *Acta Crys.* **C43**, 806–8.
Glennon, T.M., Zheng, Y.-J., Le Grand, S.M. *et al.* (1994) *J. Comput. Chem.* **15**, 1019–40.
Gress, M.A. and Jeffrey, G.A. (1977) *Acta Crys.* **B33**, 2490–5.
Ha, S.N., Giammona, A., Field, M. and Brady, J.W. (1988a) *Carbohydr. Res.* **180**, 207–21.
Ha, S.N., Madsen, L.J. and Brady, J.W. (1988b) *Biopolymers* **27**, 1927–52.
Hajduk, P.J., Horita, D.A. and Lerner, L.E. (1993) *J. Am. Chem. Soc.* **115**, 9196–201.
Harata, K. (1977) *Bull. Chem. Soc. Jpn.* **50**, 1416–24.
Hardy, B.J. and Sarko, A. (1993a) *J. Comput. Chem.* **14**, 831–47.
Hardy, B.J. and Sarko, A. (1993b) *J. Comput. Chem.* **14**, 848–57.
Head-Gordon, T., Head-Gordon, M., Frisch, M.J. *et al.* (1991) *J. Am. Chem. Soc.* **113**, 5989–97.
Hermans, J. (1989) in *Computer Simulations of Biomolecular Systems.* van Gunsteren, W.F. and Weiner, K. (eds). ESCOM, Leiden.
Hinrichs, W. and Saenger, W. (1990) *J. Am. Chem. Soc.* **112**, 2789–96.
Homans, S.W. (1990) *Biochemistry* **29**, 9110–18.
Hwang, M.-J., Ni, X., Waldman, M. *et al.* (1998) *Biopolymers*, **45**, 435–68.

Imberty, A. and Pérez, S. (1988) *Carbohydr. Res.* **181**, 41–55.
Imberty, A. and Pérez, S. (1995) *Protein Engineering* **8**, 699–709.
Jeffrey, G.A. and Huang, D.-B. (1990) *Carbohydr. Res.* **206**, 173–82.
Jeffrey, G.A. and Huang, D.-B. (1993) *Carbohydr. Res.* **247**, 37–50.
Jeffrey, G.A. and Park, Y.J. (1972) *Acta Crys.* **B28**, 257–67.
Jeffrey, G.A. and Sundaralingam, M. (1985) *Adv. Carbohydr. Chem. Biochem.* **43**, 203–421.
Jensen, J.H. and Gordon, M.S. (1991) *J. Am. Chem. Soc.* **113**, 7917–24.
Joshi, N.V. and Rao, V.S.R. (1979) *Biopolymers* **18**, 2993–3004.
Koehler, J.E.H., Saenger, W. and van Gunsteren, W.F. (1987a) *Eur. Biophys. J.* **15**, 197–210.
Koehler, J.E.H., Saenger, W. and van Gunsteren, W.F. (1987b) *Eur. Biophys. J.* **15**, 211–24.
Koehler, J.E.H., Saenger, W. and van Gunsteren, W.F. (1988a) *Eur. Biophys. J.* **16**, 153–68.
Koehler, J.E.H., Saenger, W. and van Gunsteren, W.F. (1988b) *J. Mol. Biol.* **203**, 241–50.
Kouwijzer, M.L.C.E., van Eijck, B.P., Kroes, S.J. and Kroon, J. (1993) *J. Comput. Chem.* **14**, 1281–9.
Lamba, D., Segre, A.L., Glover, S. *et al.* (1990) *Carbohydr. Res.* **208**, 215–30.
Lemieux, R.U., Bock, K., Delbaere, L.T.J. *et al.* (1980) *Can. J. Chem.* **58**, 631–53.
Lemieux, R.U., Pavia, A.A., Martin, J.C. and Watanabe, K.A. (1969) *Can. J. Chem.* **47**, 4427–39.
Liu, Q. and Brady, J.W. (1996) *J. Am. Chem. Soc.* **118**, 12276–86.
Mackie, W., Sheldrick, B., Akrigg, D. and Pérez, S. (1986) *Int. J. Biol. Macromol.* **8**, 43–51.
Mathlouthi, M. (1981) *Carbohydr. Res.* **91**, 113–23.
Mathlouthi, M. and Luu, D.V. (1980) *Carbohydr. Res.* **81**, 203–12.
Mathlouthi, M., Luu, C., Meffroy-Biget, A.M. and Luu, D.V. (1980) *Carbohydr. Res.* **81**, 213–23.
McCain, D.C. and Markley, J.L. (1986a) *Carbohydr. Res.* **152**, 73–80.
McCain, D.C. and Markley, J.L. (1986b) *J. Am. Chem. Soc.* **108**, 4259–64.
McCammon, J.A. and Harvey, S.C. (1987) *Dynamics of Proteins and Nucleic Acids*. Cambridge University Press, Cambridge.
Niketic, S.R. and Rasmussen, K. (1977) *The Consistent Force Field: A Documentation*. Springer, Berlin.
Odelius, M., Laaksonen, A. and Widmalm, G. (1995) *J. Phys. Chem.* **99**, 12686–92.
Pérez, S. (1991) *Methods in Enzymology* **203**, 510–56.
Pérez, S. and Marchessault, R.H. (1978) *Carbohydr. Res.* **65**, 114–20.
Poppes, L. and van Halbeek, H. (1994) *Structural Biology* **1**, 215–16.
Quigley, G.J., Sarko, A. and Marchessault, R.H. (1970) *J. Am. Chem. Soc.* **92**, 5834–9.
Ramachandran, G.N., Ramakrishnan, C. and Sasisekharan, V. (1963) *J. Mol. Biol.* **7**, 95–9.
Rao, V.S.R., Sundararajan, P.R., Ramakrishnan, C. and Ramachandran, G.N. (1967) in *Conformation in Biopolymers*. Vol. 2, G.N. Ramachandran (ed.). Academic Press, London.
Rasmussen, K. (1982) *Acta Chem. Scand., Ser. A* **36**, 323–7.
Raymond, S., Heyraud, A., Tran Qui, D. *et al.* (1995) *Macromolecules* **28**, 2096.
Scheraga, H.A. (1983) *Biopolymers* **22**, 1–14.
Schmidt, R.K. (1995) PhD. thesis, Cornell University, Ithaca, NY.
Schmidt, R.K., Karplus, M. and Brady, J.W. (1996) *J. Am. Chem. Soc.* **118**, 541–6.
Schmidt, R.K., Teo, B. and Brady, J.W. (1995) *J. Phys. Chem.* **99**, 11339–43.
Shaanan, B., Lis, H. and Sharon, N. (1991) *Science* **254**, 862–6.
Stevens, E.S. (1994) *Biopolymers* **34**, 1395–401.
Stevens, E.S. and Duda, C.A. (1991) *J. Am. Chem. Soc.* **113**, 8622–7.
Stevens, E.S. and Sathyanarayana, B.K. (1987) *Carbohydr. Res.* **166**, 181–93.
Stevens, E.S. and Sathyanarayana, B.K. (1988) *J. Am. Chem. Soc.* **111**, 4149–54.
Taga, T., Senma, M. and Osaki, K. (1972) *Acta Crys.* **B28**, 3258–63.
Takusagawa, F. and Jacobson, R.A. (1978) *Acta Crys.* **B34**, 213–18.
Thogerson, H., Lemieux, R.U., Bock, K. and Meyer, B. (1982) *Can. J. Chem.* **60**, 44–57.
Tran, V. and Brady, J.W. (1990) *Biopolymers* **29**, 961–76.
Tran, V., Buleon, A., Imberty, A. and Pérez, S. (1989) *Biopolymers* **28**, 679–90.
Tvaroska, I. (1982) *Biopolymers* **21**, 1887–97.
Tvaroska, I. (1984) *Biopolymers* **23**, 1951–60.
Tvaroska, I. and Bleha, T. (1989) *Adv. Carbohydr. Chem. Biochem.* **47**, 45–123.
Tvaroska, I. and Carver, J.P. (1995) *J. Phys. Chem.* **99**, 6234–41.
Tvaroska, I. and Kozar, T. (1980) *J. Am. Chem. Soc.* **102**, 6929–36.
Tvaroska, I., Hricovini, M. and Petrakova, E. (1988) *Carbohydr. Res.* **189**, 359–362.

Ueda, K. and Brady, J.W. (1996) *Biopolymers* **38**, 461–9.
van Eijck, B.P., Hooft, R.W.W. and Kroon, J. (1993) *J. Phys. Chem.* **97**, 12093–9.
Wiberg, K.B. and Marquez, M. (1994) *J. Am. Chem. Soc.* **116**, 2197–8.
Wiberg, K.B. and Murcko, M.A. (1989) *J. Am. Chem. Soc.* **111**, 4821–8.
Woods, R.J., Dwek, R.A., Edge, C. and Fraser-Reid, B. (1995) *J. Phys. Chem.* **99**, 3832–46.
Wooten, E.W., Edge, C.J., Bazzo, R. *et al.* (1990) *Carbohydr. Res.* **203**, 13–17.
Wyss, D.F., Choi, J.S., Li, J. *et al.* (1995) *Science* **269**, 1273–8.
Xu, Q. and Bush, C.A. (1996) *Biochemistry* **35**, 14512–20.
Zabel, V., Saenger, W. and Mason, S.A. (1986) *J. Am. Chem. Soc.* **108**, 3664–73.

7 Shapes and interactions of polysaccharide chains

SERGE PÉREZ and MILOU KOUWIJZER

7.1 Introduction

Polysaccharides form the most abundant and diverse family of biopolymers. With several hundreds of known examples they offer a great diversity of chemical structures, ranging from simple linear homopolymers to branched heteropolymers, having repeating units of up to octasaccharides. Simple polysaccharides, with a repeating structure composed of monosaccharides, are used to store energy, as in starch, glycogen, locust bean gum and guar gum. Carbohydrate functions are not limited to the storage or production of energy. Cellulose, a simple polymer of glucose, is an essential constituent of plant cell walls. It generates hard and solid elements in the form of tough fibers. The plasticity of the cell wall is further regulated via hydrated cross-linked three-dimensional networks where polysaccharides such as pectins play a key function. In marine species, carbohydrate polymers such as agar, alginates and carrageenans play a similar role. Other polysaccharides create viscous extracellular layers around bacteria. In the animal kingdom, the family of glycosaminoglycans (hyaluronate, chondroitin sulfate, dermatan sulfate etc.) plays a key role in governing the solution properties of some physiological fluids as well as participating in the structural buildup of the intercellular matrix.

In order to understand the molecular basis of the native arrangements of polysaccharides as well as relating their properties and functions to their structures, their different levels of structural organization must be determined. As with other macromolecules, the elucidation of the **primary structure** (the covalent sequence of monomeric units along with the respective glycosidic linkages) is a prerequisite. Polysaccharides may also be branched, which is a unique feature among naturally occurring macromolecules. Depending on their primary structures, polysaccharide chains adopt characteristic shapes such as ribbons, extended helices and hollow helices which characterize their **secondary structures**. Biosynthesis may also result in the formation of multiple helices. Some of these features may persist locally in the diluted state and may be directly responsible for the solution properties of some polysaccharides. Evidently, the characterization of the overall shape of the random coil in terms of end-to-end distances, persistence lengths and radius of gyration is required. Energetically favored interactions between chains of well-defined secondary structures result in ordered organizations known as **tertiary structures**. These may also result from biogenesis where polymerization and crystallization are often concomitant. A higher level of organization involving further associations between these well-structured entities will result in **quaternary structures**.

Unlike globular proteins, the length in polysaccharide chains is not fixed, but can vary, often widely, from one molecule to another, and an average molecular weight or molecular weight distribution must be considered. In polysaccharides, one end of the chain has the reducing character of monosaccharides and the other end is non-reducing. The determination of the relative orientation of such polar chains in crystalline or semicrystalline structures has been a central theme in the field of polysaccharide crystallography.

The present chapter covers some of the methods which have been used to elucidate the different structural levels of polysaccharides. A particular emphasis is given to the combination of the crystallographic and the modeling methods which yield information about the secondary and tertiary structures of polysaccharides. The significance of the information derived from such methods, along with the limit of applications, is emphasized. The structural features of the three more abundant polysaccharides, that is, cellulose, starch and chitin, are presented.

Chain–chain interactions as those found in the crystalline arrangement may also occur in solution, where they have a significant influence on the properties. When they develop over a significant length and period of time they may result in gel formation. When their nature is more transient or when they perturb significantly the surrounding water, they are the origin of viscous effects. Such systems are extensively studied by rheological methods. A summary of these methods and the results are also included in this chapter.

7.2 Ordered structures of polysaccharides

7.2.1 Structural features

7.2.1.1 Building blocks

Carbohydrate building units usually have cyclic structures. One of the ring atoms is almost always an oxygen and the remaining atoms are carbon atoms. In polysaccharides most commonly the sugar residues are either pentoses or hexoses. Ring shapes can be defined in terms of reference conformations (chair, C; twist, T; boat, B; envelope, E; skew, S) or by the so-called puckering parameters (Cremer and Pople, 1975; see also Chapter 1). In general, furanose ring conformations have two major local minima and a path of interconversion. A few examples are shown in Figure 7.1. With the exception of idoses, most of the monosaccharide units which occur in polysaccharides have a chair conformation. In the case of strained rings, a certain percentage of boat conformations may be observed.

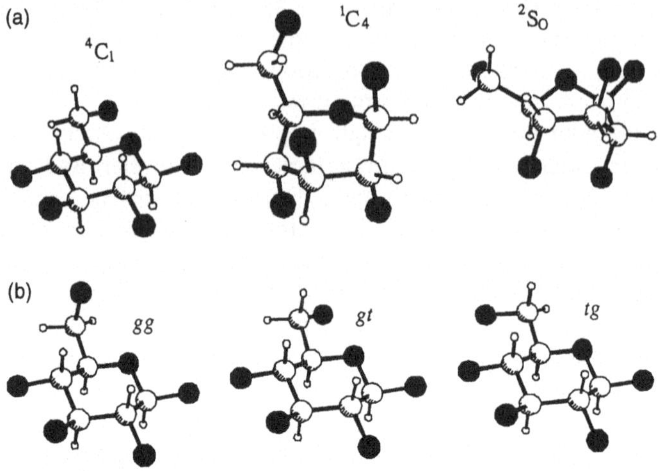

Figure 7.1 (a) Examples of conformations of pyranose rings; (b) $^4C_1$ conformation of β-D-glucopyranose with three conformations of the hydroxymethyl group.

In an effort to facilitate the construction of polysaccharides (and other carbohydrate-containing macromolecules) a carbohydrate fragment library has been created (Pérez and Delage, 1991). In its latest version, the data bank contains more than 80 monosaccharides. With the inclusion of sulfated monosaccharides, it covers more than 97% of the units which occur in oligo- and polysaccharides and in complex carbohydrates. A computerized version of the data bank is available on internet at http://www.cermav.cnrs.fr/mono.

The exocyclic primary alcohol groups can adopt a number of low-energy conformations (Figure 7.1). They are in staggered arrangements that correspond to local minima. In the case of pyranoses, primary hydroxyl groups frequently occupy two positions, avoiding interactions between O4 and O6. However, each of the secondary hydroxyl groups can rotate almost freely. All the hydroxyl groups can donate and accept protons, thereby participating in the creation of a network of hydrogen bonds. As a result of the many possible orientations of such groups, the determination and prediction of hydrogen bonding is a difficult task, which is almost never fulfilled in the structural elucidation of polysaccharides.

7.2.1.2 Junctions

Conformational analysis of the constituting disaccharides yields useful information for modeling polysaccharides. Structural information derived from crystal structure determinations of oligosaccharides has justified that, as a first approximation, the internal parameters could be divided into rigid monomeric units and flexible glycosidic linkages (Brant, 1976). Because

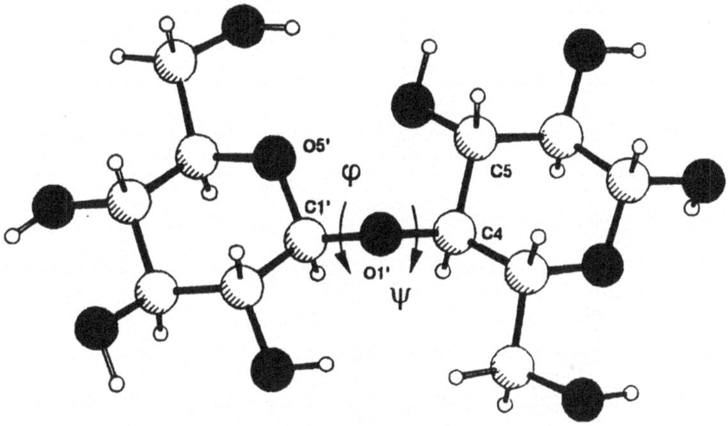

Figure 7.2 A disaccharide (cellobiose) with the glycosidic dihedrals ϕ (O5′–C1′–O1′–C4) and ψ (C1′–O1′–C4–C5) indicated. The signs of the torsion angles are given in agreement with the IUPAC–IUB Joint Commission of Biochemical Nomenclature (1996).

of the spatial separation of these rigid entities which are interposed between the flexible linkages (Figure 7.2) there is an almost total independence between successive sets of glycosidic torsion angles (Gagnaire *et al.*, 1982). For these reasons, the assessment of preferred conformations is usually performed on disaccharides by assessing the space that is available for rotation about the glycosidic junctions (Rao *et al.*, 1967). Sets of limiting distances for different pairs of atoms to be used in the construction of steric maps have been proposed. These maps show the allowed and disallowed conformations for a particular disaccharide unit. An example is shown in Figure 7.3. More precise information about the potential energy of a given conformation can be gained by using molecular mechanics methods. The total potential energy E_{tot} is partitioned into a number of discrete contributions, for example $E_{tot} = E_{nb} + E_{tor} + E_{hb} + E_{exo} + \ldots$ where E_{nb} is the energy due to non-bonded interactions, E_{tor} is the energy due to the torsional strain about the glycosidic bonds, E_{hb} is the energy from hydrogen bonds and E_{exo} is the energy due to the exoanomeric effect. Detailed descriptions of these functions have been reported, and are included in computer programs such as PFOS (Potential Functions for Oligosaccharide Structures) (Pérez, 1978; Tvaroska and Pérez, 1986) and many other computer programs.

Conformational analyses of disaccharides give different results depending on whether the residues are considered as rigid or are allowed to adjust internally at each increment in ϕ and ψ. Compared with conventional approaches where residues are considered as rigid, optimization of all the internal parameters (i.e. bond lengths, bond angles and all the torsional angles) is an important step towards more realistic information; this is obtained at the expense of computer time. Calculations of potential energy

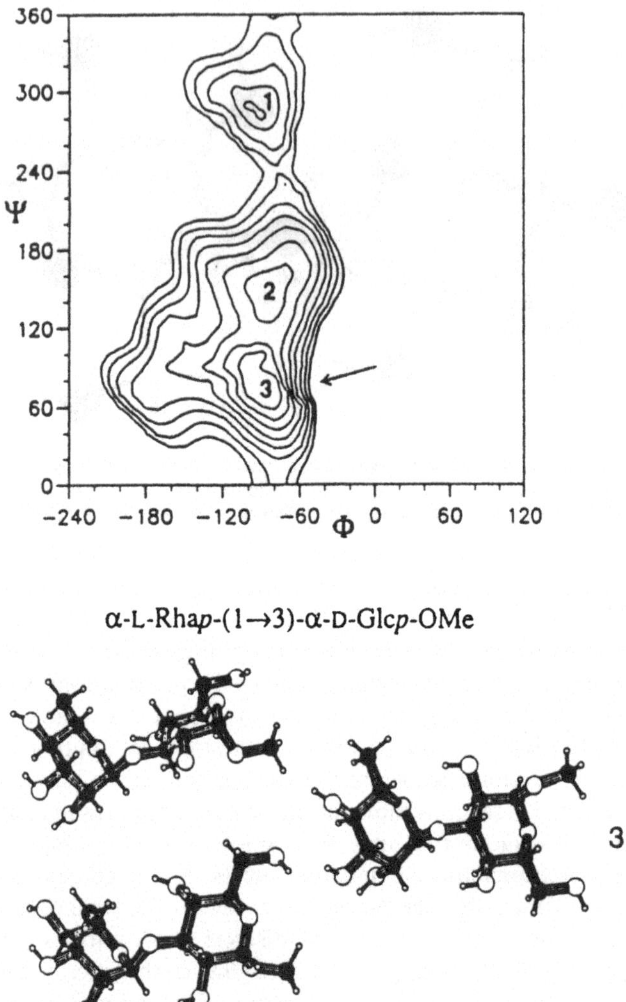

Figure 7.3 Example of a potential energy map for a disaccharide α-L-Rhap-(1 → 3)-α-D-Glcp-OCH_3 as a function of the glycosidic dihedrals, Φ and Ψ. The iso-energy contours are 1 kcal/mol (*c.* 4 kJ/mol) apart. The molecular representations corresponding to the three lowest energy arrangements are shown. Reprinted from Robijn *et al.*, 1996.

surfaces where all the internal coordinates of the molecules are relaxed and minimized through an extensive molecular mechanics scheme have been reported for several disaccharides (French, 1988; French *et al.*, 1990; Ha *et al.*, 1988; Imberty *et al.*, 1989; Tran and Brady, 1990; Tran *et al.*, 1989; see also Chapter 6). It was found that the inclusion of the relaxed principle into conformational descriptions of disaccharides does not generally alter the overall shape of the allowed low-energy regions or the position of the local minima. However, flexibility within the ring plays a crucial role. Its principal effect is to lower energy barriers of conformational transitions about the glycosidic bonds, permitting pathways among the low-energy minima. Such a procedure must be used in the investigations of polysaccharides comprising five-membered rings.

7.2.1.3 Helical parameters

When subjected to the constraints imposed by the helical symmetry of macromolecular chains, equivalent repeating units (either made of one monomer or of an oligomer) should occupy equivalent positions about the axis (Natta and Corradini, 1960). This is achieved when the ϕ, ψ torsion angles of identical links are the same in a large part of the polysaccharide chain. The secondary structure can be described by just two parameters, n and h, defined thus (Figure 7.4):

n = number of units (residues) per turn of the helix (not necessarily integral);
h = translation of each residue along the helix axis.

Other related parameters are as follows:

r = radius of helix;
P = pitch (i.e. axial translation per turn);

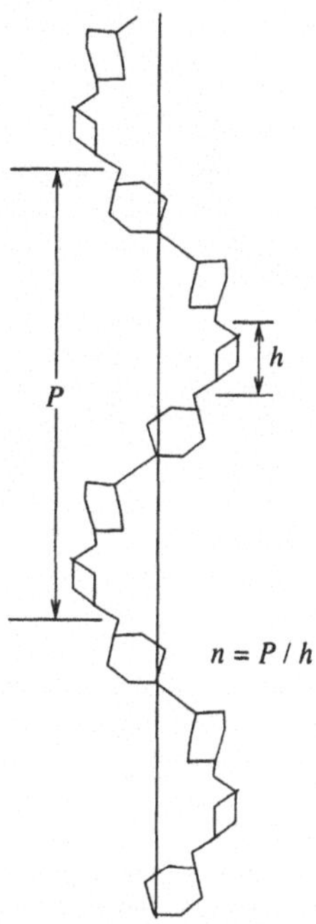

Figure 7.4 Schematic representation of a polysaccharide helix of pitch P, along with the helical parameters n (number of repeat units per helical turn) and h (axial advance of one repeat).

t = twist [i.e. angular rotation per residue about helixaxis ($= 2\pi \times h/P$)];
u = number of residues per axial repeat (must be integral);
v = number of turns per axial repeat;
c = axial repeat distance ($= u \times h = v \times P$).

The symmetry of the helix is then given as u, or u_v. The chirality of the helix is defined by the signs of n and h. Arbitrarily, positive values of n and h will designate a right-handed helix, and opposite signs will correspond to a left-handed chirality. Whenever the values $h = 0$ or $n = 2$ are intercrossed, the screw sense of the helix changes to the opposite sign. In practice, the helical parameters are readily observed on the fiber diffraction pattern, where the spacing of the layer lines provides the pitch (P) of the helical structure. The helix type (u) is deduced from the positions of the meridional reflections. In conformational analysis studies of polysaccharides that have only a monosaccharide as the repeating unit the helical parameters can be plotted as iso-n and iso-h contours as a function of torsional angles ϕ and ψ, as shown in Figure 7.5. Values of the torsional angles consistent with the observed parameters are found at the intersection of the corresponding iso-h and

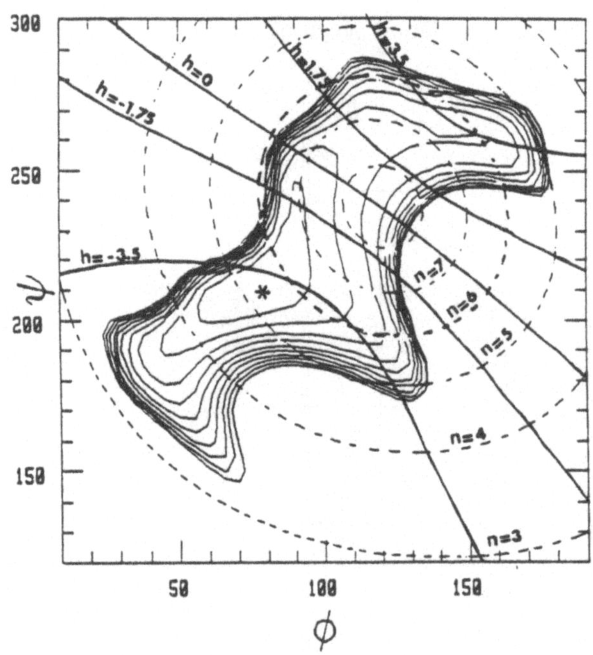

Figure 7.5 Selected iso-n and iso-h contours superimposed on the potential energy map of maltose. With respect to the relative energy minimum (*), the iso-energy contours are drawn by interpolation of 1 kcal/mol. The iso-$h = 0$ contour divides the map into two regions corresponding to right-handed and left-handed chirality. Note that a range of ϕ, ψ or h values are consistent with a given value of n.

iso-n contours. Discrimination between possible solutions is based on the magnitude of the potential energy.

Polysaccharides often crystallize in various shapes with slight variations in the geometry of the constituent monosaccharides. Therefore a thorough conformational analysis for a polysaccharide requires the use either of flexible residues or of parallel studies with several residues that span the range of residue variation. When flexibility of the residues and linkages is incorporated, the values of ϕ and ψ cannot specify n and h. Thus an alternative representation of conformational space must be used (French and Murphy, 1977; see also Chapter 1).

7.2.1.4 Multiple helices

Some polysaccharides adopt a very extended conformation, others form a rather wide, hollow helix. The latter especially is able to form multiple helices. In Figure 7.6 examples are shown of three glucan chains. Cellulose consists of β(1 → 4) linked glucopyranose units, and adopts in fibers extended, two-fold single helices (Marchessault and Sundararajan, 1983).

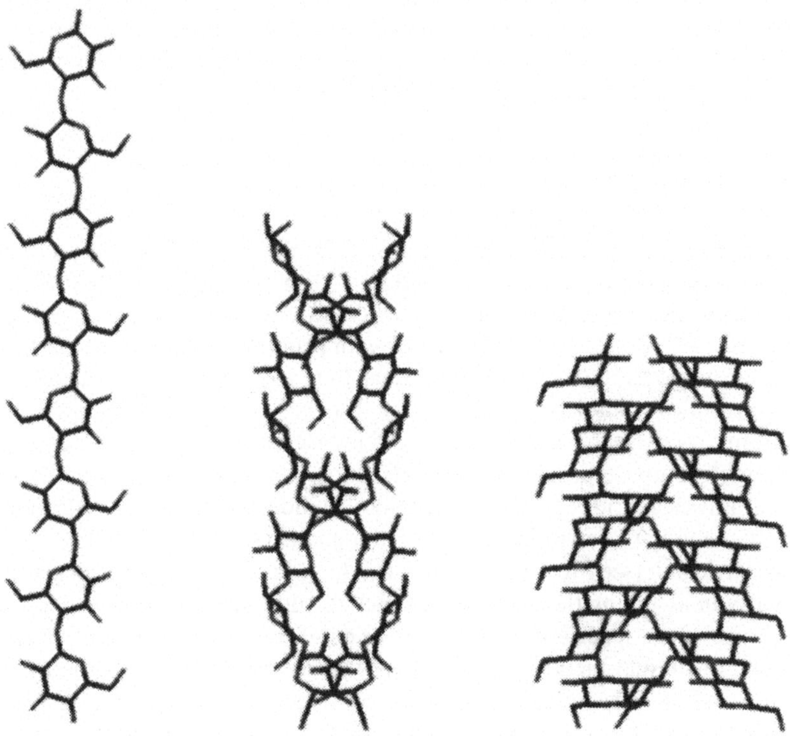

(a) (b) (c)

Figure 7.6 Schematic models of (a) a single helix formed by cellulose, (b) a double helix formed by amylose and (c) a triple helix formed by β(1 → 3)glucan.

When the links between the glucose units are not β(1 → 4), but α(1 → 4), as in amylose, double helices can be formed (Imberty *et al.*, 1988). Shorter and wider helices are observed in the triple helix formed by β(1 → 3) glucan (Deslandes *et al.*, 1980). Although the multiplicity of the helices in these examples is widely accepted, it is often not easy to determine this. For example, ι-carrageenan, agarose and xanthan are polysaccharides for which it is still not known unambiguously whether they are able to form double helices or not. The fiber diffraction patterns are not of sufficient quality to distinguish between single- and double-helical models.

The POLYS program (Engelsen *et al.*, 1996) has been written to facilitate model building of polysaccharides. The primary structure can be given in a simple ASCII syntax in accordance with the IUPAC nomenclature. This input is read and interpreted by the program, which can generate secondary and tertiary structures [making use of the previously mentioned carbohydrate fragment library (Pérez and Delage, 1991)] in the form of Cartesian coordinates, exported in many different possible formats. Intrinsic parameters such as the radius of gyration and the persistence length (Section 7.3.3) can easily be calculated; as well as the helical parameters. Recently, the program has been extended to include the calculation of multiple helices. Rotation or screw operations around the helix axis are performed to obtain double, triple, etc. helices from one single strand.

7.2.2 *Diffraction methods*

7.2.2.1 *X-ray fiber diffraction*

The most important method for the structural determination of crystalline polysaccharides is X-ray fiber diffraction. It has been observed that linear polysaccharides prefer to exist as long helices rather than as more convoluted structures. After dissolution, one usually can produce samples in which helical macromolecules are aligned with their long axes parallel or antiparallel (by convention, this is the *c* direction). Further lateral organization may occur, but rarely to the degree of a three-dimensionally ordered single crystal. Fibrous structures typically provide diffraction data of low resolution. Many helical polysaccharides yield no more than 50 independent X-ray reflections that can be used to determine the molecular geometry of the crystallographic asymmetric unit (see Figure 7.7). Other shortcomings of this approach are the difficulty in assigning the unit-cell parameters and the ambiguities regarding the choice of the space group. Also, the orientations of the individual chains (parallel or antiparallel) sometimes cannot be determined unambiguously.

Single helical structures are seen not only in the crystalline state. Some polysaccharides form in gels and more ordered states double or even triple helices. These multiple helices make the interpretation of the diffraction patterns even more difficult; the same pattern might be explained by both single- and double-helical models.

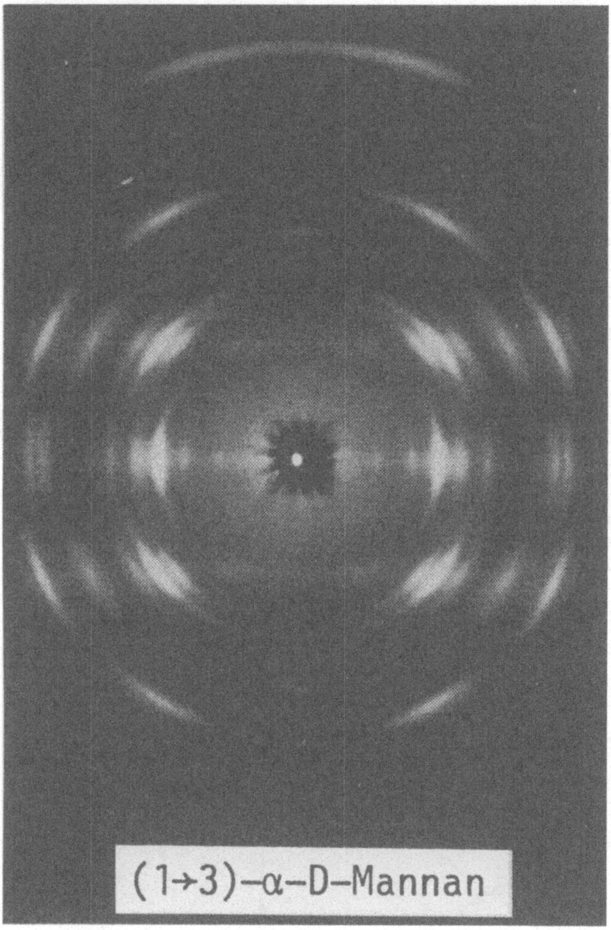

Figure 7.7 X-ray fiber diffractogram of α-(1 → 3)-D-mannan.

7.2.2.2 Electron diffraction

As with many other stereoregular polymers, simple linear polysaccharides, once dissolved and recrystallized, can yield single crystals (Figure 7.8). Polysaccharides often crystallize with the incorporation of water or other solvent molecules. In most cases, a well-defined morphology is obtained, the most common being plate-like. These thin lamellar surfaces have lateral dimensions of several micrometers for only a few tenths of an Ångström in thickness. Crystalline domains of such dimensions are well suited for examination by transmission electron microscopy both in imaging and in diffraction modes. A problem frequently encountered when studying crystalline biopolymers with the electron microscope relates to the vacuum dehydration of the specimen when inserted in the instrument column. This

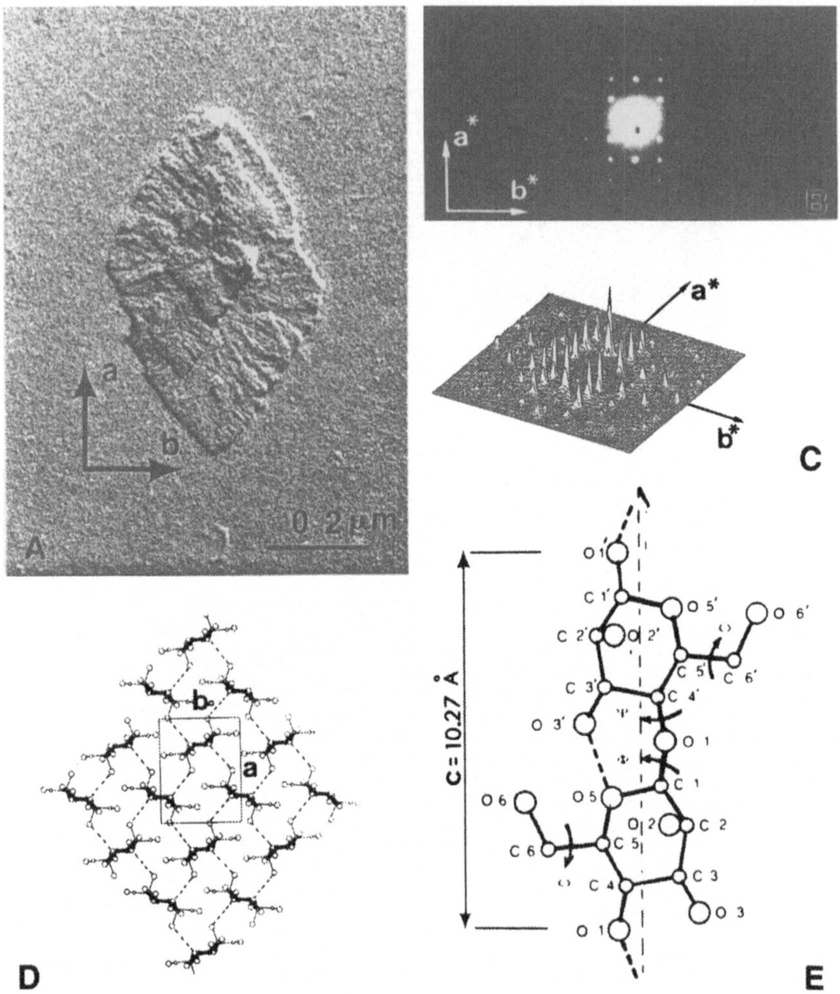

Figure 7.8 Single crystal of a polysaccharide: (A) mannan I; (B) its electron diffractogram properly orientated with respect to the crystals as in (A); (C) digitized electron diffractogram as in B after noise removal; (D) projection of the mannan I structure; (E) the mannobiose repeat unit viewed along the mannan chain axis. Reprinted from Pérez and Chanzy (1989).

is particularly critical when water or solvent is part of the crystalline structure. In such instances, total or partial decrystallization takes place, in a matter of minutes, accompanied by drastic distortion of the sample. Several methods have been developed whereby the sample is viewed either inside an hydration chamber (Hui and Parsons, 1974; Matricardi *et al.*, 1972) or quenched in a cryogenic bath prior to insertion into the electron microscope (Chanzy *et al.*, 1971; Taylor and Glaeser, 1974; Taylor *et al.*, 1975). In these instances, the observations are performed at a temperature close to that of boiling liquid nitrogen, where the water of crystallization is stable in high vacuum. With such a technique, frozen wet electron diffractograms are readily recorded.

7.2.2.3 Solving crystal structures of polysaccharides

In contrast to other macromolecules, the diffraction data that can be obtained from polysaccharides are not sufficient to permit crystal structure determination based on the data alone. A modeling technique must be used which allows the calculation of diffraction intensities from various models for comparison with the observed intensities. The first, computer-assisted and reasonably detailed structure analysis of a polysaccharide was completed more than 20 years ago. Now the number of apparently successful structure solutions of polysaccharides exceeds 100, counting all polymorphs, variants, derivatives and complexes (Chandrasekharan, 1997). The joint use of molecular modeling and diffraction techniques has been invaluable in the quantitative elucidation of crystals and molecular structure.

Providing that the data set is of sufficient quality and/or the unit-cell dimensions and space group symmetry are well assigned, the final stage of the structure determination involves the unit-cell content. A linked-atom description similar to that reported by Smith and Arnott (1978) or Zugenmaier and Sarko (1976) may be used. Optimization procedures are used to fit observed and calculated structure amplitudes (related to the intensities of the individual reflections) with simultaneous optimization of the non-bonded interactions and preservation of helix pitch and symmetry as well as ring closure. Interatomic energy functions, mimicking intermolecular non-bonded interactions (Williams, 1969), are used extensively for this purpose.

In the linked-atom least squares (LALS) procedure (Smith and Arnott, 1978) packing of the chains in the unit cell is described by the variable (μ) for orientation, and ($\bar{\omega}$) for translation along the helix axis. The procedure is aimed at minimizing the discrepancy function (Ω) which is given by:

$$\Omega = \sum W_m(|_{\mathrm{o}}F_m|^2 - K^2|_{\mathrm{c}}F_m|^2) + \sum \varepsilon_{ij} + \sum \lambda_{\mathrm{q}} G_{\mathrm{q}} \qquad (7.1)$$

The first terms seek to optimize agreement between the observed amplitude ${}_{\mathrm{o}}F_m$ and those calculated from the models ${}_{\mathrm{c}}F_m$; K is a scaling factor that places the observed amplitudes on the same scale as the calculated values; W_m is a weight and the summation is carried out over the m independent reflections. The second term provides a simple quadratic approximation to the non-bonded repulsion and is applied to all pairwise interactions between atoms i and j, when the distance between them is less than some standard values. The final term includes a set of coordinate constraints equations G_{q}, together with the initially undefined Lagrangian multipliers λ_{q}, and is used to preserve helix pitch and symmetry as well as, in the case of a flexible ring, the ring closure.

In another method (Zugenmaier and Sarko, 1976), both the monomer residue and the chain conformations are varied in such a way that all conformational and chain-packing features of the structures are optimized relative to the superimposed constraints and available data. In stereochemical refinements, the function contains terms for bond-length strain, bond-angle and conformational-angle strain and non-bonded contacts. It has the following form:

$$Y = \sum_{i=1}^{l}\left(\frac{r_i - r_{o_i}}{S_i^r}\right)^2 + \sum_{i=1}^{m}\left(\frac{\theta_i - \theta_{o_i} S_i^{\phi}}{S_i^{\theta}}\right)^2 + \sum_{i=1}^{n}\left(\frac{\phi_i - \phi_{o_i}}{S_i^{\phi}}\right)^2 + \frac{1}{W^2}\sum_{i=1,j=1}^{N} w_{ij}(d_{ij} - d_{o_{ij}}) \quad (7.2)$$

where r_i, θ_i and ϕ_i are, respectively, the bond length, bond angle and conformational angle for the l, m and n corresponding variables in the residue; with r_{o_i}, ϕ_{o_i}, θ_{o_i} and S_i^r, S_i^{ϕ}, S_i^{θ} the corresponding best (or average) values and their standard deviations. The fourth term approximates non-bonded repulsion, with d_{ij} the non-bonded distance between the atoms i and j, $d_{o_{ij}}$ the corresponding equilibrium distance and w_{ij} the weight of that contact in the function Y. The summation is over all N non-bonded distances taken pairwise. When the refinement is against observed X-ray intensity data, the function is simply the equation for any of the crystallographic R factors, the simplest choice of which is:

$$R = \sum \left| |F_o| - |F_c| \right| / \sum |F_o| \quad (7.3)$$

where F_o and F_c are the observed and calculated structure amplitudes, respectively. It is desirable to perform refinement against observed structure factors, while maintaining some stereochemical constraints. To do so, a linear combination of both functions (7.2) and (7.3) should be minimized in the form:

$$\zeta = f_x R + (1 - f_x) Y \quad (7.4)$$

where f_x is the fractional weight of the R factor in the linear combinations of R and Y.

A general scheme for defining and refining polysaccharide structures from either X-ray fiber diffraction data or electron diffraction data is given in Figure 7.9.

7.2.2.4 Placement of water molecules

On the basis of crystalline density, a putative number of water molecules is assumed, in agreement with the space group symmetry. Several strategies have been used in the location of these crystalline water molecules. In an

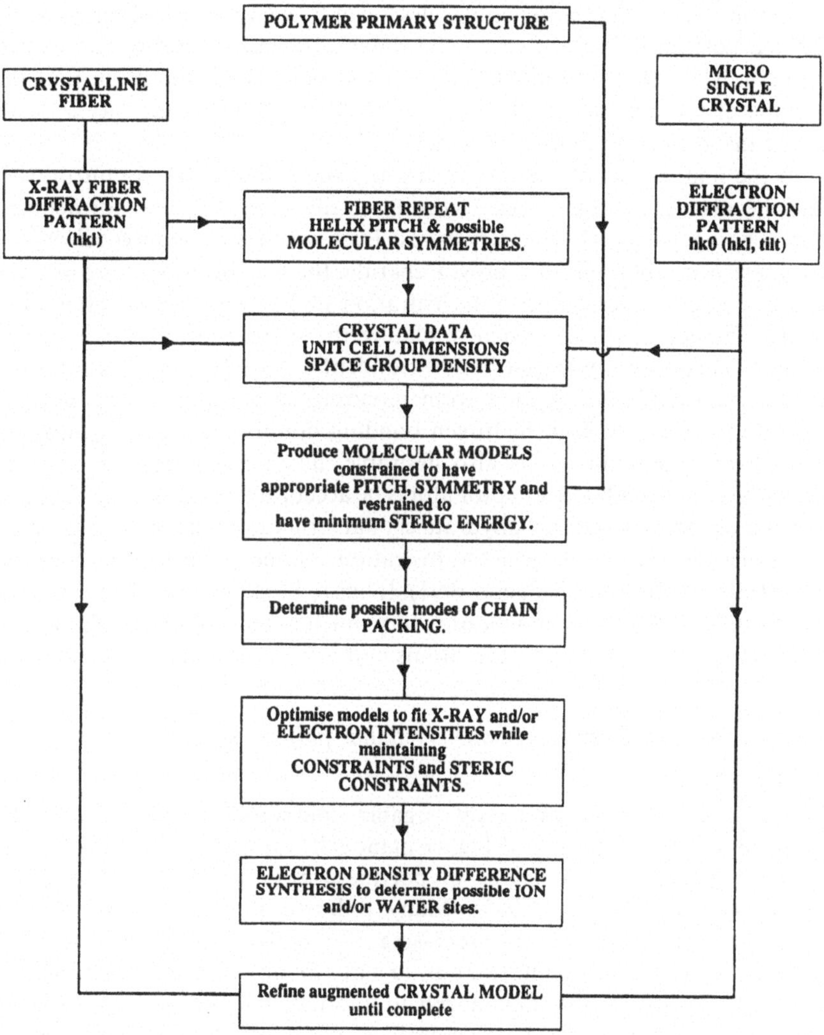

Figure 7.9 Flow chart for solving structures with X-ray fiber diffraction.

initial stage, the water positions are determined in the *ab* projection using the available *hk0* reflection intensities. The *ab* plane corresponding to the asymmetric unit is divided into small rectangular regions, and the *x,y*-coordinates of a single water molecule are refined in each such region in turn. This produces a weighted *R*-factor map (where *R* has its usual crystallographic meaning), which usually indicates the position of the water molecules. The *z* coordinate of each water molecule is then quickly refined with the *hkl* reflection intensities, starting from several trial positions.

In some instances, water molecules do not occupy crystallographically defined positions. In these cases, the water-weighted scattering factors can be used. The use of the latter factors is based on the assumption that the volume of the unit cell is filled with an electron gas whose density is equal to the mean electron density of water.

In the final stages of the refinement, the water molecules are introduced and the structure is minimized according to the general scheme for defining and refining polysaccharide structures. Even when an unequivocal crystal structure is established for a polysaccharide, the hydrogen bonding and the subsequent characterization of the hydration are likely to remain ambiguous, in the absence of direct evidence relating the orientation of the hydroxyl groups. Therefore, hydrogen bonds are identified by the O···O separation. It is a reasonable assumption from monosaccharide and oligosaccharide crystal structures that all hydrogen bonding donor functional groups will form hydrogen bonds. It is also desirable, albeit not necessary, that all potential hydrogen bond acceptor atoms be acceptors. Most of these features are derived from some carbohydrate crystal structural studies. However, as the molecules become larger, the hydration number increases as does the complexity (Jeffrey and Saenger, 1991). It has to be stated that characterizing water–polysaccharide interactions is very much hampered by the absence of information about the hydrogen atoms and is likely to remain speculative.

7.2.3 *Structural features of some crystalline polysaccharides*

7.2.3.1 *Cellulose*

Cellulose is the most abundant organic compound in nature and the most abundant compound of higher plant cell walls. It is a linear $\beta(1 \rightarrow 4)$ linked D-glucan with a degree of polymerization which is of the order of 10 000 (James *et al.*, 1985). Many different polymorphs are known from chemical treatment and processing. The native form is referred to as cellulose I. The chain conformation (a flat ribbon) remains unaffected in the conversion from one polymorph to another, but the packing patterns vary. The most important polymorph is called cellulose II, and can be obtained from the native form (irreversibly) mainly by two different processes: mercerization and regeneration. In the former process, discovered by Mercer in 1844, cellulose I is treated with a sodium hydroxide solution. The regenerating process starts with a similar treatment, but then carbon disulfide is used to produce the sodium xanthate derivative. This is dissolved, and cellulose II is regenerated from this solution by coagulation in an acid bath (Marchessault and Sundararajan, 1983).

An important difference in the two processes is that in the regeneration process the cellulose microfibrils are dissolved, and in the mercerization process they only swelled. This is of particular interest because of the packing of the cellulose chains in the polymorphs.

The determination of the different crystal structures is not straightforward, since the crystallinity of the samples is usually not very high and the diffraction patterns from different sources of cellulose may differ. Especially the structure of cellulose I is still under debate. The unit cell of native cellulose is most commonly seen as a monoclinic $P2_1$ cell with $a = 7.78$ Å, $b = 8.20$ Å, c (fiber axis) = 10.34 Å and $\gamma = 96.5°$ comprising two chains. The ribbon-like chains are arranged in sheets, stabilized by hydrogen bonds both in and between the chains. The chains are probably packed parallel. There are no hydrogen bonds between successive sheets, only van der Waals interactions (Woodcock and Sarko, 1980). More recently, cellulose I has been recognized to occur mostly as mixtures of two subclasses, Iα and Iβ. These two subclasses occur in different amounts depending on the origin of the cellulose; they have one and two chains per unit cell, respectively. The proposed unit cell dimensions are $a = 6.74$ Å, $b = 5.93$ Å, $c = 10.36$ Å, $\alpha = 117°$, $\beta = 113°$ and $\gamma = 81°$ for cellulose Iα and are $a = 8.01$ Å, $b = 8.17$ Å, $c = 10.36$ Å and $\gamma = 97.3°$ for cellulose Iβ (Sugiyama *et al.*, 1991).

The unit cell of cellulose II has generated less debate. It appears to be $P2_1$ too, with $a = 8.01$ Å, $b = 9.04$ Å, $c = 10.36$ Å and $\gamma = 117.1°$. It is likely that the chains are antiparallel, with additional stabilization (with respect to the cellulose I structure) by intersheet hydrogen bonding (Marchessault and Sundararajan, 1983). The conversion of parallel-packed chains in cellulose I to an antiparallel packing in cellulose II by the mercerization process, where no dissolvation but only swelling takes place, is very difficult to explain. Pictures of cellulose are shown in Figure 7.10.

7.2.3.2 Starch

Starch is nature's primary means of storing energy in green plants over long periods of time. X-ray patterns of native starch granules have shown several allomorphs. Generally, cereal starches give what is called the A-type, and tuber starches the B-type pattern.

The major part of the granule (often *c.* 75%) consists of amylopectin, an $\alpha(1 \rightarrow 4)$ linked glucan with $\alpha(1 \rightarrow 6)$ linked branch points. The other major macromolecular component is amylose, the linear $\alpha(1 \rightarrow 4)$ linked glucan. The three-dimensional structure of amylopectin is not yet known, but the crystalline structures of amylose are.

In A starch the glucan chains crystallize in the monoclinic space group *B*2 with cell dimensions $a = 21.24$ Å, $b = 11.72$ Å, $c = 10.69$ Å and $\gamma = 123.5°$ (Imberty *et al.*, 1988). The glucan chains form left-handed, parallel-stranded double helices, which are very compact; there is no room for water or any other molecule in the center. There are no hydrogen bonds in a single chain, but there is one between the two strands in a helix [Figure 7.11(a)]. The two double helices in the unit cell are parallel and connected through hydrogen bonds either directly or through the four water molecules in the cell. These

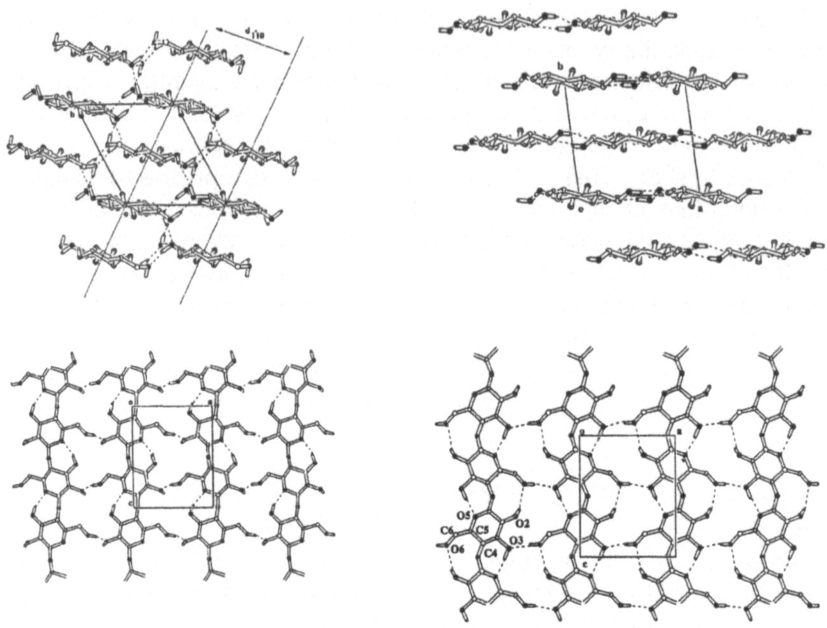

(a) (b)

Figure 7.10 Structures of (a) cellulose Iβ and (b) cellulose II viewed along the chain axis (top) and perpendicular to the chain axis (bottom). The boxes enclose the unit cells.

water molecules are buried deep in the crystal structure, and it is impossible to remove them without complete destruction of the crystalline structure.

In B starch, the chains are also organized in double helices. The structure differs in crystal packing and water content; the latter ranging from 10% to 50% (Imberty and Pérez, 1988). The unit cell is hexagonal with $a = b = 18.5$ Å and $c = 10.4$ Å. The double helices are connected through a network of hydrogen bonds that form a channel inside the hexagonal arrangement of six double helices. This channel is filled with water molecules. The structure is shown in Figure 7.11(b).

7.2.3.3 Chitin

Chitin is a naturally occurring fiber-forming polymer and is an essential component of the skeletal materials of many lower animals, most notably the arthropods. Chemically, it is a repeating polysaccharide of $\beta(1 \rightarrow 4)$ linked anhydro-2-acetamido-D-glucose. Polymorphs have been recognized, of which the α and β forms are the best characterized. Both have apparently the same 2_1 helical conformation.

The structure of the α-allomorph was determined from a fiber diffraction pattern recorded on a lobster tendon after proper deproteinization. The

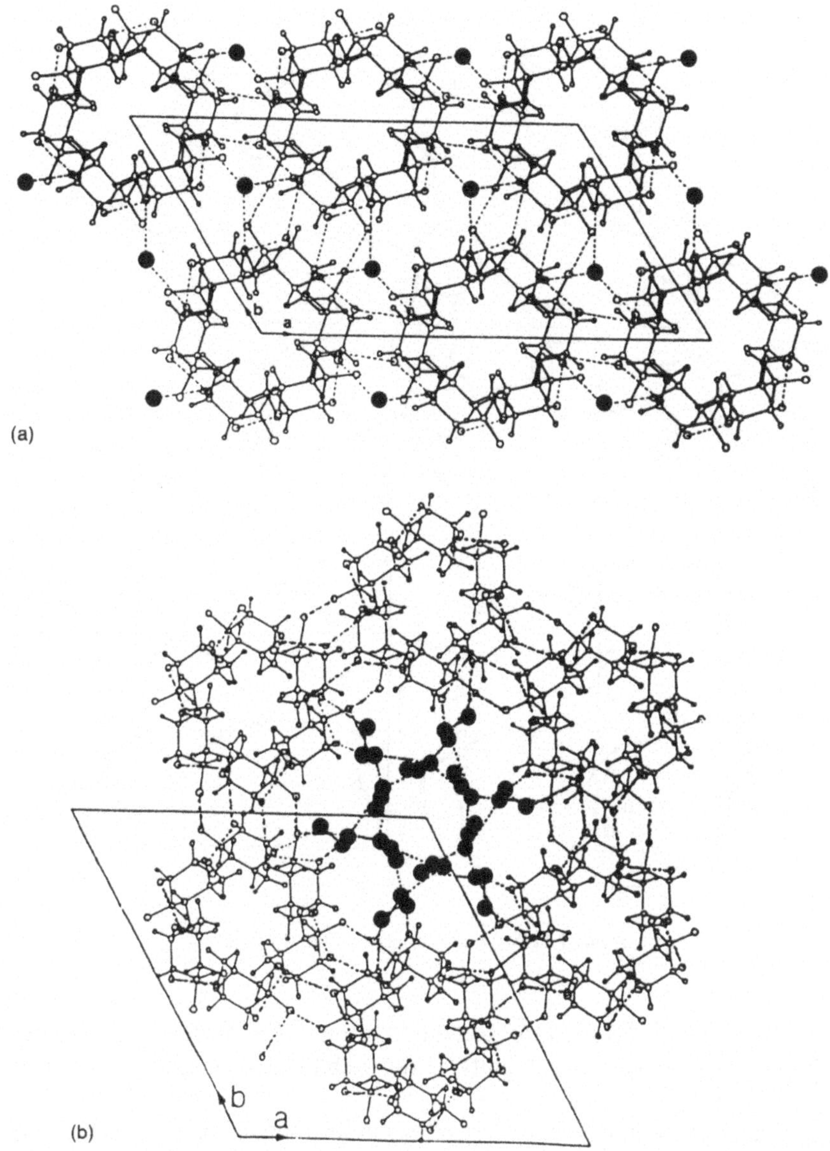

Figure 7.11 Structures of (a) A-type starch; (b) B-type amylose.

unit cell is orthorhombic with dimensions $a = 4.74$ Å, $b = 18.86$ Å and $c = 10.32$ Å. The assignment of the space group as $P2_12_12_1$ (orthorhombic) is ambiguous. Nevertheless, packing of polar chains in such a space group implies an antiparallel arrangement of the chains [Figure 7.12(a)]. The chains are so arranged to form sheets along the *a* axis; linked by N–H···O=C

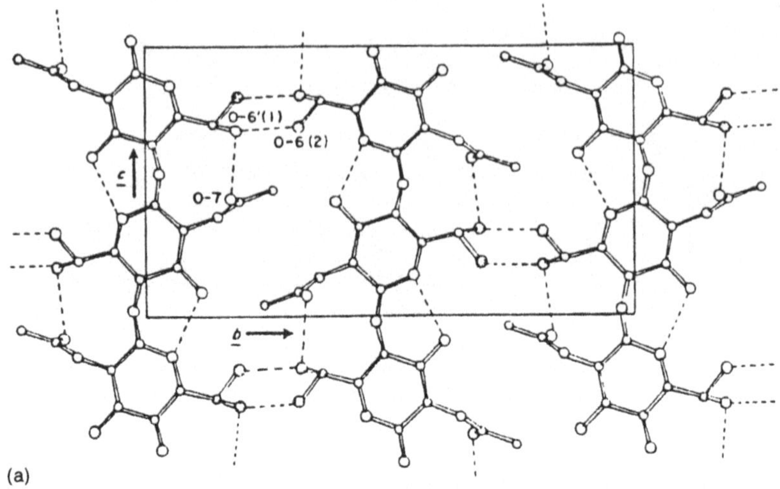

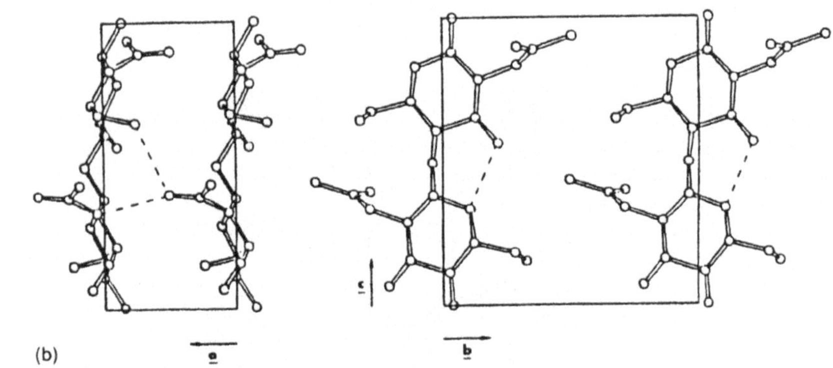

Figure 7.12 Structures of (a) α-chitin; (b) β-chitin.

hydrogen bonds between the acetamido groups. The existence of intersheet bonding can explain the stability of the α-chitin structure, in particular its inability to swell in water.

Highly crystalline β-chitin crystallizes in the monoclinic $P2_1$ space group, with cell parameters $a = 4.85$ Å, $b = 9.26$ Å, $c = 10.38$ Å (fiber axis) and $\gamma = 97.5°$; it is composed of an array of parallel chains. The structure refined for β-chitin is shown in Figure 7.12(b). The chains have a 2_1 helical conformation and are stacked along the *a* axis and linked by C=O···H–N hydrogen bonds. The primary hydroxyl groups are all hydrogen bonded to carbonyl groups in the same stack of chains. As such, there is no hydrogen bonding between adjacent stacks. Such a feature may explain why the structure is easily swollen by water molecules between the stacks of chains.

7.3 Secondary and tertiary structures of polysaccharides in solutions and gels

The ordered conformation adopted by a given polysaccharide in the solid state may not be retained in solution. Once dissolved, the polysaccharide chain tends to adopt a more or less coiled structure. In the absence of other enthalpic contributions, polymer–solvent interactions provide only a small increment to the mixing entropy, making the dissolution of a crystalline polysaccharide thermodynamically unfavorable. Actually, the disordered state results in a summation of a large number of disordered states which provides the required entropy contribution to the mixing free energy of the macromolecule. Such a dissolved random coil would fluctuate between different local and overall conformations (Figure 7.13). Theoretical treatment of the polysaccharide chains such as the worm-like model have been proposed which considers the polymer as a thin, slightly curved rod having descriptors such as the persistence length, the hydrodynamic diameter and the projection of the residue along the chain axis. The observable properties of dissolved polysaccharides normally reflect an average over the entire range of conformations accessible to the chain. Some averaged observable

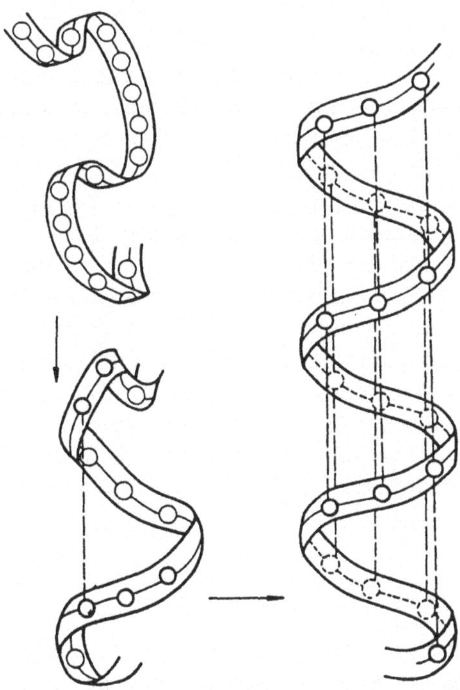

Figure 7.13 Carbohydrate chain, represented schematically as converting from random coil to helix, through intermediates in which one turn and then three turns of the helix have formed. The hydrogen bonds are shown by dotted lines (taken from Rees, 1977).

properties are: mean square radius of gyration, persistence length, dipole moment, optical activity and NMR coupling constants. Since many of the functional properties of polysaccharides are expressed either in solution or in gels, methods are required to characterize their levels of secondary and tertiary structures.

7.3.1 Chirooptical methods

The optical rotatory dispersion (ORD) and circular dichroism (CD), including ultraviolet (UV) techniques, are applicable to characterize changes in polysaccharide chains and/or their association. A polymer chromophore can absorb left and right circularly polarized lights to different extents. In CD, the differential absorption (expressed as a positive or negative quantity) is expressed as a function of the wavelength. ORD measures the differential refraction. ORD and CD are not independent and they yield in principle the same information about structure and conformation. In the investigation of polysaccharide shape, the importance of the CD effects arises from the fact that the contribution of a chromophore to the relative spectrum is influenced by its molecular neighborhood. Because the transitions associated with the conformational transitions of the backbone of a polysaccharide fall below 200 nm, special vacuum spectropolarimeters (VUCD) need to be used. Polysaccharide chain conformation can be probed under special circumstances. An additivity principle, in which optical rotation contributions are assigned to various structural and/or conformational units, is used. Therefore, the optical rotation of the overall molecule can be estimated by a summation. Empirical interpretations of optical activity have been proposed. A model based on interacting oscillators has been extended to the calculation of optical rotation of carbohydrates. The calculations refer to molecules in vacuum (Stevens and Sathyanarayana, 1987). For comparison with aqueous solution data, a solvent correction can be applied which takes into account its refractive index. Combined quantitative VUCD and ORD studies allow examination of helix–coil transitions of polysaccharides. Structurally sensitive carbohydrate transitions can be monitored directly. Results have been reported for most of the industrially important polysaccharides, such as agarose, alginate, ι-carrageeenan, hyaluronic acid and xanthan, particularly by Stevens and co-workers.

7.3.2 Nuclear magnetic resonance spectroscopy

Internuclear NMR effects are of two types: those transmitted through bonds (scalar effects such as coupling constants) and those transmitted through space [dipolar effects such as nuclear Overhauser effects (NOEs)]. Three bond coupling constants provide a source of information with regard to the torsional angles about the glycosidic linkages. As for NOEs, these through-space effects may be used to estimate internuclear distances, provided that the distance dependence of the effect is well understood. Altogether, these measurements provide structural information over a molecular distance scale but they do not yield any direct characterization of the dimension of the polysaccharide chain. Indirect measurement may provide some description of the chain dynamics and, consequently, some information about secondary structural elements may be derived. This may be achieved via the measurement of NMR relaxation of both ^{1}H and ^{13}C; it provides a useful way to characterize the motional modes of a polysaccharide chain in solution. A convenient index of the occurrence of conformational order in a polysaccharide chain is given through the use of spin–spin relaxation time T_2. This technique reflects the dissipation of the energy of the excited state to adjacent nuclei. T_2 values for polysaccharides exhibiting rapid segmental motions (milliseconds) are much larger than those exhibited by conformationally rigid segments in solution or gels (microseconds).

NMR methods have been used to characterize the gelation of polysaccharides even though they do not provide any direct information about the molecular crosslinking or the network organization. Upon gel formation, chain rigidification and immobilization follows, which is accompanied by a loss of the high-resolution NMR signal. High-resolution ^{13}C NMR of polysaccharides in the solid state has been applied to insoluble polysaccharides. The so-called cross-polarization/magic-angle-spinning (CP/MAS) technique could yield structural details on gels.

7.3.3 Light, X-ray and neutron scattering

In solution all scattering techniques measure the spherically averaged structure. Light, X-ray and neutron scattering can be treated by the same fundamental sets of equations. Scattering is due to concentration fluctuations which can be treated as a contrast between a given region and its neighbor. In the case of light scattering the contrast ($\Delta\rho$) is the polarizability or the refractive index difference between the solute and the solvent; it is the electron density in the case of X-ray scattering and the scattering length for neutron scattering. In all cases, the sample is illuminated by a collimated beam of radiation at wavelength λ, and scattered radiation at the same wavelength is measured as a function of 2θ. An important descriptor is the wave vector, $q = 4\pi \sin\theta/\lambda$. The scattering intensity is a function of the spatial organization of the macromolecules, $I(q) \simeq \Delta\rho^2 P(q)S(q)$, where $P(q)$ and $S(q)$ are related to features of the intramolecular and intermolecular organization, respectively. $P(q)$ is the form factor and $S(q)$ is the structure factor. The procedure is carried out over a wide range of polymer concentrations and wave vectors, that is, exploring the different ratios R/d, where R is the dimension of the macromolecule and d the distance between these macromolecules. Molecular weight estimates can be derived from the scattering intensity at

$q = 0$. The angular dependence of the scattering intensity for $q \rightarrow 0$ yields the radius of gyration of the sample. Measurements carried out at intermediate length scale, $a < r^{-1} < R$, where a is the local space correlation between monomers, provide information about the number of monomers inside the q^{-3} volume along with the solvent quality. The persistence length of a macromolecular chain (Lp) as well as its mass per unit length (M_l) can be determined. Seemingly, for $q^{-1} < Lp$, the radius of the chain cross-section Rc can be assessed. An illustration is shown in Figure 7.14 (Huggins and Benoit, 1994).

7.3.4 Rheology

When any real material is subjected to a stress, for example an external force, it will deform to a greater or lesser extent. For a perfect liquid, this deformation is continuous (it flows). A purely rigid solid will show almost no deformation. The measure of deformation is called strain, and the functional relation between stress (σ) and strain (γ) is a unique property of the material under investigation. Rheology is the science of this deformation and flow of matter (Darby, 1976).

Rheological experiments can be divided into two categories: small-deformation and large-deformation measurements. The former are used to probe viscoelastic properties in the linear regime; the structure is retained in these experiments. For example, the strain response of a sample to which an oscillatory stress is applied is measured. For perfect solids, stress and strain are in-phase: the greatest deformation occurs at maximal applied stress. For perfect liquids, however, the strain is at its maximum value when the rate of the applied stress is maximal too, so here stress and strain are out of phase. Polysaccharide solutions and gels show intermediate behavior. The degree of solid-like and liquid-like nature is characterized by the storage modulus G' and the loss modulus G'', respectively (Rees *et al.*, 1982). Although these quantities are almost independent of the applied frequency ω for true gels (where G' is generally much larger then G''), they are not for concentrated polysaccharide solutions. At low frequencies, flow occurs to accommodate the strain. At higher frequencies, however, the time scale is too rapid for molecular rearrangement, so the original structure is recovered substantially on release of the stress (elasticity). From the frequency dependence of G' and G'' information can be extracted about the number and the time scale of transient intermolecular associations.

The dynamic viscosity η^* is related to G', G'' and ω $\{\eta^* = [(G')^2 + (G'')^2]^{1/2}/\omega\}$ and decreases with increased frequency. Only at low frequencies, when the individual chains have sufficient time to re-entangle within the period of oscillation, does η^* remain constant. Next to the dynamic viscosity, the intrinsic viscosity $[\eta]$ is a useful parameter. It determines the molecular weight M through the relation $[\eta] = KM^{\alpha}$, where K and α are

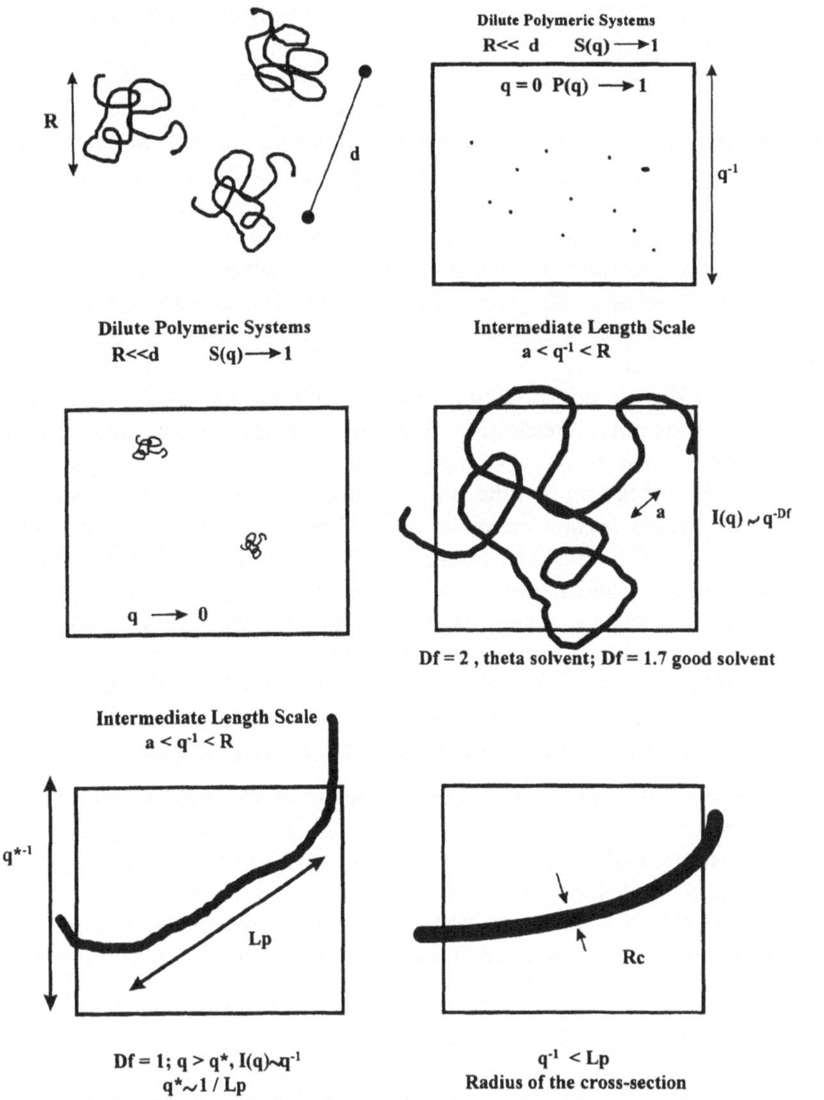

Figure 7.14 Illustration of the measurable quantities through scattering techniques for polysaccharide chains (see text for explanations). Taking into account the excluded volume parameter (∇; fractal dimension (DF) $= 1/\nabla$) $I(q) \simeq q^{-1/\nabla}$. In the case of θ solvent ($\nabla = \frac{1}{2}$, Df $= 2$), whereas ($\nabla = \frac{3}{5}$, Df $= \frac{5}{3}$).

constants. The intrinsic viscosity is determined by extrapolation from solutions of low concentration:

$$[\eta] = \lim_{c \to 0} \frac{\eta_{\text{solution}} - \eta_{\text{solvent}}}{c\eta_{\text{solvent}}}$$

Fundamentally, $[\eta]$ provides an index of random coil dimensions and hence of chain conformation (Rees *et al.*, 1982).

Large deformation experiments can be divided into those that do and those that do not rupture the gel. An example of the former class for gels is the determination of the yield stress, which is not a fundamental molecular property, unlike the above-mentioned parameters. It is the force required to fracture gel samples of given fixed geometry on compression between parallel plates. It is a valuable parameter in comparative studies of network properties in related systems (Rees *et al.*, 1982). For example, with competitive inhibition of network structure, primary structural requirements for the formation of stable interchain junctions in polysaccharide gels can be investigated (Morris *et al.*, 1980).

Two well-known non-breaking tests for polysaccharide gels are creep compliance and stress relaxation (both fundamental parameters). In the former test a constant stress is applied to the gel and the resulting strain is measured as a function of time (in creep and recoil experiments the stress is removed after a certain period of time). The creep compliance is defined as $J(t) \equiv \gamma(t)/\sigma$. In stress relaxation experiments an instantaneous strain is applied to the gel and maintained constant; now the corresponding stress as a function of time is measured. The elastic or shear modulus is defined as $G(t) \equiv \sigma(t)/\gamma$. Because of very slow changes in the stress relaxation and creep compliance with time, the measured equilibrium properties are actually 'pseudo-equilibrium' properties (Darby, 1976).

Another division of rheological experiments is given by Mitchell (1976):

- fundamental tests (for well-defined parameters such as elastic modulus or viscosity);
- empirical tests (the results obtained depend upon the geometry of the instrument employed);
- imitative tests (to measure properties under test conditions similar to those to which the material is to be subjected in practice).

7.3.5 *Marine polysaccharides*

Agarose is a linear, neutral polymer whose idealized structure is made up of alternating residues of a $1 \rightarrow 3$ linked β-D-galactopyranose and $1 \rightarrow 4$ linked 3,6-anhydro-α-L-galactopyranose (Figure 7.15). The polymer exists as random coils in dilute or hot solutions, but undergoes a conformational transition to form ordered helices in the solid state (Arnott *et al.*, 1974). The individual helices are thought to be stabilized by extensive aggregation (Morris and Norton, 1983). Already at low concentrations, agarose is able to form gels. It is assumed that impurities found in native agarose, such as the absence of the 3,6-anhydro bridge or the presence of methyl or sulfate esters, prevent the perfect ordering and introduce kinks in the helices.

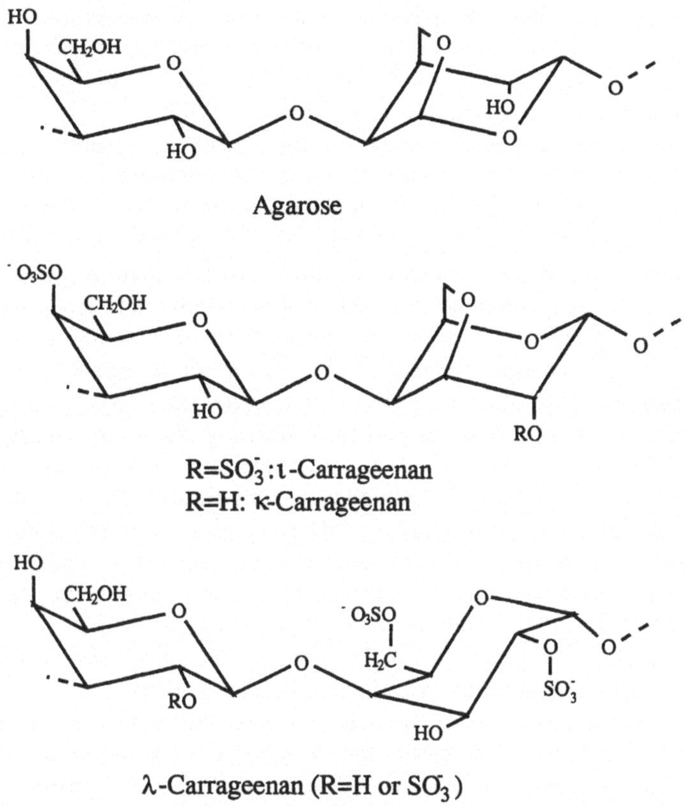

Figure 7.15 Schematic representations of the repeating units of agarose and carrageenans.

These kinks might be the cause of gel formation instead of precipitation (Clark *et al.*, 1987).

Watase and Arakawa (1968) performed stress relaxation studies of agar and agarose at a variety of concentrations, temperatures and pHs, and determined an activation energy of the main network structure in the gels of about 5.0 kcal/mol (*c.* 21 kJ/mol; constant throughout a broad pH range). Since this is the order of magnitude of the dissociation energy of hydrogen bonds, they concluded that hydrogen bonding plays a major role in the gelation mechanism.

Although the rheological implications of agarose network formation have not been investigated extensively, measurements of the shear modulus of agarose gels at various concentrations have been carried out (Clark *et al.*, 1983). It was shown that the critical concentration, C_0, below which no macroscopic gel is formed under the prevailing experimental conditions, is very low (*c.* 0.2% w/w). The storage modulus was proportional to C^n with n very much greater than 2 at concentrations close to C_0, but limiting to 2 at higher concentrations. Finally, the contribution of individual elastically active agarose network chains to G' is rather big, and appears to be caused by the shorter and stiffer segments of single agarose chains in agarose gels at higher concentrations (Clark and Ross-Murphy, 1987).

The primary structures of ι-, κ- and λ-carrageenans are shown in Figure 7.15. The main differences from agarose are the presence of a second D-galactose and sulfate groups. At high temperature, κ-carrageenan and agarose exist in solution as disordered coils, but a rigid, ordered form is found after cooling; they undergo a cooperative transition (Rees *et al.*, 1982). It is generally accepted that gels of these compounds are crosslinked by ordered aggregates of (double) helices, and that 'kinking' residues (residues with a changed geometry as a result of a missing anhydride bridge) promote formation of a three-dimensional network by limiting the length of the individual helices and thus allowing chains to participate in several different 'junction zones'. In λ-carrageenan such kinking residues predominate; no evidence of double-helix formation in the solid state has been shown, and in solution only disordered random coils are found under all conditions of temperature and ionic environment (Morris and Norton, 1983). The individual helices of agarose and κ-carrageenan are stabilized by the aggregation; the melting of aggregates and associated loss of gel structure on heating occurs at a higher temperature than does helix formation and gelation on cooling. With increasing charge density of the polysaccharide (ι-carrageenan > κ-carrageenan > agarose) the extent of aggregation and thermal hysteresis increases, the polymer concentration needed for gelation decreases and the resulting gels become increasingly brittle.

Dynamic viscosity measurements on κ-carrageenan in the presence of K^+ showed an initial maximum in the storage modulus during cooling and gel formation, the maximum being more pronounced at higher K^+ concentrations (Hermansson, 1989). It was suggested that double helices of K^+–κ-carrageenan form a fine network in the initial stage of gelation. This structure has the characteristics of a true gel, but is unstable. By lowering the temperature it is broken down partially by aggregation into ordered superstrands.

Although stabilization of ordered structures of ι-carrageenan by mono- or divalent ions has been reported (Morris *et al.*, 1983), this might be a result of impurities since native ι-carrageenan usually contains some κ-carrageenan too. Aqueous systems of pure ι-carrageenan do not show the thermal hysteresis; they appear to be at equilibrium at each temperature. This indicates that there is (almost) no interhelical aggregation, which is in contrast to the model of a coil-to-helix transition followed by the aggregation of domains of κ-carrageenan double helices. Therefore, the gelation of ι-carrageenan is not yet completely understood (Picullel *et al.*, 1992).

Despite the resemblances between agarose, ι-carrageenan and κ-carrageenan, mixtures of these polysaccharides do not always give homogeneous gels.

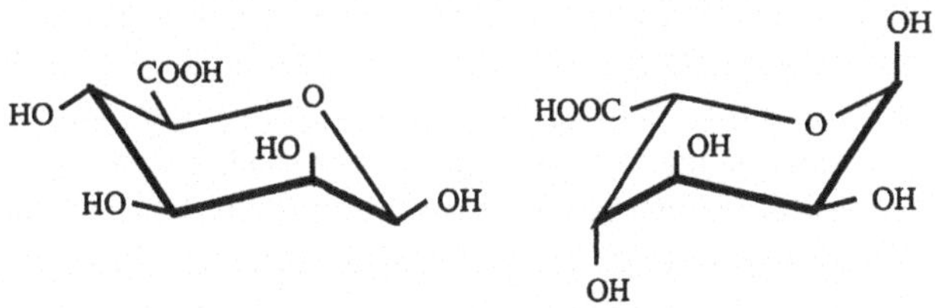

Figure 7.16 Building blocks of alginates: left, β-D-mannuronic acid, right, α-L-guluronic acid.

In a blend of agarose and κ-carrageenan investigated by Zhang and Rochas (1990) the networks of the individual polymers appeared to be conserved. Consequently, the blend was composed of interpenetrating networks interacting only by mutual entanglements. Parker *et al.* (1993) observed that mixed ι- and κ-carrageenan solutions show under certain circumstances (both Na^+ and K^+ present, in a Na^+/K^+ mole ratio of between 1 and 100) a 'two-step gelation', the ι form gelling first upon cooling. Two-step gelation is not observed for pure Na^+ or pure K^+ systems, but otherwise a phase-separated interpenetrating network is formed.

Goycoolea *et al.* (1995) investigated the rheological effect of the addition of certain plant polysaccharides to agarose and κ-carrageenan, namely, a galactomannan (locust bean gum) and a glucomannan (konjac mannan). It was already known for many years that these additions resulted in stronger gels and that they reduced the concentration needed for gelation. The rheological results confirm the model of the formation of a coupled network, in which the galactomannans (and related plant polysaccharides) are bound directly to the agarose or carrageenan double helix. Some aggregation of the algal polysaccharides is required for efficient binding, but extensive aggregation competes with heterotypic association. The heterotypic junctions appear to augment, rather than replace, the algal polysaccharide network. The galacto- or glucomannan chains suppress the back-reaction in the coil–helix transition of agarose and carrageenan, thus promoting the conformational ordering.

Alginates form strong, rigid gels with divalent cations, typically Ca^{2+} (Morris and Norton, 1983). Alginates consist of β-D-mannuronic acid and α-L-guluronic acid (Figure 7.16), occurring in either homo- or heteropolymeric sequences or blocks. Gel strength shows a progressive increase with the level of polyguluronate present, indicating a specific involvement of these sequences in interchain association. Both polymannuronate and the polymer with the heteropolymeric sequence remain soluble in the presence of Ca^{2+}, whereas polyguluronate forms stable dimers with Ca^{2+} ions according to the so-called egg-box model. Therefore, the former two sequences probably act as solubilizing, interconnecting regions between calcium polyguluronate junction zones in alginate gels (Grant *et al.*, 1973).

Smidsrød and Haug (1972) reported alginate gel rigidity studies as a function of polymer concentration, molecular weight and mannuronic (M)/guluronic (G) acid ratio. Systems rich in polyguluronate were rigid and very brittle; those rich in mannuronate gave very weak gels of high turbidity and greater extensibility. Increased deformability was ascribed to the presence of increasing amounts of MM and MG sequences between junction zones. Next to that, the modulus was found to increase with molecular weight until a degree of polymerization of about 500; then it became almost constant. The storage modulus was found to be proportional to the square of the concentration. It was shown that the rigidity modulus of gels formed by different cations is directly dependent on their ability to bind to the polyuronides by a cooperative interchain binding mechanism (Smidsrød, 1974). Such an autocooperative mechanism was found for the binding of Ca^{2+} by GG blocks, unlike MM or GM blocks. Segeren *et al.* (1974) found with dynamic viscoelastic studies that the modulus of tested alginate gels was independent of the frequency in the range 10^{-1} to 10^{1} Hz; that the ratio G'/G'' was very small and that G'' increased linearly with the temperature in the range 22°C to 49°C. Together with the data from Smidsrød, this led them to test the validity of the classical theory of rubber elasticity (this implies the presence of long-chain molecules that can form statistical coils between so-called junction zones in which there are chain associations and by which the three-dimensional network is built up). Although some data justified this theory, other data led to the rejection of it. They assumed that the apparent validity was caused by opposing effects during deformation, such as the breaking of weaker bonds and the establishment of new stronger intersegmental contacts.

7.3.6 Plant polysaccharides

Pectins are a very complex class of cell wall polysaccharides. The backbone exists of homogalacturonan (1,4-linked α-D-galacturonic acid residues) or rhamnogalacturonan (alternating 4-linked α-D-galacturonic acid and 2-linked α-L-rhamnopyranose). Many different side-chains can be attached to the rhamnose units. An important feature of the galacturonans is the degree of methylation, defined as the number of moles of esterified methanol per 100 moles of galacturonic acid.

High-methoxyl pectins form gels in the presence of sugar or polyols at low pH. The electrostatic chain repulsion is diminished by the acid (pectins with more methoxyl esters need less acid to gel; completely methoxylated pectin will gel without any acid), and the presence of sugar creates conditions of low water activity which in turn promote chain–chain rather than chain–solvent interaction. Higher molecular weight increases a gel's strength; breaking strength is increased more than rigidity (elasticity). The breaking strength gives a good correlation with the weight average molecular weight; apparently, it depends on the number of junction zones per single (long-)chain molecule. The rigidity, however, appears to depend more on the number of junction zones per unit gel volume: it correlates well with the number average molecular weight (longer-chain molecules contribute more to the weight average than to the number average molecular weight). The junction zones are stabilized both by hydrogen bonds, between undissociated carboxyl and hydroxyl groups, and by hydrophobic interactions between methoxyl groups; sucrose fortifies both types of binding forces. The esterification of hydroxyl groups with acetyl groups prevents gelation (Voragen *et al.*, 1995).

Low-methoxyl (less than 50%) pectins form gels in the presence of divalent ions such as calcium. Intermolecular junction zones between homogalacturonic regions of different chains are formed which can be compared with polyguluronate–calcium association, although it has not been shown yet that the egg-box model is valid for pectins too. The nature, distribution and amounts of substituents along the galacturonic backbone influence the resulting gel: a blockwise distribution of free carboxyl groups has a high sensitivity for calcium; this affinity is decreased by acetylation; and amidation improves the gelling ability. Next to that, temperature, pH, ionic strength and amount of calcium added also influence the gelation process (Voragen *et al.*, 1995). Braudo *et al.* (1992) investigated binding isotherms of Ca^{2+} with sodium pectate (degree of methylation, DM, 0%) and sodium pectinate (DM, 58%). They found that the binding of Ca^{2+} by the pectinate was close to that predicted for Coulombic binding, whereas the binding by the pectate was much stronger. Apparently, a cooperativety exists for pectate as a result of both the Coulombic interaction and the coordinate binding of the Ca^{2+}. This resembles the findings for alginates by Smidsrød (1974, *vide supra*).

Mixtures of pectins and alginates also form gels; the properties depend mainly on the mannuronic/guluronic acid ratio of the alginate, the degree of methoxylation of the pectin and the pH. Interaction appears only to occur when the chains are sufficiently uncharged, at low electrostatic repulsion, namely, at $pH < 4$ for high-methoxyl pectins and $pH < 2.8$ for low-methoxyl pectins (Voragen *et al.*, 1995).

The gelling behavior of galactomannan (linear 1,4-linked chains of β-D-mannopyranosyl residues to which single α-D-galactopyranosyl side-chains are linked at position 6) solutions decreases systematically with increasing galactose content. It is likely that aggregation occurs through unsubstituted regions (Morris and Norton, 1983). They do not form strong gels. Apparently, in general, 'hyperentanglement' networks are formed rather than crosslinks through specific interchain associations (Morris *et al.*, 1981). Strong gels are formed, however, when galactomannans are mixed with other types of polysaccharides, such as agarose or carrageenan (*vide supra*) or xanthan (*vide infra*) (Clark and Ross-Murphy, 1987).

Although amylose is insoluble in cold water, it can be dissolved in alkali or by heating in water to approximately 150°C. Neutralization or cooling leads to precipitation, either directly or through an unstable gel (Morris and Norton, 1983). Network formation by linear amylose chains has been followed using a combination of turbidity, X-ray diffraction and shear modulus measurements (Miles *et al.*, 1984). They found that both the shear modulus and the turbidity increased rapidly when amylose solutions were quenched from 90°C to 32°C; the recovery of crystallinity was slow and was still incomplete after many hours. Apparently, when hot amylose solutions are cooled, rapid network formation occurs through the formation of ordered junction zones based on amylose double helices. Subsequently, the aggregates of helices become bigger and develop into crystallites (Clark and Ross-Murphy, 1987).

Another component of starch, amylopectin (readily soluble in water), was rheologically investigated by Ring (1985). When solutions are cooled, it aggregates very slowly. It is suggested that the high molecular weight, and highly branched, amylopectin species slowly associate, with a concomitant crystallization of short branches situated near their surfaces. The mechanical properties of a starch gel differ from those of pure amylose or pure amylopectin gels. Those gels are based on an amylose network in which the swollen amylopectin granules are suspended (Clark and Ross-Murphy, 1987).

7.3.7 Microbial polysaccharides

Some bacterial polysaccharides, such as curdlan [a linear $\beta(1 \rightarrow 3)$glucan] and gellan (a linear, anionic polytetrasaccharide), can form strong gels like alginates and carrageenans, probably through the association of ordered junction zones (Clark and Ross-Murphy, 1987). Gellan gels in the presence of salts; a thermally reversible transition from random coils to (probably double) helices takes place with no detectable hysteresis (Lapasin and Pricl, 1995).

Xanthan has a backbone similar to cellulose, but a charged trisaccharide is attached to every second glucose residue (Figure 7.17). It can form weak gels. Generally, solutions of disordered polysaccharides flow in response to a small stress. Solutions of xanthan in the ordered-chain conformation appear to require a minimum amount of yield stress before they get freely mobile. This yield stress can be enhanced by addition of salt, which indicates a cation-mediated non-covalent association of chains into a weak but stable network structure (Morris and Norton, 1983). Combined rheo-optical and viscoelastic measurements appear to confirm that intermolecular ion and hydrogen bonds are involved (Morris *et al.*, 1983). Although the rheological behavior depends on the presence of a microgel fraction, the number and species of cations present and the ionic strength, the 'weak gel' rheological response does involve the shear breakdown of a supramolecular aggregate.

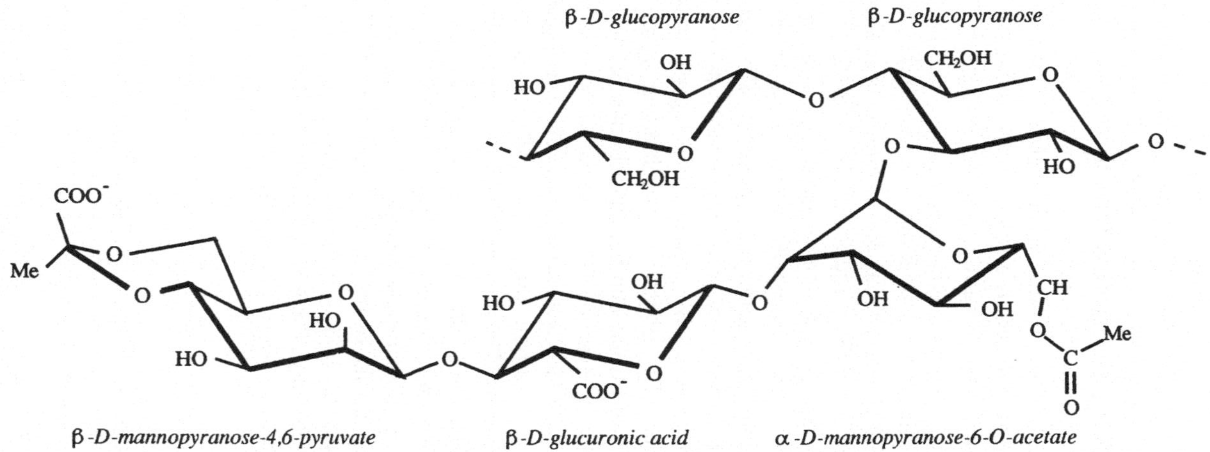

Figure 7.17 Repeating unit of xanthan.

The fast recovery of this structure implies the reformation of non-covalent intermolecular bonds (Clark and Ross-Murphy, 1987).

As was described before for agarose and carrageenan, the addition of locust bean gum (LBG) changes the gelling behavior of xanthan. When the two polysaccharides were mixed at a temperature under the helix–random-coil transition temperature of xanthan, weak elastic gels resulted which depended only slightly on the degree of galactose substitution in the galactomannan. However, when the mixtures were heated the rheological properties of the resulting strong elastic gels depended highly upon this degree of substitution (Mannion *et al.*, 1992). Lundin and Hermansson (1995) investigated the temperature dependence of mixtures of xanthan and LBG with different degrees of galactose substitution with transmission electron microscopy and small deformation viscoelastic measurements. They conclude that the supramolecular xanthan structure (xanthan superstrands) is not structurally influenced by the presence of LBG, but that the superstrands are connected by bridges of smaller LBG polymers. The binding areas on these LBG polymers are the unsubstituted mannan backbones. With a high degree of galactose substitution [a mannose (M) to galactose (G) ratio of 3], the unsubstituted parts of the LBG backbone are relatively small; the polymers are excluded from the xanthan aggregates and they can bind only to the surface of the superstrands. The polymers with a lower degree of substitution (an M to G ratio of 5) are not excluded from the aggregates as a result of their greater tendency to interact. Therefore, they can bind both to the surface of the superstrands and to the xanthan helices. When mixed cool, both types of LBG will bind to the surface of the xanthan superstrands. This leads to the same type of gels (with rheological properties completely different from pure xanthan gels). After heating, the low substituted LBG–xanthan mixture is reordered into a network with increased strength.

References

Arnott, S., Fulmer, A., Scott, W.E. *et al.* (1974) The agarose double-helix and its function in agarose gel structure. *Journal of Molecular Biology*, **90**, 269–84.

Brant, D.A. (1976) Conformational theory applied to polysaccharide structure. *Quarterly Reviews of Biophysics*, **9**, 527–96.

Braudo, E.E., Soshinsky, A.A., Yuryev, V.P. and Tolstoguzov, V.B. (1992) The interaction of polyuronides with calcium ions. 1: Binding isotherms of calcium ions with pectic substances. *Carbohydrate Polymers*, **18**, 165–9.

Chandrasekharan, R. (1997) Molecular architecture of polysaccharide helices in oriented fibres. *Adv. Carb. Chem. Biochem.*, **52**, 311–439.

Chanzy, H., Roche, E. and Vuong, R. (1971) Electron diffraction of cellulose triacetate single crystals. *Kolloid-Zeitschrift & Zeitschrift für Polymere*, **248**, 1034–5.

Clark, A.H. and Ross-Murphy, S.B. (1987) Structural and mechanical properties of biopolymer gels. *Advances in Polymer Science*, **83**, 57–192.

Clark, A.H., Richardson, R.K., Ross-Murphy, S.B. and Stubbs, J.M. (1983) Structural and mechanical properties of agar/gelatin co-gels. Small-deformation studies. *Macromolecules*, **16**, 1367–74.

Cremer, D. and Pople, J.A. (1975) Molecular orbital theory of the electronic structure of organic compounds. XXIII. Pseudorotation in saturated five-membered ring compounds. *Journal of the American Chemical Society*, **97**, 1358–67.

Darby, R. (1976) *Viscoelastic Fluids: An Introduction to Their Properties and Behavior*, Marcel Dekker, New York.

Deslandes, Y., Marchessault, R.H. and Sarko, A. (1980) Triple-helical structure of (1 → 3)-β-D-glucan. *Macromolecules*, **13**, 1466–71.

Engelsen, S.B., Cros, S., Mackie, W. and Pérez, S. (1996) A molecular builder for carbohydrates: application to polysaccharides and complex carbohydrates. *Biopolymers*, **39**, 417–33.

French, A.D. (1988) Rigid- and relaxed-residue conformational analyses of cellobiose using the computer program MM2. *Biopolymers*, **27**, 1519–25.

French, A.D. and Murphy, V.G. (1977) Intramolecular changes during polymorphic transformations of amylose. *Polymer*, **18**, 489–94.

French, A.D., Tran, V.H. and Pérez, S. (1990) Conformational analysis of a disaccharide (cellobiose) with the molecular mechanics program (MM2), in *Computer Modeling of Carbohydrate Molecules* (eds A.D. French and J.W. Brady), ACS Symposium Series **430**, American Chemical Society, Washington, DC, pp. 191–212.

Gagnaire, D., Pérez, S. and Tran, V.H. (1982) Configurational statistics of single chains of α-linked glucans. *Carbohydrate Polymers*, **2**, 171–91.

Goycoolea, F.M., Richardson, R.K., Morris, E.R. and Gidley, M.J. (1995) Effect of locust bean gum and konjac glucomannan on the conformation and rheology of agarose and κ-carrageenan. *Biopolymers*, **36**, 643–58.

Grant, G.T., Morris, E.R., Rees, D.A. *et al.* (1973) Biological interactions between polysaccharides and divalent cations: the egg-box model. *FEBS Letter*, **32**, 195–206.

Ha, S.N., Madsen, L.J. and Brady, J.W. (1988) Conformational analysis and molecular dynamics simulations of maltose. *Biopolymers*, **27**, 1927–52.

Hermansson, A.-M. (1989) Rheological and microstructural evidence for transient states during gelation of kappa-carrageenan in the presence of potassium. *Carbohydrate Polymers*, **10**, 163–81.

Huggins, J.S. and Benoit H.C. (1994) *Polymers and Neutron Scattering*, Oxford Sciences, Oxford.

Hui, S.W. and Parsons, D.F. (1974) Electron diffraction of wet biological membranes. *Science*, **184**, 77–8.

Imberty, A. and Pérez, S. (1988) A revisit to the three-dimensional structure of B-type starch. *Biopolymers*, **27**, 1205–21.

Imberty, A., Chanzy, H., Pérez, S. *et al.* (1988) New three-dimensional structures for A-type starch. *Journal of Molecular Biology*, **201**, 365–78.

Imberty, A., Tran, V.H. and Pérez, S. (1989) Relaxed potential energy surfaces of *N*-linked oligosaccharides: the mannose-α(1 → 3)-mannose case. *Journal of Computational Chemistry*, **11**, 205–16.

IUPAC–IUB Joint Commission on Biochemical Nomenclature: Nomenclature of Carbohydrates. *Pure Appl. Chem.* (1996) **68**, 1919; *Carbohydr. Res.* (1977) **297**, 1–92.

James D.W., Jr, Preiss, J. and Elbein, A.D. (1985) Biosynthesis of polysaccharides, in *The Polysaccharides, Vol. 3* (ed. G.O. Aspinall), Academic Press, New York, pp. 107–207.

Jeffrey, G.A. and Saenger, W. (1991) *Hydrogen Bonding in Biological Structures*, Springer, Berlin.

Kakudo, M. and Kasai, K. (1972) *X-ray Diffraction by Polymers*, Elsevier Publishing Company, Amsterdam, pp. 303–21.

Lapasin, R. and Pricl, S. (1995) *Rheology of Industrial Polysaccharides: Theory and Applications*, Blackie Academic & Professional, London.

Lundin, L. and Hermansson, A.-M. (1995) Supramolecular aspects of xanthan-locust bean gum gels based on rheology and electron microscopy. *Carbohydrate Polymers*, **26**, 129–40.

Mannion, R.O., Melia, C.D., Launay, B. *et al.* (1992) Xanthan/locust bean gum interactions at room temperature. *Carbohydrate Polymers*, **19**, 91–7.

Marchessault, R.H. and Sundararajan, P.R. (1983) Cellulose, in *The Polysaccharides, Vol. 2* (ed. G.O. Aspinall), Academic Press, New York, pp. 11–95.

Matricardi, V.R., Moretz, R.C. and Parsons, D.F. (1972) Electron diffraction of wet proteins. Catalase. *Science*, **177**, 268–70.

Miles, M.J., Morris, V.J. and Ring, S.G. (1984) Some recent observations on the retrogradation of amylose. *Carbohydrate Polymers*, **4**, 73–7.

Mitchell, J.R. (1976) Rheology of gels. *Journal of Texture Studies*, **7**, 313–39.

Morris, E.R. and Norton, I.T. (1983) Polysaccharide aggregation in solutions and gels, in *Aggregation Processes in Solution* (eds. E. Wyn-Jones and J. Gormally), Elsevier, Amsterdam, pp. 549–93.

Morris, E.R., Cutler, A.N., Ross-Murphy, S.B. *et al.* (1981) Concentration and shear rate dependence of viscosity in random coil polysaccharide solutions. *Carbohydrate Polymers*, **1**, 5–21.

Morris, E.R., Rees, D.A., Robinson, G. and Young, G.A. (1980) Competitive inhibition of interchain interactions in polysaccharide systems. *Journal of Molecular Biology*, **138**, 363–74.

Morris, V.J., Franklin, D. and I'Anson, K. (1983) Rheology and microstructure of dispersions and solutions of the microbial polysaccharide from *Xanthomonas campestris* (xanthan gum). *Carbohydrate Research*, **121**, 13–30.

Parker, A., Brigand, G., Miniou, C. *et al.* (1993) Rheology and fracture of mixed ι- and κ-carrageenan gels: two-step gelation. *Carbohydrate Polymers*, **20**, 253–62.

Pérez, S. (1978) *Analyse cristallographique de structures polymères: Conception et critique de nouveaux systèmes d'information*, DSc. Thesis, University of Grenoble.

Pérez, S. and Chanzy, H. (1989) Electron crystallography of linear polysaccharides, *Journal of Electron Microscopy Technique*, **11**, 280–5.

Pérez, S. and Delage, M.M. (1991) A database of three-dimensional structures of monosaccharides from molecular mechanics calculations. *Carbohydrate Research*, **212**, 253–9.

Picullel, L., Nilsson, S. and Muhrbeck, P. (1992) Effects of small amounts of kappa-carrageenan on the rheology of aqueous iota-carrageenan. *Carbohydrate Polymers*, **18**, 199–208.

Rao, V.S.R., Sundararajan, P.R., Ramakrishnan, C. and Ramachandran, G.N. (1967) Conformational studies of amylose, in *Conformation in Biopolymers. Vol. 2* (ed. G.N. Ramachandran), Academic Press, London, pp. 721–37.

Rees, D.A., (1977) *Polysaccharide Shapes* (Outline Studies in Biology Series), Chapman & Hall, London.

Rees, D.A., Morris, E.R., Thom, D. and Madden, J.K. (1982) Shapes and interactions of carbohydrate chains, in *The Polysaccharides, Vol. 1* (ed. G.O. Aspinall), Academic Press, New York, pp. 195–290.

Ring, S.G. (1985) Observations on the crystallization of amylopectin from aqueous solution. *International Journal of Biological Macromolecules*, **7**, 253–4.

Robijn, G.W., Imberty, A., van den Berg, D.J.C. *et al.* (1996) Predicting helical structures of the exopolysaccharide produced by *Lactobacillus Sake. Carbohydr. Res.*, **288**, 57–74.

Stevens, E.S. and Sathyanarayana, B.K. (1987) A semi-empirical theory of saccharide optical activity. *Carbohydr. Res.*, **166**, 181–93.

Segeren, A.J.M., Boskamp, J.V. and van den Tempel, M. (1974) Rheological and swelling properties of alginate gels. *Faraday Discussions of the Chemical Society*, **57**, 255–62.

Smidsrød, O. (1974) Molecular basis for some physical properties of alginates in the gel state. *Faraday Discussions of the Chemical Society*, **57**, 263–74.

Smidsrød, O. and Haug, A. (1972) Properties of poly(1,4-hexuronates) in the gel state. II. Comparison of gels of different chemical composition. *Acta Chemica Scandinavica*, **26**, 79–88.

Smith, P.J.C. and Arnott, S. (1978) LALS: A linked-atom least-squares reciprocal-space refinement system incorporating stereochemical restraints to supplement sparse diffraction data. *Acta Crystallographica*, **A34**, 3–11.

Sugiyama, J., Vuong, R. and Chanzy, H. (1991) Electron diffraction study on the two crystalline phases occurring in native cellulose from an algal cell wall. *Macromolecules*, **7**, 4168–75.

Taylor, K.A. and Glaeser, R.M. (1974) Electron diffraction of frozen hydrated protein crystals. *Science*, **186**, 1036–7.

Taylor, K.J., Chanzy, H. and Marchessault, R.H. (1975) Electron diffraction for hydrated crystalline biopolymers: nigeran. *Journal of Molecular Biology*, **92**, 165–7.

Tran, V.H., Buléon, A., Imberty, A. and Pérez, S. (1989) Relaxed potential energy maps of maltose. *Biopolymers*, **28**, 679–90.

Tran, V.H. and Brady, J.W. (1990) Disaccharide conformational flexibility. I. An adiabatic potential energy map for sucrose. *Biopolymers*, **29**, 961–76.

Tvaroska, I. and Pérez, S. (1986) Conformational-energy calculations for oligosaccharides: a comparison of methods and a strategy of calculation. *Carbohydrate Research*, **149**, 389–410.

Voragen, A.G.J., Pilnik, W., Thibault, J.-F. *et al.* (1995) Pectins, in *Food Polysaccharides and Their Applications* (ed. A.M. Stephen), Marcel Dekker, New York, pp. 287–339.
Watase M. and Arakawa, A. (1968) Rheological properties of hydrogels of agar-agar. III. Stress relaxation of agarose gels. *Bulletin of the Chemical Society of Japan*, **41**, 1830–4.
Williams, D.E. (1969) A method of calculating molecular crystal structures. *Acta Crystallographica*, **A25**, 464–70.
Woodcock, C. and Sarko, A. (1980) Packing analysis of carbohydrates and polysaccharides. 11. Molecular and crystal structure of native ramie cellulose. *Macromolecules*, **13**, 1183–7.
Zhang, J. and Rochas, C. (1990) Interactions between agarose and κ-carrageenans in aqueous solutions. *Carbohydrate Polymers*, **13**, 257–71.
Zugenmaier, P. and Sarko, A. (1976) Packing analysis of carbohydrates and polysaccharides. IV. A new method for detailed crystal structure refinement of polysaccharides and its application to V-amylose. *Biopolymers*, **15**, 2121–36.

8 Chemistry of polysaccharide modification and degradation

MANSSUR YALPANI

8.1 Introduction

Polysaccharides offer a vast array of primary structures and conformations. Their ubiquitous presence and diverse roles in biological systems bear testimony to the wide spectrum of functionalities. It has long been recognized that the characteristics and utility of native polysaccharides can be substantially extended by chemical modification. The alterations of parameters, such as primary structure, conformation, hydrophilicity, solubility, polyelectrolyte nature, rheology and stability may profoundly impact their biological, chemical and physical properties. Such modulations can therefore be employed to tailor functional properties. As a result, a wealth of synthetic technology has been established since the discovery by John Hyatt of the first glycan derivative, cellulose nitrate, in 1869.

A large body of knowledge is available on polysaccharide modifications [1, 2]. The useful properties of these hydrophilic polymers have been exploited for numerous new uses and have led to a proliferation of synthetic methodology in recent years. The synthetic techniques for polysaccharide modifications can be broadly classified as either non-selective or selective. The former methods introduce substituents more or less randomly along the polymer chain and within each repeat unit, whereas the latter target specific functional groups. Substituent distribution patterns in non-selectively modified derivatives have been the subject of many investigations. They are found to vary considerably, depending on the type of derivatization and the reaction parameters involved. For most ether and ester derivatives, higher equilibrium rate constants are observed for substitution at O-2 and O-6 than at O-3. A detailed discussion of these parameters is given elsewhere (Sections 8.3.1 and 8.3.2) [2].

The substitution pattern within the polymeric units can affect the macromolecular properties to the same extent as the overall uniformity of substitution [2]. The majority of the early synthetic techniques were developed for cellulose, starch and other selected glycans, and achieved non-selective modifications of polysaccharide functionalities. This approach is in large measure a consequence of limitations imposed by their native morphology and the often modest reactivity differences of the polyhydroxylated substrates. Numerous methods for selective and homogeneous modifications have been introduced in past decades in response to the growing demand for tailored glycans with controlled and uniform substitution patterns [1,2].

This chapter provides a brief survey of chemical modification and degradation methods. In view of the expansive body of existing literature, this survey is not intended to be comprehensive. Several earlier accounts provide general overviews of polysaccharide chemistry and the preparation of specific polysaccharide products [3–6], including those derived from cellulose [7–11], chitin [12] dextran [13, 14] galactomannans [15], β-glucans [16], heparin [17, 18], and starch [19, 20]. Other reviews deal with methods for the selective derivatization and degradation of polysaccharides for structure elucidations [21–23].

8.2 Non-selective chemical modifications

Most industrial polysaccharide derivatives are obtained in heterogeneous processes. Typically, the native glycan is suspended in an aqueous or other suitable medium that swells but that does not dissolve the polymer. This is followed by activation and treatment of the glycan under heterogeneous conditions with the reagent of choice. Depending on the type and degree of modification the glycan may be converted into a soluble product. When less than fully substituted products are obtained, such a process generally results in non-uniform substitution patterns, as crystalline and non-crystalline glycan regions are modified to variable extents over the course of the reaction. The analysis and functional implications of such heterogeneously modified derivatives are described in greater detail elsewhere [2].

The degree of substitution (DS) defines the average number of substituted repeat unit sites. For an anhydroglucose residue with three free hydroxyl groups, the maximum DS is 3.0. The indefinite term, molar substitution (MS), is operational for derivatizations involving reagents (e.g. alkylene oxides) that have the propensity to self-condense with residues already attached to the polymer, affording side-chains of variable lengths (Section 8.2.2.1). The average side-chain length can be assessed by the MS:DS ratio.

Major advances have been accomplished in recent years in the development of new solvents and activation methods for intractable polysaccharides. These have, in turn, facilitated significant design changes for derivatization processes [24, 25]. The activation of glycan functional groups for derivatization can be accomplished by solvent exchange, swelling with acids, metal hydroxides, salts, amines, as well as by controlled enzymatic and mechanical degradation. For cellulose, for example, numerous solvent systems are now available, which vary in their complexity, cost, dissolving power [some are limited to polymers with a low degree of polymerization (DP)] and mechanism (reversible or irreversible polymer derivatization or degradation) [26, 27].

While homogeneous modifications promise to facilitate access to various types of new and interesting products, many aspects relating to structure–activity relations remain to be investigated. The following discussion highlights some of the general features of non-selective derivation methods.

8.2.1 Acylations

8.2.1.1 Heterogeneous processes

Polysaccharide acylations can be accomplished by a wide variety of approaches, including the use of inorganic and organic acids, anhydrides and chlorides of alkyl or aryl carboxylic acids, sulfonyl chlorides and isocyanates in the presence of suitable catalysts [28]. Among the industrially important glycan esters are the acetate, butyrate, formate, propionate, nitrate, phosphate, sulfate and xanthate derivatives.

Cellulose acetates, particularly the triacetate, constitute one of the most significant types of commercial polysaccharide ester. They are prepared by treatment of the native polymer with acetic anhydride and glacial acetic acid in the presence of sulfuric acid catalyst. Cellulose acetates with lower acetyl contents are derived from the triacetate by controlled hydrolysis. The properties of the resulting acetates depend to a considerable degree on the reaction conditions [29, 30]. The most common starch acetates are low-DS products (DS 0.1–0.2) obtained from heterogeneous phase reactions. High-DS starch acetates can be prepared following alkali pretreatment or by using more stringent conditions. The acetylation of some other polysaccharides may also require such elaborate methods in order to proceed to completion [31, 32]. New acylation catalysts, such as 4-dimethylaminopyridine and 1-methyl-imidazole, have been introduced for polysaccharide modifications [33, 34]. Several mixed esters, such as cellulose acetate propionate and acetate butyrate, are also commercially available, and display unique properties [35]. Acylations of cellulose can be facilitated by pretreatment with a dehydrating agent, for example, with glacial acetic acid, prior to chemical modification. This helps to reduce the extensive hydrogen bonding network, thereby enhancing the accessibility and reactivity of the glucan.

Cellulose nitrate, one of the oldest products, is typically prepared by treatment of cotton with a sulfuric acid–nitric acid mixture. Quantitative nitration can usually not be achieved under these conditions. However, nitrations with close to theoretical yields are obtainable within 10 min in mixtures of liquid nitrogen tetroxide and hydrogen fluoride [36]. Numerous alternative nitration methods are now available [37].

The sulfation of polysaccharides can be achieved by using a variety of techniques, including concentrated sulfuric acid at low temperatures, chlorosulfonic acid in pyridine, piperidine-*N*-sulfonic acid in dimethyl sulfoxide (DMSO), and sulfur trioxide in trimethylamine (TME) or *N*,*N*-

dimethylformamide (DMF) [38–40]. The use of tetraphosphorus decasulfide can lead to sulfur- and phosphorus-containing products with sulfur and phosphorus contents of 2%–22% and 1%–10%, respectively, and S/P ratios of 0.2–5.6, depending on reaction conditions [41]. The preparation of sulfated products is of interest for heparin-like anticoagulant products, as well as antitumor and antiviral derivatives [42, 43].

Phosphorylations are most commonly accomplished with inorganic reagents, such as concentrated phosphoric acid, phosphorus oxyhalides, orthophosphoric acid, phosphorus pentoxide or sodium pyrophosphate or tripolyphosphate. Less frequently encountered reagents include acyl phosphites or similar organic reagents. Starch phosphate esters, obtained from condensation of starch with orthophosphates or tri- or tetrapolyphosphoric acid at temperatures below 60°C–70°C, are extensively used in the food industry. Phosphorus-containing cellulose derivatives find applications as flame retardants [44]. Metal-chelating phosphorylated chitin and chitosan derivatives have been obtained, using orthophosphoric acid and urea in DMF [45].

Cellulose xanthates, obtained by treatment of alkali cellulose with carbon disulfide, are employed for the production of cellophane and rayon fibers. While dry xanthate products are relatively stable, their aqueous solutions are readily susceptible to hydrolysis and oxidation [46].

Polysaccharide carbamates can be derived from condensation with isocyanates, urea or urea derivatives. Treatment of cellulose with urea at temperatures at or above the latter's melting point can be employed for the production of cellulose carbamate fibers [47]. Cellulose carbamates can also be prepared by using urea at 130°C or urea in liquid ammonia [48]. Similar amylose carbamates have been prepared with urea or urea derivatives [49]. The widely used cyanogen bromide activation method affords *N*-substituted carbamate products, as well as inert carbamate side products via cyanate ester intermediates [50].

Sulfonate esters are derived from the reaction of organic sulfonyl chlorides (e.g. benzene-, methane-, or toluene sulfonyl chloride) in the presence of tertiary amines. Sulfonic acid groups have also been introduced by employing 2-chloroethanesulfonic acid and 5-formyl-2-furansulfonic acid [51]. These derivatives can be employed as partially protected or reactive intermediates, which can undergo nucleophilic substitution reactions.

8.2.1.2 Homogeneous processes

Alternatives to many of the above heterogeneous processes have been introduced in recent years. Homogeneous processes [10, 52] for the preparation of cellulose acetates and other esters are based on the use of solvent systems, such as lithium chloride/*N*,*N*-dimethylacetamide (DMAc) [53], DMSO/paraformaldehyde (PF) [54, 55], DMAc/PF [56], DMF/PF [57], *N*-methylmorpholine *N*-oxide (MMNO), dinitrogen tetroxide-DMF, lithium chloride/DMSO [58] and sulfur dioxide-nitrosyl chloride [59, 60]. Although a number of solvents are available, only a few are sufficiently versatile to permit their use in a wide range of organic reactions. The MMNO system has been implemented commercially. In many other cases solvent reactivity, limitations in the attainable degrees of substitution and solvent cost are often of concern. In some cases, such as in the DMSO/PF system, completely new product types can be also obtained as a result of the incorporation of covalently linked functionalities originating from the solvent (e.g. polyoxymethylene residues). Cellulose acetates with varying thermal properties are obtainable by controlling the length of the methylol side-chains of cellulose in the DMSO/PF solvent system [54].The LiCl/DMAc system permits the use of mild reaction conditions and has been extensively investigated for the preparation of high DS derivatives, including esters, ethers and carbamates [27, 61]. The nuclear magnetic resonance (NMR) spectroscopic characterization of homogeneously prepared cellulose acetates and their substitution patterns has been reviewed [62].

The syntheses of various esters of chitin and chitosan, including acetate, formyl, propionyl and butyrate derivatives, have been developed in recent years [63–65]. These esters can be prepared from the corresponding acids, acid anhydrides or their mixtures in methanesulfonic acid at reduced temperatures. Highly substituted esters are often more difficult to obtain. Kurita and co-workers have developed methods for highly substituted chitosan derivatives, based on *N*-phthaloyl intermediates [66].

Water-soluble chitin phosphates with DS values up to 1.7 can be obtained by phosphorus pentoxide treatment of the native or deacetylated polymer in methane sulphonic acid solution [67]. Cellulose sulfonates have been prepared in the DMF/chloral and LiCl/DMAc solvent systems [68].

Water-soluble cellulose sulfate derivatives are obtainable under homogeneous conditions via transesterification of cellulose nitrite intermediates, using sulfurtrioxide in DMF [69]. Another recently developed method is based on dimethylol urea cross-linking of cellulose prior to esterification with sulfuric acid-propanol. This treatment reportedly affords cellulose sulfates of high DP and uniform substitution [70]. High DS carboxymethyl cellulose sulfates have been prepared via tosyl intermediates, using sulfur trioxide in pyridine or chlorosulfonic acid in DMAc [71]. An alternate procedure employs *O*-trimethylsilyl cellulose intermediates [72]. Chitosan 3,6-*O*-disulphate has been prepared using sulfur trioxide-DMF [73].

A more recently discovered class of derivatives are the hydrophobic polysaccharide esters. They are readily prepared from fatty acid chlorides or alkylamines and are of interest as surfactants, liquid crystalline materials and for various other applications [74–77]. Their tendency to self-associate into networks provides access to new supramolecular structures [78]. Thus, amphiphilic alginate derivatives exhibit interesting rheological features [75], hydroxypropyl cellulose perfluorooctanoate displays cholesteric properties [79] and tripalmitoyl–chitosan derivatives produce liposomes on ultrasonic irradiation [66]. *O*-palmitoyl and cholesterol derivatives of pullulan spontaneously form nanoparticles [80].

8.2.2 Alkylations

8.2.2.1 Heterogeneous processes

Polysaccharide alkylations are achieved on a commercial scale, by using highly reactive and inexpensive alkyl or aryl halogenides, epoxides or sulfates [81–84]. Commonly, the substitution is preceded by a conversion of the polysaccharide to an anionic derivative, for example, by treatment with alkali metal hydroxides. Typical representatives of such reaction products are the methyl, ethyl, carboxymethyl, hydroxyethyl, hydroxypropyl and alkylamino ethers. Other alkylating agents include alkyl tosylates, mesylates, triflates, phosphates and borates.

Alkylations with ionic reagents provide access to water-soluble derivatives. Carboxymethylations with chloroacetic acid are usually conducted in an aqueous suspension or in the presence of water-miscible organic solvents if higher degrees of substitution are desired. Carboxymethyl cellulose (CMC) is the most prominent representative of this class of derivatives. Other ionic derivatives include sulfoethyl, sulfopropyl and diethylaminoethyl derivatives.

A number of mixed ethers, such as hydroxyethyl hydroxypropyl cellulose, are also produced for various commercial uses. The preparation of mixed cellulose esters containing ether functions, such as epoxydated cellulose crotonate and hydroxyalkylated cellulose acetate, are known. The products retain most of the characteristics of the parent ester derivatives.

The preparation of permethylated polysaccharide derivatives is of special importance for many classical structure elucidation methods. One of the most widely used methods is the Hakamori methylation [85], which involves the generation of sodium methylsulfinyl methylide in methyl sulfoxide with sodium hydride. The carbanion has also been derived from potassium hydride, alkali metal amides [86] and butyl lithium [87]. A new methylation procedure, based on the use of methyl iodide and powdered sodium hydroxide in the sulfurdioxide–diethylamine–methylsulfoxide solvent system, is reportedly equally efficient, offering the added advantage of producing lower levels of polymer degradation [88]. Another recent modification of the Hakamori method relies on the use of 1,1,3,3-tetramethyl urea [89].

An important series of polysaccharide ethers is obtained by derivatization of alkali intermediates with epoxides, such as ethylene or propylene oxide. These substitutions are accompanied by the formation of oligomeric side-chains. Molar substitution (MS) values of commercial hydroxyethyl cellulose and hydroxypropyl cellulose derivatives, for example, may be in the range of 1.8–3.5 (DS 0.8–1.8) and 3.5–4.5 (2.2–2.8), respectively [90]. Base-catalysed addition reactions can be performed with a variety of other reagents, such as acrylamide, acrylonitrile and acrylic acid [23, 91].

An important class of glycan products is derived by grafting reactions. Graft condensation of vinyl monomers onto polymers are radical-mediated chain-transfer reactions, involving reagents such as cerium ions, Fenton's reagent or high-energy irradiation processes [92]. Ceric salts form effective redox systems in the presence of alcohols, where the oxidation and reduction produces cerous ions and transient free radical species capable of initiating vinyl polymerization, according to:

$$Ce^{4+} + RCH_2OH \leftrightarrow \text{[ceric-alcohol complex]}$$
$$\rightarrow Ce^{3+} + H^+ + RCH^{\bullet}OH \text{ (or } RCH_2O^{\bullet}\text{)}$$

Many polysaccharide graft copolymers have been prepared [93, 94]. Among the most prominent are starch co-acrylamide products that are able to absorb remarkable quantities of fluids ('super slurpers') and find use in disposable diapers and various agricultural products [95].

8.2.2.2 Homogeneous processes

Numerous homogeneous etherifications of cellulose have been reported. Thus, methyl and carboxymethyl cellulose can be prepared in DMSO/PF and MMNO/DMSO solvents [96]. A high DS, organosoluble trimethylsilyl cellulose has been prepared in DMAc/LiCl [97]. Etherification of cellulose in various other aprotic solvent systems has also been described [98].

Hydrophobic polysaccharide ether derivatives constitute a new class of products for a variety of applications, including hydrophobic chromatography [99] and surfactants [100]. Two new dextran surfactants, *N*-n-hexyl dextran aldonamide and *N*,*N*′-hexamethylene bis(dextran aldonamide) have been reported to comprise diblock (AB type) and triblock (ABA type) polymers [101]. Fluorocarbon-modified water-soluble hydroxyethyl cellulose derivatives have been reported [102]. New types of lyotropic products have been derived from triethyl and trimethyl amylose ethers [103]. Low levels of hydrocarbon residues can be incorporated into cellulose ethers to yield highly viscous, water-soluble products, which display non-Newtonian behaviour at low shear rates [104].

8.2.3 Oxidations

Partial oxidations of glycans are among the most versatile transformations, as they provide access to various intermediates and new products with altered properties. A wide range of oxidants are available, of which only a selection is described here. Most common oxidants are non-selective in their action and produce both carbonyl and carboxylate functions in varying proportions, depending on experimental parameters, such as pH, temperature, etc. Non-selective oxidations can potentially afford materials with a dozen or more types of oxidized species, including sugar residues with one or more ketone, aldehyde and carboxyl functions at different positions. Many oxidation processes are also liable to produce hydrolytically more labile derivatives.

Among the inexpensive and most widely used oxidants are hypohalide and halides. Sodium hypochlorite oxidations [105] may lead to the incorporation of carbonyl, dialdehyde and carboxylic acid functions into polysaccharides. The oxidation of cellulose with hypobromic or hypochloric acid proceeds slowly at high pH values (>11) and affords predominantly carboxylic acid groups and some aldehydic functions. At neutral pH values, maximum oxidation rates are observed, and both acidic and aldehydic functions are generated, while in acidic media aldehydic groups are mostly formed with some proportion of acidic groups. Limited depolymerization may arise from the scission of glycosidic linkages. Sodium hypochlorite is also extensively employed in the starch industry, affording the so-called 'chlorinated' starches. As in the case of cellulose, the oxidation proceeds very slowly at pH 11–13 and is fastest at or near neutral pH.

Bromine and chlorine can also be employed for polysaccharide oxidations and are known to produce mainly aldehyde and some carboxyl functions. However, these oxidations are usually accompanied by severe depolymerization. 2,3-Dicarboxy starch is obtainable by a combination of sodium hypochlorite and sodium bromide oxidation [106].

Periodic acid is the most widely used oxidant with specificity for vicinal diol functions. The resulting dialdehyde polysaccharide derivatives may serve as intermediates for a variety of products, including oxidized and reduced species, amines, dioximes, dicyanates and imine products [107]. The periodate oxidation of many other polysaccharides can be conducted in a quantitative fashion [108].

8.3 Selective chemical modifications

8.3.1 Acylations

8.3.1.1 Primary hydroxyl functions

Many heterogeneous polysaccharide esterifications, particularly those involving *p*-toluenesulfonyl (tosyl) chloride, proceed preferentially at the primary hydroxyl functions [109, 110]. The tosylation of cellulose under heterogeneous conditions occurs about six times faster at the primary than at the secondary positions [111]. The relative reactivities of the O-6, O-2 and O-3 functions in the homogeneous tosylation of partly substituted ethylcellulose and cellulose acetate were determined to be 214 : 33 : 1 and 209 : 20 : 1, respectively [112]. Similarly, the tosylation of amylose proceeds preferentially at the primary hydroxyl position, when the substitution levels are kept relatively low. Thus, exclusively 6-*O*-tosylated amylose with

DS 0.6 can be prepared, using a slight excess of tosyl chloride at 35°C for 0.5 h in pyridine, while a product with DS 0.85 contains a small proportion of secondary sulfonate functions. The application of the same methodology to cellulose leads to an even lower degree of selectivity. The tosylation of polysaccharides can be accompanied by the formation of by-products, such as 3,6-anhydro products, or, in the presence of pyridine, chlorinated derivatives or quaternary pyridinium salts [105].

An exclusively 6-*O*-acetylated amylose derivative was prepared, using a pertrimethylsilyl ether precursor and acetic anhydride–pyridine–aqueous acetic acid in carbon tetrachloride at 50°C [113]. Following removal of the trimethylsilyl ether functions, products with DS 1.0 were obtained after reaction for five days.

A preferentially C-6 modified cellulose carbamate derivative has been obtained [114]. Sulfation of chitosan with either sulfurtrioxide/dimethylformamide or chlorosulfonic acid/pyridine produces the 6-*O*-sulfated derivative in good yields [115]. The preferential O-6 sulfation of a number of polysaccharides with chlorosulfonic acid–pyridine has been demonstrated by ^{13}C-NMR [116]. Regioselective sulfation of the O-6 residues of dermatan sulfate has been achieved with sulfur trioxide-trimethylamine in DMF, when substitution levels were limited to below 50% [117].

Various amine, azide and halogenated derivatives and precursors for alkylated derivatives have been reported [118]. Thus, 6-amino-6-deoxyamylose was prepared via 6-halogenated intermediates under homogeneous conditions [119], and a 6-chloro-6-deoxy chitin derivative was prepared with *N*-chlorosuccinimide-triphenylphosphine in LiCl/DMAc [120].

8.3.1.2 Secondary hydroxyl functions

Cellulose sulfate derivatives are obtainable under homogeneous conditions by transesterification of cellulose nitrite with *N,N*-dimethylformamide-sulfur trioxide [71]. The replacement of the labile nitrite groups by half ester sulfate groups affords cellulose sulfate products with DS levels of up to 2, following removal of the residual nitrite functions by hydrolysis. In contrast to other heterogeneous sulfate ester preparations, the above products are more uniformly substituted and display water solubility at much lower DS levels (about 0.3, compared with $DS > 1.0$). The substitution occurs primarily at the secondary positions, particularly when the DS levels are kept below unity.

The preferential acetylation of cellulose at position O-2 can be accomplished for low-DS products by reaction in dimethylformamide using a cellulose trinitrate precursor [121]. Various other 2,3-di-*O*-substituted cellulose derivatives have been obtained [122]. Inulin monosuccinates, prepared by condensation with succinic anhydride in the presence of acylation catalysts, are found to be preferentially substituted at positions C-6 and C-3, rather than at C-4 [123].

A series of amylose derivatives with substituents at C-3 or C-2, such as a D-allose residue-containing product, derived from 2-*O*-benzoyl-6-*O*-trityl-amylose [124], and a 3-phosphate derivative [125], have served as enzyme substrates. The secondary hydroxyl positions of xylan and several O-6-tritylated polysaccharides have been protected by tosylation or mesylation in efforts to prepare various types of selectively derivatized products. Alternatively, O-6-tosylated starting materials can be employed to obtain products with phosphate substituents at the secondary functions [126].

8.3.2 *Alkylations*

8.3.2.1 *Primary hydroxyl functions*

The condensation of polysaccharides with triphenylmethyl (trityl) chloride proceeds generally with preference for the primary hydroxyl positions. Thus, the tritylation of cellulose occurs initially 58 times faster at O-6 than at either O-2 or O-3 [127]. However, with increasing DS levels, the substitution gradually extends to the secondary hydroxyl functions. The selectivity of the reaction can be also influenced by using a larger excess of the trityl reagent. The reactivity of cellulose towards tri-(*p*-toluenesulfonyl) methyl chloride has been examined [65]. The tolyl reagent is more reactive than is trityl chloride, and the primary hydroxyl position exhibits a 43-fold higher reactivity than the secondary hydroxyl groups.

The preparation of predominantly O-6-modified carboxymethyl cellulose can be achieved in homogeneous solution, using the MMNO/DMSO solvent system [128]. Likewise, the carboxymethylation and other alkylations of alkali chitin can be performed to afford primarily O-6 modified products [129]. New types of predominantly C-6 aldehyde-containing polysaccharides have been derived from condensation of chloroacetaldehyde dimethyl acetal and amylose, dextran, and a linear 1-3-β-D-glucan and subsequent deprotection of the corresponding masked aldehyde intermediates [130]. These products were employed in coupling reactions with amines and proteins.

6-*O*-methylamylose has been prepared by deesterification of 2,3-di-*O*-benzoyl-6-*O*-methylamylose [131]. A number of cellulose acetal derivatives ($DS < 1.0$), obtained from dihydropyran, and methyl or octadecyl vinyl ether, were shown to be predominantly substituted at the primary hydroxyl positions. Similarly, carboxymethylation of chitin to DS values of 0.8 yields mainly 6-*O*-substitution [33].

8.3.2.2 *Secondary hydroxyl functions*

A convenient approach to the modification of the secondary glycan hydroxyl functions involves the use of intermediates, whose primary positions are reversibly protected. Numerous selective modifications of secondary polysaccharide hydroxyl groups are based on strategies employing O-6 trityl intermediates, such as in the case of 2,3-dibromo-2,3-dideoxy amylose [132].

Tritylation of the primary hydroxyl functions of amylose and cellulose facilitates the selective oxidation of the O-2 hydroxyl functions.

Preferentially O-2 substituted methyl dextran derivatives can be prepared, using methyl vinyl ether modification of dextran acetates [133]. A sequence of tritylation, methylation and detritylation of dextran yields a product with methyl substituents at the secondary positions [134].

For carboxymethyl cellulose, the substituents are either uniformly distributed or are predominantly located at the secondary positions, depending on whether the reaction is conducted at 50°C or at 19°C, respectively. Alternatively, preferential carboxymethylation at the primary hydroxyl positions can be accomplished in the MMNO/DMSO solvent system [12].

8.3.3 Oxidations

8.3.3.1 Primary hydroxyl functions

One of the exciting breakthroughs in selective oxidations is based on the combined use of 2,2,6,6-tetramethyl-1-piperidinyloxy (TEMPO) and hypobromite at pH 10.5–11. Such TEMPO-mediated oxidations are found to be highly selective for the primary alcohol functions of several glycans. Thus, a 98% level of selectivity is achieved for starch and over 90% for inulin [135].

Several other, less efficient, oxidations are known. The complete conversion of the primary hydroxyl functions of α- and β-cyclodextrin can be accomplished by either catalytic oxidation (O_2/Pt) or nitrogen dioxide treatments [136]. However, the application of equivalent techniques to polysaccharides leads to the formation of either substantially depolymerized materials, or product mixtures containing both acid and aldehyde groups [137]. Secondary hydroxyl functions are also oxidized to some extent, and nitrogen is incorporated in the form of nitrites or nitrates. The oxidation of carbohydrates with nitrogen dioxide or nitrous acid is proposed to involve a nitric or nitrous acid ester intermediate [138].

The nitrogen dioxide oxidation method can be employed for the preparation of heparin analogues derived from various glycans, for which C-6 oxidation yields of between 20%–60% are obtainable [139]. For cellulose, the application of nitrogen dioxide in the gas phase or dissolved in carbon tetrachloride results in the predominant formation of D-glucuronic acid residues [140].

An improved nitrogen dioxide oxidation method for cellulose has been developed, based on the use of phosphoric acid and sodium nitrite. This technique yields products with oxidation of up to 88% of primary and about 6% secondary hydroxyl positions and offers the advantage of lower degrees of depolymerization than do the previous methods. In the case of amylose, this procedure furnishes water-soluble products with D-glucuronic acid/D-glucose ratios of 0.5–3.0 [141]. Oxidations in the presence of sodium borohydride mitigate the extent of depolymerization arising from non-specific oxidation. It has been cautioned, however, that the use of the reducing reagent could possibly give rise to β-elimination reactions. This method has also been applied to the oxidation of scleroglucan and dextran, for which quantitative oxidation of the primary hydroxyl groups was obtained [142].

The oxidation of cotton with dimethyl sulfoxide/acetic anhydride (DMSO/Ac_2O) followed by chlorous acid treatment has yielded oxidized products with 47%–62% carboxyl functions, with the balance being 2- or 3-keto functions [143].

Another preferential oxidation method for primary hydroxyl functions involves the use of oxygen and Adams catalyst and can be applied to polysaccharide branch and terminal non-reducing residues or to otherwise sterically favoured positions. This method introduces furanosyluronic and pyranosyluronic acid residues into polysaccharides. Treatment of arabinoxylan and arabinogalactan with oxygen, Adams catalyst and sodium hydrogen carbonate for 4 days and 14 days afforded the corresponding oxidized products containing 4% and 8% carboxyl functions, respectively [144, 145]. The unfavourably long reaction periods and low yields clearly limit the utility of this method. Similarly low yields are reported for the application of this catalytic oxidation method to $1 \rightarrow 4$-linked polysaccharides [146]. On the other hand, highly branched $1 \rightarrow 6$-linked polysaccharides seem to constitute better substrates for the technique. Thus, various branched and unbranched dextrans have been oxidized in yields of up to 80%–85% [147, 148].

A heparin analogue has been prepared from partially 2-trifluoroacetamide-substituted amylose by selective catalytic oxidation (O_2/Pt, $DS_{C\text{-}6}$0.46) and subsequent sulfation [149]. The use of hydrogen peroxide has been reported to facilitate the selective oxidation of the primary hydroxyl functions of amylose, but not those of cellulose [2, 10].

Selective oxidations of chitosan can be effected by chromium trioxide treatment following prior quaternization of the primary amine function with perchloric acid [150]. The bulkiness of the resulting ammonium perchlorate moiety provides the requisite protection of the O-3 function during the oxidation. A C-6 carboxyl chitosan product with a DS of 1.0 has been obtained with apparent retention of polymer integrity.

8.3.3.2 Secondary hydroxyl functions

The selective oxidation of amylose at the C-3 position has been achieved with DMSO-phosphorus pentoxide via a 2-*O*-benzoyl-6-*O*-trityl intermediate [151]. The borohydride-reduced, detritylated product contained 10% D-allose residues. In another study, a similar oxidation–reduction sequence with 6-*O*-trityl-amylose afforded a product with 26% D-allose content [152].

Selective oxidations of amylose and cellulose can be accomplished, using the DMSO/Ac_2O method. The 6-*O*-trityl derivatives of amylose and cellulose have, for instance, been oxidized with this reagent to the respective 2-oxy-6-*O*-trityl-amylose [153] and cellulose [154] derivatives. 2-Oxy-6-*O*-trityl-amylose can also be obtained by employing dicyclohexylcarbodiimide (DCC) in DMSO [155]. The oxidation of unprotected cellulose with DMSO/Ac_2O or DCC/DMSO/pyridine/trifluoroacetic acid affords a mixture of 2-oxy-, 3-oxy- and 2,3-dioxy-cellulose [156]. More recently, it was observed that oxidation of unprotected cellulose with DMSO/Ac_2O in the DMSO/PF solvent system affords exclusively 3-oxycellulose, owing to the reversible formation of O-6 and O-2 hydroxymethyl and poly(oxymethylene)ol side chains [157]. For unprotected amylose the same oxidation procedure yields also 3-oxy-amylose, with minor amounts (10%) of the 2-oxy-product. The application of the DMSO/Ac_2O oxidation method to dextran-2,4-diphenyl boronate resulted in the formation of products with carbonyl functions at C-2 (8%) and C-3 (11.5%), but not at C-4 [158].

Mild bromine oxidation methods have been developed that facilitate the attainment of relatively high degrees of selectivity for certain polysaccharides under carefully controlled conditions [155]. Thus, oxidation of amylose with dilute aqueous bromine at pH 6–7 introduces only carbonyl functions, while similar oxidations at pH 8 produce equimolar quantities of carboxyl and carbonyl groups. Treatment of amylopectin at pH 6–8, on the other hand affords a 2:1 ratio of carbonyl to carboxyl functions [159].

Mild bromine oxidations of Sepharose are selective for the hydroxyl group at C-4 [160], but afford mixed oxidation products of cellulose and curdlan [161]. For dextran, the method results in oxidation at C-2 and C-4, with minor amounts of C-3 glycosylose and acidic ring cleavage products [162]. In a modification of this method, in which oxidations of dextran are conducted in the presence of borate ions, the formation of dicarboxylic acid components can be significantly suppressed [163]. Oxidations of xylan in the presence of metaborate ions yield a product oxidized at C-2 and C-3, but not at C-4 [164]. The mechanism of the bromine oxidation is believed to involve the abstraction of the methine hydrogen by bromine. It is suggested that 1,3-diaxial steric hindrance is responsible for the fact that, unlike sodium nitrite oxidations, bromine oxidations of dextran do not take place to any significant degree at C-3 [156].

8.4 Modification of carbonyl functions

8.4.1 Esterifications

The esterification of carboxylic-acid-containing polysaccharides can be accomplished using acid-catalysed alcoholysis, diazomethane or alkylene oxides, or via acid chloride or other types of intermediates. The modification of polyuronides, such as alginic acid, with methanolic hydrogen chloride often affords only partially esterified products, even under severe reaction conditions. Thus the ambient temperature esterification of alginic acid and pectic acid for periods of 1–13 days furnishes the corresponding methyl ester derivatives with yields of 21%–60% and 1%–12%, respectively [165]. Higher esterification yields are obtainable if the reaction is preceded by an activation or pretreatment step, such as by regeneration of the polymer, partial neutralization of the carboxylic acid groups, or by solvent exchange using glacial acetic acid in order to reduce the extent of hydrogen bonding.

The preparation of methyl esters can be conveniently affected using diazomethane. The reactions proceed with good efficiencies and relatively little attendant degradation. Thus, esterification yields of up to 92% have been obtained for alginic acid, although this was accompanied by substantial (up to 34%) degrees of methylations at the secondary hydroxyl groups [166]. Diazomethane esterifications of pectins have also been performed yielding products with varying degrees of esterification (0%–95%) and essentially constant degrees of polymerization [167].

The esterification of alginic acid in aqueous medium with alkylene oxides, including butylene, ethylene, pentylene or propylene oxide, is performed on an industrial scale, leading to products such as propylene glycol alginate, with 50%–80% conversion yields [168]. The carboxylic acid functions of polyuronides have been protected by propionation [169], and 6-carboxycellulose has been esterified with epichlorohydrin [170].

8.4.2 Amidations

Carbodiimide-mediated couplings are widely employed for the preparation of amides. The use of water-soluble carbodiimides, such as *N*-ethyl-*N*-dimethylaminopropyl carbodiimide (EDC) has been extensively documented in the literature for the modification of alginic acid, xanthan gum, a dextran carboxylic acid derivative and several glycosaminoglycans, including dermatan sulfate, chondroitin sulfates, heparin and hyaluronic acid [171, 172]. Alternatively, amidations have been performed in organic media, using DCC [173]. The course, efficiency and specificity of the coupling reaction have been monitored by inclusion of fluorescent, nitroxide or radioisotope probes. It has been demonstrated that the EDC activation of glycosaminoglycans results in the formation of *O*-acylisourea intermediates [174].

The preparation of polysaccharide amides by direct condensation of their carboxyl groups with amines has been accomplished for alginic acid, CMC and xanthan gum [175–177]. Alcoholic ammonia treatment of pectate has afforded pectinic acid amides. [178]

8.4.3 Aminations

Selectively oxidized glucan derivatives can be transformed into the corresponding amines by oximation and subsequent reduction with lithium aluminum hydride [179]. A 2-amino-2-deoxy-amylose derivative (DS 0.8) has been obtained with retention of the D-gluco configuration, but with substantial depolymerization and the formation of minor proportions of residues containing keto functions. The equivalent 2-amino-2-deoxy-cellulose derivative prepared by this method had DS 0.4. Other aminodeoxy polysaccharides have been synthesized via azide or hydrazide intermediates [180].

The reductive amination method, using sodium cyanoborohydride, is now widely employed for the preparation of amine derivatives. Thus, 2-amino-2-deoxy derivatives of amylose and cellulose have been prepared from the corresponding 6-*O*-trityl-2-oxy-precursors with essentially quantitative conversions being obtained for the latter [181, 182]. Similarly, 3-amino-3-deoxy derivatives of both polymers can be derived, in addition to a series of products bearing substituted amines at position C-2 or C-3, respectively. The method can also be applied to the synthesis of amine derivatives of various other oxidized polysaccharides, including alginic acid, galactomannans [183], and xanthan gum [184].

8.5 Modification of amines

8.5.1 N-*acylations*

Selective *N*-acylations of a number of aminopolysaccharides and glycosaminoglucuronans can be performed, using various alkyl and aryl carboxylic acid anhydride derivatives, haloanhydride derivatives and saturated and unsaturated dibasic anhydrides. Thus, chitosan can be efficiently *N*-acylated without concomitant hydroxyl modifications, using anhydrides (2–3-fold excess) in organic acid media, while mixed *N*- and *O*-acylated products are obtained under similar conditions, using larger anhydride to amine ratios (10-fold excess) [185]. In an alternative approach, *N*-, *O*-haloacylated chitosan products with DS 1.2–3.0 are prepared and subsequently *O*-deactylated by alkali treatment to obtain the exclusively *N*-haloacylated chitosan derivatives (DS 1.0) [186]. Quantitative or near quantitative acylations of the amine functions have been accomplished at ambient temperatures within a few minutes [187].

The acylation of *N*-desulfated heparin amine functions has been described, using acid anhydrides or halides, including a number of fatty acyl derivatives, in the presence of amines, catalysts, or anion exchange resins [188, 189].

8.5.2 N-*alkylations*

Water-soluble, quaternary alkyl ammonium chitosan derivatives with DS 0.5–0.8 are prepared by using methyl iodide and pyridine or triethylamine

[190]. A low-acetyl chitosan has been transformed into *N*-trimethylammonium iodide chitosan derivative, using methyl iodide in the presence of sodium hydroxide [191]. Products with DS 0.25 and higher were water soluble. The condensation of chitosan with formaldehyde has been used to obtain *N*-dimethyl [192] or *N*-methylene chitosan derivatives [193]. The synthesis of *N*-methyl *O*-hydroxyethyl chitosan has also been described [194].

The primary amine functions of chitosan can be efficiently modified with a wide range of aliphatic and aromatic carbonyl compounds, including formaldehyde, glutaraldehyde, unsaturated alkyl aldehydes and carbonyl-containing carbohydrates, etc., to form *N*-alkylidene and *N*-arylidene derivatives [195, 196]. Chitosan Schiff's-base derivatives can be prepared as reversible protection of the amine functions in *O*-alkylation reactions.

The syntheses of chitosan imine products are generally observed to be less efficient than those of analogous derivatives with stable amine linkages. Products of the latter type have been obtained from a series of aromatic aldehydes, for example, salicylaldehyde, anthraldehyde [197], *m*-fluorobenzaldehyde [198], carbonyl-containing saccharides and other types of aldehydes, such as nitroxide spin labels [199].

Similar modifications are described for selectively aminated cellulose, dextran and guar gum derivatives. Various carboxylic acid and lactone derivatives have been condensed with chitosan to afford the corresponding amides. Aldonic acid lactones of cellobiose, glucose and maltose were coupled to chitosan salts in organic solvents to afford water-insoluble products (DS 0.8–1.0) [200], while related branched chitosan products (DS 0.77), derived from condensation with D-glucoheptonic acid-δ-lactone in aqueous acidic acid–methanol, were water soluble. Carbodiimide-mediated amidation of a Sepharose carboxylic acid derivative has been performed with heparin. Similar EDC couplings of chitosan have been reported, involving several types of acid derivatives [192].

8.6 Polysaccharide degradations

The structural integrity of polysaccharides is of critical importance for many applications. At the same time there is a growing realization that valuable new products may be obtained from glycan depolymerization products. Among the numerous uses of oligosaccharides are their use as food additives, antibodies, vaccines, plant growth regulators and synthetic antigens. Many immunological properties are modulated by the molecular weight of glycans [201, 202]. This section highlights various chemical polysaccharide degradation techniques. Additional methods, including enzymatic, thermal, ultrasonic, mechanical, lyophilization-mediated [203], radiation-induced [204], metal-catalysed [205, 206] and radical-catalysed [207, 208] depolymerizations are described elsewhere [2, 21–23].

The various depolymerization techniques are mediated by differing mechanisms and can consequently be employed to obtain potentially distinct products. Thus chemical and enzymatic depolymerizations of a given polysaccharide often afford two distinct and complementary series of oligosaccharides. The depolymerization kinetics also vary depending on the method. Thus, for hyaluronic acid, ultrasonic depolymerization was found to follow a linear relation between the inverse of molecular weight and depolymerization time, whereas heating and ultraviolet (UV) irradiation obeyed non-linear relations [200]. A similar linear relation between the inverse of molecular weight and depolymerization time has been observed in random depolymerizations of single-stranded polymers, for example, alginate, hyaluronic acid, hydroxyethyl cellulose, carboxymethyl cellulose and carrageenan [209, 210].The depolymerization of multiple-stranded polymers, on the other hand, deviates from linear kinetics, and follows power law relations. The power law exponent is an indicator of the multiplicity of the polymer strandedness and kinetic analysis can therefore serve as a conformational tool [206, 211].

The random, partial depolymerization of polysaccharides can be accomplished by various chemical techniques. Hydrolysis with inorganic acids is the most frequently used method, in which H^+ ions catalyse glycosidic cleavage in the presence of water. Additional methods include acid-catalysed alcoholysis, acetolysis, autohydrolysis and solid superacids. In free radical depolymerizations, $^{\bullet}OH$ radicals are believed to cause polymer chain cleavage.

Differences in the hydrolytic stability of carbohydrate residues and glycosidic linkages can often be exploited for selective polysaccharide degradations [212]. Thus, pentose and furanose units are more labile than are hexose and pyranose units, respectively, and α-linked residues are less stable than are β-linked residues. Furthermore, some saccharide residues may be prone to acid hydrolysis, but stable under conditions of acetolysis, thereby allowing for complementary fragmentation patterns [213, 214]. A number of other examples of selective degradations will be presented below. In general, however, the development of the hydrolytic conditions for the preparation of a given oligosaccharide fraction is based on an empirical basis.

While acid hydrolysis is generally assumed to be a random process, non-random processes have been observed for a number of polysaccharides. Thus, the hydrolysis of amylopectin at 50°C was shown to involve a two-step mechanism, which produced well-defined products with molecular weights of 3×10^5 and 6×10^4 daltons, respectively [215]. Treatment of dextran with low concentrations of acid (0.33 M HCl) at 95°C leads to the preferential cleavage of the $1 \rightarrow 3$-α-linkages rather than the $1 \rightarrow 6$-α-linkages, that is, to debranching [216].

High-resolution size-exclusion chromatography has facilitated the efficient preparation of acid hydrolysis-derived cellooligomers with DP 3–8 [217].

Poly-(galacturonate) fragments (DP 25) with a narrow molecular weight range can be derived from pectin [218], and dextran oligosaccharides (M_w 1000 daltons) from dextran [219]. Various other types of oligosaccharide fragments of alginic acid [220], carrageenans [221], cellulose [222], chitosan [223], dextran [224], and heparin [225] are obtainable by partial acid hydrolysis.

The Smith periodate degradation of λ-carrageenan (M_w 12×10^4 daltons) yielded fragments with lower molecular weights (M_w 6×10^3–68×10^3 daltons) [226]. Agarose sulphate chain segments were similarly prepared [227]. Other partial depolymerizations of agarose [228] and heparin [229] have been reported.

Glycosaminoglycan-derived oligosaccharides are of considerable interest and have been extensively investigated [230]. Thus, treatment of hyaluronate with 10% 0.1 M HCl in DMSO produces a series of even-numbered (di- to octa-) oligosaccharides containing *N*-acetyl-D-glucosamine reducing end groups, which could subsequently be transformed into a series of odd-numbered (mono- to penta-) oligosaccharides with terminal D-glucuronic acid residues. Degradation of desulfated chondroitin-6-sulfate in 10% aqueous DMSO affords a series of di- to octasaccharides with *N*-acetyl-D-glucosamine residue reducing end groups, while a similar treatment of dermatan sulfate yields even-numbered (di- to octa-) oligosaccharides with *N*-acetyl-D-galactosamine residues [231].

The nitrous acid-mediated deaminative cleavage of chitosan can be performed in 20% acetic acid and affords acetylated oligosaccharides containing terminal anhydromannitol (anManOH) residues in 83% yield [232]. This method is also applicable to the preparation of heparin [233], dermatan sulfate, keratan sulfate and chondroitin-6-sulfate fragments [234]. Low molecular weight heparins (3000–9000 daltons) are of special pharmaceutical interest and are prepared by fractionation or chemical or enzymatic hydrolysis of commercial heparin [235]. Peroxide and deaminative cleavage are employed for this purpose. Nitric acid treatment of heparin affords a mixture of di- and tetrasaccharides [236]. The deaminative cleavage of equatorially oriented 2-amino-2-deoxyglycosidic bonds by nitrous acid is a stoichiometric process that proceeds mainly via an intermediate diazonium ion with concomitant ring contraction. Instead of participation of electrons of the O–C1 bond, electrons of the C3–C4 bond may be involved, resulting in the formation of 2-*C*-formyl-pentofuranosides in an alternate non-lytic pathway that may account for up to 25% of the product (Scheme 8.1) [21, 237]. Variations of the deamination reaction permit the selective cleavage of either *N*-substituted (*N*-sulfated) or *N*-unsubstituted residues [238, 239]. Deamination of heparin, followed by borohydride reduction, has also yielded a series of oligosaccharides with DP 1–6 [240]. Depolymerizations of glycosaminoglycans and their methyl esters can alternatively be performed by using diazomethane-nitrous acid treatments to afford oligosaccharides

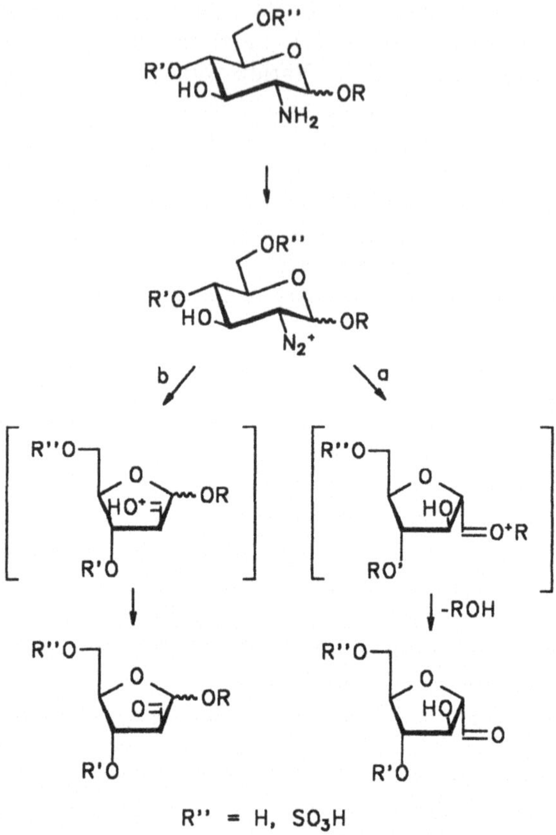

Scheme 8.1 Nitrous acid deamination of glycosaminoglycans, resulting in (a) formation of 2,5-anhydro-D-mannose; (b) 2-deoxy-2-*C*-formyl-pentafuranosides.

bearing 4,5-unsaturated uronic acid residues at the non-reducing termini [226].

The depolymerization of polysaccharides with hydrogen fluoride proceeds more rapidly and at lower temperatures than with other acids [241]. The effectiveness of HF solvolysis is based on the high efficiency with which HF cleaves glycosidic linkages and the high solubility of polysaccharides in liquid HF. A number of instrumental designs have been proposed for the convenient handling of HF reactions [242]. The treatment of amylose, cellulose and D-glucose with anhydrous hydrogen fluoride affords equilibrium mixtures of α-D-glucopyranosylfluoride and oligomeric reversion products, with the former predominating in dilute solutions and the latter upon evaporation of hydrogen fluoride [243].

Treatment of aqueous chitosan suspensions with hydrogen peroxide at 70°C transformed the polymer (M_w 240 000 daltons) into low molecular weight products (M_w 12 000 daltons) in 4 h [244]. Low molecular weight chitosan (M_w 12 000 daltons) has also been prepared via glycosidic cleavage of chitosan (M_w 225 000 daltons), using aqueous perborate solutions [245].

References

1. Yalpani, M. (1985) *Tetrahedron*, **41**, 2957–3020.
2. Yalpani, M. (1988) *Polysaccharides, Syntheses, Modifications and Structure/Property Relations*, Elsevier, Amsterdam.
3. Beck, R.H.F., Fitton, M.G. and Kricheldorf, H.R. (1992) in *Handbook of Polymer Synthesis, Part B*, Kricheldorf, H.R. (ed.), Marcel Dekker, New York, pp. 1517–77.
4. Whistler, R.L. (ed.) (1973) *Industrial Gums*, 2nd edn, Academic Press, New York.
5. Meltzer, Y.L. (1981) *Water-soluble Polymers, Developments Since 1978*, Noyes Data Corporation, Park Ridge, NJ.
6. Davidson, R.L. (ed.) (1980) *Handbook of Water-soluble Gums and Resins*, McGraw-Hill, New York.
7. Krässig, H.A. (1993) *Cellulose, Structure, Accessibility and Reactivity*, Gordon & Breach, New York.
8. Gilbert, R.D. (ed.) (1994) *Cellulosic Polymers, Blends and Composites*, Hanser, Munich.
9. Fengel, D. (1994) *Macromol. Symp.*, **83**, 311–23.
10. Philipp, B.J.M.S. (1993) *Pure Appl. Chem.*, **A30**, 703–14.
11. Vigo, T.L. (1985) in *Encyclopedia of Polymer Science and Engineering*, Mark, H.F., Bikales, N.M., Overberger, C.G. and Menges, G. (eds), 2nd edn., vol. 3, Wiley Interscience, New York, pp. 124–39.
12. Skjak-Braek, G., Anthonson, T. and Sandford, P. (eds) (1989) *Chitin and Chitosan*, Elsevier, Amsterdam.
13. Alsop, R.M. (1983) *Progr. Ind. Microbiol.*, **18**, 1–44.
14. Harris, M.J. and Yalpani, M. (1985) in *Partitioning in Aqueous Two-phase Systems*, Walter, H., Brooks, D.E. and Fisher, D. (eds), Academic Press, New York, pp. 590–626.
15. Seaman, J.K. (1980) in *Handbook of Water-soluble Gums and Resins*, Davidson, R.L. (ed.), McGraw-Hill, New York, pp. 6/1–6/19.
16. Stone, B.A. and Clarke, A.E. (1992) *Chemistry and Biology of 1-3-β-Glucans*, La Trobe University Press, La Trobe.
17. Lane, D.A. and Lindahl, U. (eds) (1989) *Heparin, Chemical and Biological Properties, Clinical Applications*, CRC Press, Boca Raton.
18. Lundblad, R.L., Brown, W.V., Mann, K.G. and Roberts, H.R. (eds) (1981) *Chemistry and Biology of Heparin*, Elsevier, Amsterdam.
19. Van Breynum, G.M.A. and Roels, J.A. (eds) (1985) *Starch Conversion Technology*, Marcel Dekker, New York.
20. Whistler, R.L., BeMiller, J.N. and Paschall, E.F. (eds) (1984) *Starch, Chemistry and Technology*, 2nd edn, Academic Press, New York.
21. Aspinall, G.O. (1982) in *The Polysaccharides*, Aspinall, G.O. (ed.), Academic Press, New York, vol. 1, pp. 36–131.
22. Pigman, W. and Horton, D. (eds) (1980) *Carbohydrates, Chemistry and Biochemistry*, 2nd edn, Academic Press, New York, vols. 2A, 1978, and 2B, 1980.
23. Burchard, W. (ed.) (1985) *Polysaccharide*, Springer, Berlin.
24. McCormick, C.L., Lichatowich, D.K., Pelezo, J.A. and Anderson, K.W. (1980) *ACS Symp. Ser.*, **121**, 371–80.
25. Aiba, S., Izume, M., Mioura, N. and Fujiwara, Y. (1985) *Br. Polym. J.*, **17**, 38–40.
26. Johnson, D.C. (1985) in *Cellulose Chemistry and its Applications*, Nevell, T.P. and Zeronian, S.H. (eds), Wiley, New York, pp. 181–201.
27. Dawsey, T.R. (1994) in *Cellulosic Polymers, Blends and Composites*, Gilbert, R.D. (ed.), Hanser, Munich, pp. 157–72.

28. Bogan, R.T. and Brewer, R.J. (1985) in *Encyclopedia of Polymer Science and Engineering*, 2nd edn., vol. 3, Mark, H.F., Bikales, N.M., Overberger, C.G. and Menges, G. (eds), Wiley Interscience, New York, pp. 158–81.
29. Doyle, S., Pethrick, R.A., Harris, R.K. *et al.* (1986) *Polymer*, **27**, 19–24.
30. Tanghe, L.J., Genung, L.B. and Mench, J.W. (1963) *Methods Carbohydr. Chem.*, **3**, 193–8.
31. Tokura, S., Nishi, N., Somorin, O. and Noguchi, J. (1980) *Polym. J.*, **12**, 695–700.
32. Carson, J.F. and Maclay, W.D. (1946) *J. Am. Chem. Soc.*, **68**, 1015–17.
33. Vermeersch, J. and Schacht, E. (1985) *Macromol. Chem.*, **187**, 125–31.
34. Muller, G., Chiron, G. and Levesque, G. (1985) *Polymer Bull.*, **15**, 1–5.
35. Gray, D.J. (1983) *Appl. Polym. Sci. Appl. Polym. Symp.*, **37**, 179–92.
36. Nissan, A.H. and Hunger, G.K. (1965) in *Encyclopedia of Polymer Science and Technology*, vol. 3, Mark, H.F., Gaylord, N.G. and Bikales, N.M. (eds), Wiley Interscience, New York, pp. 131–226.
37. Wint, R.W. and Shaw, K.G. (1985) *ACS Symp. Ser.*, **285**, 1073–99.
38. Yamamoto, N., Takayama, K., Honma, K. *et al.* (1991) *Carbohydr. Polym.*, **14**, 53–63.
39. Hatanaka, K., Kurihara, Y., Uryu, T. *et al.* (1991) *Carbohydr. Res.*, **214**, 147–54.
40. Nishi, N., Maekita, Y., Nishimura, S. *et al.* (1987) *Int. J. Biol. Macromol.*, **9**, 109–11.
41. Yalpani, M. (1992) *Carbohydr. Polym.*, **19**, 35–9.
42. De Clercq, E. (1993) in *Carbohydrates and Carbohydrate Polymers, Analysis, Biotechnology, Modification, Antiviral, Biomedical and Other Applications*, Yalpani, M. (ed.), ATL Press, Mt. Prospect, pp. 87–100.
43. Uryu, T., Kaneko, Y., Yoshida, T. *et al.* (1993) in *Carbohydrates and Carbohydrate Polymers, Analysis, Biotechnology, Modification, Antiviral, Biomedical and Other Applications*, Yalpani, M. (ed.), ATL Press, Mt. Prospect, pp. 101–15.
44. Langley, J.D., Drews, M.J. and Barker, R.M. (1980) *J. Appl. Polym. Sci.*, **25**, 243–62.
45. Sakaguchi, T., Horikoshi, T. and Nakajima, A. (1981) *Agric. Biol. Chem.*, **45**, 2191.
46. Federowicz, M., Palasinski, M. and Tomasik, P. (1984) *Acta Aliment. Polon.*, **10**, 163–78.
47. Selin, J.-F., Huttunen, J., Turunen, O. *et al.* (1985) *US. Pat.* 4,530,999.
48. Selin, J.-F., Huttenen, J.I. and Turunene, O.T. (1985) in *Cellulose and its Derivatives*, Kennedy, J.F., Phillips, G.O., Wedlock, D.J. and Williams, P.A. (eds), Ellis Horwood, Chichester, Sussex, pp. 521–5.
49. Yoshida, M. and Yuen, S. (1976) *US Pat.* 3,997,398.
50. Wilchek, M. and Miron, T. (1986) *J. Chromatogr.*, **357**, 315–17.
51. Muzzarelli, R.A.A. (1992) *Carbohydr. Polym.*, **19**, 231–6.
52. Nehls, I., Wagenknecht, W., Philipp, B. and Stscherbina, D. (1994) *Prog. Polym. Sci.*, **19**, 29–78.
53. McCormick, C.L. and Sheu, T.S. (1981) in *Macromolecular Solutions, Solvent–Property Relationships in Polymers*, Seymour, R.B. and Stahl, G.A. (eds), Pergamon, New York, pp. 101–7.
54. Miyagi, Y., Shiraishi, N., Yokota, T. *et al.* (1983) *J. Wood Chem. Technol.*, **3**, 59–78.
55. Seymor R.B. and Johnson, E.L. (1978) *J. Polym. Sci. Polym. Chem. Edn.*, **16**, 1–11.
56. Leoni, R. and Baldini, A. (1982) *Carbohydr. Polym.*, **2**, 298–301.
57. Terbojevich, M., Cosani, A., Carrao, C. and Leoni, R. (1985) *Carbohydr. Polym.*, **5**, 351–65.
58. Petrus, L., Gray, D.G. and BeMiller, J.N. (1995) *Carbohydr. Res.*, **268**, 319–23.
59. Shiraishi, N., Katayama, T. and Yokota, T. (1978) *Cell. Chem. Technol.*, **12**, 429–43.
60. Miyamoto, T., Sato, Y., Shibata, T. *et al.* (1984) *J. Polym. Sci. Polym. Chem. Edn.*, **22**, 2363–70.
61. McCormick, C.L. and Callais, P.A. (1987) *Polymer*, **28**, 2317–23.
62. Kamide, K. and Saito, M. (1994) *Macromol. Symp.*, **83**, 233–71.
63. Nishi, N., Noguchi, J., Tokura, S. and Shiota, H. (1979) *Polym. J.*, **11**, 27–32.
64. Kaifu, K., Nishi, N., Komai, T. *et al.* (1981) *Polym. J.*, **13**, 241–5.
65. Kaifu, K., Nishi, N. and Komai, T. (1981) *J. Polym. Chem., Polym. Chem. Edn.*, **19**, 2361–3.
66. Kurita, K., Nishimura, S.-I., Ishi, S. *et al.* (1993) in *Carbohydrates and Carbohydrate Polymers, Analysis, Biotechnology, Modification, Antiviral, Biomedical and other Applications*, Yalpani, M. (ed.), ATL Press, Mt. Prospect, pp. 218–26.
67. Nishi, N., Ebina, A., Nishimura, S. *et al.* (1986) *Int. J. Biol. Macromol.*, **8**, 311–17.
68. Ishii, T., Ishizu, A. and Nakano, J. (1977) *Carbohydr. Res.*, **59**, 155–63.
69. Schweiger, R.G. (1979) *Carbohydr. Res.*, **70**, 185–98.

70. Petropavlovskii, G.A., Vasil'eva, G.G. and Kuvaldin, K.V. (1985) *Zh. Prikl. Khim.*, **58**, 421–3.
71. Vogt, S., Heinze, T., Rottig, K. and Klemm, D. (1995) *Carbohydr. Res.*, **266**, 315–20.
72. Wagenknecht, W., Nehls, I., Stein, A. *et al.* (1992) *Acta Polymer*, **43**, 266–9.
73. Hirano, S., Hasegawa, M. and Kinugawa, J. (1991) *Int. J. Biol. Macromol.*, **13**, 316–7.
74. Landoll, L.M. (1982) *J. Polym. Sci.*, **20**, 443.
75. Sinquin, A., Hubert, P. and Dellacherie, E. (1993) *Langmuir*, **9**, 3334–7.
76. Suzuki, M., Mikami, T., Matsumoto, T. and Suzki, S. (1977) *Carbohydr. Res.*, **53**, 223–9.
77. Hirano, S. and Ohe, Y. (1975) *Carbohydr. Res.*, **41**, C1–C2.
78. Braunmühl, V., Jonas, G. and Stadler, R. (1995) *Macromolecules*, **28**, 17–24.
79. Guittard, F., Yamagishi, T., Cambou, A. and Sixou, P. (1994) *Macromolecules*, **27**, 6988–90.
80. Akiyoshi, K., Nagai, K., Nishikawa, T. and Sunamoto, J. (1992) *J. Chem. Lett.*, 1727.
81. Greenway, T. M. (1994) in *Cellulosic Polymers, Blends and Composites*, Gilbert, R.D. (ed.), Hanser, Munich, pp. 173–88.
82. Stetzer, G.I. and Klug, E.D. (1980) in *Handbook of Water-soluble Gums and Resins*, Davidson, R.L. (ed.), McGraw-Hill, New York, pp. 4/1–4/28.
83. Just, E.K. and Majewicz, T.G. (1985) in *Encyclopedia of Polymer Science and Engineering*, vol. 3, Mark, H.F., Bikales, N.M., Overberger, C.G. and Menges, G. (eds), Wiley Interscience, New York, pp. 226–69.
84. Felcht, U.-H. (1985) in *Cellulose and its Derivatives*, Kennedy, J.F., Phillips, G.O., Wedlock, D.J. and Williams, D.J. (eds), Wiley, New York, pp. 273–84.
85. Hakamori, S. (1964) *J. Biochem. (Tokyo)*, **55**, 205–8.
86. Exner, J.H. and Steiner, E.C. (1974) *J. Am. Chem. Soc.*, **96**, 1782–7.
87. Blakeney, A.B. and Stone, B.A. (1985) *Carbohydr. Res.*, **140**, 319–24.
88. Needs, P.W. and Selvendran, R.R. (1993) *Carbohydr. Res.*, **245**, 1–10.
89. Narui, T., Takahashi, K., Kobayashi, M. and Shibata, S. (1982) *Carbohydr. Res.*, **103**, 293–5.
90. Savage, A.B. (1971) in *Cellulose and Cellulose Derivatives*, vol. 5, Bikales, N.M. and Segal, L. (eds) Wiley, New York, pp. 785–809.
91. Compton, J. (1963) *Methods Carbohydr. Chem.*, **3**, 317.
92. Battaerd, H.A.J. and Tregear, G.W. (1967) *Graft Copolymers*, John Wiley, New York.
93. Shah, S.B., Patel, C.P. and Trivedi, H.C. (1995) *Carohydr. Polym.*, **26**, 61–7.
94. Fanta, G.F. (1974) in *Block and Graft Polymerization*, vol. 1, Ceresa, R.J. (ed.), Wiley, London, 4.
95. Fanta, G.F., Burr, R.C. and Doane, W.M. (1979) *J. Appl. Polym. Sci.*, **24**, 2015–23.
96. Baker, T.J., Schroeder, L.R. and Johnson, D.C. (1981) *Cell. Chem. Technol.*, **15**, 311–20.
97. Schempp, W., Krause, Th., Seilfried, U. and Koura, A. (1984) *Papier*, **38**, 607–10.
98. Isogai, A., Ishizu, A. and Nakano, J. (1984) *J. Appl. Polym. Sci.*, **29**, 2097–109.
99. Genieser, H.-G., Gabel, D. and Jastorff, B. (1981) *J. Chromatogr.*, **215**, 235–42.
100. Landoll, I.M. (1982) *J. Polym. Sci., Polym. Chem. Edn.*, **20**, 443–55.
101. Zhang, T. and Marchant, R.E. (1994) *Macromolecules*, **27**, 7302–8.
102. Hwang, F.S. and Hogen-Esch, T.E. (1993) *Macromolecules*, **26**, 3156–60.
103. Zugenmaier, P. and Voihsel, M. (1993) *Makromol. Chem. Rapid Commun.*, **5**, 245–53.
104. Gelman, R.A. and Barth, H.G. (1986) *ACS Symp. Ser.*, **235**, 101–10.
105. Santacesaria, E., Trulli, F., Brussani, G.F. *et al.* (1994) *Carbohydr. Polym.*, **23**, 35–46.
106. Besemer, A.C. and van Bekkum, H. (1994) *Starch/Stärke*, **46**, 95–101.
107. Fujimoto, M., Fukami, K., Tsuji, K. and Nagase, T. (1980) *US Pat.* 4,186,024.
108. Nevell, P.T. (1963) *Methods Carbohydr. Chem.*, **3**, 164–85.
109. Pacsu, E. (1963) *Methods Carbohydr. Chem.*, **3**, 251–9.
110. Roberts, H.J. (1964) *Methods Carbohydr. Chem.*, **4**, 299–301.
111. Mahoney, J.F. and Purves, C.B. (1942) *J. Am. Chem. Soc.*, **64**, 9.
112. Heuser, E., Heath, M. and Shockley, W.H. (1950) *J. Am. Chem. Soc.*, **72**, 670.
113. Horton, D. and Lehman, J.H. (1978) *Carbohydr. Res.*, **61**, 553–6.
114. Ekman, K., Eklund, V., Fors, J. *et al.* (1986) in *Cellulose, Structure, Modification and Hydrolysis*, Young, R.A. and Rowell, R.M. (eds), Wiley Interscience, New York, pp. 131–48.
115. Naggi, A.M., Torri, G., Compagnoni, T. and Casu, B. (1986) in *Chitin in Nature and Technology*, Muzzarelli, R., Jeuniaux, C., Gooday, G.W. (eds), Plenum Press, New York, pp. 371–7.

116. Doctor, V.M. and Esho, D. (1983) *Carbohydr. Res.*, **121**, 312–15.
117. Ludwig-Baxter, K.G., Liu, Z. and Perlin, A.S. (1991) *Carbohydr. Res.*, **214**, 245–56.
118. Ball, D.H., Wiley, B.J. and Reese, E.T. (1992) *Can. J. Microbiol.*, **38**, 324–7.
119. Cimecioglu, A.L., Ball, D.H., Kaplan, D.L. and Huan, S.H. (1994) *Macromolecules*, **27**, 2917–22.
120. Sakamoto, M., Tseng, H. and Furuhata, K. (1994) *Carbohydr. Res.*, **265**, 271–80.
121. Mansson, P. and Wesfelt, L. (1980) *Cellulose Chem. Technol.*, **14**, 13–17.
122. Kondo, T. (1994) *J. Polym. Sci. B, Polym. Phys.*, **32**, 1229–36.
123. Vermeersch, J. and Schacht, E. (1986) *Makromol. Chem.*, **187**, 125–31.
124. Braun, P.J., French, D. and Robyt, J.F. (1985) *Carbohydr. Res.*, **141**, 265–71.
125. Takeda, Y., Hizukuri, S., Ozono, Y. and Suetake, M. (1983) *Biochim. Biophys. Acta*, **749**, 302–11.
126. Jain, R.K., Lal, K. and Bhantnagar, H.L. (1986) *Thermochim. Acta*, **97**, 99–114.
127. Green, J.W. (1963) *Methods Carbohydr. Chem.*, **3**, 327–31.
128. Nicholson, M.D., Johnson, D.C. and Haigh, F.C. (1976) *Appl. Polym. Symp.*, **28**, 931.
129. Nishiura, S., Nishi, N., Tokura, S. *et al.* (1986) *Carbohydr. Res.*, **146**.
130. Bogwald, J., Seljelid, R. and Hoffman, J. (1986) *Carbohydr. Res.*, **148**, 101–7.
131. Weill, C.E. and Bratt, M. (1967) *Carbohydr. Res.*, **4**, 230–8.
132. Horton, D. and Meshreki, M.H. (1975) *Carbohydr. Res.*, **40**, 345–52.
133. DeBelder, A.N. and Norrman, B. (1968) *Carbohydr. Res.*, **8**, 1–6.
134. Rees, D.A., Richardson, N.G., Wright, N.J. and Hirst, E. (1969) *Carbohydr. Res.*, **9**, 451–62.
135. de Noy, A.E.J., Besemer, A.C. and van Bekkum, H. (1994) *Recl. Trav. Chim. Pays-Bas*, **113**, 165–6.
136. Casu, B., Scovenna, G., Cifonelli, A.J. and Perlin, A.S. (1968) *Carbohydr. Res.*, **63**, 13.
137. Pigman, W.W., Browning, B.L., McPherson, W.H. *et al. J. Am. Chem. Soc.*, **71**, 2200.
138. Hoffman, J., Larm, O., Larsson, K. and Scholander, G. (1985) *Acta Chem. Scand., Ser. B*, **39**, 513–15.
139. Hoffman, J., Larm, O., Larsson, K. *et al.* (1982) *Carbohydr. Polym.*, **2**, 115–21.
140. Allen, T.C. and Cuculo, J.A. (1973) *J. Polym. Sci., Macromol. Rev.*, **7**, 189–262.
141. Painter, T.J., Cesaro, A., Delben, F. and Paoletti, S. (1985) *Carbohydr. Res.*, **140**, 61–8.
142. Cesaro, A., Delben, F., Painter, T.J. and Paoletti, S. (1985) in *New Developments in Industrial Polysaccharides*, Crescenzi, V., Dea, I.C.M. and Stivala, S.S. (eds), Gordon & Breach, New York, pp. 307–15.
143. Snyder, S.L., Vigo, T.L. and Welch, C.M. (1974) *Carbohydr. Res.*, **34**, 91.
144. Aspinall, G.O. and Nicolson, A. (1960) *J. Chem. Soc.*, 2503–7.
145. Aspinall, G.O. and Cairncross, I.M. (1960) *J. Chem. Soc.*, 3998–4000.
146. Heyns, K. and Beck, M. (1962) cited as unpublished results in Heyns, K. and Paulson, H., *Adv. Carbohydr. Chem. Biochem.*, **17**, 194.
147. Abbott, D., Bourne, E.J. and Weigel, H. (1966) *J. Chem. Soc.*, 827–8.
148. Miyah, H., Misaki, A. and Toru, M. (1973) *Carbohydr. Res.*, **31**, 277–87.
149. Horton, D. and Just, E. (1973) *Carbohydr. Res.*, **30**, 349–57.
150. Horton, D. and Just, E. (1973) *Carbohydr. Res.*, **29**, 173–9.
151. Braun, P.J., French, D. and Robyt, J.F. (1985) *Carbohydr. Res.*, **141**, 265–71.
152. Kondo, Y. and Takeo, K. (1976) *Carbohydr. Res.*, **52**, 232–4.
153. Wolfram, M.L. and Wang, P.Y. (1970) *Carbohydr. Res.*, **12**, 109.
154. Defaye, J., Driguez, H. and Gadelle, A. (1976) *Appl. Polym. Symp.*, **28**, 955.
155. Horton, D. and Usui, T. (1978) in *Carbohydrate Sulfates*, Schweiger, R.G. (ed.), *ACS Symp. Ser.*, **77**, 95–112.
156. Bredereck, K. (1967) *Tetrahedron Lett.*, 695–8.
157. Bosso, C., Defaye, J., Gadelle, A., Wong, C.C. and Pederson, C. (1982) *J. Chem. Soc. Perkin Trans.* I, 1579–85.
158. DeBelder, A.N., Lindberg, B. and Svensson, S. (1968) *Acta Chem. Scand.*, **22**, 949–52.
159. Ziderman, I. and Bel-Ayche, J. (1973) *Carbohydr. Res.*, **27**, 341–52.
160. Einarsson, M., Forsberg, B., Larm, O. *et al.* (1981) *J. Chromatogr.*, **215**, 45–53.
161. Andersson, L.O., Hoffman, J., Hohner, E. *et al.* (1982) *Thromb. Res.*, **28**, 741–7.
162. Larm, O., Larsson, K., Scholander, E. *et al.* (1981) *Carbohydr. Res.*, **91**, 13–20.
163. Augustinsson, B. and Scholander, E. (1984) *Carbohydr. Res.*, **126**, 162–4.

164. Augustinsson, B., Larm. O., Scholander, E. and Nahar, N. (1986) *Carbohydr. Res.*, **152**, 305–9.
165. Jansen, E.F. and Jang, R.J. (1946) *J. Am. Chem. Soc.*, **68**, 1475–7.
166. Lukas, H.J. and Stewart, W.T. (1940) *J. Am. Chem. Soc.*, **62**, 1070–4.
167. Plashchina, I.G., Semenova, M.G., Braudo, E.E. and Tolstoguzov, V.B. (1985) *Carbohydr. Polym.*, **5**, 159–79.
168. Steiner, A.B. and McNeely, W.H. (1950) *US Pat.* 2,494,912.
169. Wakamoto Pharmaceutical Co. (1980) *Jpn. Kokai Tokkyo Koho.* 81 79,101 (*Chem. Abstr.*, **95**, 175,786, 1981).
170. Sikorski, R.T. and Kokocinski, J. (1982) *Pol. Pl.* 113,337 (*Chem. Abstr.*, **98**, 162,680, 1982).
171. Ogamo, A., Matsuzaki, K., Uchiyama, H. and Nagasawa, K. (1982) *Carbohydr. Res.*, **105**, 69–85.
172. Funahashi, M., Matsumoto, I. and Seno, N. (1982) *Anal. Biochem.*, **126**, 414–21.
173. Yalpani, M. and Brooks, D.E. (1985) *J. Polym. Sci., Polym Chem. Edn.*, **23**, 1396–405.
174. Inoue, Y. and Nagaswa, K. (1982) *Carbohydr. Res.*, **111**, 113–25.
175. Yalpani, M. and Abdel-Malik, M.M. (1990) *US Pat.*, 4,963,664.
176. Yalpani, M. and Abdel-Malik, M.M. (1993) in *Carbohydrates and Carbohydrate Polymers, Analysis, Biotechnology, Modification, Antiviral, Biomedical and other Applications*, Yalpani, M. (ed.), ATL Press, Mt. Prospect, pp. 235–45.
177. Yalpani, M. and Hall, L.D. (1981) *Can. J. Chem.*, **59**, 3105–19.
178. Lockwood, J. (1972) *Food Process Ind.*, 47.
179. Teshirogi, T., Yamamoto, H., Sakamoto, M. and Tonami, H. (1980) *Sen-I Gakkaishi*, **36**, T501–T505.
180. Wolfram, M.L., Kato, H., Taha, M.I. *et al.* (1967) *J. Org. Chem.*, **32**, 3086–9.
181. Yalpani, M., Hall, L.D., Defaye, J. and Gadelle, A. (1984) *Can. J. Chem.*, **62**, 260–2.
182. Yalpani, M. (1985) *US Pat.* 4,531 000.
183. Yalpani, M. and Hall, L.D. (1982) *J. Polym. Sci., Polym. Chem. Edn.*, **20**, 3399–420.
184. Yalpani, M. (1987) *US Pat.* 4,683,298.
185. Yamaguchi, R., Arai, Y., Itoh, T. and Hirano, S. (1981) *Carbohydr. Res.*, **88**, 172–5.
186. Hirano, S. and Kondo, Y. (1982) *J. Chem. Soc. Jpn.*, 1622.
187. Kurita, K., Ichikawa, H., Ishizeki, S. *et al.* (1982) *Macromol. Chem.*, **183**, 1161–9.
188. Wolfram, M.L. and Montgomery, R. (1950) *J. Am. Chem. Soc.*, **72**, 2859.
189. Hirano, S. and Ohashi, W. (1977) *Carbohydr. Res.*, **59**, 285–8.
190. Nud'ga, L.A., Plisko, E.A. and Danilov, S.N. (1973) *Zh. Obshch. Khim.*, **43**, 2756–60 (2733–2736 in trans.).
191. Domard, A., Rinaudo, M. and Terrassin, C. (1986) *Int. J. Biol. Macromol.*, **8**, 105–7.
192. Nikolayev, A.F., Prokonov, A.A., Shulina, E.S. and Chudnova, W.M. (1985) *Zh. Prekl. Khim.* (*Leningrad*), **58**, 1916–17.
193. Hirano, S., Matsuda, N., Miura, O. and Tanaka, T. (1979) *Carbohydr. Res.*, **71**, 344–8.
194. Okimasu, S. (1956) *Bull. Agr. Chem. Soc. Jpn.*, **20**, 29–33.
195. Hirano, S. and Takeuji, M. (1983) *Int. J. Biol. Macromol.*, **5**, 373–6.
196. Yalpani, M. and Hall, L.D. (1984) *Macromolecules*, **17**, 272–81.
197. Yalpani, M. and Hall, L.D. (1981) *Can. J. Chem.*, **59**, 2934–9.
198. Hall, L.D. and Yalpani, M. (1981) *Carbohydr. Res.*, **91**, C1–C4.
199. Yalpani, M. and Hall, L.D. (1984) *Can. J. Chem.*, **62**, 975–80.
200. Pfannemüller, B. and Emmerling, W.N. (1984) *Stärke*, **35**, 298–303.
201. Hirsch, J. and Levine, M.N. (1992) *Blood*, **79**, 1–17.
202. Makela, D.G., Peterfly, F., Outschoorn, J.G. *et al.* (1984) *Scand. J. Immunol.*, **19**, 541–50.
203. Doherty, M.M., Hughes, P.J., Kim, S.R. *et al.* (1994) *Int. J. Pham.*, **111**, 205–11.
204. Rehakova, M., Bakos, D., Soldan, M. and Vizarova, K. (1994) *Int. J. Biol. Macromol.*, **16**, 121–4.
205. Smidsrød, O., Haug, A. and Larson, B. (1963) *Acta Chem. Scand.*, **17**, 1473–4.
206. Samal, R., Satrusallya, S.C., Sahoo, P.K. *et al.* (1984) *Colloid Polym. Sci.*, **262**, 939–47.
207. Simkovic, I. (1986) *J. Macromol. Sci., Rev. Macromol. Chem. Phys.*, **C26**, 67–80.
208. Gilbert, B.C., King, D.M. and Thomas, C.B. (1984) *Carbohydr. Res.*, **125**, 217–35.
209. Haug, A., Larsen, B. and Smidsrød, O. (1963) *Acta Chem. Scand.*, **17**, 1466–8.
210. Hjerde, T., Kristiansem, T.S., Stokke, B.T. *et al.* (1994) *Carbohydr. Polym.*, **224**, 265–75.

211. Christensen, B.E., Smidsrød, O. and Stokke, B.T. (1993) in *Carbohydrates and Carbohydrate Polymers, Analysis, Biotechnology, Modification, Antiviral, Biomedical and Other Applications*, Yalpani, M. (ed.), ATL Press, Mt. Prospect, pp. 166–973.
212. BeMiller, J.N. (1967) *Adv. Carbohydr. Chem.*, **22**, 25–108.
213. Aspinall, G.O. (1977) *Pure Appl. Chem.*, **45**, 1105–34.
214. Aspinall, G.O. (1982) in *The Polysaccharides*, vol. 1, Aspinall, G.O. (ed.), Academic Press, New York, pp. 35–131.
215. Salemis, Ph. and Rinaudo, M. (1984) *Polym. Bull.*, **12**, 283–5.
216. Kroner, K.H., Hustedt, H. and Kula, M.-R. (1982) *Biotechnol. Bioeng.*, **24**, 1015.
217. Hamacher, K., Schmid, G., Sahm, H. and Wandrey, Ch. (1985) *J. Chromatogr.*, **319**, 311–18.
218. Powell, D.A., Morris, E.R., Gidley, M.J. and Rees, D.A. (1982) *J. Mol. Biol.*, **155**, 517–31.
219. Richter, A.W., Ingelman, B. and Granath, K. (1982) *Abstracts XIth Intern. Carbohydrate Symposium*, Vancouver.
220. Haug, A., Larsen, B. and Smidsrød, O. (1967) *Acta Chem. Scand.*, **21**, 691–704.
221. Rees, D.A. and Welsh, E.J. (1977) *Angew. Chem. Intern. Edn. Engl.*, **16**, 214–42.
222. Miller, G.L. (1963) *Methods Carbohydr. Chem.*, **3**, 134–8.
223. Tsukada, S. and Inoue, Y. (1981) *Carbohydr. Res.*, **88**, 19–38.
224. Jeremic, K., Ilic, L. and Jovanovic, S. (1985) *Eur. Polym. J.*, **21**, 537–40.
225. Gordon, D.L., Linhardt, R. and Adams, H.P. (1990) *Clin. Neuropharm.*, **13**, 522–8.
226. Rees, D.A., Williamson, F.B., Frangou, S.A. and Morris, E.R. (1982) *Eur. J. Biochem.*, **122**, 71–9.
227. Norton, I.T., Goodall, D.M., Austen, K.R.J. *et al.* (1986) *Biopolymers*, **25**, 1009–29.
228. Dea, I.C.M., McKinnon, A.A. and Rees, D.A. (1972) *J. Mol. Biol.*, **68**, 153–72.
229. Casu, B. (1984) *Nouv. Rev. Fr. Hematol.*, **26**, 211–19.
230. Inoue, Y. and Nagasawa, K. (1985) *Carbohydr. Res.*, **141**, 99–110.
231. Nagaswa, K., Ogamo, A., Ishihara, H. and Yoshida, K. (1984) *Carbohydr. Res.*, **131**, 301–14.
232. Hirano, S., Kondo, Y. and Fujii, K. (1985) *Carbohydr. Res.*, **144**, 338–41.
233. Lindahl, U. and Axelsson, O. (1971) *J. Biol. Chem.*, **246**, 74–82.
234. Hopwood, J.J. and Miller, V.J. (1983) *Carbohydr. Res.*, **122**, 227–39.
235. Atha, D.H., Coxon, B., Reipa, V. and Gaiglas, A.K. (1995) *J. Pharm. Sci.*, **84**, 360–64.
236. Bienkowski, M.J. and Conrad, H.E. (1985) *J. Biol. Chem.*, **260**, 356–65.
237. Erbing, C., Lindberg, B. and Svensson, S. (1973) *Acta Chem. Scand.*, **27**, 3699–704.
238. Lindahl, U. and Axelsson, O. (1971) *J. Biol. Chem.*, **246**, 74–82.
239. Perlin, A.S., Ng Ying Kin, N.M., Bhattacharjee, S.S. and Johnson, L.F. (1973) *Can. J. Chem.*, **50**, 2437–41.
240. Huckerby, T.N., Sanderson, P.N. and Niedusynski, I.A. (1986) *Carbohydr. Res.*, **154**, 15–27.
241. Bock, C., Pederson, C., Defaye, J. and Gadelle, A. (1991) *Carbohydr. Res.*, **216**, 141–8.
242. Bergamaschi, B. and Hedges, J. (1995) *Carbohydr. Res.*, **267**, 115–26.
243. Defaye, J., Gadelle, A. and Pedersen, C. (1982) *Carbohydr. Res.*, **110**, 217–27.
244. Hiroi, O. and Kawahata, K. (1979) *Jpn. Kokai Tokkyo Koho* 79,148,890 (*Chem. Abstr.*, **92**, 148883).
245. Kurahashi, I., Yabe, H., Kawamura, Y. and Seo, H. (1986) *Jpn. Kokai Tokkyo Koho JP* 61,40,303 (*Chem. Abstr.*, **105**, 81083).

9 Carbohydrate–protein interactions

LOUIS T.J. DELBAERE and LATA PRASAD

9.1 General features

The common aspects of receptor sites in proteins are the presence of a significant number of hydrophilic and hydrophobic (often aromatic) residues which make up the carbohydrate combining site; usually there are adjustments of less than 2 Å in the positions of certain atoms of amino acid residues upon formation of the carbohydrate complex. This is illustrated in Figure 9.1 with the structure of the fourth lectin isolated from *Griffonia simplicifolia* (GS4) and the structure of the complex of GS4 with the methyl glycoside of the Lewis b blood group determinant, α-L-Fuc(1 → 2)β-D-Gal(1 → 3)[α-L-Fuc(1 → 4)]β-D-GlcNAc (Delbaere *et al.*, 1993). All of the illustrations in this chapter have been carried out with the use of the software SETOR (Evans, 1993). The sugar residues of Lewis b are denoted as (*a*), β-D-GlcNAc; (*b*), β-D-Gal; (*c*), α-L-Fuc(1 → 4); (*d*), α-L-Fuc(1 → 2).

Inhibition studies on complexes of GS4 with Lewis b analogues indicated that there are three key hydroxyl groups essential for complex formation

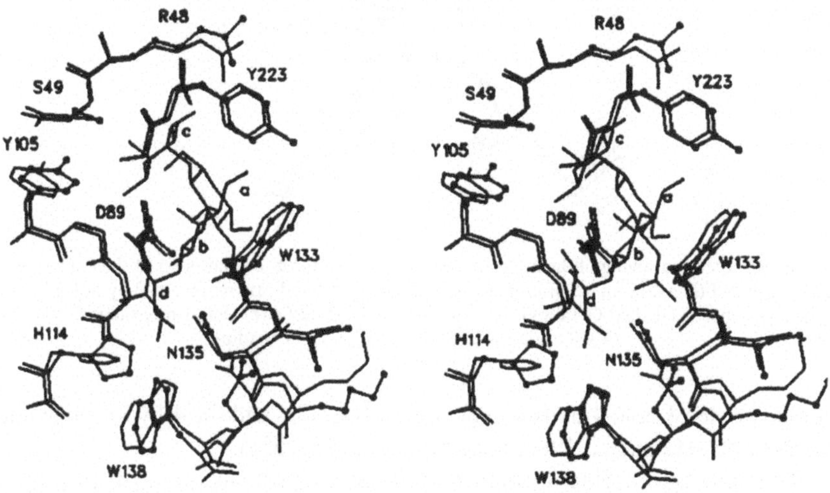

Figure 9.1 Stereo drawing of the non-hydrogen atoms of the carbohydrate-binding site of native GS4 (ball and stick) and the complex (stick bonds) of GS4 with the Lewis b human blood group determinant. The sugar residues of Lewis b are denoted as: *a*, β-D-GlcNAc; *b*, β-D-Gal; *c*, α-L-Fuc(1 → 4); *d*, α-L-Fuc(1 → 2).

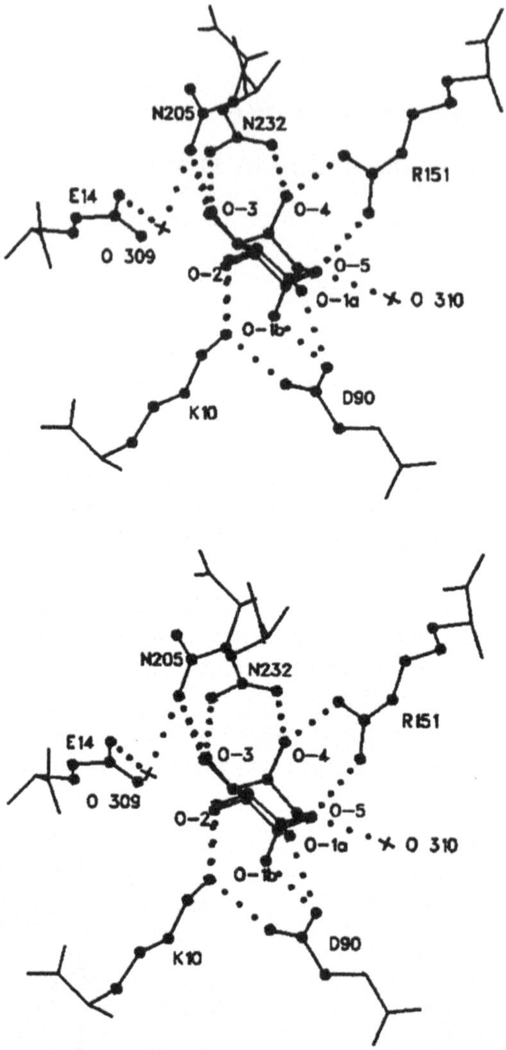

Figure 9.2 Stereo drawing of the hydrogen bonds between the α- or β-anomer of the L-arabinose substrate and the L-arabinose-binding protein. Hydrogen bonds are represented by dotted lines. The α-anomer is labelled O-1a and the β-anomer is labelled O-1b. Water molecules O 309 and O 310 are intimately involved in sugar binding.

which are the 3- and 4-hydroxyl groups of the *b*Gal and the 4-hydroxyl group of the *c*Fuc (Lemieux, 1989; Spohr *et al.*, 1985, 1986).

The monosaccharide–protein complexes of proteins which transport a specific sugar have the sugar completely buried and isolated from contact with bulk solvent, for example L-arabinose binding protein (Quiocho, 1988), as shown in Figure 9.2.

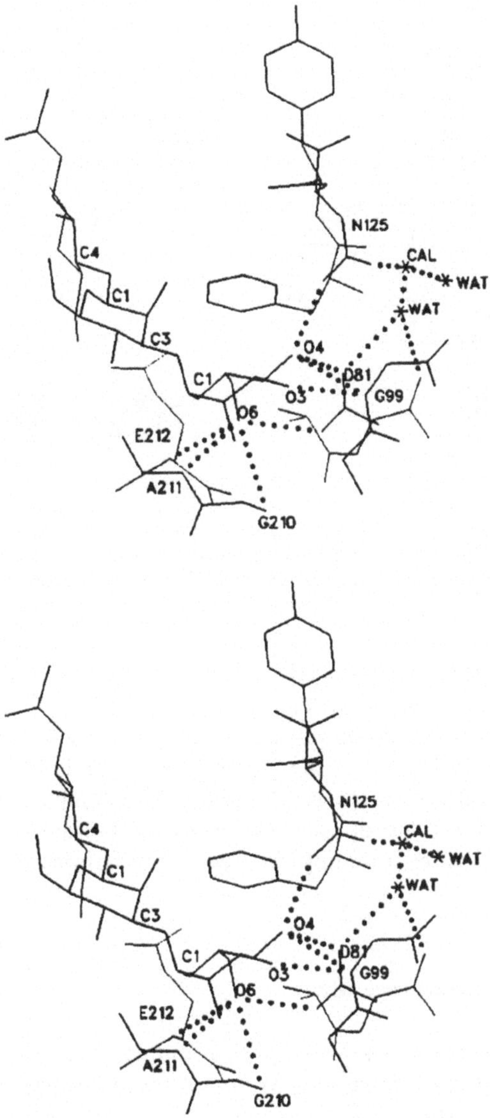

Figure 9.3 Stereo illustration of the interaction of the trisaccharide residue bound to *Lathyrus ochrus* (LOLI); the α-Man residue of the trisaccharide is the only sugar which is extensively bound to the protein. Hydrogen bonds are denoted by dotted lines.

Oligosaccharides often have one monosaccharide residue extensively bound to the protein whereas others have additional contacts (Bourne *et al.*, 1990). The complex of α-D-Man(1 → 3)β-D-Man(1 → 4)β-D-GlcNAc with isolectin I from *Lathyrus ochrus* (LOLI) in Figure 9.3 provides a good

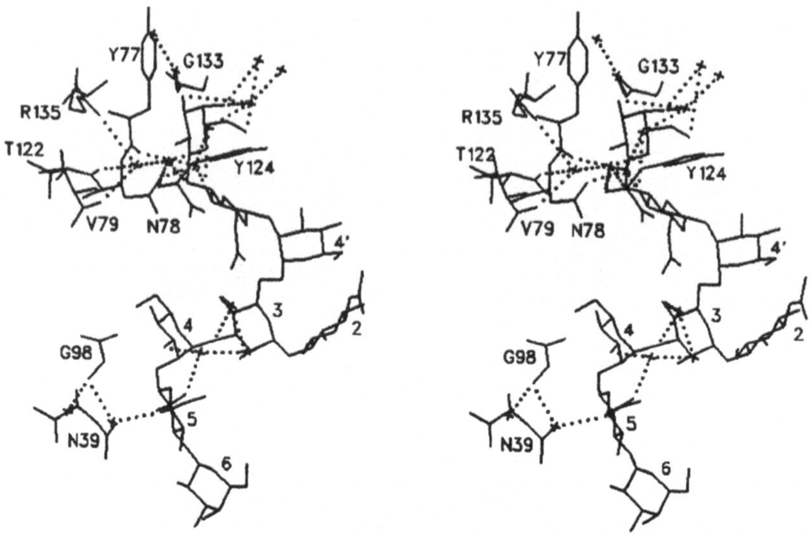

Figure 9.4 Stereo diagram which illustrates the polar interactions for the water molecules that bridge the octasaccharide and *Lathyrus ochrus* (LOLI) in the complex. Hydrogen bonds are shown as dotted lines.

illustration of this aspect in that the α-Man residue is the only sugar which is comprehensively bound to the lectin.

In some cases of oligosaccharide binding there are many interactions with one sugar residue and the other sugars are linked to the protein via water molecules (Bourne *et al.*, 1992). The complex of LOLI with a biantennary octasaccharide of the *N*-acetyllactosamine type provides an example of water bridging between the carbohydrate and protein (Figure 9.4).

9.2 Basis of structural specificity

Carbohydrate-binding sites have a shape complementary to the sugar; both polar and non-polar interactions are involved in binding carbohydrate to protein. Hydrogen bonds (including some ionic interactions) provide more directional specificity than do van der Waals contacts, but both are important for this example of molecular recognition. The binding of methoxy-α-mannose to the mannose-specific *Galanthus nivalis* lectin (Hester *et al.*, 1995) is illustrated in Figure 9.5. The 2-hydroxyl group interacts with the side chains of Asp 91 and Asn 93; the 3-hydroxyl group and 4-hydroxyl group interact with Gln 89 and Tyr 97, respectively. In addition, there are non-polar van der Waals interactions with the side chain of Val 95 and the C3 and C4 ring atoms and polar van der Waals interactions between the

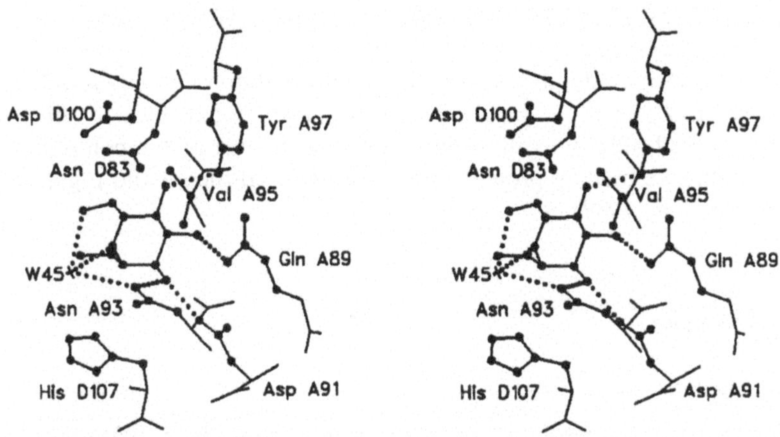

Figure 9.5 Stereo view of Me-α-Man bound to *Galanthus nivalis* lectin. Hydrogen bonds are shown as dotted lines. In addition to hydrophilic interactions there are also non-polar van der Waals contacts between the side chain of Val95 and the C3 and C4 atoms of the Man residue.

side chains of Asn 83, Asp 100, His 107 and the 1-methoxy methyl and 6-hydroxyl groups.

Water sometimes plays a pivotal role in binding carbohydrate to a protein in forming a bridge between the sugar and protein as shown in Figure 9.6 (Cygler *et al.*, 1991) in the structure of the complex of an oligosaccharide with an antibody Fab fragment. This water molecule accepts a hydrogen bond from the 4-hydroxyl group of the abequose residue at the bottom of the binding pocket and donates a hydrogen bond to the ring O5 oxygen atom. In addition, the water molecule accepts a hydrogen bond from the NE2 of His 35H and donates a bifurcated bond to the main chain carbonyl

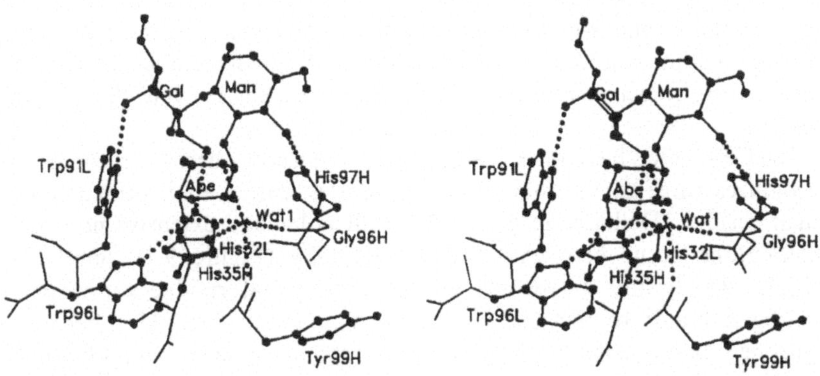

Figure 9.6 Stereo diagram of the complex of the α-D-Gal[α-D-Abe]α-D-Man epitope bound to the Se155-4 Fab fragment. Hydrogen bonds are illustrated as dotted lines. Note the key interactions of the water molecule Wat1 in forming a bridge between the sugar and the protein.

groups of Gly 96H and Tyr 99H, where the H suffix refers to the heavy chain of the Fab fragment.

Also, water may stabilize a particular conformation of an oligosaccharide (Bourne *et al.*, 1992; see Figure 9.4) or water may contribute to substrate specificity and affinity as illustrated in the structures of the complexes of the L-arabinose-binding protein with L-arabinose, D-fucose and D-galactose (Quiocho *et al.*, 1989).

9.3 Examples

9.3.1 Lectins

Lectin structure has recently been extensively reviewed (Rini, 1995) and some typical examples will be mentioned herein. LOLI has four ordered water molecules located in the binding site and these are displaced when a complex with sugar is formed (Bourne *et al.*, 1990). In the LOLI/trisaccharide complex, there is one Man residue bound tightly in a binding pocket and the other Man residue and a GlcNAc residue are linked to the protein by a channel of water molecules (Figure 9.3). The LOLI/octasaccharide complex also has a corresponding Man residue in the binding pocket whereas the other seven sugar residues are linked to the protein by way of water molecules (Figure 9.4); in this case it appears that the conformation of the octasaccharide is stabilized by hydrogen bonds with waters within the octasaccharide and by hydrogen bonds between the octasaccharide and the protein (Bourne *et al.*, 1992).

The EcorL complex with lactose has only the Gal residue and the glycosidic bond completely resolved in the electron density map; electron density for the Glc residue is barely visible because it is likely disordered (Shaanan *et al.*, 1991). As illustrated in Figure 9.7, there are strong hydrogen bonds between the 3- and 4-hydroxy groups of the galactose and the side chains of Asp 89 and Asn 133. In addition there are hydrophobic interactions between the galactose and the side chains of Ala 88, Tyr 106, Phe 131 and Ala 218.

The GS4/Lewis b complex is an example of a lectin where the receptor site recognizes three of the four monosaccharides, rather than primarily one sugar residue (Delbaere *et al.*, 1993). Both polar and extensive non-polar interactions occur between Lewis b and GS4 (Figure 9.1). Similar to EcorL, the 3- and 4-hydroxyl groups of the Gal residue form strong hydrogen bonds to the side chains of Asp 89 and Asn 135. There is a hydrogen bond between the 3-hydroxyl group of *b*Gal and Gly 107N. There are also hydrogen bonds between the 4-hydroxyl group of *c*Fuc and Ser 49N, Ser 49OG and Gly 222N. The 3-hydroxyl group of *c*Fuc is hydrogen bonded to Arg 48NH2 and Gly 222O. The 2-hydroxyl group of *d*Fuc is

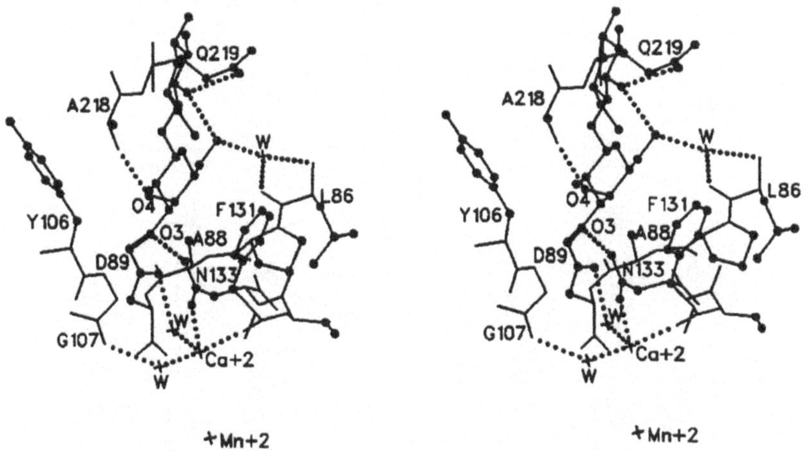

Figure 9.7 Stereo illustration of the complex of lactose with EcorL. Hydrogen bonds are shown as dotted lines. Only the 3- and 4-hydroxyl groups of the galactosyl residue make hydrogen bonds to protein residues; there are hydrophobic interactions between the hydrocarbon face of the galactosyl residue and the side chain of Phe 131.

hydrogen bonded to Asn 135ND2 and Trp 138NE1 and the 3-hydroxyl group of *d*Fuc hydrogen bonds to Trp 138NE1. In addition, several hydrophobic interactions occur between the tetrasaccharide and six aromatic amino acid residues, Tyr 105, Phe 108, His 114, Trp 133, Trp 138 and Tyr 223. In comparison with the native GS4 structure, several amino acid residues move by 1–2 Å in order to optimize the polar and non-polar interactions in the complex.

The complex of *Galanthus nivalis* lectin/methyl α-D-mannoside has several hydrogen bonds involved in the carbohydrate–protein interactions as well as van der Waals contacts, but no stacking interactions with an aromatic side chain of an amino acid residue (Hester *et al.*, 1995).

Galectins bind β-galactoside and have many hydrogen bonds and the structures of two complexes of oligosaccharides with bovine galectin-1 have a stacking of the side chain of Trp 68 with the hydrophobic face of the Gal ring (Bourne *et al.*, 1994; Figure 9.8). There are also 10 waters which form hydrogen bond bridges between sugar and lectin and one of these water molecules is illustrated in Figure 9.8.

9.3.2 Antibodies

The structure of the Se155-4 Fab fragment/dodecasaccharide complex revealed three of the 12 sugars bound to the antigen-binding site (Cygler *et al.*, 1991; Figure 9.6). The abequose (3,6-dideoxygalactose) residue was completely buried in the complex whereas the Gal and Man residues are on the surface of the protein. Aromatic amino acids Trp and His and Tyr form stacking

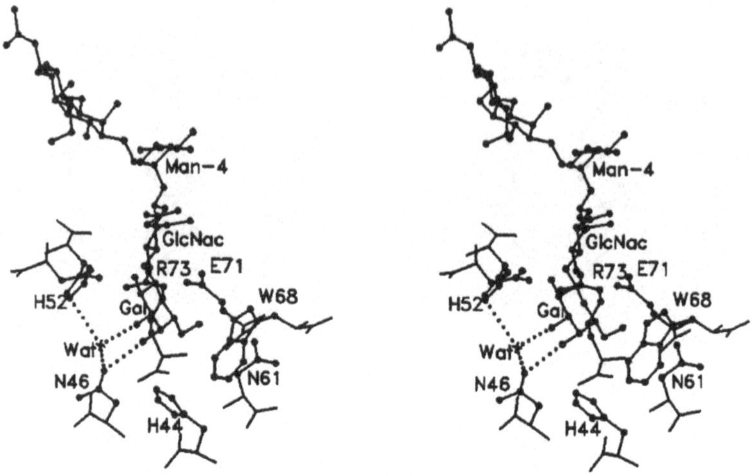

Figure 9.8 Stereo view of the complex of a biantennary octasaccharide and bovine galectin-1. Hydrogen bonds are denoted as dotted lines. There are also non-polar van der Waals contacts between the hydrocarbon face of the galactosyl residue and the side chain of Trp 68.

interactions with the hydrophobic faces of the sugars and also form hydrogen bonds to the sugar atoms. One water molecule is located at the bottom of the binding pocket and it forms essential hydrogen bonds and contributes importantly to the specificity of the carbohydrate–protein interaction.

9.4 Enzymes

The hen-egg-white lysozyme (HEWL) structure was the first enzyme whose X-ray structural analysis was carried out. Structures of complexes of disaccharides, trisaccharides and tetrasaccharides bound to HEWL and additional model building have revealed the various interactions which are formed between substrates and enzyme and have provided insight into the mechanism of catalysis (Cheetham *et al.*, 1992; Ford *et al.*, 1974; Strynadka and James, 1991). Recently, the structure of di-, tri- and tetrasaccharide complexes with rainbow trout lysozyme (RBTL) have provided new valuable information (Karlsen and Hough, 1995). RBTL is very homologous in primary, secondary and tertiary structure to HEWL. Several sites of binding for the various monosaccharide residues of an oligosaccharide have been identified and interpreted from these studies. Lysozyme structural work provides a general example for the binding of carbohydrates and the catalysis of relevant reactions by enzymes. The oligosaccharide units bound to lysozyme were either *N*-acetylglucosamine (NAG) polymers or alternating *N*-acetylmuramic acid (NAM) and NAG units. Figure 9.9 illustrates the binding of a tetra-NAG-oligosaccharide to RBTL in sites A to D.

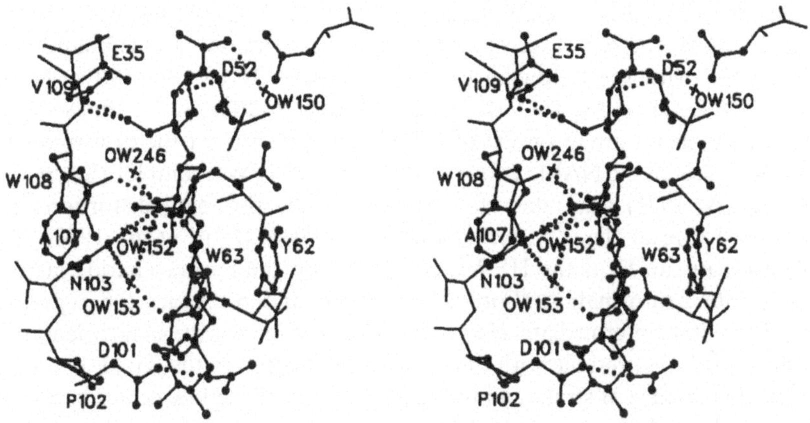

Figure 9.9 Stereo drawing of the interaction between the tetra-*N*-acetylglucosamine residues bound to the sites A to D in rainbow trout lysozyme. Hydrogen bonds are denoted as dotted lines and water molecules are illustrated as crosses.

In site A, there is a single hydrogen bond from the N2-acetamido group of the first NAG residue to Asp 101OD2.

In site B there is a hydrogen bond from O6 of the pyranosyl ring to Asp 101OD2 and a hydrogen bond from O7 of the acetamido group of the second NAG residue to Asn 103ND2. In addition, there are extensive hydrophobic interactions between the hydrophobic face of the second NAG residue and the phenyl ring of Tyr 62. There are similar interactions in the structure of the complex of NAM–NAG–NAM with HEWL, where the hydrogen bond from Asn 103ND2 is to O10 of the lactyl group on O3 of the first NAM and the hydrophobic interactions are between the hydrophobic face of the first NAM and the indole group of Trp 62 of HEWL. Complexes with both RBTL and HEWL also have hydrogen bonds between water molecules and the O6 and O7 atoms of the pyranosyl ring in this site.

Site C is specific for NAG, and both RBTL and HEWL bind this sugar residue very similarly (Cheetham *et al.*, 1992; Karlsen and Hough, 1995; Strynadka and James, 1991). Hydrogen bonds occur between the N2 and O7 of the 2-acetamido group and Ala 107O and Asn 59N, respectively. There is a hydrogen bond between O3 of the NAG residue and Trp 63NE1. Hydrophobic interactions occur between the methyl group at C2 of the 2-acetamido group and the side chains of Trp 108, Ile 58 and Val 98 (Ile 98 in HEWL). The complex with HEWL has an additional hydrogen bond from O6 of the NAG residue to Trp 62NE1.

Site D in the RBTL complex binds a NAG residue which has the chair conformation and has the following hydrogen bonds: O6 to Glu 35OE1, O6 to Val 109N, O5 and O1 to Asp 52OD2, O1 to Asn 46OD1 and a water molecule to N2 of the acetamido group. In the HEWL complex with NAM–NAG–NAM, the NAM residue has a sofa conformation and is bound very differently, with the following hydrogen bonds: O1 to Glu 35OE1, N2 to Asp 52OD2, O1 to Asn 46OD1, O6 to Val 109N and two water molecules to O10 of the lactyl group on O3 of NAM. The presence of the distorted sofa conformation of the NAM residue in site D of the HEWL complex (Strynadka and James, 1991) was taken as evidence for the proposed distortion of a pyranosyl ring in this site during enzyme catalysis (Ford *et al.*, 1974). The δ-lactone in site D of the HEWL complex with the tetrasaccharide lactone has a sofa conformation (Ford *et al.*, 1974) and resembles the supposed transition-state intermediate. However, this δ-lactone has the same orientation as does the NAG residue with the chair conformation in the RBTL complex, so the importance of strain and the exact position of the D saccharide residue has been brought into question (Karlsen and Hough, 1995).

In any case, the interactions that occur in the complexes of lysozyme with oligosaccharides are typical of those that occur in other enzyme-inhibitor/substrate complexes. For example, in the X-ray structure of a covalent catalytic intermediate of a β-glycosidase there are many hydrogen bonding interactions between polar groups of the disaccharide and the enzyme as well as hydrophobic interactions between hydrophobic regions of the sugar residues and the side chains of three tryptophan residues of the enzyme (White *et al.*, 1996).

References

Bourne, Y., Bolgiano, B., Laio, D.-I. *et al.* (1994) Crosslinking of mammalian lectin (galectin-1) by complex biantennary saccharides. *Nature Structural Biology*, **1**, 863–70.

Bourne, Y., Rougé, P. and Cambillau, C. (1990) X-ray structure of a (α-Man(1-3)β-Man(1-4)GlcNAc)-lectin complex at 2.1 Å resolution. *Journal of Biological Chemistry*, **265**, 18161–5.

Bourne, Y., Rougé, P. and Cambillau, C. (1992) X-ray structure of a biantennary octasaccharide–lectin complex refined at 2.3 Å resolution. *Journal of Biological Chemistry*, **267**, 197–203.

Cheetham, J.C., Artymiuk, P.J. and Phillips, D.C. (1992) Refinement of an enzyme complex with inhibitor bound at partial occupancy Hen egg-white lysozyme and tri-*N*-acetylchitotriose at 1.75 Å. *Journal of Molecular Biology*, **224**, 613–28.

Cygler, M., Rose, D.R. and Bundle, D.R. (1991) Recognition of a cell-surface oligosaccharide of pathogenic *Salmonella* by an antibody Fab fragment. *Science*, **253**, 442–445.

Delbaere, L.T.J., Vandonselaar, M., Prasad, L. *et al.* (1993) Structures of the lectin IV of Griffonia *simplicifolia* and its complex with the Lewis b human blood group determinant at 2.0 Å resolution. *Journal of Molecular Biology*, **230**, 950–65.

Evans, S.V. (1993) SETOR: hardware-lighted three-dimensional solid model representations of macromolecules. *Journal of Molecular Graphics*, **11**, 134–8.

Ford, L.O., Johnson, L.N., Machin, P.A. *et al.* (1974) Crystal structure of a lysozyme–tetrasaccharide lactone complex. *Journal of Molecular Biology*, **88**, 349–71.

Hester, G., Kaku, H., Goldstein, I.J. and Wright, C.S. (1995) Structure of mannose-specific snowdrop (*Galanthus nivalis*) lectin is representative of a new plant lectin family. *Nature Structural Biology*, **2**, 472–9.

Karlsen, S. and Hough, E. (1995) Crystal structures of three complexes between chito-oligosaccharides and lysozyme from the rainbow trout. How distorted is the NAG sugar in site D? *Acta Crystallographica* section D, **51**, 962–78.

Lemieux, R.U. (1989) The origin of the specificity in the recognition of oligosaccharides by proteins. *Chemical Society Reviews*, **18**, 347–74.

Quiocho, F.A. (1988) Molecular features and basic understanding of protein–carbohydrate interactions: the arabinose-binding protein–sugar complex. *Current Topics in Microbiology and Immunology*, **139**, 135–48.

Quiocho, F.A., Wilson, D.K. and Vyas, N.K. (1989) Substrate specificity and affinity of a protein modulated by bound water molecules. *Nature*, **340**, 404–7.

Rini, J.M. (1995) Lectin structure. *Annual Reviews of Biophysics and Biomolecular Structure*, **24**, 551–77.

Shaanan, B., Lis, H. and Sharon, N. (1991) Structure of a legume lectin with an ordered *N*-linked carbohydrate in complex with lactose. *Science*, **254**, 862–6.

Spohr, U., Back, M. and Lemieux, R.U. (1986) *Abstracts of the XIIIth International Carbohydrate Symposium*, August 10–15, 1986, p. 353, Cornell University Press, Ithaca, NY.

Spohr, U., Hindsgaul, O. and Lemieux, R.U. (1985) Molecular recognition II. The binding of the Lewis b and Y human blood group determinants by the lectin IV of *Griffonia simplicifolia*. *Canadian Journal of Chemistry*, **63**, 2644–52.

Strynadka, N.C.J. and James, M.N.G. (1991) Lysozyme revisited: crystallographic evidence for distortion of an *N*-acetylmuramic acid residue bound in site D. *Journal of Molecular Biology*, **220**, 401–24.

White, A., Tull, D., Johns, K. *et al.* (1996) Crystallographic observation of a covalent catalytic intermediate in a β-glycosidase. *Nature Structural Biology*, **3**, 149–54.

Index

Note: page numbers in **bold** refer to figures

ab initio calculations 237
N-Acetylneuraminic acid **70**, 72
Acid scavenger 158
Adam's catalyst 305
Adiabatic energy map 243, 251
Agar 108, 283
Agarose 282, **283**
Alginic acid 285, 308
  esterification 306
D-Allose 58, **60,**
  5-amino-5-deoxy- 80, **81**
  2,3,4-tri-*O*-benzyl-5,6-isopropylidene- **65**
  6-deoxy- 54, **55**
  3,6-dideoxy 99, **100**
  5-thio- **77**
D-Altrose
  5-thio- 77, **78**
L-Altrose
  carba-α-pentaacetate 83, **86**
  carba-β-pentaacetate 82, **83**
AMBER 16, 245
AM1 calculation 238
Amylose **275**, 288
  acetate 302
  carbamate 297
  oxidation 305, 306
  tosylate 301
Amylopectin 273, 288
  oxidation 306
Angyal, S.J. 6, 11, 29
Anomeric effect 7, 15
Anomeric carbene 177
Anomeric radical 176
Anomers 7, **8,** 36
Antibody Fab fragment **323**
Anticoagulant 107, 111, 113, 114, 120, 132
Anti-thrombin 111, 120, 121, 130, 132
Anti-tumour 113, 127
Anti-viral 111, 113, 132
D-Arabinose, carba-α- 88, **88**
L-Arabinose, β- **4**
L-Arabinose-binding protein **320**, 324
L-Arcanose **66**, 67

Block synthesis 162
Boltzmann weighting 12, 36, 246, 247
Borate complexation 6
Böseken, J. 6
Bromine oxidation 306

Calicheamycin 102
Carba-monosaccharides 81
Carrageenans 109, 134, **283**, 284
Carbodiimides 307
Cellobiose 237, **261**
Cellulose 231, 232, 272, **274**
  acetate 296, 298
  carbamate 297, 302
  carboxymethyl 299, 300, 303
  esters 296
  nitrate 296, 302
  nitrite 302
  oxidation 304, 305, 306
  phosphate 297
  sulphate 132, 298, 302
  tosylate 301
  xanthate 297
C-glycoside 175
Chair form 4, 8
  see also Conformation
CHARMM 16, 37, 241, 245, 246
Chitin 274, **276**
  carboxymethyl 303
  esters 298
  sulphate 132
Chitosan
  derivatives 309, 313
  esters 298
  oxidation 305
  sulphate 132, 133, 302
Chiro-optical methods 236, 278
Chlorosulphonic acid 129, 130
Chondroitin sulphate 119, 311
Combinatorial synthesis 216
Computer modelling 12
  see also Density functional method; Force fields; Molecular dynamics; Multiple minimum problem; Quantum mechanics
Conformation
  acyclic carbohydrates 17
  boat 23, 140
  chair 4, 8, **260**
  cycloamyloses 34

Conformation (*cont'd*)
disaccharides 260
envelope 22
exocyclic groups 32, 33, 37, 260
flexibility 3, **4**, 10, 34
free energy 251, 252
half-chair 23
instability factors 6
monosaccharides 5, 18, 259
multiple minima 4
oligosaccharide 228
pseudorotation 4, 24, 27, 36
polysaccharides 228, 263
representation 23
Altona–Sundaralingham system 24
Cremer–Pople system 25
puckering parameter 26
seven-membered rings 26
skew 23
sulphates 140
twist 22
Conformational wheel 24, **25**
Consecutive synthesis 162
Consensus sequence 188, 190
Convergent synthesis 162, 195
Crystalline structure
databases 11
disorder 3, 10
packing effects 35
Cyclodextrins 34, 230, 304
Cuprammonium reaction 6
Curdlan 288
oxidation 306
sulphate 133
Cystic fibrosis 125

L-Daunosamine 50, **51**
Degree of substitution (DS) 295
Density functional method 15
Dermatan sulphate 117, **118**, 302, 311
Dextran 304
aldonamides 300
oxidation 305, 306
sulphate 132
D-Digitoxose 49, **50**
Dipole–dipole interaction energies 22
DISCOVER 240
Distortion energies 19–21
DMSO oxidation 305
DCC 306

Electron diffraction 229, 267
Energy functions 240
Energy surfaces 28
Enzymatic synthesis 178, 213
Exo-anomeric effect 230
Extensin 189

Fibre diffraction 264
Fischer E. 187
glycosidation 152
Food products 110
Force fields 16, 240
D-Fructose
carba-β-furanose 89, **89**
carba-β-pyranose 90, **90**
α-furanose **27**
5-thiofuranose **79**
Fucans 110
Fucoidan 111, **112**
L-Fucose
5-thio- 77, **78**

GAG *see* Glycosaminoglycan
Galactomannan 285, 287, 308
D-Galactose
5-amino-5-deoxy- 80
3,6-anhydro- **108**, 134
carba-α- 82, **84**
2-deoxy- **49**
α-pyranosyl- 107, **108**
L-Galactose
3,6-anhydro- **108**, 136
α-pyranosyl- 107, **108**
6-sulphate 135
Gauche effect 8
D-Glucose
anomers 7, 36
2-amino-2-deoxy-carba-β-pentaacetate 83, **87**
carba-β-pentaacetate 82, **83, 87**
2-deoxy-2-acetamido-5-thio- **76**
(5RS)-hydroxyphosphinyl- 81, **82**
L-Glucose **48,** 56, **58**
Glycals 104
D-Glyceraldehyde conformation 29, **29**
Glycogenin 189
Glycoprotein
N-linked 188
O-linked 188, **189**
Glycosaminoglycans 114–24, **115**, 311
Glycosyl acceptor 151
Glycosylamine 173, 192, 194, 198, 215
N-Glycosylaminoacid 174, 188, 192
N-Glycosyl asparagine 188, 192
Glycosyl donor 151, 164
2-acetamido 159
acetate 152, 175
1,2-anhydro 156, 157, 175
2-azido 159, 204
2-*N,N*-diacetyl 159
fluoride 154, 181
glycal 156, 175, 197, 219
halides 153, 175, 193, 200, 206, 207, **208**
oxazoline 193, **194, 206**
pentenyl glycoside 155

Glycosyl donor (*cont'd*)
  phosphate 156
  2-phthalimido 159
  selenoglycoside 156
  sulfoxide 156, 164
  thioglycoside 154, 161, 164, 207, 215
  trichloroacetimidate 154, 160, 175, 202, 203, 209
  vinyl glycoside 156, 164
  xanthogenate 209
Glycosyl hydrolase 181
Glycosyl transferase 178, 213
Grafting reactions 300
GROMOS 16, 240, 245

Heparan sulphate 120
Heparin 120, 311
  analogue 305
Hexoses
  5-amino-5-deoxy- 79
  6-deoxy- 54
  2,6-dideoxy- 49, 52, 63
  L- 48
  2,3,6-trideoxy-3-amino-L- 50
HIV 111,
HSEA force field 240, 245
Hyaluronic acid 311
Hydrogen bonding 31, 33, 35, 234, 235, 248, 323, 327
Hydrogen fluoride 312
Hydrogen peroxide 313

α-Idopyranose **28, 31**
Iduronic acid 67, **68, 119**, 132
Infrared spectroscopy 241
  sulphates 138
Isomaltose **231**
Inositol
  L-*chiro*- 82, **84**
Internal delivery 177

KDO **72**, 73
Keratan sulphate 117

Latent–active acceptor–donor pairs **165**
Lectins 190, 324
  EcorL 234
  galectins 325
  GS4 324
  Galanthus nivalis 322
  Griffonia simplicifolia 319
  Lathyrus ochrus **321, 322**
Lewis[a] trisaccharide 196
Lewis b determinant **319, 324**
Lewis[X] antigen 211, 213
Locust bean gum 290
L-selectin 111
Lysozyme 326

Maltose **229**, 237, **242, 244**, 246, **247**
Mannan I 268
D-Mannose
  5-amino-5-deoxy- 80
  β-linkages 210
  4-deoxy-α- 61, **61**
  2-deoxy-2-acetamido-5-thio- **76**
Mannotriose 233
Mass spectrometry
  sulphates 138
Melezitose 233
Methyl dextran 304
Methylation analysis 139
Microbial polysaccharides 288
Microheterogeneity 191
MM3 16, 29, 31, 240
Molar substitution (MS) 295
Molecular dynamics 13, 17, 36, 246, 248
Molecular mechanics 13, 15, 240, 245, 261
Monosaccharide 1
  acyclic form 2, 38
  asymmetric synthesis 48
  carbonyl form 2
  descent of series 72
  dynamic nature 3
  halogenation 98
  higher carbon 65, 68, 69, **70**
  hydrate 2, 38
  representation 8
  sulphates 129
  synthons
    allyl-tri-butylstannanes **68, 69**
    2-acetylfuran **51**
    4-benzhydryloxy-(*E*)-2-buten-1-ol 47, **48**
    1,4-dihydroxybut-2-ene **49**
    1,3-dioxan-5-one **57**
    2,3-isopropylidene-D-glyceraldehyde 62, **65, 66**
    ethyl-2-furancarboxylate **53**
    4-hydroxyhepta-1,5-diene **50**
    methylcyclopentadiene **52**
    7-oxabicyclo[2,2,1]hept-5-en-2-one 59, **60**
    quebrachitol 82, **84**
    (−)-quinic acid 90, **90**
  structure library 260
  5-thio- 75
Multiple helices 265
Multiple minimum problem 14, 236, 243
Mutarotation 2, 37
  catalysis 38
  in metabolism 40

Nanoparticles 299
Neocarrabiose 248, **249, 250**
Neoglycoproteins 221
Neutron scattering 279

N-glycoside *see* Glycosylamine
Nitrous acid 123, 311
Nitrogen dioxide 304
Nojirimycin 79, **80**
Nod factors **127**
NMR spectroscopy 11, 278
  ^{13}C substitution 12
  chemical shifts 12
  coupling constants 11, 235
  Karplus equation 11, 236
  NOE measurements 11, 236, 278
  oligosaccharides 235
  relaxation times 279
  sulphates 137
Nystose 233

Oligosaccharide
  conformation 228
  by degradation 123, 311
  crystal structures 233
  hydration 247, 324
  mapping 123
  NMR 235
  N-linked 234
  sulphates 125, **131**
  synthesis 150, 207
D-Olivose 49
L-Olivomycose 67, **67**
Optical rotation, ORD 12, 37, 237, 278
Orthoester 158
Oxazolidine 159

Panose 233
PCILO calculation 239
Pectin 286
  esterification 307
Periodic acid 301, 311
Phenylchlorosulphate 129, 133
PIMM 16
Polysaccharide
  acylation 296, 301, 308
  alkylation 299, 303, 308
  amidation 307
  amination 307
  carbamates 297
  conformation 228
  degradation 309
  esters 298, 306
  ethers 299, 300
  fibre diffraction 234, 266
  graft copolymers 300
  helical parameters 263
  hydration 270
  nanoparticles 299
  NMR spectroscopy 278
  oxidation 300, 304
  phosphates 297
  POLYS program 266
  primary structure 258
  quaternary structure 259
  rheology 280
  secondary structure 259, 277
  solvents 297
  sulphates 107, 131, 296
  sulphonates 297
    classification 107, **108**
  tertiary structure 259, 277
  tosylates 297
  tritylation 304
Potential of mean force (PMF) 252, **253**
Promoter 151
  BF_3-etherate 160
  DMTST 155
  IDCP 155
  mercury bromide 152, 173
  mercury cyanide 152
  methylsulphenyl triflate 209
  methyl triflate 154
  NBS 173
  NIS 155, 173
  silver silicate 210
Proteoglycans 116, 189, 199
Protecting groups 91, 161, 165
  acetamido 171
  acetate 157, 166, 167
  allyl 169, 200, 202, 213, 219
  allyloxycarbonyl 194, 202, 207
  azido 171
  benzoate 157, 167, 201, 202
  O-benzyl 95, 166, 168, 200, 215
  4,6-*O*-benzylidene 95, 166, 168, 215
  benzyloxycarbonyl 201
  tert-butyl ester 194, 221
  chloroacetate 169
  bis-dihydropyran 92
  diphenylmethyl ether 91
  dispoke acetal 91
  9-fluorenylmethoxycarbonyl 201, 206, 217, 219
  isopropylidene 166, 168
  levulinate 169
  p-methoxybenzyl 169, 196
  4,6-(4-methoxybenzylidene) 195
  phthalimido 171
  silyl 169, 170, 219
  N-trichloroethoxycarbonyl 207
  trityl 170
Pseudosugars 73

Quebrachitol 82, **84**
Quantum mechanics 14, 237

Raffinose 233
Ramachandran maps 237, 239, 240, **242**, 243, 251
Reeves, R.E. 6

Relaxed energy map 242
RGD peptide 216
Rheology 280
D-Ribose
  β-methyl-2,3-isopropylidene 58, **59**
  4-*C*-methyl- 54
  propane-1,3-diyl dithioacetal 62, **65**
L-Ribose 60, **63**
Rigid rotation energy map 241

SAMP/RAMP procedure 55
Saponins **127**
Scattering techniques 279, **281**
S-Glycoside 172
Sharpless, K.B. 47
Sialyl Le^x 180, 213, 214
Sialyl T_N antigen 208, 209
Solid phase synthesis 156, 195, 198, 213, 216
Solvent effects 34, 35
  continuum model 35
  on monosaccharide equilibrium 33
Solvents for polysaccharides 297
Stachyose 233
Starch 273
Sucrose 233, 234, 249
Sulphates 107
  analysis 137
  conformation 140
  cyclic 98, 133
  chemical stability 134
  cleavage 134
  diesters 133
  dye binding 139
  glycolipids 126
  glycoprotein 124
  glycosides 128
  infrared spectroscopy 138
  mass spectrometry 138
  NMR spectroscopy 137
  oligosaccharides 123, 124, 125, 127
  polysaccharides 108–24, 131, 296
  synthesis 128, 296
  vs phosphates 128
Sulphomucins 125
Sulphur trioxide 129, 130, 131, 132, 133
SYBYL 16

T-antigen 204, 207
L-Talose, 6-deoxy- **54**
TEMPO 304
T_N-antigen 204, 209, 218
Torsion angles **229**
α,α-Trehalose 233
β,β-Trehalose 231

UV/Visible spectroscopy
  sulphates 123, 138

Vibrational spectroscopy 11

WWW 138

Xanthan 288, **289**, 290, 308
X-ray diffraction 9, 232, 234, 266
  R-factor 10, 271

The manufacturer's authorised representative in the EU is Springer Nature Customer Service Centre GmbH, Europaplatz 3, 69115 Heidelberg, Germany. If you have any concerns regarding our products, please contact ProductSafety@springernature.com

Printed and bound by CPI Group (UK) Ltd, Croydon, CR0 4YY
22/09/2026
02220945-0014